Drought and Heat Stress in Agriculture

Rachana Dubey • Kumari Shubha
Amitava Rakshit • Sanjeev Kumar
Anup Das
Editors

Drought and Heat Stress in Agriculture

Implications, Mitigation and Policy Approaches

Springer

Editors
Rachana Dubey
Division of Crop Research
ICAR Research Complex
for Eastern Region
Patna, Bihar, India

Kumari Shubha
Division of Crop Research
ICAR Research Complex
for Eastern Region
Patna, Bihar, India

Amitava Rakshit
Department of Soil Science
and Agricultural Chemistry
Institute of Agricultural Science, BHU
Varanasi, Uttar Pradesh, India

Sanjeev Kumar
Division of Crop Research
ICAR Research Complex
for Eastern Region
Patna, Bihar, India

Anup Das
ICAR Research Complex
for Eastern Region
Patna, Bihar, India

ISBN 978-981-96-5734-6 ISBN 978-981-96-5735-3 (eBook)
https://doi.org/10.1007/978-981-96-5735-3

This Springer imprint is published by the registered company Springer Nature Singapore Pte Ltd.
The registered company address is: 152 Beach Road, #21-01/04 Gateway East, Singapore 189721, Singapore

Dedicated to our parents for their relentless devotion and being architects of our happiness.

Preface

Predictions indicate a projected global temperature rise of 7 °C by the end of the century. Important crops' productivity will be greatly impacted, and usual ecosystem processes will be compromised. Drought conditions are brought on by a lack of water due to a decrease in rainfall and an increase in the frequency of dry spells. Frequently, drought is combined by heat for worse impacts. The agricultural sector is highly dependent on climatic conditions and is particularly vulnerable to the impacts of drought and heat stress. For instance, it has been demonstrated that crop yields in Africa, Asia, and the Middle East are reduced by 15–35% when temperatures rise by 3–4 °C. The competition for environmental resources brought on by human population increase may be further accelerated by the obvious limits. The book *Drought and Heat Stress in Agriculture: Implications, Mitigation and Policy Approaches* provides a comprehensive overview of the latest research and innovative strategies for building agricultural resilience in the face of these challenges. The book explores the impacts of climate change on agriculture and how farmers can adapt their practices to overcome the dual threat of drought and heat stress on cereals, pulses, oilseeds, vegetables, and other allied sectors. One of the key themes of the book is the importance of smart water management in building agricultural resilience to drought and heat stress. Efficient water management plays a crucial role in plant growth, water retention, and nutrient cycling. By adopting agronomic options such as cover cropping and reduced tillage, farmers can enhance soil health and build resilience to drought and heat stress. Another important theme of the book is the role of community collaboration in building agricultural resilience. In the face of drought and heat stress, communities can come together to share resources and knowledge, and to develop collective solutions to the challenges they face. The book also emphasizes the importance of livestock management in building agricultural resilience. Heat stress can have a significant impact on the health and productivity of livestock, leading to reduced milk production and lower meat quality. By adopting practices such as shade provision and heat-tolerant breeds, farmers can minimize the impacts of heat stress on their livestock. The book explores the role of AI-assisted solutions in enhancing agricultural resilience to drought and heat stress. With advances in machine learning and remote sensing, farmers can now collect and analyze vast amounts of data to make informed decisions about crop management and irrigation. Finally, this book also highlights climate-smart agricultural policies refer to those policies that address the challenges posed by climate change while promoting sustainable agricultural practices and the socioeconomic impacts of climate change-induced drought and heat stress that are significant and can affect food security, livelihoods, and economic growth. These impacts can be particularly severe in developing countries, where agriculture is a primary source of income and food security. Future prospects and opportunities for addressing drought and heat stress in agriculture and allied sectors include the development and adoption of climate-smart agricultural practices, the use of drought-resistant crops and technologies, and the implementation of sustainable water management practices. There are also opportunities for public-private partnerships to invest in research and development of climate-smart technologies, such as precision agriculture, that can help farmers adapt to the changing climate. Additionally, there is an opportunity to increase awareness of the impacts of climate change on agriculture and to promote policy interventions that promote sustainable and resilient agricultural practices. By taking action to address drought and heat stress, agriculture can continue to provide food security and economic opportunities for future generations. In summary, *Drought and Heat Stress in Agriculture: Implication, Mitigation and Policy Approaches* is a valuable resource for farmers, researchers, policymakers, and anyone concerned about the future of our food systems in a changing climate. By adopting the innovative strategies and practices outlined in the book, farmers can turn the challenges of drought and heat stress into opportunities for greater agricultural productivity and sustainability.

Patna, Bihar, India — Rachana Dubey
Patna, Bihar, India — Kumari Shubha
Varanasi, Uttar Pradesh, India — Amitava Rakshit
Patna, Bihar, India — Sanjeev Kumar
Patna, Bihar, India — Anup Das

Acknowledgments

We extend our thanks to all the valued select group of contributors who produced updated science ideas from the unique perspective of being involved in both education and ground truth research. We also wish to acknowledge the help rendered by Indian Council of Agricultural Research (ICAR), New Delhi, and Institute of Eminence (IOE) Cell, Banaras Hindu University, in preparing this book. We gratefully acknowledge the support of the production team led by Nupur Verma and Muthuneela Muthukumar.

Last but not least, our deepest appreciation to our kind, affectionate, and compassionate family members' openhanded support and encouragement during the preparation this book.

Rachana Dubey
Kumari Shubha
Amitava Rakshit
Sanjeev Kumar
Anup Das

Contents

About the Editors and Contributors

Editors

Rachana Dubey is working as Scientist (Environmental Science) at ICAR Research Complex for Eastern Region, Patna. She did her master's and PhD at Indian Agricultural Research Institute. Her research area includes managing heat and drought stress in cereals, carbon sequestration, greenhouse gas quantification, and total ecosystem services valuation. She has done quantification of greenhouse gas emissions from rice-based systems in Indo-Gangetic Plains. She has received young scientist award from Department of Science Technology-Science and Engineering Research Board (DST-SERB) with a project on optimizing soil organic carbon stock in rice-based systems. Dr. Rachana is measuring greenhouse fluxes from long-term experiments under Cereal Systems Initiative for South Asia (CSISA) and has quantified the total economic value of different rice-based ecosystems of middle Indo-Gangetic plains. She has done her master's in mitigation of greenhouse gases in agriculture and her doctorate in mitigation of terminal heat stress in wheat. She is a recipient of the DST Inspire fellowship, the Best Student of the Division of Environmental Sciences, Indian Agricultural Research Institute (IARI), National Environmental Science Academy young scientist award, and many others. Dr. Rachana has published around 40 research papers, 10 book chapters, and many technical bulletins and manuals. Additionally, she is a peer reviewer of many high-impact journals.

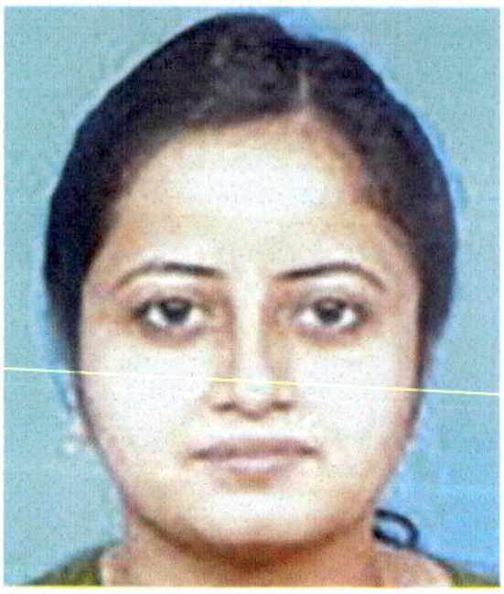

Kumari Shubha is a Scientist specializing in vegetable breeding, with a focus on abiotic and biotic stress. She is currently associated with the prestigious ICAR-Research Complex for Eastern Region in Patna, India. Her academic journey has been marked by exceptional achievements, having obtained both her MSc and PhD in Vegetable Science from G. B. Pant University of Agriculture and Technology (GBPUA&T), Pantnagar, Uttarakhand, and ICAR-IARI, New Delhi, India, respectively. Especially, she was a gold medalist during her graduation and was honored as a recipient of the DST Inspire Fellow award. Her research endeavors are aimed at advancing the field of vegetable breeding by developing climate-resilient varieties of dolichos bean and yardlong bean. She made significant contributions during her tenure at ICAR-National Bureau of Plant Genetic Resources (ICAR-NBPGR), New Delhi, where she dedicated her efforts to conserving and maintaining numerous biotic and abiotic stress-tolerant accessions of vegetable crops. Beyond her scientific achievements, she is an accomplished editor, having lent her expertise to the publication of two books. She is also an author of more than 20 research papers and reviews, showcasing her impactful contributions to the scientific community. Furthermore, she serves as a distinguished reviewer for peer-reviewed journals of renowned publishers such as Springer and Elsevier, adding to her extensive involvement in shaping the progress of research in her field.

Amitava Rakshit an IIT-Kharagpur alumnus, is presently a Professor at the Department of Soil Science and Agricultural Chemistry at Institute of Agricultural Sciences, Banaras Hindu University. Dr. Rakshit worked in the Department of Agriculture, Government of West Bengal, in administration and extension roles. He has visited Scandinavia, Europe, Africa, the UAE, and Bangladesh pertaining to his research work and presentation. He was awarded with The World Academy of Sciences (TWAS) Nxt Fellow (Italy), Biovision Nxt Fellow (France), Darwin Now Bursary (British Council, UK), Young achiever award, Shastri Indo-Canadian Institute award, Award of Excellence in Soil Science Education and Research by ISSS, New Delhi and Best Teacher's Award at UG and PG level by Banaras Hindu University-Indian Council of Agricultural Research (BHU-ICAR), state level by Uttar Pradesh Academy of

Agricultural Sciences, and national level by All India Agricultural Students Association in 2009, 2012, 2018, and 2019, respectively. He is serving as a review college member of British Ecological Society, London, since 2011, member of Global Forum on Food Security and Nutrition of FAO, Rome, and Commission on Ecosystem Management of International Union for Conservation of Nature (ICUN). He has published 178 research papers, 55 book chapters, 28 popular articles, and 3 manuals, and authored 40 books for Springer, Elsevier, and CRC Press.

Sanjeev Kumar is working as Principal Scientist & Head, Division of Crop Research at ICAR Research Complex for Eastern Region, Patna, Bihar, India. He has obtained his graduation and post-graduation degree from G. B. Pant University of Agriculture and Technology, Pantnagar, Uttarakhand, India in the year 1994 and 1996, respectively, and further received PhD degree (Agronomy) from Dr Rajendra Prasad Central Agricultural University, Pusa, Samastipur, Bihar, in 2005. His research interests include farming systems, conservation agriculture, maize agronomy, weed control, GHG emissions, water management, and crop production technologies. He has published 62 numbers of research papers in high National Academy of Agricultural Sciences (NAAS) rating journals of international and national reputes. He has completed 14 research projects (national/international) as principal investigator and 18 research projects as project co-investigator. Apart from this, he also published 3 books, 36 popular articles, 12 technical bulletins, etc. He is recipient of many national awards, viz., Fellow of Indian Society of Agronomy, New Delhi, Fellow of SURE, Varanasi (UP), Rajeev Gandhi Gyan-Vigyan Pustak Lekhan Puraskar, Ministry of HRD, New Delhi, Young Scientist Award from Shyam-Madhavi Trust, best poster, best paper presentation awards, best AICRP-IFS Centre Awards, etc. He has developed a number of climate smart integrated farming system models for irrigated, lowlands, and uplands with high adoption rate. He also developed and standardized technology for winter maize transplanting and raising of rice mat nursery for the eastern region of India. He is an editor of many national journals and a member of different professional and national societies in India. He is an expert member of the promotion of IFS in Bihar through JEEVIKA (Govt. of Bihar), a member of Institute Research Council at MGIFRI, Motihari, an expert member for development of IFS models at different locations in Bihar and Jharkhand (ATARI, Patna), and a member of Institute Management Committee (IMC) of ATARI, Patna.

Anup Das is presently serving as Director, ICAR Research Complex for Eastern Region, Patna, Bihar. His extensive research works led to development of package of practices for no-till production of pulses and oilseeds in rice and maize fallow areas; organic production packages of 33 important crops on cropping system approach; standardization of Modified System of Rice Intensification and Raised and Sunken Bed land configuration, watershed-based land use and IFS models for climate resilience, enhancing crop and water productivity, produce quality, soil properties, and C-stocks under different land uses and management practices. He has conceptualized and executed Integrated Organic Farming System in a cluster approach for safe and healthy food production. Dr. Das is associated with release of one Soybean variety (Umiam Sb-1) and three maize varieties (Megha maize 1, Megha maize 2, and Mizo mimpui). He has guided/co-guided 17 master's and 2 PhD students and taught 6 courses to PG students. He has published 220 research articles, 74 books, compendium and technical bulletins, 75 book chapters, 35 popular articles/extension leaflets with total citation of 5200 and h-index 40. He received several national awards like ICAR-Lal Bahadur Shastri outstanding young scientist award, Swami Sahajanand Saraswati outstanding extension scientist award, Fakhruddin Ali Ahmed award, Inter-disciplinary team research award, and First Dr. H.K. Jain – CAU, Imphal Award, among many others. Dr. Das is a recipient of the prestigious Fellowship of National Academy of Agricultural Sciences (2019), Indian Society of Agronomy (2017), National Academy of Biological Sciences (2021), Indian Association of Hill Farming (IAHF) (2017), and Society for Biotic and Environmental Research (2020). He has served as Vice president, Secretary, and Joint Secretary of IAHF; Co-Chief Editor, *Indian Journal of Hill Farming*; Editor and Councilor, *Indian Journal of*

Agronomy and *Indian Journal of Soil Conservation*; President, Society for Advancement of Agricultural Innovations and organizing secretary of one international and five national conferences besides serving as Chairman/Co-chairman/member/reviewer of various scientific sessions, task forces, committees, and 40 national and international journals.

Contributors

Abhilash Division of Natural Resource and Management, ICAR—Central Island Agricultural Research Institute, Port Blair, Andaman and Nicobar Islands, India

Anurag Ajay International Maize and Wheat Improvement Centre (CIMMYT), Patna, India

Leibniz Institute for Agricultural Development in Transition Economies (IAMO), Halle (Saale), Germany

B. N. Aloo Department of Biological Sciences, University of Eldoret, Eldoret, Kenya

Md Asadullah MBA (Sustainable Management), Nalanda University, Rajgir, Bihar, India

P. S. Basavaraj ICAR-National Institute of Abiotic Stress Management, Malegaon, Baramati, Pune, Maharashtra, India

Kuntal Bera Department of Seed Science and Technology, Uttar Banga Krishi Vishwavidyalaya, Cooch Behar, West Bengal, India

Department of Botany, Raiganj University, Raiganj, West Bengal, India

Debarati Bhaduri Crop Production Division, ICAR Central Rice Research Institute, Cuttack, Odisha, India

Preetam Biswas Department of Agronomy, Bidhan Chandra Krishi Viswavidyalaya, Mohanpur, Nadia, West Bengal, India

Preetam Biswas Department of Seed Science and Technology, Uttar Banga Krishi Vishwavidyalaya, Cooch Behar, West Bengal, India

K. M. Boraiah ICAR-National Institute of Abiotic Stress Management, Malegaon, Baramati, Pune, Maharashtra, India

R. Roy Burman Division of Agricultural Extension, Indian Council of Agricultural Research,, New Delhi, India

Aranab Chakraborty Leibniz Institute for Agricultural Development in Transition Economies (IAMO), Halle (Saale), Germany

Sozial-nachhaltige Transformation, Institute for Advanced Studies, Vienna, Austria

Kajal Mog Chaudhuri Department of Seed Science and Technology, Uttar Banga Krishi Vishwavidyalaya, Cooch Behar, West Bengal, India

Sangram Chavan ICAR-National Institute of Abiotic Stress Management, Baramati, Maharashtra, India

Adarsh Dangwal Department of Agronomy, G B Pant University of Agriculture & Technology, Pantnagar, Uttarakhand, India

Anup Das ICAR Research Complex for Eastern Region, Patna, Bihar, India

Rajkumar Dhakar Division of Agricultural Physics, ICAR—Indian Agricultural Research Institute, New Delhi, India

G. K. Dinesh SRM College of Agricultural Sciences, Baburayanpettai, Elapakkam, Tamil Nadu, India

Rachana Dubey Division of Crop Research, ICAR Research Complex for Eastern Region, Patna, Bihar, India

Puspendu Dutta Department of Seed Science and Technology, Uttar Banga Krishi Vishwavidyalaya, Cooch Behar, West Bengal, India

Hanamant M. Halli ICAR-National Institute of Abiotic Stress Management, Malegaon, Baramati, Pune, Maharashtra, India

C. B. Harisha ICAR-National Institute of Abiotic Stress Management, Malegaon, Baramati, Pune, Maharashtra, India

S. K. Mosharaf Hossain International Rice Research Institute, Varanasi, Uttar Pradesh, India

Shikha Jain ICAR-Indian Agricultural Research, New Delhi, India

Kalyan Jana Department of Agronomy, Bidhan Chandra Krishi Viswavidyalaya, Mohanpur, Nadia, West Bengal, India

Alok Singh Jayara Department of Agronomy, G B Pant University of Agriculture & Technology, Pantnagar, Uttarakhand, India

Neha Joshi ICAR—National Institute of Agricultural Economics and Policy Research, New Delhi, India

Prabhat Kishore ICAR—National Institute of Agricultural Economics and Policy Research, New Delhi, India

V. V. S. Jaya Krishna Department of Agronomy, Bidhan Chandra Krishi Viswavidyalaya, Nadia, West Bengal, India

Abhishek Kumar Division of Crop Research, ICAR Research Complex for Eastern Region, Patna, Bihar, India

Ujjwal Kumar ICAR Research Complex for Eastern Region, Patna, Bihar, India

Kavita Kumari Crop Production Division, ICAR-Central Rice Research Institute, Cuttack, Odisha, India

Meenu Kumari ICAR-RCER, Farming System Research Centre for Hill and Plateau Region, Ranchi, Jharkhand, India

Kaushalendra Kumar Department of Animal Nutrition, Bihar Animal Sciences University, Bihar Veterinary College Campus, Patna, Bihar, India

Rajeew Kumar Department of Agronomy, G B Pant University of Agriculture & Technology, Pantnagar, Uttarakhand, India

Rakesh Kumar Division of Crop Research, ICAR Research Complex for Eastern Region, Patna, Bihar, India

Sanjeev Kumar Division of Crop Research, ICAR Research Complex for Eastern Region, Patna, Bihar, India

Santosh Kumar Division of Crop Research, ICAR Research Complex for Eastern Region, Patna, Odisha, India

Manimala Mahato Crop Production Division, ICAR-Central Rice Research Institute, Cuttack, West Bengal, India

Sanjit Maiti Veterinary Extension Education, ICAR-National Dairy Research Institute, Karnal, India

Srijani Maji AICRP on MULLaRP, Directorate of Research, Bidhan Chandra Krishi Viswavidyalaya, Nadia, West Bengal, India

Subhasmita Mohapatra International Rice Research Institute, Varanasi, Uttar Pradesh, India

Sanchita Mondal Survey, Selection and Mass Production Unit, Directorate of Research, Bidhan Chandra Krishi Viswavidyalaya, Nadia, West Bengal, India

Anirban Mukherjee ICAR Research Complex for Eastern Region, Patna, Bihar, India

Kanu Murmu Department of Agronomy, Bidhan Chandra Krishi Viswavidyalaya, Mohanpur, Nadia, West Bengal, India

Sidhu Murmu Department of Agronomy, Bidhan Chandra Krishi Viswavidyalaya, Nadia, West Bengal, India

Bigyananda Mutum Department of Seed Science and Technology, Uttar Banga Krishi Vishwavidyalaya, Cooch Behar, West Bengal, India

Anirban Nath International Rice Research Institute, Varanasi, Uttar Pradesh, India

Swati Nayak International Rice Research Institute, Varanasi, Uttar Pradesh, India

Ravikant Nirala Department of Livestock Production Management, Bihar Animal Sciences University, Bihar Veterinary College Campus, Patna, Bihar, India

Priyanka Pandey Department of Molecular Biology and Genetic Engineering, G B Pant University of Agriculture & Technology, Pantnagar, Uttarakhand, India

Devesh Kumar Pant ICAR—National Institute of Agricultural Economics and Policy Research, New Delhi, India

Ratneswar Poddar AICRP on MULLaRP, Directorate of Research, Bidhan Chandra Krishi Viswavidyalaya, Nadia, West Bengal, India

Jai Prakash ICAR-Indian Agricultural Research, New Delhi, India

Abhijit Rai Department of Plant Sciences, College of Agriculture and Bioresources, University of Saskatchewan, Saskatoon, SK, Canada

Alka Rani Division of Agricultural Physics, ICAR—Indian Agricultural Research Institute, New Delhi, India

Banda Sainath ICAR Research Complex for Eastern Region, Patna, Bihar, India

K. Sammi Reddy ICAR-National Institute of Abiotic Stress Management, Malegaon, Baramati, Pune, Maharashtra, India

Arup Sarkar Department of Agronomy, Bidhan Chandra Krishi Viswavidyalaya, Mohanpur, Nadia, West Bengal, India

Sayantan Sarkar Texas A&M AgriLife Research and Extension Center, Corpus Christi, Temple, TX, USA

Kamal Sarma DLFM, ICAR- Research Complex for Eastern Region, Patna, Bihar, India

Sramana Sen Sarma Department of Agronomy, Bidhan Chandra Krishi Viswavidyalaya, Mohanpur, Nadia, West Bengal, India

Vinay Kumar Sehgal Division of Agricultural Physics, ICAR—Indian Agricultural Research Institute, New Delhi, India

E. Senthamil Department of Agronomy, University of Agricultural Sciences, Dharwad, Karnataka, India

Sudhanshu Shekhar Krishi Vigyan Kendra Ramgarh (ICAR-RCER, Patna), Mandu, Jharkhand, India

Shivani Division of Crop Research, ICAR Research Complex for Eastern Region, Patna, Bihar, India

Kumari Shubha Division of Crop Research, ICAR Research Complex for Eastern Region, Patna, Bihar, India

Anshuman Singh ICAR-Central Institute for Subtropical Horticulture, Lucknow, Uttar Pradesh, India

Jaspreet Singh Germplasm Exploration Evaluation and Conservation Division, ICAR National Bureau of Fish Genetic Resources, Lucknow, India

Varsha Singh International Rice Research Institute, Varanasi, Uttar Pradesh, India

R. R. K. Sinha Department of Livestock Production Management, Bihar Animal Sciences University, Bihar Veterinary College Campus, Patna, Bihar, India

S. V. Bangararaju Tatipudi ICAR—National Institute of Agricultural Economics and Policy Research, New Delhi, India

M. G. Vinay Department of Agronomy, GKVK, University of Agricultural Sciences, Bangalore, Karnataka, India

T. Vishal Department of Biotechnology, Graphic Era (Deemed to be University), Dehradun, Uttarakhand, India

O. J. Were Department of Seed, Crop and Horticultural Sciences, University of Eldoret, Eldoret, Kenya

Part I

Understanding the Impacts of Drought and Heat Stress on Agriculture

Climate Change and the Incidence of Drought and Heat Stress in Agriculture

1

Rachana Dubey, Kumari Shubha, Abhishek Kumar, G. K. Dinesh, Sanjeev Kumar, and Anup Das

Abstract

One of the main global concerns is the gradual rise in temperature brought on by human activity. Plant development is negatively impacted by the resulting heat (HS) and drought stress (DS), endangering ecosystem quality, and global food security at jeopardy. Plant growth, physiological processes, and yield are all impacted by HS and DS due to the physical damage, physiological disruptions, and biochemical changes that happen at distinct stages of growth. Thus, a deeper comprehension of how plants respond to HS/DS has practical ramifications for developing defence mechanisms, mitigating tactics, and recognizing the distinctions between HS/DS and the accompanying drought stress. To create HS/DS-tolerant genotypes in crops grown for food or feed, conventional breeding, biotechnological, and molecular methods are utilized. The molecular mechanisms behind HS/DS are now better understood thanks to recent advancements in omics methods. To increase plant resilience to HS, smart crop practice management is still beneficial. Proper timing of seeding, priming the seed, treating the seed with bacteria, controlling water and nutrients, applying osmo-protectants exogenously, and preserving soil moisture are crucial strategies for enhancing plant behavior in the critical habitat situations brought about by climate change and warming of the planet.

Keywords

Agriculture · Climate change · Drought · Heat · Mitigation

R. Dubey (✉) · K. Shubha · A. Kumar · S. Kumar · A. Das
Division of Crop Research, ICAR Research Complex for Eastern Region, Patna, Bihar, India
e-mail: Rachana.dubey@icar.gov.in

G. K. Dinesh
SRM College of Agricultural Sciences, Baburayanpettai, Tamil Nadu, India

R. Dubey et al. (eds.), *Drought and Heat Stress in Agriculture*,
https://doi.org/10.1007/978-981-96-5735-3_1

1.1 Impact of Climate Change in Agriculture

Increases in greenhouse gas (GHG) emissions are expected to cause Earth's temperature to rise by 2.5–4.5 °C by the year 2100 (A. Farooq et al., 2023). Reduced crop yields could arise from plants absorbing net carbon inefficiently because of rising atmospheric carbon dioxide levels. Future food security is called into question by this. Water shortages, changes in soil quality, and biodiversity loss are all direct effects of climate change that make it difficult to produce food sustainably for the world's expanding population. The Intergovernmental Panel on Climate Change (IPCC) predicts that while climate change may have both good and negative effects in northern latitudes, it will consistently and negatively affect crop productivity in low-latitude countries. However, climate change may make some high-latitude areas more suited for crops, sustained agricultural production increases in these areas may be limited by soil quality and water availability. Climate models predict that, in comparison to a scenario with a climate that is stable, global rice and wheat yields could decrease by 17% by 2050 under the most heated scenario (A. Farooq et al., 2023). It is critical to address this issue by making effective use of the resources at hand and identifying the steps that must be taken to control future climate risks in agricultural output. Carbon farming is crucial for reducing greenhouse gas emissions, improving soil health, and increasing farmers' income. This practice has the potential to transform agriculture, making it a solution to climate change and contributing to achieving sustainable development goals. It is our responsibility to inform farmers about the existence and advantages of carbon credit programs. The Government of India's initiative to incentivise farmers who adopt these carbon farming practices can act as a catalyst for change. In the medium term, the priority should be to establish an adequate enabling environment to support these farmers and scale the initiative. The engagement of stakeholders, particularly organizations that facilitate farmers in this process, should also be intensified. In the long run, carbon trading in agriculture can be a productive and profitable proposition for farmers. Global warming and other climatic effects are the outcome of human activities that have led to harmful levels of CO_2 accumulation and other greenhouse gases in the atmosphere over the past century (Chaudhry & Sidhu, 2022). Global agricultural productivity and food security are being significantly impacted by the increasing frequency of extreme weather events, including heat waves, droughts, strong precipitation, flooding, and changes in freezing patterns (Dempewolf et al., 2014). Feeding a growing human population while reducing production losses and the environmental effects of climate change is agriculture's biggest challenge. Developing sustainable strategies for crop adaptation to shifting climatic circumstances and increasing crop productivity are essential to ensuring global food security (St. Clair & Lynch, 2010). Abiotic stress episodes are growing increasingly frequent, causing considerable yield losses in crops worldwide (Chaudhry & Sidhu, 2022). By establishing a global target to keep warming to less than 2 °C, with efforts to keep it to 1.5 °C over preindustrial levels, the Paris Agreement seeks to mitigate the hazards associated with climate change. A thorough and reliable scientific analysis of the effects of varying degrees of warming on local climates in various places is necessary to support multilevel policy development. The hazards and effects of varying degrees of warming on various sectors have recently been quantitatively examined in several studies (Sun et al., 2019). Dryland regions, for instance, are extremely vulnerable to the effects of global warming. Dry regions are expected to warm 40–45% more than wet regions under a global warming scenario of 2.0 °C. Should the rate of global warming rise from 1.5 to 2.0 °C, drylands would face increased risks of adverse impacts on agriculture, hydrology, drought, and health. For both 1.5 and 2.0 °C warming targets, it is anticipated that the danger of drought will rise considerably in Mediterranean regions and central Europe relative to the Southwestern and Central Great Plains regions of the United States. The risk is significantly increased by the incremental temperature increase of 0.5 °C (Lehner et al., 2017). A 20–40% reduction in area exposures to hazardous extreme precipitation events and a 36% (22–46%) reduction in population changes could be achieved with a 0.5 °C drop in warming (W. Zhang et al., 2018). It is therefore imperative to understand the consequences of these in agriculture to manage and cope accordingly. In this chapter, we will understand the impact of heat, drought, and combined stress on agricultural crops.

Abiotic stressors, such as heat, cold, and drought, are important limiting factors that have an impact on crop yield both quantitatively and qualitatively. Climate change and the pressures of an ever-increasing human population are predicted to make these risks even more serious. In 2022, there were 8 billion people on the planet; according to the United Nations' most recent estimates, that number might rise to approximately 8.5 billion by 2030 and 9.7 billion by 2050. Environmental change makes it extremely difficult for agriculture to meet the world's expanding food needs. Regrettably, stressful conditions affect most of the agricultural land, which can typically lower yields by more than 60% as compared to record yields. (Kopecká et al., 2023).

The consequences of climate change have had a major impact on agricultural production in the past few decades. The increased impact of climate change on agriculture over the past few years has drawn more attention to mitigation techniques. During their developmental stages, agricultural crops usually go through a growth cycle that is extremely sensitive to environmental influences. Agricultural crops are grown in environments that are frequently unfavorable and change frequently. Crop performance during the course of production can be negatively impacted by stress at any stage of growth. Crop stress is the term used to describe the adverse effects that continuously shifting environmental conditions—which are frequently detrimental to crops' growth and development—have on them. Reducing stressors to the lowest feasible level is crucial for optimum crop growth and development (Oshunsanya et al., 2019).

The four main abiotic stressors that limit agricultural productivity worldwide are salinity, heavy metals, drought, and extremely high or low temperatures. Globally, abiotic stressors like temperature, water availability, radiation, and nutrient deficits significantly reduce agricultural yields. It has been observed that agricultural yield can be considerably impacted by various pressures, either separately or in

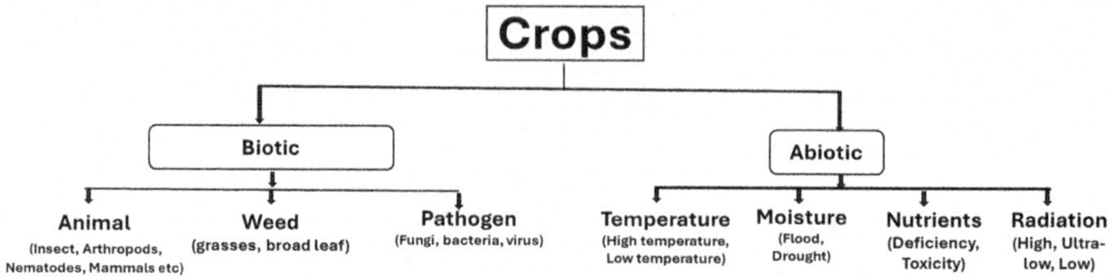

Fig. 1.1 Different types of biotic and abiotic stresses in crop plants. (Oshunsanya et al., 2019)

combination. Figure 1.1 presents a comparison of the abiotic and biotic stresses commonly observed in agricultural crops.

1.2 Impact of Drought and Heat Stress on Crops

Transient episodes of high temperatures that surpass ideal growth conditions are the source of HS in plants, which causes irreparable harm to cellular components, plant development, and plant functioning. Reactive oxygen species overproduction, liquidation, and damage to the plasma membrane (PM) are some of the events that produce this harm (Zenda et al., 2022). An imbalance between soil water input and evapotranspiration flux is the cause of this situation. A thorough multilocation study emphasises how temperature affects yields. The development of drought-prone areas is facilitated by these unfavorable conditions, which in turn affects significant agricultural productivity and plant growth (Lamaoui et al., 2018). Heat and drought have a greater combined impact when considered jointly, and hence it is crucial to do so. Heat and drought have a greater combined effect than when they are examined separately, and thus it is crucial to take both into account simultaneously. In contrast to other characteristics, such as biotic pressures, which are usually characterized by monogenic resistance, abiotic challenges, such as heat and drought, are usually controlled by several genes and have more complex underlying mechanisms. Furthermore, the study is made more difficult by the fact that many biotic and/or abiotic stressors frequently have an additive influence on the response to heat and drought (Lamaoui et al., 2018). Global warming of 1.5 °C is likely to be reached or surpassed between 2021 and 2040, with a strong probability (>50%) that this warming will continue (IPCC, 2021). Heat waves will therefore be more common and linger longer over most terrestrial regions. Global food security will be impacted because of the substantial effects this will have on crop growth and production in cereal grain crops, including maize, wheat, rice, barley, and sorghum (IPCC, 2021). Heat and drought can drastically diminish crop yields and productivity, which lowers farmers' profits. For example, yield reductions of up to 40% and 21%, respectively, were noted for maize and wheat when exposed to a water loss of roughly 40% (Daryanto et al., 2016).

The challenges created by high temperatures harming crops like rice are revealed by the recent increase in air temperature. Rising temperatures have caused rice production in China, a major rice-growing nation, to fall (Hassan et al., 2021a). High temperatures in the late stages of rice growth decrease yield due to lower seed production from pollen sterility. Between 1980 and 2000, a 1% increase in temperature in China led to a decrease in wheat production of more than 10%. By 2050, China's rainfed wheat yield would decrease by 5–7%, while India might lose 0.45 tons/ha because of a 0.5% rise in temperature. High temperatures also pose a threat to rice production in the Philippines. Fifteen percentage of the world's wheat is produced in the Indo-Gangetic Plains, but by 2050, drought and high temperatures are predicted. Countries that grow rice, wheat, and maize in central and eastern Asia, North America, and the Indian subcontinent are predicted to be most affected by heat stress (Hassan et al., 2021a).

Elevated temperatures have the potential to decrease wheat grain yield, quantity, and quality. Cantered on anthesis is when wheat is most vulnerable. It has been determined that increased development, decreased photosynthesis, and direct effects on reproductive systems are the causes of the observed susceptibility of wheat yields to high temperatures. The link between high temperatures during flowering and fewer grains, which significantly lowers grain output, is supported by a large body of experimental data. It is widely acknowledged that wheat close to blooming can withstand temperatures as high as 31 °C without seeing a decrease in grain number. The crop's hydration state, genetics, and developmental stage all affect how sensitive it is to temperature. Reduced grain counts might result from high-temperature occurrences (>30 °C), notably during the time around anthesis (5 days before to 2 days after), with the 20 days before and 10 days after being the most sensitive (Mitchell et al., 1993). The effects of high temperatures around anthesis on pollen fertility or sterile grains are mostly responsible for the fall in grain number. Apart from the direct detrimental effects of heat stress on pollen productivity and grain abortion, higher temperatures accelerate crop development rates, which causes anthesis and the emergence of double ridges to occur more quickly, resulting in fewer spikelets per spike and grains per spikelet (Eyshi Rezaei et al., 2015).

The length of time and rate of grain filling determine the ultimate weight of wheat grains. The amount of assimilate produced by photosynthesis in leaves and stems, as well as the remobilization of stored carbohydrates and nitrogen-containing substances inside these organs, all have an impact on this process. These compounds are then transported to the ear and grains. It is important to note that this process is governed by objective factors and not subjective evaluations. Wheat yield is quite vulnerable to heat stress, which reduces photosynthesis during grain filling at high temperatures. For example, 75 wheat cultivars in Australia showed a 23–37% decrease in individual grain weight in response to brief flashes of high temperatures above 35 °C (during grain filling). By reducing the time needed for grain filling, lowering photosynthetic rates, and directly blocking the endosperm cells' ability to produce starch, temperatures above 34 °C can lower final grain weights. It is important to note that this effect is caused by the mentioned factors and is not a subjective evaluation. Raising the temperature by 5 °C above 20 °C shortens the grain filling period of wheat by 12 days. With a 5 °C increase in temperature above 20 °C, wheat grain filling time is shortened by 12 days. Furthermore, during the early stages of grain growth, grain weight decreases at 38 °C; in wheat, this fall is 13% greater in heat-susceptible cultivars than in heat-tolerant cultivars. According to reports, 16–21 °C is the ideal temperature range for grain filling durations (Eyshi Rezaei et al., 2015). Al-Khatib and Paulsen (1984) concluded that heat stress (35/25 °C) during wheat grain growth speeds up senescence because it stops photosynthesis. When temperatures rise to 35 °C, wheat leaf senescence rates dramatically increase. In contrast to delayed senescence under ideal temperature treatments of 26 °C/14 °C and 24 °C/16 °C, greater daily temperature fluctuations (34 versus 22 °C) under heat stress circumstances hastened the senescence of flag leaves in wheat (Eyshi Rezaei et al., 2015).

The two primary reasons of decreased kernel quantity in maize under heat stress during flowering are probably decreased biomass absorption and kernel abortion. The lower kernel number and final yield were explained by the lowered resource usage efficiency that resulted from applying heat stress (30–36 °C) during silking. This also caused a decrease in plant and ear growth rates. Biomass partitioning to the ear during silking was unaffected by these temperatures at anthesis or silking (Eyshi Rezaei et al., 2015). The sole effects of heat stress on partitioning occurred during anthesis, when temperatures were high. This reaction was temporary because nearly all assimilate partitioning to the ear occurred when heat stress was removed. Reduced assimilation rates during kernel formation and/or kernel abortion can lead to fewer kernels, which in turn lowers yield indices when heat stress occurs. These sink limited situations are important physiological processes in the response of maize yield components. When combined, these findings highlight how crucial the source–sink relationship is in assessing how high temperatures close to anthesis affect grain output. Although there is a decrease in both kernel number and biomass assimilation, if the decreased kernel numbers lead to limited sink availability, the recovery of resource utilization and biomass output after heat stress is eliminated may result in a lower harvest index (Eyshi Rezaei et al., 2015).

Drought during the reproductive phase of maize and wheat resulted in a higher yield loss than drought during the vegetative phase ($P < 0.05$). The higher yield was probably caused by the increased amount of water that was available during the vegetative period. In the reproductive phase, maize was significantly more vulnerable to drought than wheat, despite wheat's equal sensitivity in the vegetative and reproductive stages. Plants show differing levels of drought sensitivity during different phases of their life cycle. When they germinate, for example, most of seed plants, including grains, start to lose their ability to withstand drought. Establishing seeds is essential, as is the growth of crown roots and the first green leaf area. Drought in the vegetative phase is generally regarded as reparable due to water shortfall (Daryanto et al., 2016). By causing stomatal closure and preventing photosynthesis, the stress restricted the synthesis of carbohydrates and, consequently, the division and growth of cells. In certain grain species, like rice and maize, drought during the reproductive stage can result in ovule abortion and pollen sterility. This implies that providing water later on will not be able to undo the impact of drought on yield during this stage. A major delay in silking brought on by drought might hinder the crucial synchronization between silking and anthesis, lowering maize production. This finding clarifies why contemporary maize hybrids have a lower anthesis-silking interval (ASI). Shortening the ASI has been the main goal of breeding initiatives designed to create drought-tolerant maize cultivars. Actually, shorter ASI (and tassel size in temperate maize) has helped newer hybrids of maize produce more grain (Daryanto et al., 2016).

During the vegetative period, wheat is susceptible to water stress, which can lead to decreased leaf area, number of leaves per plant, size, and longevity. Furthermore, it has been established that early-season dryness, which occurs 4 weeks prior to anthesis, drastically lowers yield by reducing the number of grains per ear. Drought can have an irreversible effect on the size of the wheat head, specifically the number of spikelets per spike, and the growth of the roots if it continues until the middle of the season, between jointing and anthesis. This is due to the fact that late-emerging tillers do not increase yield. Yield can be significantly reduced by a decline in photosynthetic sink and nutrient absorption capability. Therefore, having characteristics that enable soil moisture conservation during early growth is especially crucial for wheat in arid regions where water loss from soil evaporation is substantial. Characteristics for soil moisture conservation by crop include thick crop establishment to reduce weed competition for water and shade the soil, as well as wider but thinner leaves to boost radiation and ground cover-use efficacy (Daryanto et al., 2016; Wang et al., 2012).

Under high-yielding circumstances, sink capacity—which is established by the combination of high spikelet density and grain weight—is linked to rice production. The variety establishes each characteristic's relative importance. Higher yielding Indica varieties are linked to source activity, which includes faster growth rates and a quicker translocation of nonstructural carbohydrates to grains in the early stages of blooming. Reduced grain weight, altered flowering dynamics, and decreased seed set appear to be related to the effects of high temperatures on rice yield formation. Rice is susceptible to high temperatures during the day and at night (Eyshi Rezaei et al., 2015). According to TAO et al. (2008), changes in flowering dynamics brought on by daytime heat stress cause an earlier peak flowering and a general shortening of the anthesis period. Depending on their daily peak flowering time, varieties with the same high temperature sensitivities displayed different rates of seed sterility (Ishimaru et al., 2010). By making sure their flowers did not open during the hottest parts of the day, those that opened earlier in the day were able to prevent the detrimental impacts of high temperatures on seed sterility. It should be noted that stigma is less susceptible to heat stress than pollen grains. In gradient tunnels, the mean temperature increases of 5 °C (from an average of 22–28 °C) resulted in a decrease in both pollen production and pollen viability. Reduced pollen production and viability, along with a 20% fall in pollen germination percentages, were the outcomes of increasing overnight temperatures from 27 to 32 °C. Reduced seed set resulted from pollen germination being adversely influenced by elevated daytime temperatures up to 37.5 °C. However, the decrease in pollen germination

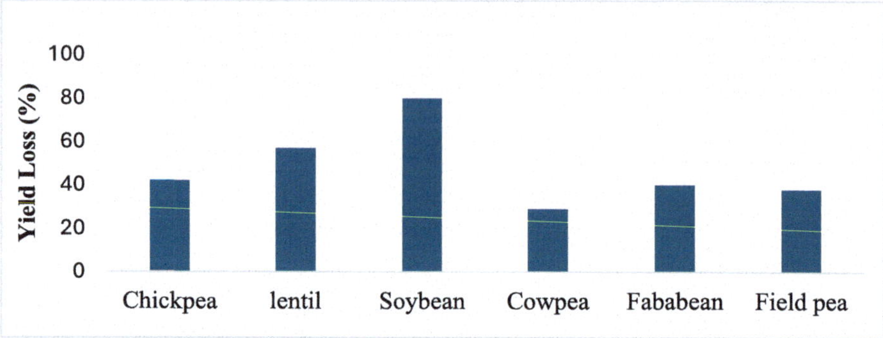

Fig. 1.2 Yield loss under drought and heat stress in cool season legumes. (Kumari et al., 2021)

was not sufficient to explain the larger fall in seed set at higher temperatures (40 °C) (Eyshi Rezaei et al., 2015).

Under heat stress, crops may exhibit a variety of visual symptoms, such as senescence, sunburn, leaf discoloration, and leaf burning (Vollenweider & Günthardt-Goerg, 2005). The duration of exposure to high temperatures is a crucial factor. High temperatures significantly impact on the reproductive growth. A temperature of 30°C or higher can negatively affect pollen viability, shedding, germination, growth, and elongation. Compared to other legume crops like lentils, faba beans and field peas, chickpeas have an enhanced temperature threshold for heat tolerance. When the temperature rises above 30 °C, chickpeas experience stress. A decreased seed set in common beans, an early onset of the reproductive stage, and an impact on floral physiology could arise from changing the night and day temperatures during sporogenesis to 27/32 °C (Kumari et al., 2021). Figure 1.2 lists the yield loss for several cool-season food legumes under heat and draught stress.

Reactive oxygen species (ROS) are created when plants are subjected to high temperatures, which causes oxidative damage. Through their effects on proteins, lipids, and membranes, these ROS can impair cell function. Membrane instability, which has been found to be more prevalent in field peas under drought stress conditions than under normal circumstances, can result from this damage. Membrane instability can also affect chloroplasts, which in turn affects photosynthesis and assimilate transportation. Under stress, plants reduce yields and quality by closing their stomata to avoid water loss. This response is triggered by lower leaf water potential or atmospheric humidity. However, excessive closure can lead to oxidative damage and hinder assimilation of CO_2. Nutrients are absorbed by roots along with water, but stress can impede their movement and retard growth. The means by which that stomata react to ABA during a drought is another noteworthy finding.

The impact of heat and drought is particularly severe in underdeveloped nations, especially for the agriculturally based economy. Agriculture employs almost 40% of the labor force. Crop productivity is drastically decreased during droughts, which has an immediate effect on the economy and farmers' livelihoods. Despite the complexity of plant responses to heat and drought and the lack of significant progress in reducing these stresses, there are still ways to increase plant production in these environments.

1.3 Management Options to Mitigate Heat and Drought Stress

It is suggested that addressing abiotic pressures requires both appropriate cultural practices and genetic advancements. While many cultural methods have been used for a long time to deal with abiotic challenges, the use of genetic tools is a relatively new addition to this activity. Breeders have historically placed a great deal of emphasis on creating high-yielding cultivars, and there is no question that such types thrived in environments free from stress (Fahad et al., 2017). However, breeding for stress resistance should also be prioritized considering shifting climatic conditions when plants are more vulnerable to abiotic stress. Recently, studies have been conducted to enhance plants' ability to withstand stress by conventional and molecular breeding techniques. We will discuss below a few management options to combat heat and drought stress in crops:

1.3.1 Heat Stress Escape and Avoidance Mechanisms

Two temporal frameworks—short-term avoidance or adaptation and long-term phenological and morphological changes—allow plants to tolerate high temperatures (HS) through a variety of methods. The short-term mechanisms include transpiration-induced cooling, leaf rolling or a shift in leaf position, and the makeup of lipidic membranes. Larger vessels and a higher density of leaf stomata and hair are examples of long-term morphological alterations brought on by HS (Hassan et al., 2021a). As an avoidance mechanism certain crops complete their life cycle sooner which although results in yield loss but it is a crucial strategy in plant ecology to avoid heat/drought stress. Under stress, plants can also change their phenotype in certain ways. Plants shrink, close their stomata, stop transpiring water, and have more xylem channels in their roots and shoots. Mesophyll cells are harmed by HS, which also makes membranes more permeable. Under HS, plants also experience subcellular alterations, such as notable adjustments in the thylakoids' structural arrangement, which have a large impact on the process of photosynthesis (Zhang et al., 2006).

1.3.2 Breeding Approaches

Using the breeding material under the desired conditions and identifying the lines that perform better is the most efficient way to create heat-tolerant cultivars. The best-performing cultivars are identified by using various morphophysiological traits as markers of heat tolerance. A plant's ability to withstand heat stress is demonstrated by increased production of defensive chemicals and little harm to the photosynthetic apparatus (Hassan et al., 2021a, b). Breeding methods that focus on the many physiological traits associated with the canopy structure of plants must be used to speed up the process of stress (heat/drought) adaptation. Stay-green qualities, delayed senescence, improved photosynthetic efficiency, decreased respiration, a better harvest index, and improved reproductive traits are some of these attributes. Therefore, physiological trait-based breeding methods are an essential strategy for creating cultivars that can withstand heat and drought stress while sustaining yield (Fahad et al., 2017). Modern breeding methods using quantitative trait loci (QTL) mapping have not been widely used for heat stress tolerance. There are wheat QTLs for a number of traits related to heat tolerance, such as leaf senescence and grain filling time. Farooq et al. (2011) noted a number of QTLs found for heat tolerance in wheat during the reproductive stage. Chromosomes 1B and 5A were found to have the QTLs associated with the length of time grains fill in wheat at high temperatures. Through the transfer of desired genes, transgenic techniques entail adjustments in both the qualitative and quantitative features (Lamaoui et al., 2018). Transferring desired genes to alter both qualitative and quantitative characteristics is known as transgenic methods (Ashraf, 2010). The majority of the focus has been on designing the genes that encode growth regulators, appropriate solutes, and antioxidants involved in stress tolerance. Successfully altering the genes encoding the two enzymes (choline monooxygenase and beta aldehyde dehydrogenase) that cause higher plants to express glycine betaine has demonstrated the potential to produce drought-resistant crops (Lamaoui et al., 2018).

1.3.3 Agronomic Strategies

The detrimental impacts of drought and heat stress may be mitigated by crop management practices such as planting windows, irrigation, crop residues and mulching, soil management and culture techniques, and choosing more suitable crop varieties.

To achieve the best yields under heat and drought conditions, choosing the right planting window is a nonfinancial best management option. Crop output under stress (HS/DS) is increased when the right planting/sowing timing is chosen to prevent heat and drought during later growth stages, such as flowering and grain filling. Furthermore, the timing of sowing affects crop growth and development thus choosing the right time to plant is crucial to maximizing production in challenging circumstances. Several authors have demonstrated that early sowing maximizes wheat production, but later sowing significantly reduces yield due to HS at later growth stages (Dubey et al., 2020). It has been shown that delayed sowing lowers crop production and has an adverse effect on all of the yield components. In actuality, delayed sowing produces test weight, productive tillers, and kernels with lower weights. Consequently, an understanding of the optimum sowing time in specific areas experiencing heat stress is essential to attain optimal yields. When heat stress (HT) occurs during the anthesis and grain filling periods of late-planted crops, crop yield is drastically decreased. This is explained by reduced grain filling rate and pollen sterility (Hassan et al., 2021b). For Indo-Gangetic belts early planted crops are protected from heat and drought stress, but late planted crops lose 20% more of their output than early planted crops when the temperature rises even for a few days. Consequently, one effective way to counteract heat stress is to plant early so that the grain-filling period occurs in a cooler climate (Dubey et al., 2020). Crop productivity is significantly impacted by mineral nutrients. However, low soil fertility and an unbalanced delivery of essential nutrients reduce crop productivity worldwide. Globally, it has been reported that about 60% of soils lack sufficient nutrients. Thus, to obtain good yields, adequate nutrient availability is required. Plant integrity is preserved by optimal nutrient availability, which supports multiple physiological processes. N and Mg are crucial nutrients for photosynthesis; P is required for energy and nucleic acid synthesis; and K is required for stomata control and enzyme activation (Hassan et al. 2021a, b).

The protective effects of different nutrients against abiotic stressors have been documented by several researchers. Among these, potassium (K) and silicon (Si) are recognized for enhancing crop resistance to salinity. By raising the concentration of antioxidant enzymes in plant cells, the application of nutrients including nitrogen (N), potassium (K), calcium (Ca), and magnesium (Mg) has also been shown to lessen toxicity to ROS (Hassan et al., 2021a, b). On the other hand, several studies have shown that applying fertilizer has no appreciable effect on drought stress. Since water is essential to the movement and metabolism of these nutrients, it follows that an ideal soil moisture level is needed (Hassan et al., 2021b). The usage of plant growth regulators, which are known to have a role in the response to stress, has also received attention. Salicylic acid, cytokinin, and ABA are among the plant growth chemicals that have been shown to be important for drought resistance. Treatments with plant growth regulators can dramatically raise the water potential and chlorophyll content in situations when there is water stress (Zhang et al., 2006). Another effective tool to combat heat and drought stress is seed priming for enhancing crop stand establishment. It significantly enhances crop stand and growth, seed germination, and seedling emergence. Utilizing KCl and $CaCl_2$ in the priming of seed greatly enhances crop yield components such as growth and dry matter production. It has been demonstrated that seed priming in normal and low vigor seeds increases root length and root and shoot growth. According to Farooq et al. (2008), seed priming with a variety of chemicals also enhances crop performance under abiotic stresses (Cicer arietinum L.), seed priming produced maximum yield under semi-arid circumstances by promoting optimal germination, improved development, and early flowering. Because seed priming promoted early blooming (8–10 days), maturity (9–10 days), and a lower chance of stress incidence during flowering, the most rice and wheat were able to stand stress with their avoiding mechanism (Hassan et al., 2021a, b).

Conserving soil moisture is one of the crucial techniques that might work for plant production in high-stress (HS/DS) environments. Grain filling time and rate, as well as final grain size, depend on the soil moisture status. Mulching is a key method for preserving the moisture and temperature balance of the soil system, but there are other strategies to preserve soil moisture as well. Straw mulching reduces evaporation, it considerably conserves soil moisture. Additionally, it enhances the efficiency of water and nitrogen usage, saving soil moisture, and improving overall plant performance under stress conditions. Conservation agriculture especially zero tillage is a very good way to not only conserve soil moisture but it can also prepone sowing windows by reducing field preparation time (Komarek et al., 2021). In the HS, effective water management is essential for maximizing crop yields. Many field crops can tolerate high temperatures (HT) of up to 40 °C if they receive enough water; however, insufficient water dehydrates plant leaves, which significantly lowers yield. The reasons for the lower production at high temperature lies in the fact that water stressed plants try to save water by closing their stomata; as a result, evaporative cooling considerably diminishes, and without cooling, leaf temperature reaches up to 50 °C. Water-stressed plants attempt to conserve water by closing their stomata, which significantly reduces evaporative cooling and causes leaf temperatures to rise above 50 °C in the absence of cooling. This is the explanation for reduced yields at high temperatures. Plant physiological activities are significantly inhibited at such HT, and, therefore, production is reduced. As a result, it is reasonable to infer that the best practices for water management involve timing irrigation and applying water using suitable irrigation techniques (micro irrigation). With the added benefit of reducing the adverse impacts of HS, irrigation timing based on the essential growth stages, water availability, and environmental variables will allow for maximum yield (Hassan et al., 2021a, b). For instance, compared to no irrigation, one additional watering at the grain filling stage of wheat under heat stress increased yield recovery by 7–8% (Dubey et al., 2020).

1.4 Conclusion

Global agricultural production is at risk of being curtailed by the existing heat and drought stress which is pertinent to the looming challenges of climate stress. Plants respond to HS/DS in a variety of ways, exhibiting significant alterations in growth, morphology, and functionality. The most significant losses in the final product are determined, particularly during the reproductive stages (anthesis-seed filling). To survive a wide range of intricate abiotic challenges, plants have evolved sophisticated adaptation mechanisms. Significant advancements in our understanding of these intricate features in higher plants have been made possible by the development of new technologies like genetic transformation and genomics. However, additional field validation of goods or prototypes is necessary for the commercial use of successful research outcomes. Agronomic techniques that can assist reduce production penalties in the field during heat and drought stress include conservation agriculture, seed priming, extra irrigation, nitrogen management at grain filling stages, and the use of Osmo protectants. All of these steps will surely assist to lessen the negative effects of abiotic stressors and improve plant production and food security under the current climate change and global warming scenarios.

References

Al-Khatib, K., & Paulsen, G. M. (1984). Mode of high temperature injury to wheat during grain development. *Physiologia Plantarum, 61*(3), 363–368. https://doi.org/10.1111/j.1399-3054.1984.tb06341.x

Ashraf, S. (2010). Land suitability analysis for wheat using multicriteria. *Research Journal of Biological Sciences, 5*(9), 601–605.

Chaudhry, S., & Sidhu, G. P. S. (2022). Climate change regulated abiotic stress mechanisms in plants: A comprehensive review. *Plant Cell Reports, 41*(1), 1–31. https://doi.org/10.1007/s00299-021-02759-5

Daryanto, S., Wang, L., & Jacinthe, P.-A. (2016). Global synthesis of drought effects on maize and wheat production. *PLoS One, 11*(5), e0156362. https://doi.org/10.1371/journal.pone.0156362

Dempewolf, H., Eastwood, R. J., Guarino, L., Khoury, C. K., Müller, J. V., & Toll, J. (2014). Adapting agriculture to climate change: A global initiative to collect, conserve, and use crop wild relatives. *Agroecology and Sustainable Food Systems, 38*(4), 369–377. https://doi.org/10.1080/21683565.2013.870629

Dubey, R., Pathak, H., Chakrabarti, B., Singh, S., Gupta, D. K., & Harit, R. C. (2020). Impact of terminal heat stress on wheat yield in India and options for adaptation. *Agricultural Systems, 181*, 102826. https://doi.org/10.1016/j.agsy.2020.102826

Eyshi Rezaei, E., Webber, H., Gaiser, T., Naab, J., & Ewert, F. (2015). Heat stress in cereals: Mechanisms and modelling. *European Journal of Agronomy, 64*, 98–113. https://doi.org/10.1016/j.eja.2014.10.003

Fahad, S., Bajwa, A. A., Nazir, U., Anjum, S. A., Farooq, A., Zohaib, A., Sadia, S., Nasim, W., Adkins, S., Saud, S., Ihsan, M. Z., Alharby, H., Wu, C., Wang, D., & Huang, J. (2017). Crop production under drought and heat stress: Plant responses and management options. *Frontiers in Plant Science, 8*. https://doi.org/10.3389/fpls.2017.01147

Farooq, M., Bramley, H., Palta, J. A., & Siddique, K. H. M. (2011). Heat stress in wheat during reproductive and grain-filling phases. *Critical Reviews in Plant Sciences, 30*(6), 491–507. https://doi.org/10.1080/07352689.2011.615687

Farooq, A., Farooq, N., Akbar, H., Hassan, Z. U., & Gheewala, S. H. (2023). A critical review of climate change impact at a global scale on cereal crop production. *Agronomy, 13*(1), 162. https://doi.org/10.3390/agronomy13010162

Farooq, M., Basra, S. M. A., Rehman, H., & Saleem, B. A. (2008). Seed priming enhances the performance of late sown wheat (*Triticum aestivum* L.) by improving chilling tolerance. *Journal of Agronomy and Crop Science, 194*(1), 55–60. https://doi.org/10.1111/j.1439-037X.2007.00287.x

Hassan, M. U., Chattha, M. U., Khan, I., Chattha, M. B., Barbanti, L., Aamer, M., Iqbal, M. M., Nawaz, M., Mahmood, A., Ali, A., & Aslam, M. T. (2021a). Heat stress in cultivated plants: Nature, impact, mechanisms, and mitigation strategies—A review. *Plant Biosystems - An International Journal Dealing with all Aspects of Plant Biology, 155*(2), 211–234. https://doi.org/10.1080/11263504.2020.1727987

Hassan, M. U., Chattha, M. U., Khan, I., Chattha, M. B., Barbanti, L., Aamer, M., Iqbal, M. M., Nawaz, M., Mahmood, A., Ali, A., & Aslam, M. T. (2021b). Heat stress in cultivated plants: Nature, impact, mechanisms, and mitigation strategies—A review. In *Plant biosystems* (Vol. 155(2), pp. 211–234). Taylor and Francis Ltd. https://doi.org/10.1080/11263504.2020.1727987

IPCC. (2021). Summary for policymakers. In *Climate change 2021—The physical science basis* (pp. 3–32). Cambridge University Press. https://doi.org/10.1017/9781009157896.001

Ishimaru, T., Hirabayashi, H., Ida, M., Takai, T., San-Oh, Y. A., Yoshinaga, S., Ando, I., Ogawa, T., & Kondo, M. (2010). A genetic resource for early-morning flowering trait of wild rice Oryza officinalis to mitigate high temperature-induced spikelet sterility at anthesis. *Annals of Botany, 106*(3), 515–520. https://doi.org/10.1093/aob/mcq124

Komarek, A. M., Thierfelder, C., & Steward, P. R. (2021). Conservation agriculture improves adaptive capacity of cropping systems to climate stress in Malawi. *Agricultural Systems, 190*, 103117. https://doi.org/10.1016/j.agsy.2021.103117

Kopecká, R., Kameniarová, M., Černý, M., Brzobohatý, B., & Novák, J. (2023). Abiotic stress in crop production. *International Journal of Molecular Sciences, 24*(7), 6603. https://doi.org/10.3390/ijms24076603

Kumari, V. V., et al. (2021). Drought and heat stress in cool-season food legumes in sub-tropical regions: Consequences, adaptation, and mitigation strategies. *Plants, 10*(6), 1038. https://doi.org/10.3390/plants10061038

Lamaoui, M., Jemo, M., Datla, R., & Bekkaoui, F. (2018). Heat and drought stresses in crops and approaches for their mitigation. *Frontiers in Chemistry, 6*. https://doi.org/10.3389/fchem.2018.00026

Lehner, F., Coats, S., Stocker, T. F., Pendergrass, A. G., Sanderson, B. M., Raible, C. C., & Smerdon, J. E. (2017). Projected drought risk in 1.5°C and 2°C warmer climates. *Geophysical Research Letters, 44*(14), 7419–7428. https://doi.org/10.1002/2017GL074117

Mitchell, R. A. C., Mitchell, V. J., Driscoll, S. P., Franklin, J., & Lawlor, D. W. (1993). Effects of increased CO_2 concentration and temperature on growth and yield of winter wheat at two levels of nitrogen application. *Plant, Cell & Environment, 16*(5), 521–529. https://doi.org/10.1111/j.1365-3040.1993.tb00899.x

Oshunsanya, S. O., Nwosu, N. J., & Li, Y. (2019). Abiotic stress in agricultural crops under climatic conditions. In *Sustainable agriculture, forest and environmental management* (pp. 71–100). Springer. https://doi.org/10.1007/978-981-13-6830-1_3

St. Clair, S. B., & Lynch, J. P. (2010). The opening of Pandora's Box: Climate change impacts on soil fertility and crop nutrition in developing countries. *Plant and Soil, 335*(1–2), 101–115. https://doi.org/10.1007/s11104-010-0328-z

Sun, Q., Miao, C., Hanel, M., Borthwick, A. G. L., Duan, Q., Ji, D., & Li, H. (2019). Global heat stress on health, wildfires, and agricultural crops under different levels of climate warming. *Environment International, 128*, 125–136. https://doi.org/10.1016/j.envint.2019.04.025

Tao, L.-X., Tan, H.-J., Wang, X., Cao, L.-Y., Song, J., & Cheng, S.-H. (2008). Effects of high-temperature stress on flowering and grain-setting characteristics of Guodao 6. *Acta Agronomica Sinica, 34*(4), 609–674. https://doi.org/10.1016/S1875-2780(08)60027-9

Vollenweider, P., & Günthardt-Goerg, M. S. (2005). Diagnosis of abiotic and biotic stress factors using the visible symptoms in foliage. *Environmental Pollution, 137*(3), 455–465. https://doi.org/10.1016/j.envpol.2005.01.032

Wang, L., D'Odorico, P., Evans, J. P., Eldridge, D. J., McCabe, M. F., Caylor, K. K., & King, E. G. (2012). Dryland ecohydrology and climate change: Critical issues and technical advances. *Hydrology and Earth System Sciences, 16*(8), 2585–2603. https://doi.org/10.5194/hess-16-2585-2012

Zenda, T., Wang, N., Dong, A., Zhou, Y., & Duan, H. (2022). Reproductive-stage heat stress in cereals: Impact, plant responses and strategies for tolerance improvement. *International Journal of Molecular Sciences, 23*(13), 6929. https://doi.org/10.3390/ijms23136929

Zhang, J., Jia, W., Yang, J., & Ismail, A. M. (2006). Role of ABA in integrating plant responses to drought and salt stresses. *Field Crops Research, 97*(1), 111–119. https://doi.org/10.1016/j.fcr.2005.08.018

Zhang, W., Zhou, T., Zou, L., Zhang, L., & Chen, X. (2018). Reduced exposure to extreme precipitation from 0.5 °C less warming in global land monsoon regions. *Nature. Communications, 9*(1), 3153. https://doi.org/10.1038/s41467-018-05633-3

Impact of Abiotic Stress on Cereal Crop Production

2

Preetam Biswas, Bigyananda Mutum, Kuntal Bera, Kajal Mog Chaudhuri, Srijani Maji, Ratneswar Poddar, Sanchita Mondal, and Puspendu Dutta

Abstract

Changing climatic pattern is an unambiguous and unavoidable global issue arising out of rapid industrialization with the progress of civilization. Unpredicted but frequent occurrence of various abiotic stresses has increased to several folds due to global climate change. Moreover, abiotic stresses may also aggravate biotic stresses and lead to exponential damage. These stresses have a huge detrimental bearing on crop production, food and nutritional security throughout the globe. The severity of different abiotic stress like drought, heat and cold stress, soil salinity, etc., vary with crops and their phenological phases, such as the effects on C_3 plants differ from that on C_4 plants or cereals like rice and wheat

P. Biswas
Department of Agronomy, Bidhan Chandra Krishi Viswavidyalaya, Mohanpur, Nadia, West Bengal, India

B. Mutum · K. M. Chaudhuri · P. Dutta (✉)
Department of Seed Science and Technology, Uttar Banga Krishi Vishwavidyalaya, Cooch Behar, West Bengal, India
e-mail: puspendu@ubkv.ac.in

K. Bera
Department of Seed Science and Technology, Uttar Banga Krishi Vishwavidyalaya, Cooch Behar, West Bengal, India

Department of Botany, Raiganj University, Raiganj, West Bengal, India

S. Maji · R. Poddar
AICRP on MULLaRP, Directorate of Research, Bidhan Chandra Krishi Viswavidyalaya, Nadia, West Bengal, India

S. Mondal
Survey, Selection and Mass Production Unit, Directorate of Research, Bidhan Chandra Krishi Viswavidyalaya, Nadia, West Bengal, India

R. Dubey et al. (eds.), *Drought and Heat Stress in Agriculture*,
https://doi.org/10.1007/978-981-96-5735-3_2

suffer more from high temperatures during their reproductive phase. The cereals, however, have a wide range of adaptation mechanisms to cope with these abiotic stresses. Activation of a complex stream of molecular networks or accumulation of different metabolites is initiated as an immediate response to stress. A proper understanding of these mechanisms along with a thorough knowledge on crop biology and weather phenomena would be helpful in the development of effective mitigation options against these abiotic stresses. Better crop performance under adverse conditions would also be ensured with the aid of genetic engineering and breeding of cereal crops or novel methods of stimulating crop growth through the use of plant growth-promoting rhizobacteria. Further, a proper integration of the above techniques along with improved agronomic practices will go a long way in meeting increasing food demands even under the global climate change scenario.

Keywords

Cereals · Abiotic stress · Climate change · C_3 plants · C_4 plants

2.1 Introduction

Cereals are grasses belonging to the family Poaceae, which are globally cultivated for their edible portion of the caryopsis, i.e., endosperm, germ, and bran. Cereals are generally rich sources of protein, carbohydrates, and dietary fiber; however, their nutrient density varies significantly depending on the crop species. Moreover, with the recent advancements in biofortification techniques, vitamin and micronutrient-rich cereal grains have great potential to reduce malnutrition, especially in low-income countries where they are widely consumed due to their low cost and are an integral component of daily diets (Garg et al., 2018; Shahzad et al., 2021). The principal staple grains that are grown most extensively worldwide are wheat, rice, maize, barley, and millets. Of these, wheat, rice, and maize account for 75% of global cereal production. According to the OECD-FAO Agricultural Outlook 2023–2032, the expected yearly rise of world consumption of wheat, rice, maize, and coarse grains in the next decade is 1.1%, 1.1%, 1.2%, and 0.8%, respectively, with Asian countries accounting for nearly half of the projected demand increase. However, many governments have imposed restrictions on the conversion of forests or pasture lands for agricultural use in order to maintain environmental stability and biodiversity thus constraining land availability for meeting the rise in demand. Instead, intensive cropping of the currently existing arable land along with heavy input use and improved varieties are expected to be the possible strategies for meeting the population-driven demand growth (OECD-FAO Agricultural Outlook 2023–2032, 2023). However, various abiotic stresses affect over 90% of the world's rural areas, making it difficult to harness the genetic yield potential of the crops (Pandey et al., 2020). Abiotic stress alone can result in crop production losses of up to 40% (Bailey-Serres et al., 2019). Therefore, if abiotic stresses are not adequately managed, they could pose a grave threat to global food security.

Moisture (drought and waterlogging), extreme thermal regimes, and soil degradation due to salinity, acidity, heavy metal toxicity, and nutrient deficiency are the major abiotic stresses affecting cereal cultivation worldwide. A recent study reported that wheat is more vulnerable to drought, salinity, and high-temperature stress than biotic stresses, indicating their severe negative impacts on the crop's cellular and physiological responses eventually resulting in compromised yield (Abhinandan et al., 2018). A meta-analysis from 15 countries showed that drought stress reduced wheat grain yield by 57.32%, with continuous and terminal droughts accounting for 83.60% and 26.34% yield losses, respectively (Wan et al., 2022). Salinity is another major abiotic stress which affects 23% of arable land and is also responsible for yearly 1–2% reduction of cultivable land (Alqahtani et al., 2018). Osmotic stress under saline conditions results in Na^+ and Cl^- accumulation in the cell sap and tissues leading to nutrient imbalance which adversely affects germination and crop metabolic activity (Hussain et al., 2019b). In fact, at the cellular level, crops interpret both salinity and drought stress as water deprivation, which results in similar physiological and biochemical responses (Tavili et al., 2011). For instance, both drought and salinity stress compromise crop productivity by adversely affecting plant water status, cell turgidity and membrane stability, photosynthetic rate, and biomass accumulation, which eventually disrupts source–sink partitioning ultimately leading to early maturity (EL Sabagh et al., 2021; Gunasekera & Ratnasekera, 2023). Heavy metal toxicity (Srivastava et al., 2017), micronutrient deficiency (Shukla et al., 2021), and soil acidification (Huang et al., 2023) are also responsible for destabilizing global cereal production. These issues are growing at an alarming rate because of intensive cropping, exhaustive land use, and injudicious application of inorganic NPK fertilizers without organic amendments. For instance, the principal foods that cause arsenic poisoning in humans are grains like rice, wheat, barley, and maize. However, rice, which is a major source of caloric intake in south and southeast Asia, can accumulate several folds more arsenic in its grain and shoot compared to wheat and maize (Upadhyay et al., 2020) thus posing a great threat to global food security (Mawia et al., 2021).

Climate change-associated increase in abiotic stress due to altered natural environment in the forms of frequent episodes of heat waves, extreme temperatures, increased incidence of drought stress, and frequent occurrence of extreme weather events like storms and flooding are also being reported globally (Eckardt et al., 2023). According to Saud et al. (2022), by 2050, China's irrigated and rainfed rice yield is expected to decrease by 7% and 6%, respectively, due to an increase in heat and drought stress episodes, heavy rainfall, and floods brought on by climate change and global warming. Moreover, crops are frequently subjected to several abiotic stressors simultaneously under field conditions. Prolonged heat stress especially in rainfed areas often results in both drought and osmotic stresses at the same time because soil moisture evaporation elevates soil salt concentrations (Abhinandan et al., 2018). A study conducted in China on temporal changes in the occurrence of combined heat and drought stress on maize found that the frequency of severe combined stress events has increased between 1961 and 2020 (Yu et al., 2023). Combined heat + drought stress has much more damaging effects on the growth and yield traits of maize than individual heat and drought stress at the pre-flowering stage, as apparent in a recent study (Yousaf et al., 2022). The authors observed significantly higher production of oxidative stress markers, viz., malondialdehyde and H_2O_2, with a marked increase in total superoxide dismutase activity in combined stress conditions. Abiotic stresses like heat and drought cause over production of reactive oxygen species (ROS), which results in membrane degradation due to lipid peroxidation, chloroplast, and peroxisome autophagy, ultimately leading to cell death (Thussagunpanit et al., 2015; Das & Roychoudhury, 2014). Therefore, understanding the underlying mechanisms of morphological, physiological, molecular, and genetic responses of cereals to individual stress and stress combinations is of paramount importance for the development of climate-resilient cereals to ensure global food security (Bhusal et al., 2023). Hence, in this chapter, an attempt has been made to summarize the recent research advancements focusing on crop response and their tolerance and adaptation to major abiotic stresses in cereals. Research insights on mitigation strategies for developing sustainable food production systems to meet the demand of a surging global population in the face of a changing climate have also been explored.

2.2 Different Abiotic Stress on Cereals

2.2.1 Heat Stress

Crop exposure to elevated temperatures beyond optimum range, even for a brief period, can cause irreversible damage to their growth and functions and is referred to as heat stress. The Intergovernmental Panel on Climate Change estimates that if temperature increases at the current rate, it will likely hit 1.5 °C between 2030 and 2052 (Masson-Delmotte et al., 2022). Frequent episodes of heat shocks are presently the main global limiting factor for crop productivity due to rising air temperatures. Zhao et al. (2017) predicted 6.0%, 3.2%, and 7.4% global yield reductions in wheat, rice, and maize, respectively, for every 1 °C rise in temperature. The damaging effects of heat stress are largely dependent on the duration of the stress period, heat intensity, crop species, and phenological stage during which the crop is exposed to the stress (Kopecká et al., 2023). Heat stress-induced genetic upregulation triggers ROS production such as singlet oxygen ($^1O_2$), superoxide anion radicals (O^{2-}), hydrogen peroxide (H_2O_2), and hydroxyl radicals (OH^-), eventually resulting in lipid peroxidation, cell membrane disintegration, and cell death (Pospíšil, 2016; El-Sappah et al., 2022). Moreover, heat stress-induced oxidative stress during the vegetative stage adversely affects photosynthetic capacity due to inhibited Rubisco activase activity (Perdomo et al., 2017) and hampered electron transport to the PSII acceptor side from the oxygen-evolving complex (Essemine et al., 2017; Xu et al., 2021), which eventually results in poor biomass accumulation. A strong antioxidative defense mechanism, comprising various enzymatic and nonenzymatic components, scavenges ROS to maintain cellular homeostasis and prevent oxidative damage under elevated temperatures (Awasthi et al., 2015; Yang et al., 2022). However, exposure to extreme heat stress can impair antioxidant enzyme activity. For instance, compromised sodium oxide dismutase and catalase activity were responsible for the over-accumulation of ROS in rice anthers, resulting in poor pollen viability and floret fertility (Zhao et al., 2018). Also, elevated temperatures during early reproductive stage alter phytohormone levels, such as decrease in active cytokinins, gibberellin A1, and indole-3-acetic acid, but increased concentration of abscisic acid and bound cytokinins in young panicles of rice (Wu et al., 2016). In fact, all three major cereals, viz., rice, wheat, and maize are more susceptible to heat stress during the reproductive stage than the vegetative stage (Thomason et al., 2018; Arshad et al., 2017; Teng et al., 2022). In wheat, rice, and maize, heat stress during anthesis or post-pollination inhibits assimilate translocation to grains hence reducing grain weight and quality (Ullah et al., 2022; Zhang et al., 2018b; Niu et al., 2021). Furthermore, heat stress during the grain-filling stage causes early senescence, which also can be ascribed to oxidative damage contributing to the loss of photosynthetic capacity of the leaves (Qu et al., 2023).

2.2.2 Cold Stress in Cereals

Low temperature stress is a commonly found abiotic inhibitor of plant growth and productivity sometimes even affecting the geographical distribution of many crop plants. Cold stress in cereals often elicits a dynamic response, this being a function of the duration and extent of prevailing low temperatures, the type of crop and its growth stages. Depending on the ambient temperature during the stress period, the stress may be either called chilling stress (<20 °C) or freezing stress (<0 °C). Temperate crops like wheat, while not tolerant to freezing temperatures, can manage chilling stress quite well. Moreover, their tolerance to freezing temperatures is improved following chilling stress, a process known as cold acclimation. However, subtropical and tropical crops like rice and maize are sensitive to even chilling stress (Ritonga & Chen, 2020; Megha et al., 2018). The effect of cold stress on plants may be two pronged: by a direct effect of low temperature or due to dehydration of cells and tissues via freezing of cellular water (Janmohammadi et al., 2015). Low temperature may result in inhibition of thylakoid electron transport which results in an over-energized state of the photosynthetic reaction centers. This in turn causes overproduction of reactive oxygen species (ROS), causing photodamage, which ultimately hampers photosynthesis. Activities of scavenging enzymes responsible for balancing the ROS are also depressing thus further aggravating the situation (Li et al., 2014). Cold stress also produces a plethora of other changes in the plants as a means of coping with abnormal situations. This can range from changes in plant hormonal balances, such as an increase in the level of abscisic acid, to changes in gene expressions and proteome (Miura & Furumoto, 2013; Janmohammadi et al., 2015). Further, low temperature stress leads to reduction in leaf area, dry matter production as well as grain filling rate and grain number which ultimately reduces the yield (Ji et al., 2017). It was found that spring freeze (reduction in 8 °C from ambient temperature) resulted in decreased gas exchange rates and reduction of maximum quantum efficiency of photosystem II in wheat, and this caused a yield reduction of about 5–14% (Li et al., 2015). Ji et al. (2017) also reported a decrease in grain yield per plant of wheat under low temperatures during booting and jointing stages, the former being more sensitive.

2.2.3 Drought Stress

Drought can be defined as a meteorological phenomenon in which water scarcity for an extended period due to lack of rainfall or high evaporation rates leads to soil water deficit condition, and thus unable to meet crop demands (Kebede et al., 2019; Gunasekera & Ratnasekera, 2023). Drought episodes are only expected to increase due to the temperature spike and rainfall shortage brought on by climate change. This significant abiotic stressor is limiting agricultural productivity worldwide; the detrimental effects are particularly pronounced in rainfed cereals (Hussain et al., 2023), and the length and severity of the spell determine the extent of yield losses (Wu et al., 2019). A recent study assessing crop sensitivity to water deficit revealed that maize and wheat yields were more vulnerable to drought stress than rice (Hendrawan et al., 2023). Wheat and maize yields are expected to decrease by 20.6% and 39.3%, respectively, with a 40% reduction in available soil moisture (Daryanto et al., 2016). However, a different meta-analysis found that the drought-induced yield losses in wheat (27.5%) and rice (25.4%) showed a similar pattern (Zhang et al., 2018a). This differential drought sensitivity among the cereals may be attributed to the fact that the recovery of photosynthetic rate in C_4 grasses is slower than C_3 grasses upon rewatering after drought stress (Ripley et al., 2010). Furthermore, drought stress promotes the development of male flowers over female inflorescence in diclinous monoecious crops like maize due to limited water availability, which lowers fertilization and compromises production (Sangoi & Salvador, 1998).

Suboptimal water availability under prolonged water-stressed conditions adversely affects photosynthetic and assimilation rates due to damaged photosystem I and II reaction center and inhibitory effect on electron transport (Jia et al., 2020). Moreover, stomatal closure under drought stress to check transpirational loss is also responsible for reducing CO_2 absorption and photosynthesis (Sheoran et al., 2022). Reduced crop dry matter accumulation under water-limited conditions may be ascribed to significant morphological changes such as rolled leaves, poor leaf area development, increased root growth at the cost of shoot growth, and stunted plant height (Schoppach et al., 2016; Yang et al., 2022). Altered crop-soil water status under drought hampers nutrient uptake by crops due to poor diffusion of nutrients to the root surface and their decreased mobilization within the crop (da Silva et al., 2011; Kulczycki et al., 2022). ROS accumulation under drought stress adversely affects membrane stability due to lipid peroxidation and electrolyte leakage (Hasanuzzaman et al., 2020). Furthermore, during prolonged dry spells, a compromised antioxidant system leads to decreased ROS scavenging, which causes oxidative damage and eventually cell death (Cruz de Carvalho, 2008).

2.2.4 Salinity Stress

Salt stress can be described as a situation with excess salts in soil solution, or expressed with electrical conductivity of at least 4 desi Siemens/m at 25 °C (Ngara & Ndimba, 2014). Worldwide, about 20% or more soil is salt-affected, greatly hindering crop production, productivity, and quality of the produce, which is further aggravated under the climate change scenario (Seleiman et al., 2022). In cereals, salt stress is a serious global concern affecting deeply the biochemical and physiological levels that in turn reflects on the morphological and phenotypic levels (Kumar et al., 2022). Generally, cereal crops can be considered as glycophytes (sensitive to salt stress) but they differ in their capacity to overcome this stress via different mechanisms, for example, rice is more sensitive compared to wheat and barley (Shelden & Roessner, 2013). Moreover, the extent of damage due to salinity stress is a function of the timing and duration of the stress, the crop growth stage, and also the climatic conditions (Shahverdi et al., 2018). Wheat and maize are particularly vulnerable to salt stress during the early development stage but salt stress at the later stages in maize causes poor setting of kernel and reduced grain weight. Rice on the other hand is sensitive to salt stress both at the seedling as well as panicle initiation to flowering stages (Kumar et al., 2022). The first stage of high salt concentration causes osmotic stress which is basically the inability of plants to uptake adequate water. Adequate cell water ensures proper turgidity required for cell growth and elongation and thus under salt stress the plant growth processes are hampered. The second stage in the salinity stress is the ionic stress which arises due to toxic concentrations of Na^+ ions in the plant tissues causing low yield or even plant death (EL Sabagh et al., 2021). Also, high sodium ion in soil causes an imbalance in the ion ratios which means low uptake of essential nutrient elements by plants, especially K^+ ions, resulting in nutrient deficiencies. Reduced root growth in plants due to salinity-induced low overall growth implies a loss in the capacity of plants to uptake nutrients which further aggravates the condition (Ngara & Ndimba, 2014). Salt stress causes a drastic reduction in photosynthesis, effecting stomatal closure, chlorophyll impairment, and breakdown of enzymatic machinery and chloroplast structure (Kumar et al., 2022). Salinity also induces stomatal closure reducing CO_2 intake and fixation. This acts as a forerunner for the development of ROS which causes oxidative stress. This results in damage to major cellular constituents including lipids, proteins, and nucleic acids (Seleiman et al., 2022).

2.2.5 Heavy Metal Stress

Heavy metal stress means the adverse effects of excessive heavy metal concentration in soil on crop growth, physiology, and metabolism of crops resulting in poor-quality produce. Heavy metals such as zinc (Zn), copper (Cu), nickel (Ni), and cobalt (Co) are essential for plants but become toxic at high concentrations. However, lead (Pb), cadmium (Cd), mercury (Hg), and arsenic (As) are nonessential and toxic at low concentrations also (Zhao et al., 2022). Cereals absorb heavy metals through their roots from the contaminated soil and water and translocate to different plant parts. Uptake and translocation depend on several factors such as soil pH, organic matter content (Hou et al., 2019), metal concentration in the soil, as well as presence of minerals that have an antagonistic effect against heavy metals (Mondal et al., 2020) and other metals competing for the absorption sites on the root surface. Heavy metal can disrupt the plant's physiological and metabolic activities by increasing the production of reactive oxygen species (ROS) (Zhao et al., 2021). Like the response of plants to other stressors, excessive production of ROS causes oxidative damage to plant cells, reduced stomatal conductance, CO_2 assimilation, and photosynthetic rate and impaired nutrient uptake leading to reduced biomass accumulation, stunted growth, chlorosis, reduction in yield, and nutritional quality (Sachdev et al., 2021). Different crops respond differently to heavy metal stress (Majeed et al., 2019). Among the cereals, rice has a higher affinity toward heavy metals and accumulates in higher concentration compared to other cereals (Khanam et al., 2020). Maize is comparatively tolerant to elevated heavy metal concentration and can be grown without interrupting the growth and production of the crop (Ruiz-Huerta et al., 2022). Millets can be a better alternative for cultivation in the heavy metals contaminated zones compared to other cereals (Asopa et al., 2017) as ROS scavenging enzymes are higher in millets that help in mitigating oxidative damage when they are exposed to heavy metal stress. Table 2.1 shows the major impacts of heavy metal stress in various cereal crops.

Besides this cereals are also facing other abiotic stress like waterlogging, high light intensity, elevated CO_2 level, etc. Flooding conditions due to excessive rainfall or over-irrigation reduce the oxygen concentration in the system that hampers different physiological and metabolic activities like CO_2 assimilation, stomatal conductance, nutrient uptake, photosynthesis rate, etc. (Loreti & Perata, 2020; Hasanuzzaman et al., 2017). Researchers have shown that due to water logging conditions, there is a huge yield loss in different cereals like wheat and barley (Celedonio et al., 2018), rice (Mei et al., 2017), and maize (Tian et al., 2019). If there is high light flux photo inhibition may occur (Takagi et al., 2019).

2.2.6 Combined Stresses

In the field, crops are frequently subjected to a combination of abiotic stressors. Studying a single stress at a time is not sufficient to predict how plants will respond to combined stress since plant responses are distinct. High evaporation rates increase salt concentrations in coastal, arid, and semi-arid areas, causing combined salinity and drought stress (Hailu & Mehari, 2021). When drought and salinity occur together, the negative impacts on root and shoot biomass, photosynthetic efficiency, and ionic adjustment are higher than when they occur separately because of an

Table 2.1 Major impacts of heavy metal stress in different cereal crops

Crop	Heavy metals	Major impact	References
Rice	As	(i) Exert severe oxidative stress in rice (ii) Arsenic accumulation in grains (iii) Compromise quality and quantity of rice grain yield	Jan et al. (2023); Islam et al. (2016)
Wheat	Cu and Pb	(i) Negative impact on seed germination. (ii) Cu influenced more than Pb in inhibition of seed germination.	Jańczak-Pieniążek et al. (2022)
	Ni	(i) Affects photosynthetic activity and respiration, hinders mineral intake, and interrupts the nitrogen metabolic pathway (ii) Oxidative damage in plants by disordering the antioxidant immune system through production of harmful ROS (iii) Reduced absorption of essential minerals and nutrients like iron, copper, manganese, and zinc by plants, resulting in nutrient deficiencies	Atta et al. (2024)
Maize	Cd	(i) Increase oxidative stress through the accumulation of Cd (ii) Reduced maize growth and biomass	Rahman et al. (2023)
Sorghum	Hg and Cd	(i) Hamper seed germination and seedling growth	Ertekin et al. (2020)
Finger millet, pearl millet, oats	Pb	(i) Reduced seed germination and seedling growth, chlorophyll content Reduced catalase activity, higher lipid peroxidation, and proline content	Jatav et al. (2021)

impaired antioxidant system failing to avert oxidative damage (Angon et al., 2022). The synergistic interaction of these two stresses resulted in significant yield losses in wheat due to reduced biomass accumulation, root growth and development, root hydraulic conductivity, and nutrient imbalance in shoots (Fu et al., 2023; Paul et al., 2019). However, their interaction may not always be additive to their individual impact (Cao et al., 2023) as apparent in maize seedling, wherein the co-occurrence of these two stresses alleviated the deleterious effects of drought stress to some extent (Yao et al., 2016). Therefore, cereals respond to these two abiotic inhibitors in a complex way that largely depends on crop species, age, size and cultivar, and length and severity of the stresses (Ma et al., 2020). Similarly, the co-occurrence of heat and drought stress also significantly affects yield and yield traits due to reduced biomass, photosynthetic capacity, stomatal conductance, chlorophyll functions, and grain numbers (Mahrookashani et al., 2017; Prasad et al., 2011). Qaseem et al. (2019) reported that wheat yield was reduced by 56.47% under combined heat and drought stress, with the extent of losses being 53.05% and 44.66% under heat and drought stress, respectively. The yield losses under combined stress were primarily due to their inhibitory effect on metabolism and reserve mobilization to the sink. The negative effect of one stress is usually aggravated by the other, with combined effects being more devastating than the additive effects of either of them (Giri, 2019; Prasad et al., 2011).

2.3 Relation of Abiotic Stress with Growth Stages

The effect of different stress stimuli on the crop plants manifests starting from the molecular level and progressively integrates into the morphological level affecting the entire plant. This effect is dependent on the intensity and duration of the stress factor as well as the time at which they appear during the life cycle of a crop. Some crops may face an early-stage stress but later recover in the upcoming phenological stages resulting in little to no effect on the productive potential. However, during prolonged stress conditions, some crops may not have noticeable effects at the early stage but when continued up to the later stages might cause irreversible damage (Feller & Vaseva, 2014). The effect of certain stress conditions is more pronounced at specific growth stages. Sorghum is expected to have higher yield reduction when drought stress occurs during booting or flowering stages. Similarly in crops like wheat, the reproductive stage is more prone to heat stress compared to the vegetative stage (Calanca, 2017). The differential response to stress during different phenological stages in cereal crops have been shown in Table 2.2 and it has been discussed in brief under three broad categories: at the germination and seedling stage, the vegetative stage, and at the reproductive stage.

2.3.1 Stress at Germination and Seedling Stage

Proper germination and seedling growth are essential for optimum crop stand establishment that ultimately dictates the yield. Germination being very sensitive to stress, any deviations in the immediate environment during this stage results in complete rewiring of the plant mechanisms from the very beginning which are then bound to have some effect throughout its life cycle. Delay in germination may also result in terminal stress due to unsuitable atmospheric conditions, such as higher temperatures, further reducing yield and quality. The primary effects of stress factors like high temperature, drought, and salinity are disruption of seed germination causing reduced stand establishment (Kumar, 2020). Moisture is essential for seed germination as it helps in the activation of several enzymes involved in the process (Rane et al., 2021). In rice, drought stress leads to an increase in the concentration of polyethylene glycol. There is also a decrease in the hypocotyl length, root-shoot biomass, and many other associated changes which results in the reduction of seed germination capability (Akram et al., 2019). High soil salinity results in abnormal ratios of Na^+ and Cl^- ions with the other cations and anions of the soil, respectively. This leads to build-up of toxicity symptoms and unbalanced nutrient uptake (Rane et al., 2021). Moreover, higher osmotic potential leads to reduced water uptake which is reflected as lowered leaf water potential in the crop seedlings, reduced evapotranspiration, and stomatal conduction which in turn causes reduced root and shoot growth (Maiti & Satya, 2014). Some crops like rice are more susceptible to damage due to salinity in the early seedling stages whereby the NaCl toxicity results in build-up of reactive oxygen species that results in membrane damage (Wani & Sah, 2014; Maiti & Satya, 2014). Salinity stress at the germination stage may cause

Table 2.2 Impact of different abiotic stresses on different crops

Abiotic stress	Crop	Major impact/damage on crop plant	References
Heat	Wheat (*Triticum aestivum L.*)	Lowering seed formation due to disturbance in fertilization process because of low fertile pollen formation	Fahad et al., (2017)
		Reduction of stomatal conductance because of high transpiration rate	Balla et al. (2019)
	Maize (*Zea mays L*)	Occurrence of male sterility due to abortion of pore in male reproductive organs	Begcy et al. (2019)
		Increase in grain abortion ratio and lowing of yield attributes parameter	Wang et al. (2020)
		Reduction of pollen viability, pollen number, and fertilization activity	Siebers et al. (2017); Wang et al. (2019)
	Rice (*Oryza sativa L*)	Reduction in yield parameters like number of seeds per spike, chaffy grain formation, and grain quality	Wu et al. (2020)
	Sorghum (*Sorghum bicolor* L. Moench.)	Hampering pollen tube growth and viability and fertilization process	Chiluwal et al. (2020)
Drought	*Wheat* (*Triticum aestivum* L.)	Reduction of salicylic acid and zeatin concentration in spikes and leaves, respectively	Shokat et al. (2021)
	Rice (*Oryza sativa L*)	Alteration of osmotic potential, antioxidant activities, and chlorophyll content	Sohag et al. (2020)
		Reduction of relative water content, nitrogen metabolism	Farooq et al. (2009); Xu et al. (2015)
	Zea mays L	Increase of enzyme like lipid peroxidation and accelerating antioxidant activity	Ali et al. (2020)
		Increase in the activities of catalase, glutathione peroxidase, and superoxide dismutase	Usmani et al. (2020)
		Development of abnormal embryo and reduction in kernel numbers	Liu et al. (2015); Wang et al. (2019)
	Sorghum (*Hordeum vulgare* L.)	Hampered plant growth by reducing nutrient uptake	Askarnejad et al. (2021)
		Reduction of transpiration rate and stomatal conductance	Yuping et al. (2017)
	Barley (*Hordeum vulgare* L.)	Reduction of shoot-to-root biomass ratio and stomata conductance	Bista et al. (2020)

(continued)

Table 2.2 (continued)

Abiotic stress	Crop	Major impact/damage on crop plant	References
Cold	Rice (*Oryza sativa L*)	Unbalancing various biochemical, molecular, and biological process of rice	Hsu and Hsu (2019)
		Cold-induced severe membrane damage, mechanical injury, and metabolic dysfunction through ice crystallization	Yadav (2010)
		Reduction in the concentration of chlorophyll	Reddy et al. (2021)
	Maize (*Zea mays L*)	Lowering photosynthesis activity, nitrogen metabolism, and dry matter accumulation	Zhang et al. (2020)
		Senescence of leaves	Bhusal et al. (2021)
	Sorghum (*Hordeum vulgare* L.)	Lowering CO_2 uptake by stomata conductance and photosynthesis activity	Ortiz et al. (2017)
		Reduction of chlorophyll formation and accumulation of harmful reactive oxygen species	Guo et al. (2018)
	Wheat (*Triticum aestivum L.*)	Hampering respiration process, enzymatic activity, and deterioration of grain fillings	Golizadeh and Kumleh (2019)
		Delay in anthesis during booting stage	Xiao et al. (2021)
	Barley (*Hordeum vulgare* L.)	Decrease in membrane stability and increase lipid peroxidation and electrolyte leakage	Valizadeh-Kamran et al. (2017)
Salinity	Wheat (*Triticum aestivum* L.)	Accelerate production of proline, antioxidant enzymes, and hampered leakage of electrolyte.	Ilyas et al. (2020)
	Rice (*Oryza sativa* L.)	Decreased in concentration of gibberellic acid during seed germination	Liu et al. (2018)
		Increase production of lipid peroxidation and hydrogen peroxide in plant leaves	Mekawy et al. (2018)
		Reduce grain yield due to malformation of spikelets, occurrence of sterility in panicle, and improper fertilization process results in poor grain development	Lv et al. (2020)
		Creation of osmotic stress, nutritional imbalance, and ion toxicity	Ghosh et al. (2016)
	Maize (*Zea mays* L.)	Increased soluble salt uptake (Na^+) and reduced plant growth and development	Ali et al. (2021); Munns and Tester (2008)
		Damage of subcellular organelles and cell membranes	Iqbal et al. (2020)
	Sorghum (*Sorghum bicolor* L. Moench.)	Reduction in seed germination, seedling growth	Dehnavi et al. (2020)
		Decrease in photosynthesis, stomatal conductance, alteration of CO_2 concentration	Sui et al. (2015)
	Barley (*Hordeum vulgare* L.)	Reduction of chlorophyll concentration while increase in proline and hydrogen peroxide concentration	Ferchichi et al. (2018); Ebrahim et al. (2020)

drastic yield reduction, especially in other sensitive crops like wheat (Abhinandan et al., 2018), maize, and pearl millet (Maiti & Satya, 2014). Optimum temperature during germination is equally important. Any rise in temperature above the base temperature results in an increased rate of germination until the optimum temperature is reached. Thereafter, any further heat will be detrimental to the developing seed and may lead to cell death causing damage to the embryo as in wheat (Hussain et al., 2019a). Similarly, it was found in an experiment that increasing temperature over 35 °C resulted in decreased seed germination in rice, maize, and sorghum, followed by a complete inhibition of germination in maize and rice at 41 °C (Akman, 2009).

2.3.2 Stress at Vegetative Growth Stage

Stress occurring during the major part of the vegetative growth stage hampers the normal physiology of the plants via various mechanisms like high oxidative stress, reduced photosynthesis, and so on. Under drought and heat stress, for example, reduced internal CO_2 concentration as well as enzyme activity greatly hinder photosynthetic processes, and as such the plants decrease its growth processes like tillering and shoot growth as a means of adaptive response against the prevailing stress (Kumar, 2020). Drought at the early growth stages results in poor tillering and crop stand whereas the mid-stage of growth produces low dry matter, less effective tillers, and grain production (Hossain et al., 2021). Drought results in loss in turgor pressure which is linked to lowered rate of leaf expansion and leaf growth in the cereal crops. It also results in damage to cells and tissues via oxidative stress, leading to low chlorophyll and carotenoid contents, reduced stomatal conductance during the drought period suggesting a decrease in photosynthesis which in turn reduces crop yield. In addition, it may result in delayed or complete inhibition of flowering (Akram et al., 2019; Maiti & Saitya, 2014). Drought stress at the vegetative stage in cereals like rice, maize, and pearl millet can lead to yield loss to the tune of 20–50% or sometimes more (Rane et al., 2021). Deviation of temperature from the optimum in both directions is harmful. Cold stress results in stunting of plants and a general bushy appearance, yellowing of leaves, and hastening of maturity. Similarly, heat stress results in the degradation of enzyme activity resulting in impaired photosynthesis which implies growth retardation and low yield. In addition, maturity comes early as the growing degree days required for maturity is achieved much earlier (Hussain et al., 2019a; Kang et al., 2017) (Table 2.3).

2.3.3 Stress at Reproductive Growth Stage

A crop facing stress at the early stages may still have the chance to recover in the upcoming growth stages. However, any fluctuations in the environmental conditions during the reproductive stages are critical as any growth impairment during this stage will directly affect the yield and quality of the produce. Crops are susceptible

Table 2.3 Major impact and yield loss due to different abiotic stresses

Abiotic stress	Crop	Growth Stage	Major impact	Yield loss	References
Heat stress	Rice	Reproductive stage	(i) Affect panicle initiation and spikelet development, resulting in deformed floral organs, and reduced spikelet number and size, (ii) while HS at the (iii) later reproductive developmental stages, such as gametogenesis and anthesis, results in spikelet sterility and decreased yields	9–13.8%	Xu et al. (2021)
	Wheat	Anthesis and grain filling stage	Increase thylakoid membrane damage, loss of chlorophyll and decrease photosynthetic rate, seed set percentage, number of grains per spike and individual grain weight	29% and 44%	Djanaguiraman et al. (2020)
	Wheat	Reproductive stage	Reduction in pollen functionalities, resulting in reduced thousand seed weight, spike length, flag leaf area.	21%	Alghabari et al. (2021)
Cold stress	Wheat	Booting	Starch synthesis enzymes activity (AGPase, SSS, GBSS, and SBE) decreased and affect starch synthesis and dry matter accumulation. Reduction of number of grains per spike and the thousand grain weight also decreased.	–	Zhang & Zhao (2021)
Drought stress	Maize	Flowering and grain filling	Increased the flowering days, days to maturity, anthesis silk interval, decreased the leaf number, abnormal expression of secondary stress responsive traits, loss of normal root architecture, leading to yield loss	34.28–66.15% in non-drought tolerant and 38.48–55.95% in drought tolerant varieties	Sah et al. (2020)
	Finger millet	Ear head emergence to grain filling	Decreased physiological and yield attributing traits compromise the grain yield	36.6%	Krishna et al. (2021)
Combined heat and drought stress	Maize	Grain filling stage	(i) Decreased stomatal conductance, increased leaf temperature (ii) Reduced kernel size and grain yield	78.9%	Bheemanahalli et al. (2022)
		Flowering to grain filling stage	Delayed silking, increased anthesis-silking interval, and reduced kernel set, resulted in yield loss	˃50%	Nelimor et al. (2019)
Combined stress of climate and soil As	Rice	–	(i) Grain As content increase by two-fold (ii) Loss of yield and nutritional quality due to climate-induced changes in soil-As pattern and plant response	39%	Muehe et al. (2019)

to heat stress at every growth stage but the major damages are done when high temperatures coincide with the reproductive phase. Even a slight increase in temperature can be detrimental. High temperatures result in defective pollen development, stigma and style deformities and these hamper the fertilization process thus producing unfertilized embryo or lead to embryo abortions (Rane et al., 2021). Further rise in temperatures leads to increased production of reactive oxygen species, accompanied by high ethylene production. As a result, the length of the grain filling period is shortened leading to advanced maturity and low test weight (Kang et al., 2017). Drought has deleterious effects on cereals like rice, wheat, maize, as well as millets like sorghum and pearl millet during this phase, particularly affecting the grain filling stage. Grain filling in cereals depends on certain sucrose to starch metabolizing enzymes, namely sucrose synthase (SuSase), soluble starch synthase (SSS), ADP-glucose pyrophosphorylase (AGPase), and starch branching enzymes (SBE). Dry period during and after anthesis negatively affects the functioning of these enzymes and as such a decline in SuSase activity or inhibition of AGPase may result in low grain filling rate and end of grain development, respectively (Mahla et al., 2017; Akram et al., 2019). In maize, the resultant low kernel weight due to drought following silking can cause yield losses to the tune of 20–30%. In addition, drought sometimes delays silking thus increasing the period between anthesis and silking (Kumar, 2020). Moreover, drought-induced low photosynthesis might also lead to early leaf senescence (Jeyasri et al., 2021). Lowering of temperatures below 20 °C in cereals following heading has greater yield loss potential as it leads to decreased spikelet fertility along with low crop growth rate, low radiation use efficiency, and leaf area (Maiti & Satya, 2014). It also disrupts pollen tube elongation, pollen sterility, delayed and distorted flowering, and also affects sink–source distribution thus affecting the grain filling process (Abhinandan et al., 2018; Hussain et al., 2018). Apart from affecting yield, stress during these stages may also affect the quality of produce. For example, during grain filling stages, water stress and high temperatures differentially regulate the activity of granule-bound starch synthase activity which decreases or increases the amylose/amylopectin ratio in the starch, respectively (Kumar, 2020). Salt stress during spikelet development also leads to a decrease in kernel number and weight (Abhinandan et al., 2018).

2.4 Crops Tolerance Against Abiotic Stress

Plants employ various strategies to overcome abiotic stress, such as escape, avoidance, or tolerance. Escape involves fleeing from stressful conditions either by shedding leaves or adjusting growth patterns. Avoidance entails reversible physiological changes, including decreased water loss and increased water absorption, to counteract adverse environmental conditions (Maroco et al., 2000). Tolerance, on the other hand, involves coordinated physiological and biochemical changes at the cellular and molecular level to withstand stress (Chaves et al., 2003). Morphological changes are also a common response to stress, such as the extension of roots to access water or the development of a more compact growth form. These changes may involve modifications in the concentrations of mRNA and protein levels, leading to various physiological, biochemical, and molecular changes in the plants to combat different types of abiotic stress (Ingram et al., 1980).

2.4.1 Morphological Changes

Under stressful conditions, plants may exhibit morphological changes to mitigate the effects of abiotic stress. One such change is the extension of roots to access water in the soil. As a result, roots experience an increase in length, which facilitates water absorption and persistence under stressful conditions. The increased root biomass in abiotic stress conditions contributes to this ability to absorb sufficient water. Shoot length is typically higher in abiotic stress conditions due to the efficient transpiration and translocation mechanisms. In contrast, water-deficient plants show a comparatively shorter shoot length because they need to overcome water and nutrient deficit conditions caused by drought. The shoot dry weight, which depends on the inner mass and tillers of the plant, may also vary in response to abiotic stress. Irrigated plants can accumulate water inside tissues due to a sufficient amount of soil water whereas water deficient cannot do so due to which inner mass decreases in case of water stress plants (Yadav et al., 2016).

2.4.2 Physiological Changes

Plants confronting abiotic stress undergo various physiological modifications at the molecular, cellular, and whole-plant levels. These modifications enable plants to adapt and respond to environmental stresses. The key physiological parameters that provide insight into plant stress are as follows.

2.4.2.1 Relative Water Content (RWC)

RWC is a measure of a leaf's water content relative to its maximum water-holding capacity. It is a reliable indicator of water deficit and heat stress, as it reflects the decline in leaf water content. RWC is closely related to leaf water potential, which is another measure of plant water status. The osmotic adjustment mechanism, a crucial process for maintaining cell hydration, is also influenced by RWC. The rate of transpiration and water content in leaves are directly influenced by water absorption from roots to leaves. In well-watered plants, the rate of transpiration is high, resulting in elevated water content in leaves, while in water-deficit plants, the rate of transpiration is low, leading to depleted water content in leaves (Maroco et al., 2000).

2.4.2.2 Relative Stress Injury (RSI)

RSI is a measure of the relative injury caused to plants under stressful conditions. It provides an assessment of the injury caused to plants under stressful conditions. When plants are exposed to stressful conditions, they activate genes that enable them to become resistant to external factors. This leads to the development of abiotic stress tolerance, which is the result of cooperation between physiological and biochemical changes at the cellular and molecular levels. These modifications may involve stiffer cell walls, osmotic adjustments, or smaller cells. Rapid changes in mRNA and protein concentrations also contribute to the development of stress tolerance in plants (Yadav et al., 2016).

2.4.2.3 Water Use Efficiency (WUE)

Water use efficiency (WUE) is a fundamental parameter that plays a critical role in determining the productivity of plants under conditions of limited water availability. In agronomic terms, WUE is defined as the ratio of the total dry matter produced or harvested to the amount of water utilized. From a physiological standpoint, however, WUE can be defined as the ratio of the fixed carbon rate to the transpired water rate. Water use efficiency is the relationship between water use and crop production rate, and it is typically measured in terms of biomass generated per unit of transpiration. In circumstances of limited water availability, the ability to generate more biomass is essential for achieving higher crop yields. Furthermore, plants may simultaneously experience multiple stresses, such as drought and salinity, which are among the most significant factors limiting crop productivity and yield worldwide.

To improve drought resistance in plants, various WUE mechanisms can be employed to maintain water levels or drought-tolerant crop varieties can be cultivated (Cattivelli et al., 2008; Yadav et al., 2016).

2.4.2.4 Osmotic Adjustment

Osmotic adjustment (OA) is the process by which intercellular solutes are elevated in response to water stress, allowing turgor to be maintained at a lower water potential. OA has been widely recognized as a primary mechanism for plant adaptation to drought, as it promotes tissue metabolic activity and enables regeneration, although the efficiency of this mechanism varies significantly between plant genotypes. OA has been linked to plant efficiency in arid conditions in several species, including sorghum, wheat, and oilseed brassicas. High levels of ions can inhibit cytosolic enzymes in plant cells, which can have a significant impact on their function. During osmotic adjustment, ion accumulation appears to be restricted to the vacuoles, where ions are kept separate from the cytosol or subcellular organelles. Because of this ion compartmentation, other solutes such as sugar alcohol, amino acid proline must be assembled in the cytoplasm in order to preserve the cell's water potential balance (Singh et al., 2017).

2.4.3 Biochemical Changes

Under stress, crops experience a number of cellular and molecular-level biochemical modifications.

2.4.3.1 Chlorophyll and Carotenoid

The content of chlorophyll and carotenoids is dependent on ATP, photosynthetic reactions, and NAD. Chlorophyll cannot capture sunlight directly, so it receives sunlight with the aid of carotenoid, an accessory pigment, and transfers it to photosystem I and photosystem II, which converts light energy into chemical energy in the form of ATP and NADPH. With the help of end products of photosystems and fixed carbon dioxide, plants produce glucose. Therefore, in wheat, the higher the carotenoid content in the chloroplast, the more sunlight will be captured, resulting in more chlorophyll (Farooq et al., 2009).

2.4.3.2 Starch

Starch also plays a crucial role in enabling the plant's response to abiotic stresses such as water deficit, salinity, or extreme temperatures. When photosynthesis is limited under stress conditions, plants tend to use starch as an energy source. The adverse effects of stress on plants can be reduced by releasing compatible solutes, osmoprotectants, derived sugars, and other metabolites to promote plant growth (Krasensky & Jonak, 2012).

2.4.3.3 Amino Acid

Under stress conditions, amino acids such as proline and arginine play a significant role in controlling osmotic pressure. Proline acts as an osmoprotectant, and its accumulation can lead to improved synthesis of cells and reduced degradation. This behavior of higher accumulation of proline is due to the expression of the gene encoding pyrroline-5-carboxylate synthase. Additional evidence for proline's defensive function was discovered in transgenic crops, where proline overproduction improves tolerance to osmotic stress. In addition to proline, the response to osmotic stress also involves other amino acids. Arginine was found to function as a compatible solute to enhance stress tolerance in leaves. The enzyme involved in arginine biosynthesis is increased under hyperosmotic conditions. In addition, osmotic stress in sunflower and wheat causes enhanced expression of asparagine synthase genes. Glutamine synthase overexpression enhances tolerance of osmotic stress in rice. These findings indicate that changes in osmotic pressure-induced amino acid levels may be due to modified gene expression encoding the enzymes engaged in their metabolism (Ashraf & Foolad, 2007).

2.4.4 Molecular Changes

2.4.4.1 Late Embryogenesis Abundant Proteins (LEA)

Late embryogenesis abundant proteins (LEA) are a group of elevated molecular weight proteins that are highly prevalent during early embryonic development and accumulate in response to water stress during seed dehydration. LEA proteins are classified into different groups, with group 3 proteins believed to participate in the sequestration of focused ions between these groups during cell dehydration. Group 1 LEA proteins are thought to possess increased water-binding ability, while group

5 LEA proteins are presumed to play a role in ion homeostasis during water loss (Galau et al., 1987).

2.4.4.2 Detoxifying Genes

In addition to LEA proteins, certain detoxifying genes help plants combat abiotic stress. By increasing tolerance to stress and accumulating appropriate solutes, plants can protect themselves from damage. Moreover, the accumulation of reactive oxygen species (ROS) can be scavenged to maintain protein structures and functions. The genes responsible for activating these enzymes—ascorbate peroxidase, glutathione peroxidase, and glutathione reductase—have been shown to play a role in various abiotic stress responses (Diamentet al., 2001).

2.4.4.3 Heat Shock Protein Genes

A characteristic biological response to abiotic stress in all species is the increase in the transcription of a set of genes in response to heat exposure or other abiotic stress. This response is mediated by the heat shock transcription factor (HSF), which exists as a monomeric, non-DNA binding form in unstressed cells. Upon exposure to stress, HSF transforms into a trimeric shape that can bind heat shock gene promoters, leading to the stimulation of genes encoding thermal shock proteins (Hsps). This is one of the most noticeable responses in organisms subjected to high molecular temperatures (Vierling, 1991).

2.4.5 In Reaction to Abiotic Stress, Various Genetic Mechanisms Begin in the Crops

2.4.5.1 ABA Pathway

The ABA pathway is activated in response to stress, with abiotic stress conditions triggering the accumulation of the hormone abscisic acid (ABA). Two pathways are activated in plants under osmotic stress, the ABA-dependent pathway and the ABA-independent pathway. The ABA-dependent pathway involves a group of transcription factors known as ABRE binding protein/ABRE binding factors (AREB/ABFs), which play a critical role in the expression of ABA-dependent genes. The dehydration-response element/C-repeat (DRE/CRT) and DREB2 transcription factors are key factors in the expression of ABA-independent genes in response to osmotic stress. The expression of AREB1/ABF2, AREB2/ABF4, and ABF3 is continuously increased in vegetative tissues under drought and salinity conditions. Overexpression studies have demonstrated that these three AREB/ABFs act as useful signals from ABA regulators. The AREB/ABF transcription factors trigger the expression of genes involved in abiotic stress response and tolerance (Shinozaki & Yamaguchi-Shinozaki, 2007).

2.4.5.2 Cold Stress Pathway

The CBF/DREB1 homologs have been identified in various species. CBF/DREB1 binds to CRT/DRE cis-elements (A/GCCGAC) in the promoter region of COR genes to regulate the expression of COR genes, which belong to the transcription factor family ERF/AP2 (ethylene-responsive factor/APETALA2). The CRT/DRE cis-acting elements are involved in the expression of the RD29A gene, which is believed to play a role in abiotic stress response and tolerance (Yamaguchi-Shinozaki & Shinozaki, 2006).

2.4.5.3 SOS Pathway

The SOS signaling pathway includes three significant enzymes, SOS1, SOS2, and SOS3. The SOS1 protein encodes for a plasma membrane Na+/H+ anti-porter. This protein is involved in cell-level regulation of Na + efflux and facilitates long-distance transport of Na + from the root to the shoot (Shi et al., 2000). The overexpression of SOS1 is linked to plant salt tolerance. Salt stress-stimulated signals from Ca2+ activate the SOS2 gene, which encodes a serine/threonine kinase (Liu, 2000). This protein comprises a well-established catalytic N-terminal domain and a regulatory C-terminal domain, the former of which is involved in the SOS stress signaling pathway. Specifically, it is the third protein in this pathway, known as the SOS3 protein, and it is a myristylated calcium-binding protein that contains an N-terminal myristylation site. This site plays a crucial role in salt tolerance. The C-terminal regulatory domain of the SOS2 protein contains an FISL motif, also known as the NAF domain, which is a 21-amino acid sequence that helps to interact with the calcium-binding SOS3 protein. The interaction between these two proteins results in the activation of the SOS2 kinase, which then phosphorylates the SOS1 protein. This enhances the SOS1 protein's initially identified yeast transportation activity. The SOS1 protein has a long cytosolic C-terminal tail composed of a putative nucleotide binding motif and an auto-inhibitory domain, which is approximately 700 amino acids long. The target site for SOS2 phosphorylation is this auto-inhibitory domain. In relation to salinity tolerance, the SOS2–SOS3 complex regulates the trafficking of membrane vesicles, pH homeostasis, and functions of the vacuole (Shi et al., 2000). There is a significant increase in intracellular calcium levels with an increase in sodium concentration, which in turn promotes the binding of the SOS3 protein. Calcium, in conjunction with the SOS proteins, regulates the homeostasis of intracellular sodium. The SOS3 protein then interacts with and activates the SOS2 protein, releasing its self-inhibition. The resulting SOS2–SOS3 complex is then transported to the plasma membrane where it reacts with SOS1. The phosphorylated SOS1 protein leads to improved sodium efflux, reducing sodium toxicity (Gupta & Huang, 2014).

2.5 Mitigation Options for Abiotic Stress in Crops

Abiotic stress factors such as drought, salinity, high temperatures, and nutrient deficiencies significantly impact crop productivity, posing a substantial threat to global food security. Addressing these challenges requires a multifaceted approach that combines traditional agricultural practices, innovative technologies, and sustainable management strategies. In this discussion, we explore various mitigation options for abiotic stress in crops, highlighting both current practices and emerging technologies.

2.5.1 Crop Diversification and Rotation

Crop diversification involves cultivating a variety of crops in a given area, while rotation involves changing the type of crops grown in a specific field over time. These practices can help mitigate abiotic stress by breaking pest and disease cycles, enhancing soil fertility, and reducing the risk of single-crop failure due to specific stress factors. Additionally, diverse crop rotations can exploit the varying stress tolerances of different crops, promoting resilience in the face of changing environmental conditions.

Crop diversification and rotation are important practices for reducing crop stress and improving soil health. By growing different crops in rotation, farmers can break pest and disease cycles, allowing the soil to recover nutrients specific to each crop (Krupinsky et al., 2006). Rotating between crops with different root structures and nutrient needs also promotes more diverse soil microbiology and prevents nutrient depletion (Lin, 2011). Some key crop rotations include alternating cereals with legumes, which fix nitrogen, or alternating shallow-rooted crops like lettuce with deep-rooted crops like potatoes (Larkin et al., 2011). Diversifying crops across both space and time increases biodiversity above and below the soil surface, allowing for more complex ecological interactions that increase resilience to stressors like pests, diseases, and weather fluctuations (Smith et al., 2007). Overall, thoughtfully planned crop rotations and diversity are critical for improving soil quality, managing pests sustainably, and reducing reliance on agricultural chemicals.

2.5.2 Breeding for Stress Tolerance

Conventional breeding and modern biotechnological approaches play a crucial role in developing crops with improved stress tolerance. Breeding programs aim to identify and incorporate genes associated with stress resistance into existing crop varieties. Genetic engineering techniques, such as CRISPR-Cas9, enable precise modifications to enhance specific stress-related traits, leading to the development of stress-resistant crop varieties. This approach has shown success in creating drought-resistant, salinity-tolerant, and heat-resistant crops.

Plant breeding efforts are increasingly focused on developing crop varieties with improved tolerance to various stresses, including biotic and abiotic factors. Key targets include tolerance to drought, heat, flooding, salinity, and resistance to pests and diseases (Cattivelli et al., 2008). Traditional breeding approaches like hybridization, as well as modern molecular techniques, are utilized to introduce desirable traits from donor lines into high-yielding varieties (Ashraf, 2010). For example, genes related to drought tolerance have been identified in rice and introduced into susceptible varieties, conferring the ability to maintain yield despite water shortage

(Venuprasad et al., 2009). Stress-tolerant varieties allow crops to better cope with suboptimal growing conditions, reducing crop losses. However, breeding for stress tolerance is a long-term effort limited by the complex genetic control of these traits. Continued research and multidisciplinary collaboration are important to enable further genetic improvements in staple crops to feed the world's growing population in the face of climate change.

2.5.3 Precision Agriculture

Precision agriculture utilizes technology to optimize resource use and enhance crop management. Remote sensing, GPS technology, and data analytics enable farmers to monitor field conditions in real time. This information can be used to adjust irrigation schedules, nutrient applications, and planting practices to mitigate the effects of abiotic stress. Precision agriculture not only improves resource efficiency but also helps farmers make informed decisions based on the current state of their crops.

Precision farming technologies allow for improved monitoring and management of crop stress through advanced sensors, data analytics, and targeted intervention. For example, sensor networks can map microclimate variability within fields, revealing zones of inadequate moisture or areas prone to heat stress (Bramley, 2009). Crop models and prediction algorithms can then determine optimal irrigation, fertilizer, and other inputs needed for each zone to minimize stress (Gebbers & Adamchuk, 2010). Robotic or aerial systems can also enable timely pesticide or growth regulator application only where and when needed to prevent crop losses. Overall, precision agriculture aims to provide crops with precise conditions to optimize growth, limit stress exposure, and reduce excessive chemical use. However, implementing these technologies requires significant investment and skilled management. While research continues on making precision farming more accessible, these tools have already demonstrated significant potential to increase productivity while reducing environmental impact.

2.5.4 Water Management Practices

Water scarcity is a critical abiotic stress factor, and efficient water management practices are essential for mitigating its impact. Techniques such as drip irrigation, rainwater harvesting, and soil moisture sensors help optimize water use. Additionally, the development of drought-tolerant crop varieties and water-saving technologies, such as precision irrigation, contributes to sustainable water management in agriculture.

Efficient irrigation and water management are critical to reducing crop stress in arid regions and during drought conditions. Key practices include drip irrigation, which precisely delivers water to plant roots, reducing losses from evaporation or runoff (Bar-Yosef, 1999). Irrigation scheduling based on soil moisture sensors, weather data, and crop models helps determine optimal timing and amounts (Jones,

2004). Drainage and runoff collection systems can also capture rainwater for supplemental irrigation. Improving soil health through reduced tillage or cover crops helps soil retain moisture (Palm et al., 2014). Additionally, mulches and windbreaks slow soil evaporation, and laser land leveling provides uniform irrigation. Adopting drought-tolerant varieties further reduces water needs. While these practices require investment, studies show significant water savings and yield improvements from improved irrigation management, especially when combined with appropriate technologies. Water management will become increasingly vital for food security considering extensive irrigation needs of staple crops and growing water scarcity globally.

2.5.5 Soil Amendment and Conservation

Healthy soil is fundamental for crop resilience against abiotic stress. Practices such as cover cropping, organic farming, and the application of organic and inorganic soil amendments contribute to improved soil structure, nutrient availability, and water retention. Conservation tillage, which minimizes soil disturbance, helps prevent erosion and enhances soil health, particularly in regions prone to drought and high temperatures.

Maintaining healthy soils through amendment and conservation practices is key to minimizing crop stress. Adding organic matter like compost or cover crops improves soil structure and nutrient retention, increasing moisture availability for plants (Diacono & Montemurro, 2010). Balanced fertilization provides nutrients to support plant growth and defense mechanisms under stress. Reduced tillage and avoiding erosion/compaction helps preserve soil biodiversity and long-term productivity (Hobbs et al., 2008). Crop rotations with legumes or mixed species cover crops can suppress disease and assemble beneficial microbiota (McDaniel et al., 2014). Integrated pest management utilizes minimal, targeted pesticide use to control pathogens without disrupting beneficial organisms. Overall, building soils high in organic matter and biological activity increases resilience to various stresses. While transitioning to regenerative practices can be labor and knowledge intensive, long-term field studies demonstrate the agronomic and economic benefits (Reeve et al., 2020). Protecting soils is a key foundation for sustainable, resilient crop production.

2.5.6 Biofertilizers and Microbial Inoculants

Biofertilizers containing beneficial microorganisms, such as nitrogen-fixing bacteria and mycorrhizal fungi, can enhance nutrient availability and uptake by plants. These microbial inoculants promote plant growth, improve stress tolerance, and contribute to overall soil health. The use of biofertilizers is an environmentally friendly approach to enhancing crop resilience against nutrient deficiency and other stress factors.

Biofertilizers and microbial inoculants are emerging agricultural amendments that can enhance crop health and resilience to stress. Many contain beneficial microbes like nitrogen-fixing rhizobia or arbuscular mycorrhizal fungi that form symbiotic relationships with plant roots (Adesemoye & Kloepper, 2009). These microbes facilitate nutrient acquisition, induce disease resistance, and increase tolerance to drought or salinity in inoculated plants (Reddy et al., 2002). Specific microbial strains can also be selected that protect against particular soil-borne diseases through competitive exclusion of pathogens (Spence et al., 2014). While application methods are still being optimized, research demonstrates significant potential for biofertilizers to reduce reliance on synthetic chemicals and improve crop vigor under stress. However, their efficacy depends on proper strain selection, quality production, and adequate field testing for consistency across variable environments (Malusá et al., 2012). When effectively implemented, microbial amendments provide an ecological approach to enhance productivity and crop resilience.

2.5.7 Climate-Smart Agriculture

Climate-smart agriculture is an integrated approach that addresses the challenges of climate change and abiotic stress. It involves the adoption of sustainable practices, resilient crop varieties, and technologies that optimize resource use. Strategies such as agroforestry, agroecology, and the use of climate-resilient crops contribute to the overall climate-smart agriculture framework, promoting long-term sustainability in the face of changing environmental conditions.

Climate-smart agriculture involves farming practices tailored to maintain productivity despite climate change stresses like drought, heat, and extreme weather. Key techniques include use of short-cycle crop varieties, shifting planting dates, and improved irrigation to align production with altered growing conditions (Khatri-Chhetri et al., 2020). Agroforestry systems and mixed crop-livestock farming enhance biodiversity and microclimate moderation. Soil and water conservation measures like terracing prevent erosion and runoff during heavy rain. Crop diversification, smallholder farmer collaboration, and climate-informed advisories also promote resilience. While implementing climate-smart strategies entails overcoming socioeconomic barriers, research by CGIAR and others demonstrates feasibility and benefits like higher, stabler yields plus mitigated climate impact (Steenwerth et al., 2014). As global warming exacerbates crop stresses, widespread adoption of tailored practices and policies will be critical to meeting twenty-first century food security challenges.

2.5.8 Advanced Monitoring and Early Warning Systems

Deploying advanced monitoring systems, including satellite imagery, drones, and sensor networks, helps detect early signs of abiotic stress. Early warning systems provide farmers with timely information, allowing them to take proactive measures to mitigate the impact of stress on crops. Timely interventions, such as adjusting irrigation or applying stress-mitigating treatments, can significantly improve crop resilience.

Advanced technologies for real-time monitoring and early stress detection allow farmers to rapidly identify and respond to crop threats. Satellite imagery, aerial drones, and proximal sensors analyze variables like canopy temperature, moisture, and vegetation indices to pinpoint problem areas in fields (Sullivan et al., 2007). Networked in-field sensors and Internet of Things devices also track microclimate and soil data at high resolutions, signaling irrigation or fertilization needs (Camargo & Smith, 2009). Complex simulation models ingest diverse datasets to forecast pest and disease outbreak risks days to weeks in advance (De Wolf & Isard, 2007). When linked to advisory services, these surveillance tools provide timely, precise recommendations to mitigate losses. However, effective implementation requires regular system calibration, farmer training, and installation costs. Overall, next-generation monitoring and early warning systems have potential to significantly increase resilience to multiple crop stressors with near real-time, preventative field management.

2.5.9 Greenhouse and Controlled Environment Agriculture

Greenhouse and controlled environment agriculture provide an enclosed environment where temperature, humidity, and other factors can be precisely controlled. These systems offer protection against extreme weather conditions and allow year-round cultivation. While resource-intensive, these methods are particularly effective in mitigating temperature-related stress and ensuring consistent crop production.

Greenhouse and controlled environment agriculture allows for precise regulation of conditions to optimize crop health and limit stress exposure. Enclosed structures maintain ideal temperature, light, humidity, and CO_2 levels while excluding pests (Ozbek et al., 2020). Hydroponic or aeroponic systems provide balanced nutrition and prevent soil-borne diseases. Environmental control systems can also administer targeted supplemental lighting, cooling, moisture, or nutrients in response to sensor data to maintain optimal growth conditions. While controlled agriculture requires significant investment in infrastructure compared to open field production, it maximizes yields while eliminating uncontrolled external stresses. This technology enables local food production in challenging climates and shows potential to sustainably intensify agriculture to meet rising global food demand.

2.5.10 Public Policies and Farmer Education

Government policies that support sustainable agricultural practices, provide incentives for adopting stress-tolerant crop varieties, and invest in research and development are crucial for mitigating abiotic stress. Farmer education programs can help disseminate knowledge about best practices, technologies, and the importance of crop diversification to enhance resilience against abiotic stress.

Agricultural extension programs that support farmer education are key to driving large-scale adoption of stress-mitigating practices. Subsidies and financial incentives can promote use of resilient technologies like drip irrigation or integrated pest management that require significant upfront investment (Stoecker et al., 2020). Crop insurance programs tailored to small farms help farmers survive catastrophic loss events and recover after disasters. Field schools, demonstration plots, and mobile apps connect farmers to context-specific guidance, equipment, and advisories to implement stress-reducing strategies (Steenwerth et al., 2014). Public agriculture research institutions also drive development of stress-tolerant varieties and sustainable farming protocols. While transitioning systems is challenging, cooperative extension paired with supportive policies enables farmers to obtain the knowledge, materials, and financial security needed to adopt productive, ecologically resilient practices.

2.6 Conclusion

In conclusion, mitigating abiotic stress in crops requires a comprehensive and integrated approach that combines traditional wisdom with cutting-edge technologies. Crop diversification, breeding for stress tolerance, precision agriculture, water management practices, soil conservation, biofertilizers, climate-smart agriculture, advanced monitoring systems, controlled environment agriculture, and supportive public policies are essential components of a holistic strategy. As the global climate continues to change, the implementation of these mitigation options becomes increasingly crucial for sustaining agricultural productivity and ensuring food security for a growing population.

References

Abhinandan, K., Skori, L., Stanic, M., Hickerson, N. M. N., Jamshed, M., & Samuel, M. A. (2018). Abiotic stress signaling in wheat—An inclusive overview of hormonal interactions during abiotic stress responses in wheat. *Frontiers in Plant Science, 9*, 734. https://doi.org/10.3389/fpls.2018.00734

Adesemoye, A. O., & Kloepper, J. W. (2009). Plant–microbes interactions in enhanced fertilizer-use efficiency. *Applied Microbiology and Biotechnology, 85*(1), 1–12.

Akman, Z. (2009). Comparison of high temperature tolerance in maize, rice and sorghum seeds by plant growth regulators. *Journal of Animal and Veterinary Advances, 8*(2), 358–361.

Akram, R., Fahad, S., Masood, N., Rasool, A., Ijaz, M., Ihsan, M. Z., ... & Nasim, W. (2019). Plant growth and morphological changes in rice under abiotic stress. In: *Advances in rice research for abiotic stress tolerance*. Woodhead publishing, pp. 69–85.

Alghabari, F., Shah, Z. H., Elfeel, A. A., & Alyami, J. H. (2021). Biochemical and physiological responses of thermostable wheat genotypes for agronomic yield under heat stress during reproductive stages. *Agron, 11*(10), 2080. https://doi.org/10.3390/agronomy11102080

Ali, M., Afzal, S., Parveen, A., Kamran, M., Javed, M. R., Abbasi, G. H., Malik, Z., Riaz, M., Ahmad, S., Chattha, M. S., Ali, M., Ali, Q., Uddin, M. Z., Rizwan, M., & Ali, S. (2021). Silicon mediated improvement in the growth, ionhomeostasis by decreasing Na+ uptake in maize (*Zea mays* L.) cultivars exposed to salinity stress. *Plant Physiology and Biochemistry, 158*, 208–218. https://doi.org/10.1016/j.plaphy.2020.10.040

Ali, Q., Javed, M. T., Haider, M. Z., Habib, N., Rizwan, M., Perveen, R., Ali, S., Alyemeni, M. N., El-Serehy, H. A., & Al-Misned, F. A. (2020). α-Tocopherol foliar spray, translocation mediates growth, photosynthetic pigments, nutrient uptake, oxidative defense in maize (Zea mays L.) under drought stress. *Agronomy, 10(9):1235*. https://doi.org/10.3390/agronomy10091235

Alqahtani, M. D. S., Roy, S. J., & Tester, M. A. (2018). *Increasing salinity tolerance of crops*. Springer.

Angon, P. B., Tahjib-Ul-Arif, M., Samin, S. I., Habiba, U., Hossain, M. A., & Brestic, M. (2022). How do plants respond to combined drought and salinity stress?—A systematic review. *Plants, 11*(21), 2884.

Ashraf, M. (2010). Inducing drought tolerance in plants: Recent advances. *Biotechnology Advances, 28*(1), 169–183.

Ashraf, M. F. M. R., & Foolad, M. R. (2007). Roles of glycine betaine and proline in improving plant abiotic stress resistance. *Environmental and Experimental Botany, 59*(2), 206–216.

Arshad, M. S., Farooq, M., Asch, F., Krishna, J. S., Prasad, P. V., & Siddique, K. H. (2017). Thermal stress impacts reproductive development and grain yield in rice. *Plant Physiology and Biochemistry, 115*, 57–72.

Askarnejad, M. R., Soleymani, A., & Javanmard, H. R. (2021). Barley (*Hordeum vulgare* L.) physiology including nutrient uptake affected by plant growth regulators under field drought conditions. *Journal of Plant Nutrition, 44*(15), 1–17. https://doi.org/10.1080/01904167.2021.1889593

Asopa, P. P., Bhatt, R., Sihag, S., Kothari, S. L., & Kachhwaha, S. (2017). Effect of cadmium on physiological parameters of cereal and millet plants—A comparative study. *Int Phytoremediation, 19*(3), 225–230.

Atta, N., Shahbaz, M., Farhat, F., Maqsood, M. F., Zulfiqar, U., Naz, N., Ahmed, M. M., Hassan, N. U., Mujahid, N., Mustafa, A. E., Elshikh, M. S., & Chaudhury, T. (2024). Proline-mediated redox regulation in wheat for mitigating nickel-induced stress and soil decontamination. *Scientific Reports, 14*(1), 456. https://doi.org/10.1038/s41598-023-50576-5

Awasthi, R., Bhandari, K., & Nayyar, H. (2015). Temperature stress and redox homeostasis in agricultural crops. *Frontiers in Environmental Science, 3*, 11. https://doi.org/10.3389/fenvs.2015.00011

Bailey-Serres, J., Parker, J. E., Ainsworth, E. A., Oldroyd, G. E., & Schroeder, J. I. (2019). Genetic strategies for improving crop yields. *Nature, 575*(7781), 109–118.

Balfagón, D., Zandalinas, S. I., & Gómez-Cadenas, A. (2019). High temperatures change the perspective: Integrating hormonal responses in citrus plants under co-occurring abiotic stress conditions. *Physiologia Plantarum, 165*(2), 183–197.

Balla, K., Karsai, I., Bónis, P., Kiss, T., Berki, Z., Horváth, Á., Mayer, M., Bencze, S., & Veisz, O. (2019). Heat stress responses in a large set of winter wheat cultivars (*Triticum aestivum* L.) depend on the timing, duration of stress. *PLoS One, 14*(9), e0222639. https://doi.org/10.1371/JOURNAL.PONE.0222639

Bar-Yosef, B. (1999). Advances in Fertigation. *Advances in Agronomy, 65*, 1–77. https://doi.org/10.1016/S0065-2113(08)60910-4

Begcy, K., Nosenko, T., Zhou, L. Z., Fragner, L., Weckwerth, W., & Dresselhaus, T. (2019). Male sterility in maize after transient heat stress during the tetrad stage of pollen development. *Plant Physiology, 181*(2), 683–700. https://doi.org/10.1104/PP.19.00707

Bheemanahalli, R., Vennam, R. R., Ramamoorthy, P., & Reddy, K. R. (2022). Effects of post-flowering heat and drought stresses on physiology, yield, and quality in maize (*Zea mays* L.). Plant. *Stress, 6*, 100106. https://doi.org/10.1016/j.stress.2022.100106

Bhusal, B., Poudel, M. R., Rishav, P., Regmi, R., Neupane, P., Bhattarai, K., Maharjan, B., Bigyan, K. C., & Acharya, S. (2021). A review on abiotic stress resistance in maize (*Zea mays* L.): Effects, resistance mechanisms and management. *Journal of Biology and Today's World, 10*(2), 1–3.

Bhusal, N., Sharma, P., Kumar, R. R., & Sareen, S. (2023). Editorial: Multiple abiotic stresses: Molecular, physiological, and genetic responses and adaptations in cereals. *Frontiers in Plant Science, 14*, 1146326. https://doi.org/10.3389/fpls.2023.1146326
Bista, D. R., Heckathorn, S. A., Jayawardena, D. M., & Boldt, J. K. (2020). Effect of drought, carbon dioxide on nutrient uptake, levels of nutrient-uptake proteins in roots of barley. *American Journal of Botany, 107*(10), 1401–1409. https://doi.org/10.1002/ajb2.1542
Bramley, R. (2009). Lessons from nearly 20 years of precision agriculture research, development, and adoption as a guide to its appropriate application. *Crop and Pasture Science, 60*(3), 197–217.
Calanca, P. P. (2017). Effects of abiotic stress in crop production. Quantification of climate variability, adaptation and mitigation for agricultural sustainability, 165–180.
Camargo, A. & Smith, J. (2009). An image-processing based early pest/disease detection system for greenhouse crops. *Biosystems Engineering, 104*(3), 292–304.
Cao, H., Ding, R., Kang, S., Du, T., Tong, L., Zhang, Y., Chen, J., & Shukla, M. K. (2023). Drought, salt, and combined stresses in plants: Effects, tolerance mechanisms, and strategies. *Advances in Agronomy, 178*, 107–163. https://doi.org/10.1016/bs.agron.2022.11.004
Cattivelli, L., Rizza, F., Badeck, F. W., Mazzucotelli, E., Mastrangelo, A. M., Francia, E., ... & Stanca, A. M. (2008). Drought tolerance improvement in crop plants: an integrated view from breeding to genomics. *Field Crops Research, 105*(1–2), 1–14.
Celedonio, R. P., De, S., Abeledo, L. G., & Miralles, D. J. (2018). Physiological traits associated with reductions in grain number in wheat, barley under waterlogging. *Plant and Soil, 429*(1), 469–481. https://doi.org/10.1007/S11104-018-3708-4
Chaudhry, S., & Sidhu, G. P. S. (2021). Climate change regulated abiotic stress mechanisms in plants: A comprehensive review. *Plant Cell Reports, 41*, 1–31.
Chaves, M. M., Maroco, J. P., & Pereira, J. S. (2003). Understanding plant responses to drought—from genes to the whole plant. *Functional Plant Biology, 30*(3), 239–264.
Chiluwal, A., Bheemanahalli, R., Kanaganahalli, V., Boyle, D., Perumal, R., Pokharel, M., Oumarou, H., & Jagadish, S. V. K. (2020). Deterioration of ovary plays a key role in heat stress-induced spikelet sterility in sorghum. *Plant, Cell & Environment, 43*(2), 448–462. https://doi.org/10.1111/PCE.13673
Cruz de Carvalho, M. H. (2008). Drought stress and reactive oxygen species: Production, scavenging and signaling. *Plant Signaling & Behavior, 3*(3), 156–165.
da Silva, E. C., Nogueira, R. J. M. C., da Silva, M. A., & de Albuquerque, M. B. (2011). Drought stress and plant nutrition. *Plant Stress, 5*(1), 32–41.
Daryanto, S., Wang, L., & Jacinthe, P. A. (2016). Global synthesis of drought effects on maize and wheat production. *PloS One, 11*(5), e0156362.
Das, K., & Roychoudhury, A. (2014). Reactive oxygen species (ROS) and response of antioxidants as ROS-scavengers during environmental stress in plants. *Frontiers in Environmental Science, 2*, 53.
De Wolf, E.D. & Isard, S.A. (2007). Disease cycle approach to plant disease prediction. *Annual Review Phytopathology, 45*, 203–220.
Dehnavi, A. R., Zahedi, M., Ludwiczak, A., Perez, S. C., & Piernik, A. (2020). Effect of salinity on seed germination, seedling development of sorghum (*Sorghum bicolor* (L.) Moench) genotypes. *Agronomy, 10*(6), 859. https://doi.org/10.3390/AGRONOMY10060859
Diacono, M., & Montemurro, F. (2010). Long-term effects of organic amendments on soil fertility. *Sustainable Agriculture, 2*, 761–786.
Djanaguiraman, M., Narayanan, S., Erdayani, E., & Prasad, P. V. (2020). Effects of high temperature stress during anthesis and grain filling periods on photosynthesis, lipids and grain yield in wheat. *BMC Plant Biology, 20*, 1–12.
Ebrahim, F., Arzani, A., Rahimmalek, M., Sun, D., & Peng, J. (2020). Salinity tolerance of wild barley *Hordeum vulgares* sp. spontaneum. *Plant Breeding, 139*(2), 304–316. https://doi.org/10.1111/PBR.12770
Eckardt, N. A., Cutler, S., Juenger, T. E., Marshall-Colon, A., Udvardi, M., & Verslues, P. E. (2023). Focus on climate change and plant abiotic stress biology. *The Plant Cell, 35*(1), 1–3.

EL Sabagh, A., Islam, M. S., Skalicky, M., Ali Raza, M., Singh, K., Anwar Hossain, M., Hossain, A., Mahboob, W., Iqbal, M. A., Ratnasekera, D., & Singhal, R. K. (2021). Salinity stress in wheat (*Triticum aestivum* L.) in the changing climate: Adaptation and management strategies. *Frontiers in Agronomy, 3*, 661932.

El-Sappah, A. H., Rather, S. A., Wani, S. H., Elrys, A. S., Bilal, M., Huang, Q., Dar, Z. A., Elashtokhy, M., Soaud, N., Koul, M., & Mir, R. R. (2022). Heat stress-mediated constraints in maize (*Zea mays*) production: Challenges and solutions. *Frontiers in Plant Science, 13*, 879366.

Ertekin, E. N., Ertekin, İ., & Bilgen, M. (2020). Effects of some heavy metals on germination and seedling growth of sorghum. *Kahramanmaraş Sütçü İmam Üniversitesi Tarım ve Doğa Dergisi, 23*(6), 1608–1615.

Essemine, J., Xiao, Y., Qu, M., Mi, H., & Zhu, X. G. (2017). Cyclic electron flow may provide some protection against PSII photoinhibition in rice (Oryza sativa L.) leaves under heat stress. *Journal of Plant Physiology, 211*, 138–146.

Fahad, S., Bajwa, A. A., Nazir, U., Anjum, S. A., Farooq, A., Zohaib, A., Sadia, S., Nasim, W., Adkins, S., Saud, S., Ihsan, M. Z., Alharby, H., Wu, C., Wang, D., & Huang, J. (2017). Crop production under drought and heat stress: Plant responses and management options. *Frontiers in Plant Science, 8*, 1147. https://doi.org/10.3389/fpls.2017.01147

Farooq, M., Wahid, A., Kobayashi, N. S., Fujita, D. B., & Basra, S. M. (2009). Plant drought stress: Effects, mechanisms and management. In *Sustainable agriculture* (pp. 153–188). Springer. https://doi.org/10.1007/978-90-481-2666-8_12

Feller, U., & Vaseva, I. I. (2014). Extreme climatic events: impacts of drought and high temperature on physiological processes in agronomically important plants. *Frontiers in Environmental Science, 2*, 39.

Ferchichi, S., Hessini, K., Dell'Aversana, E., D'Amelia, L., Woodrow, P., Ciarmiello, L. F., Fuggi, A., & Carillo, P. (2018). *Hordeum vulgare, Hordeum maritimum* respond to extended salinity stress displaying different temporal accumulation pattern of metabolites. *Functional Plant Biology, 45*(11), 1096–1109. https://doi.org/10.1071/FP18046

Fu, Y., Li, P., Mounkaila Hamani, A. K., Wan, S., Gao, Y., & Wang, X. (2023). Effects of single and combined drought and salinity stress on the root morphological characteristics and root hydraulic conductivity of different winter wheat varieties. *Plants, 12*(14), 2694.

Galau, G.A., Bijaisoradat, N., & Hughes, D.W. (1987). Accumulation kinetics of cotton late embryogenesis- abundant mRNAs and storage protein mRNAs: Coordinate regulation during embryogenesis and the role of abscisic acid. *Developmental Biology, 123*(1):198–212

Garg, M., Sharma, N., Sharma, S., Kapoor, P., Kumar, A., Chunduri, V., & Arora, P. (2018). Biofortified crops generated by breeding, agronomy, and transgenic approaches are improving lives of millions of people around the world. *Frontiers in Nutrition, 5*, 12. https://doi.org/10.3389/fnut.2018.00012

Gebbers, R. & Adamchuk, V. I. (2010). Precision agriculture and food security. *Science, 327*(5967), 828–831.

Ghosh, B., Md, N. A., & Gantait, S. (2016). Response of rice under salinity stress: A review update. *Journal of Rice Research, 4*, 2.

Giri, A. (2019). *Wheat improvement for heat and drought stress tolerance* (Doctoral dissertation).

Golizadeh, F., & Kumleh, H. H. (2019). Physiological responses, expression changes of fatty acid metabolism–related genes in wheat (*Triticum aestivum*) under cold stress. *Plant Molecular Biology Reporter, 37*(3), 224–236. https://doi.org/10.1007/S11105-019-01150-9

Gunasekera, D., & Ratnasekera, D. (2023). Advancement in mitigating the effects of drought stress in wheat. In *Abiotic stresses in wheat* (pp. 297–311). Academic Press. https://doi.org/10.1016/B978-0-323-95368-9.00023-0

Guo, W., Zheng, B., Potgieter, A. B., Diot, J., Watanabe, K., Noshita, K., Jordan, D. R., Wang, X., Watson, J., Ninomiya, S., & Chapman, S. C. (2018). Aerial imagery analysis–quantifying appearance and number of sorghum heads for applications in breeding and agronomy. *Frontiers in Plant Science, 9*, 1544.

Gupta, B., & Huang, B. (2014). Mechanism of salinity tolerance in plants: Physiological, biochemical, and molecular characterization. *International Journal of Genomics*, 701596. https://doi.org/10.1155/2014/701596

Hailu, B., & Mehari, H. (2021). Impacts of soil salinity/sodicity on soil-water relations and plant growth in dry land areas: A review. *Journal of Natural Science Research, 12*, 1–10.

Hasanuzzaman, M., Mahmud, J., Al-Nahar, K., Anee, T. I., Inafuku, M., Oku, H., & Fujita, M. (2017). Responses, adaptation, ROS metabolism in plants exposed to water logging stress. In *Reactive oxygen species, antioxidant systems in plants: Role, regulation under abiotic stress* (pp. 257–281). Springer. https://doi.org/10.1007/978-981-10-5254-5_10

Hasanuzzaman, M., Bhuyan, M. B., Zulfiqar, F., Raza, A., Mohsin, S. M., Mahmud, J. A., Fujita, M., & Fotopoulos, V. (2020). Reactive oxygen species and antioxidant defense in plants under abiotic stress: Revisiting the crucial role of a universal defense regulator. *Antioxid, 9*(8), 681.

Hendrawan, V. S. A., Komori, D., & Kim, W. (2023). Possible factors determining global-scale patterns of crop yield sensitivity to drought. *PLoS One, 18*(2), e0281287.

Hobbs, P. R., Sayre, K., & Gupta, R. (2008). The role of conservation agriculture in sustainable agriculture. *Philosophical Transactions of the Royal Society B: Biological Sciences, 363*(1491), 543–555.

Hossain, A., Skalicky, M., Brestic, M., Maitra, S., Ashraful Alam, M., Syed, M. A., ... & Islam, T. (2021). Consequences and mitigation strategies of abiotic stresses in wheat (Triticum aestivum L.) under the changing climate. *Agronomy, 11*(2), 241.

Hou, S., Zheng, N., Tang, L., Ji, X., & Li, Y. (2019). Effect of soil pH and organic matter content on heavy metals availability in maize (*Zea mays* L.) rhizospheric soil of non-ferrous metals smelting area. *Environmental Monitoring and Assessment, 191*, 1–10.

Hsu, C. H., & Hsu, Y. T. (2019). Biochemical responses of rice roots to cold stress. *Botanical Studies, 60*(1), 1–12. https://doi.org/10.1186/S40529-019-0262-1

Huang, K., Li, M., Li, R., Rasul, F., Shahzad, S., Wu, C., Shao, J., Huang, G., Li, R., Almari, S., & Hashem, M. (2023). Soil acidification and salinity: The importance of biochar application to agricultural soils. *Frontiers in Plant Science, 14*, 1206820. https://doi.org/10.3389/fpls.2023.1206820

Hussain, S., Khan, A. A., Shakoor, A., Goheer, A., Qadir, T., Khan, M. M., & Hussain, Z. (2018). Effect of cold and heat stress on different stages of wheat: a review. *Journal of Global Innovations in Agricultural and Social Sciences, 6*(4), 123–128.

Hussain, H. A., Men, S., Hussain, S., Chen, Y., Ali, S., Zhang, S., Zhang, K., Li, Y., Xu, Q., Liao, C., & Wang, L. (2019a). Interactive effects of drought, heat stresses on morpho-physiological attributes, yield, nutrient uptake, oxidative status in maize hybrids. *Scientific Reports, 9*(1), 1–12. https://doi.org/10.1038/s41598-019-40362-7

Hussain, S., Shaukat, M., Ashraf, M., Zhu, C., Jin, Q., & Zhang, J. (2019b). Salinity stress in arid and semi-arid climates: Effects and management in field crops. *Climate Change and Agriculture, 13*, 201–226.

Hussain, S., Wang, J., Asad Naseer, M., Saqib, M., Siddiqui, M. H., Ihsan, F., Xiaoli, C., Xiaolong, R., Hussain, S., & Ramzan, H. N. (2023). Water stress memory in wheat/maize intercropping regulated photosynthetic and antioxidative responses under rainfed conditions. *Scientific Reports, 13*(1), 13688.

Ilyas, N., Mazhar, R., Yasmin, H., Khan, W., Iqba, S., El-Enshasy, H., & Dailin, D. J. (2020). Rhizobacteria isolated from saline soil induce systemic tolerance in wheat (*Triticum aestivum* L.) against salinity stress. *Agronomy, 10*(7), 989. https://doi.org/10.3390/agronomy10070989

Iqbal, S., Hussain, S., Qayyaum, M. A., & Ashraf, M. (2020). The response of maize physiology under salinity stress and its coping strategies. In *Plant stress physiology*. Intech Open.

Islam, S., Rahman, M. M., Islam, M. R., & Naidu, R. (2016). Arsenic accumulation in rice: Consequences of rice genotypes and management practices to reduce human health risk. *Environment International, 96*, 139–155.

Jan, R., Asif, S., Asaf, S., Du, X. X., Park, J. R., Nari, K., Bhatta, D., Lee, I. J., & Kim, K. M. (2023). Melatonin alleviates arsenic (As) toxicity in rice plants via modulating antioxidant defense system and secondary metabolites and reducing oxidative stress. *Environmental Pollution, 318*, 120868.

Jańczak-Pieniążek, M., Cichoński, J., Michalik, P., & Chrzanowski, G. (2022). Effect of heavy metal stress on phenolic compounds accumulation in winter wheat plants. *Molecules, 28*(1), 241.

Janmohammadi, M., Zolla, L., & Rinalducci, S. (2015). Low temperature tolerance in plants: Changes at the protein level. *Phytochemistry, 117*, 76–89.

Jatav, P. K., Verma, R., Kothari, S. L., Jain, R., & Kachhwaha, S. (2021). Relative morpho-physiological responses of millets and oats against lead toxicity. *Environmental and Experimental Botany, 192*, 104658.

Jeyasri, R., Muthuramalingam, P., Satish, L., Pandian, S. K., Chen, J. T., Ahmar, S., Wang, X., Mora-Poblete, F., & Ramesh, M. (2021). An overview of abiotic stress in cereal crops: Negative impacts, regulation, biotechnology, integrated omics. *Plants, 10*(7), 1472. https://doi.org/10.3390/PLANTS10071472

Ji, H., Xiao, L., Xia, Y., Song, H., Liu, B., Tang, L., Cao, W., Zhu, Y., & Liu, L. (2017). Effects of jointing and booting low temperature stresses on grain yield and yield components in wheat. *Agricultural and Forest Meteorology, 243*, 33–42.

Jia, Y., Xiao, W., Ye, Y., Wang, X., Liu, X., Wang, G., Li, G., & Wang, Y. (2020). Response of photosynthetic performance to drought duration and re-watering in maize. *Agronomy, 10*(4), 533.

Jones, H.G. (2004). Irrigation scheduling: advantages and pitfalls of plant-based methods. *Journal of Experimental Botany, 55*(407), 2427–2436.

Kang, J. S., Singh, H., Singh, G., Kang, H., Kalra, V. P., & Kaur, J. (2017). Abiotic stress and its amelioration in cereals and pulses: a review. *International Journal of Current Microbiology and Applied Sciences, 6*(3), 1019–1045.

Kebede, A., Kang, M. S., & Bekele, E. (2019). Advances in mechanisms of drought tolerance in crops, with emphasis on barley. *Advances in Agronomy, 156*, 265–314.

Khanam, R., Kumar, A., Nayak, A. K., Shahid, M., Tripathi, R., Vijayakumar, S., Bhaduri, D., Kumar, U., Mohanty, S., Panneerselvam, P., & Chatterjee, D. (2020). Metal (loid) s (As, Hg, Se, Pb and Cd) in paddy soil: Bioavailability and potential risk to human health. *Science of the Total Environment, 699*, 134330.

Khatri-Chhetri, A., Aggarwal, P.K., Joshi, P. K., & Vyas, S. (2020). Farmers' prioritization of climate-smart agriculture (CSA) technologies. *Agricultural Systems, 181*, 102810.

Kopecká, R., Kameniarová, M., Černý, M., Brzobohatý, B., & Novák, J. (2023). Abiotic stress in crop production. *International Journal of Molecular Sciences, 24*(7), 6603.

Krasensky, J., & Jonak, C. (2012). Drought, salt, and temperature stress-induced metabolic rearrangements and regulatory networks. *Journal of Experimental Botany, 63*(4), 1593–1608.

Krishna, S. S., Reddy, Y. N., & Kumar, R. R. (2021). Assessment of traits for grain yield under drought in finger millet. *Plant Physiology Reports, 26*, 84–94.

Krupinsky, J. M., Tanaka, D. L., Merrill, S. D., Liebig, M. A., & Hanson, J. D. (2006). Crop sequence effects of 10 crops in the northern Great Plains. *Agricultural Systems, 88*(2), 227–254.

Kulczycki, G., Sacała, E., Chohura, P., & Załuska, J. (2022). Maize and wheat response to drought stress under varied sulphur fertilisation. *Agronomy, 12*(5), 1076.

Kumar, S. (2020). Abiotic stresses and their effects on plant growth, yield and nutritional quality of agricultural produce. *International Journal of Food Science and Agriculture, 4*, 367–378.

Kumar, P., Choudhary, M., Halder, T., Prakash, N. R., Singh, V., Vineeth, V. T., Sheoran, S., Longmei, N., Rakshit, S., & Siddique, K. H. (2022). Salinity stress tolerance and omics approaches: Revisiting the progress and achievements in major cereal crops. *Heredity, 128*(6), 497–518.

Larkin, R.P., Honeycutt, C.W., & Olanya, O.M. (2011). Management of Verticillium wilt of potato with diseasesuppressive green manures and as biological soil amendments. *Plant Disease, 95*(5), 568–576.

Li, X., Cai, J., Liu, F., Dai, T., Cao, W., & Jiang, D. (2014). Cold priming drives the sub-cellular antioxidant systems to protect photosynthetic electron transport against subsequent low temperature stress in winter wheat. *Plant Physiology and Biochemistry, 82*, 34–43.

Li, X., Pu, H., Liu, F., Zhou, Q., Cai, J., Dai, T., Cao, W., & Jiang, D. (2015). Winter wheat photosynthesis and grain yield responses to spring freeze. *Agronomy Journal, 107*(3), 1002–1010.

Lin, B.B. (2011). Resilience in agriculture through crop diversification: adaptive management for environmental change. *BioScience, 61*(3), 183–193.

Liu, J. (2000). The Arabidopsis thaliana SOS2 gene encodes a protein kinase that is required for salt tolerance. *Proceedings of the National Academy of Sciences, 97*(7):3730–3734

Liu, T., Gu, L., Dong, S., Zhang, J., Liu, P., & Zhao, B. (2015). Optimum leaf removal increases canopy apparent photosynthesis, 13C-photosynthate distribution and grain yield of maize crops grown at high density. *Field Crops Research, 170*, 32–39.

Liu, L., Xia, W., Li, H., Zeng, H., Wei, B., Han, S., & Yin, C. (2018). Salinity inhibits rice seed germination by reducing α-amylase activity via decreased bioactive gibberellins content. *Frontiers in Plant Science, 9*, 275. https://doi.org/10.3389/fpls.2018.00275

Loreti, E., & Perata, P. (2020). The many facets of hypoxia in plants. *Plants, 9*(6), 745. https://doi.org/10.3390/PLANTS9060745

Lv, C., Huang, Y., Sun, W., Yu, L., & Zhu, J. (2020). Response of rice yield and yield components to elevated [CO2]: Asynthesis of updated data from FAC Eexperiments. *European Journal of Agronomy, 112*, 125961.

Ma, Y., Dias, M. C., & Freitas, H. (2020). Drought and salinity stress responses and microbe-induced tolerance in plants. *Frontiers in Plant Science, 11*, 591911.

Mahla, R., Madan, S., Kaur, V., Munjal, R., Behl, R. K., & Midathala, R. (2017). Activities of sucrose to starch metabolizing enzymes during grain filling in late sown wheat under water stress. *Journal of Applied and Natural Science, 9*(1), 338–343. https://doi.org/10.31018/jans.v9i1.1193

Mahrookashani, A., Siebert, S., Hüging, H., & Ewert, F. (2017). Independent and combined effects of high temperature and drought stress around anthesis on wheat. *Journal of Agronomy and Crop Science, 203*(6), 453–463.

Maiti, R. K., & Satya, P. (2014). Research advances in major cereal crops for adaptation to abiotic stresses. *GM Crops & Food, 5*(4), 259–279.

Majeed, A., Muhammad, Z., & Siyar, S. (2019). Assessment of heavy metal induced stress responses in pea (Pisum sativum L.). *Acta Ecologica Sinica, 39*(4), 284–288.

Malusá, E., Sas-Paszt, L., & Ciesielska, J. (2012). Technologies for beneficial microorganisms inocula used as biofertilizers. *The Scientific World Journal*, 2012:491206. https://doi.org/10.1100/2012/491206

Maroco, J. P., Pereira, J. S., & Chaves, M. M. (2000). Growth, photosynthesis and water-use efficiency of two C4Sahelian grasses subjected to water deficits. *Journal of Arid Environments, 45*(2), 119–137.

Masson-Delmotte, V., Zhai, P., Pörtner, H. O., Roberts, D., Skea, J., & Shukla, P. R. (2022). *Global Warming of 1.5 C: IPCC special report on impacts of global warming of 1.5 C above pre-industrial levels in context of strengthening response to climate change, sustainable development, and efforts to eradicate poverty*. Cambridge University Press.

Mawia, A. M., Hui, S., Zhou, L., Li, H., Tabassum, J., Lai, C., Wang, J., Shao, G., Wei, X., Tang, S., & Luo, J. (2021). Inorganic arsenic toxicity and alleviation strategies in rice. *Journal of Hazardous Materials, 408*, 124751.

McDaniel, M.D., Tiemann, L.K., & Grandy, A.S. (2014). Does agricultural crop diversity enhance soil microbial biomass and organic matter dynamics? A meta-analysis. *Ecological Applications, 24*(3), 560–570.

Megha, S., Basu, U., & Kav, N. N. (2018). Regulation of low temperature stress in plants by microRNAs. *Plant, Cell & Environment, 41*(1), 1–15.

Mei, J., Wang, W., Peng, S., & Nie, L. (2017). Seed pelleting with calcium peroxide improves crop establishment of direct-seeded rice under waterlogging conditions. *Scientific Reports, 7*(1), 1–12. https://doi.org/10.1038/s41598-017-04966-1

Mekawy, A. M. M., Abdelaziz, M. N., & Ueda, A. (2018). Apigenin pre treatment enhances growth, salinity tolerance of rice seedlings. *Plant Physiology and Biochemistry, 130*, 94–104. https://doi.org/10.1016/j.plaphy.2018.06.036

Miura, K., & Furumoto, T. (2013). Cold signaling and cold response in plants. *International Journal of Molecular Sciences, 14*(3), 5312–5337.

Mondal, S., Dutta, P., Bandopadhyay, P., & Maji, S. (2020). Arsenic contamination in major food crops: Issues and mitigation in Indian subcontinent perspective. In *Agronomic crops* (Stress response tolerance) (Vol. 3, pp. 209–234).

Muehe, E. M., Wang, T., Kerl, C. F., Planer-Friedrich, B., & Fendorf, S. (2019). Rice production threatened by coupled stresses of climate and soil arsenic. *Nature Communications, 10*(1), 4985.

Munns, R., & Tester, M. (2008). Mechanisms of salinity tolerance. *Annual Review of Plant Biology, 59*, 651–681.

Nelimor, C., Badu-Apraku, B., Tetteh, A. Y., & N'guetta, A. S. (2019). Assessment of genetic diversity for drought, heat and combined drought and heat stress tolerance in early maturing maize landraces. *Plants, 8*(11), 518.

Ngara, R., & Ndimba, B. K. (2014). Understanding the complex nature of salinity and drought-stress response in cereals using proteomics technologies. *Proteom, 14*(4–5), 611–621.

Niu, S., Du, X., Wei, D., Liu, S., Tang, Q., Bian, D., Zhang, Y., Cui, Y., & Gao, Z. (2021). Heat stress after pollination reduces kernel number in maize by insufficient assimilates. *Front Gene, 12*, 728166.

OECD-FAO. (2023). *Agricultural outlook 2023–2032*. OECD/FAO.

Ortiz, D., Hu, J., & Salas-Fernandez, M. G. (2017). Genetic architecture of photosynthesis in Sorghum bicolor under non-stress, cold stress conditions. *Journal of Experimental Botany, 68*(16), 4545–4557. https://doi.org/10.1093/JXB/ERX276

Ozbek, H. F., Topal, A., Dumanoglu, Y., Solgon, V., & Bay, K. (2020). Controlled environment agriculture: plant protection and plant health. *GAIA Scientia, 2020* (Special Issue 1), 84–94.

Palm, C., Blanco-Canqui, H., DeClerck, F., Gatere, L., & Grace, P. (2014). Conservation agriculture and ecosystem services: An overview. *Agriculture, Ecosystems & Environment, 187*, 87–105.

Pandey, P., Srivastava, S., Pandey, A. K., & Dubey, R. S. (2020). Abiotic-stress tolerance in plants-system biology approach. In *Plant life under changing environment* (pp. 577–609). Academic Press.

Paul, K., Pauk, J., Kondic-Spika, A., Grausgruber, H., Allahverdiyev, T., Sass, L., & Vass, I. (2019). Co-occurrence of mild salinity and drought synergistically enhances biomass and grain retardation in wheat. *Frontiers in Plant Science, 10*, 501.

Perdomo, J. A., Capó-Bauçà, S., Carmo-Silva, E., & Galmés, J. (2017). Rubisco and rubisco activase play an important role in the biochemical limitations of photosynthesis in rice, wheat, and maize under high temperature and water deficit. *Frontiers in Plant Science, 8*, 490.

Pospíšil, P. (2016). Production of reactive oxygen species by photosystem II as a response to light and temperature stress. *Frontiers in Plant Science, 7*, 1950.

Prasad, P. V. V., Pisipati, S. R., Momčilović, I., & Ristic, Z. (2011). Independent and combined effects of high temperature and drought stress during grain filling on plant yield and chloroplast EF-Tu expression in spring wheat. *Journal of Agronomy and Crop Science, 197*(6), 430–441.

Qaseem, M. F., Qureshi, R., & Shaheen, H. (2019). Effects of pre-anthesis drought, heat and their combination on the growth, yield and physiology of diverse wheat (Triticum aestivum L.) genotypes varying in sensitivity to heat and drought stress. *Scientific Reports, 9*(1), 6955.

Qu, L., Gu, X., Li, J., Guo, J., & Lu, D. (2023). Leaf photosynthetic characteristics of waxy maize in response to different degrees of heat stress during grain filling. *BMC Plant Biology, 23*(1), 469.

Rahman, S., Ahmad, I., & Nafees, M. (2023). Mitigation of heavy metal stress in maize (Zea mays L.) through application of silicon nanoparticles. *Biocatalysis and Agricultural Biotechnology, 102757*. https://doi.org/10.1016/j.bcab.2023.102757

Rane, J., Singh, A. K., Kumar, M., Boraiah, K. M., Meena, K. K., Pradhan, A., & Prasad, P. V. (2021). The adaptation and tolerance of major cereals and legumes to important abiotic stresses. *International Journal of Molecular Sciences, 22*(23), 12970.

Reddy, M.S., Kumar, S., Babita, K., & Reddy, M.S. (2002). Biosolubilization of poorly soluble rock phosphates by Aspergillus tubingensis and Aspergillus niger. *Bioresource Technology, 84*(2), 187–189.

Reddy, K. R., Seghal, A., Jumaa, S., Bheemanahalli, R., Kakar, N., Redoña, E. D., Wijewardana, C., Alsajri, F. A., Chastain, D., Gao, W., Taduri, S., & Lone, A. A. (2021). Morpho-physiological characterization of diverse rice genotypes for seedling stage high-, low-temperature tolerance. *Agronomy, 11*(1), 112. https://doi.org/10.3390/AGRONOMY11010112

Reeve, J. R., Hoagland, L. A., Villalba, J. J., Carr, P. M., Atucha, A., Cambardella, C., Davis, D. R., & Delate, K. (2020). Organic farming promotes ecosystem services. *Nature Sustainability, 3*(1), 54–60.

Ripley, B., Frole, K., & Gilbert, M. (2010). Differences in drought sensitivities and photosynthetic limitations between co-occurring C3 and C4 (NADP-ME) Panicoid grasses. *Annals of Botany, 105*(3), 493–503.

Ritonga, F. N., & Chen, S. (2020). Physiological and molecular mechanism involved in cold stress tolerance in plants. *Plants, 9*(5), 560.

Ruiz-Huerta, E. A., Armienta-Hernández, M. A., Dubrovsky, J. G., & Gómez-Bernal, J. M. (2022). Bioaccumulation of heavy metals and As in maize (Zea mays L) grown close to mine tailings strongly impacts plant development. *Ecotoxicology, 31*(3), 447–467.

Sachdev, S., Ansari, S. A., Ansari, M. I., Fujita, M., & Hasanuzzaman, M. (2021). Abiotic stress and reactive oxygen species: Generation, signaling, and defense mechanisms. *Antioxid, 10*(2), 277.

Sah, R. P., Chakraborty, M., Prasad, K., Pandit, M., Tudu, V. K., Chakravarty, M. K., Narayan, S. C., Rana, M., & Moharana, D. (2020). Impact of water deficit stress in maize: Phenology and yield components. *Scientific Reports, 10*(1), 2944.

Sangoi, L., & Salvador, R. J. (1998). Maize susceptibility to drought at flowering: A new approach to overcome the problem. *Ciência Rural, 28*, 699–706.

Saud, S., Wang, D., Fahad, S., Alharby, H. F., Bamagoos, A. A., Mjrashi, A., Alabdallah, N. M., AlZahrani, S. S., AbdElgawad, H., Adnan, M., & Sayyed, R. Z. (2022). Comprehensive impacts of climate change on rice production and adaptive strategies in China. *Frontiers in Microbiology, 13*, 926059.

Schoppach, R., Taylor, J. D., Majerus, E., Claverie, E., Baumann, U., Suchecki, R., Fleury, D., & Sadok, W. (2016). High resolution mapping of traits related to whole-plant transpiration under increasing evaporative demand in wheat. *Journal of Experimental Botany, 67*(9), 2847–2860.

Seleiman, M. F., Aslam, M. T., Alhammad, B. A., Hassan, M. U., Maqbool, R., Chattha, M. U., Khan, I., Gitari, H. I., Uslu, O. S., Roy, R., & Battaglia, M. L. (2022). Salinity stress in wheat: Effects, mechanisms and management strategies. *Phyton, 91*(4), 0031–9457.

Shahverdi, M. A., Omidi, H., & Tabatabaei, S. J. (2018). Plant growth and steviol glycosides as affected by foliar application of selenium, boron, and iron under NaCl stress in Stevia rebaudiana Bertoni. *Industrial Crops and Products, 125*, 408–415.

Shahzad, R., Jamil, S., Ahmad, S., Nisar, A., Khan, S., Amina, Z., Kanwal, S., Aslam, H. M. U., Gill, R. A., & Zhou, W. (2021). Biofortification of cereals and pulses using new breeding techniques: Current and future perspectives. *Frontiers in Nutrition, 8*, 721728. https://doi.org/10.3389/fnut.2021.721728

Shelden, M. C., & Roessner, U. (2013). Advances in functional genomics for investigating salinity stress tolerance mechanisms in cereals. *Frontiers in Plant Science, 4*, 123.

Sheoran, S., Kaur, Y., Kumar, S., Shukla, S., Rakshit, S., & Kumar, R. (2022). Recent advances for drought stress tolerance in maize (*Zea mays* l.): Present status and future prospects. *Frontiers in Plant Science, 13*, 872566.

Shi, H., Ishitani M., Kim, C., Zhu, J-K. (2000). The Arabidopsis thaliana salt tolerance gene SOS1 encodes a putative Na+/H+ antiporter. *Proceedings of the National Academy of Sciences, 97*(12), 6896–6901.

Singh, A., Sharma, M. K., & Sengar, R. S. (2017). Osmolytes: proline metabolism in plants as sensors of abiotic stress. *Journal of Applied & Natural Science, 9*(4), 2079–2092. https://doi.org/10.31018/jans.v9i4.1492

Shinozaki, K., & Yamaguchi-Shinozaki, K. (2007). Gene networks involved in drought stress response and tolerance. *Journal of Experimental Botany, 58*(2), 221–227.

Shokat, S., Großkinsky, D. K., & Liu, F. (2021). Impact of elevated CO2 on two contrasting wheat genotypes exposed to intermediate drought stress at anthesis. *Journal of Agronomy and Crop Science, 207*(1), 20–33. https://doi.org/10.1111/jac.12442

Shukla, A. K., Behera, S. K., Prakash, C., Tripathi, A., Patra, A. K., Dwivedi, B. S., Trivedi, V., Rao, C. S., Chaudhari, S. K., Das, S., & Singh, A. K. (2021). Deficiency of phyto-available sulphur, zinc, boron, iron, copper and manganese in soils of India. *Scientific Reports, 11*(1), 19760.

Siebers, M. H., Slattery, R. A., Yendrek, C. R., Locke, A. M., Drag, D., Ainsworth, E. A., Bernacchi, C. J., & Ort, D. R. (2017). Simulated heat waves during maize reproductive stages alter reproductive growth but have no lasting effect when applied during vegetative stages. *Agriculture, Ecosystems and Environment, 240*, 162–170.

Smith, R.G., Menalled, F.D., & Robertson, G.P. (2007). Temporal yield variability under conventional and alternative management systems. *Agronomy Journal, 99*(6), 1629–1634.

Sohag, A. A. M., Tahjib-Ul-Arif, M., Polash, M. A. S., Belal-Chowdhury, M., Afrin, S., Burritt, D. J., Murata, Y., Hossain, M. A., & Afzal-Hossain, M. (2020). Exogenous glutathione-mediated drought stress tolerance in rice (Oryza sativa L.) is associated with lower oxidative damage, favorable ionic homeostasis. *Iranian Journal of Science and Technology, Transaction A, Science, 44*(4), 955–971. https://doi.org/10.1007/s40995-020-00917-0

Srivastava, V., Sarkar, A., Singh, S., Singh, P., De Araujo, A. S., & Singh, R. P. (2017). Agroecological responses of heavy metal pollution with special emphasis on soil health and plant performances. *Frontiers in Environmental Science, 5*, 64.

Spence, C., Alff, E., Johnson, C., Ramos, C., Donofrio, N., Sundaresan, V., & Bais, H. (2014). Natural ricerhizospheric microbes suppress rice blast infections. *BMC Plant Biology, 14*(1), 130.

Steenwerth, K.L., Hodson, A.K., Bloom, A.J., Carter, M.R., Cattaneo, A., Chartres, C.J., ... & Jackson, L.E. (2014). Climate-smart agriculture global research agenda: scientific basis for action. *Agriculture & Food Security, 3*(1), 1–39.

Stoecker, B. J., Bernhardt, E. S., & Millet, D. B. (2020). Benefits and trade-offs of public subsidies for agricultural best management practices to improve water quality. *Proceedings of the National Academy of Sciences, 117*(27), 15581–15588.

Sui, N., Yang, Z., Liu, M., & Wang, B. (2015). Identification and transcriptomic profiling of genes involved in increasing sugar content during salt stress in sweet sorghum leaves. *BMC Genomics, 16*(1), 1–18.

Sullivan, D. G., Fulton, J. P., Shaw, J. N., & Bland, G. (2007). Evaluating the sensitivity of an unmanned thermal infrared aerial system to detect water stress in a cotton canopy. *Transactions of the ASABE, 50*(6), 1963–1969.

Takagi, D., Ihara, H., Takumi, S., & Miyake, C. (2019). Growth light environment changes the sensitivity of photosystem I photo inhibition depending on common wheat cultivars. *Frontiers in Plant Science, 10*, 686. https://doi.org/10.3389/FPLS.2019.00686

Tavili, A., Zare, S., Moosavi, S. A., & Enayati, A. (2011). Effects of seed priming on germination characteristics of Bromus species under salt and drought conditions. *American-Eurasian Journal of Agricultural & Environmental Sciences, 10*(2), 163–168.

Teng, L. I., Zhang, X. P., Qing, L. I. U., Jin, L. I. U., Chen, Y. Q., & Peng, S. U. I. (2022). Yield penalty of maize (Zea mays L.) under heat stress in different growth stages: A review. *Journal of Integrative Agriculture, 21*(9), 2465–2476. https://doi.org/10.1016/j.jia.2022.07.013

Thomason, K., Babar, M. A., Erickson, J. E., Mulvaney, M., Beecher, C., & MacDonald, G. (2018). Comparative physiological and metabolomics analysis of wheat (Triticum aestivum L.) following post-anthesis heat stress. *PLoS One, 13*(6), 0197919.

Thussagunpanit, J., Jutamanee, K., Kaveeta, L., Chai-arree, W., Pankean, P., Homvisasevongsa, S., & Suksamrarn, A. (2015). Comparative effects of brassinosteroid and brassinosteroid mimic on improving photosynthesis, lipid peroxidation, and rice seed set under heat stress. *Journal of Plant Growth Regulation, 34*, 320–331.

Tian, L., Li, J., Bi, W., Zuo, S., Li, L., Li, W., & Sun, L. (2019). Effects of waterlogging stress at different growth stages on the photosynthetic characteristics, grain yield of spring maize (*Zea mays* L.) under field conditions. *Agricultural Water Management, 218*, 250–258. https://doi.org/10.1016/J.AGWAT.2019.03.054

Ullah, A., Nadeem, F., Nawaz, A., Siddique, K. H., & Farooq, M. (2022). Heat stress effects on the reproductive physiology and yield of wheat. *Journal of Agronomy and Crop Science, 208*(1), 1–17.

Upadhyay, M. K., Majumdar, A., Suresh Kumar, J., & Srivastava, S. (2020). Arsenic in rice agro-ecosystem: Solutions for safe and sustainable rice production. *Frontiers in Sustainable Food Systems, 4*, 53.

Usmani, M. M., Nawaz, F., Majeed, S., Shehzad, M. A., Ahmad, K. S., Akhtar, G., Aqib, M., & Shabbir, R. N. (2020). Sulfate-mediated drought tolerance in maize involves regulation at physiological, biochemical levels. *Scientific Reports, 10*(1), 1–13. https://doi.org/10.1038/s41598-020-58169-2

Valizadeh-Kamran, R., Toorchi, M., Mogadam, M., Mohammadi, H., & Pessarakli, M. (2017). Effects of freeze, cold stress on certain physiological, biochemical traits in sensitive, tolerant barley (*Hordeum vulgare*) genotypes. *Journal of Plant Nutrition, 41*(1), 102–111. https://doi.org/10.1080/01904167.2017.1381730

Venuprasad, R., Dalid, C.O., Del Valle, M., Zhao, D., Espiritu, M., Cruz, M.T.S., Amante, M., Kumar, A., & Atlin, G.N. (2009). Identification and characterization of large-effect quantitative trait loci for grain yield under lowland drought stress in rice using bulk-segregant analysis. *Theoretical and Applied Genetics, 120*(1), 177–190.

Vierling, E. (1991). The roles of heat shock proteins in plants. *Annual Review of Plant Physiology and Plant Molecular Biology, 42*(1):579–620

Wani, S. H., & Sah, S. K. (2014). Biotechnology and abiotic stress tolerance in rice. *Journal of Rice Research, 2*(2), e105.

Wan, C., Dang, P., Gao, L., Wang, J., Tao, J., Qin, X., Feng, B., & Gao, J. (2022). How does the environment affect wheat yield and protein content response to drought? A meta-analysis. *Frontiers in Plant Science, 13*, 896985.

Wang, H. Q., Liu, P., Zhang, J. W., Zhao, B., & Ren, B. Z. (2020). Endogenous hormones inhibit differentiation of young ears in maize (*Zea mays* L.) under heat stress. *Frontiers in Plant Science, 11*, 533046. https://doi.org/10.3389/FPLS.2020.533046

Wang, Y., Tao, H., Tian, B., Sheng, D., Xu, C., Zhou, H., Huang, S., & Wang, P. (2019). Flowering dynamics, pollen, and pistil contribution to grain yield in response to high temperature during maize flowering. *Environmental and Experimental Botany, 158*, 80–88.

Wu, C., Cui, K., Tang, S., Li, G., Wang, S., Fahad, S., Nie, L., Huang, J., Peng, S., & Ding, Y. (2020). Intensified pollination, fertilization ameliorate heat injury in rice (*Oryza sativa* L.) during the flowering stage. *Field Crops Research, 252*, 107795. https://doi.org/10.1016/J.FCR.2020.107795

Wu, C., Cui, K., Wang, W., Li, Q., Fahad, S., Hu, Q., Huang, J., Nie, L., & Peng, S. (2016). Heat-induced phytohormone changes are associated with disrupted early reproductive development and reduced yield in rice. *Scientific Reports, 6*(1), 34978.

Wu, L., Feng, L., Li, Y., Wang, J., & Wu, L. (2019). A yield-related agricultural drought index reveals spatio-temporal characteristics of droughts in Southwestern China. *Sustainability, 11*(3), 714.

Xiao, L., Liu, B., Zhang, H., Gu, J., Fu, T., Asseng, S., Liu, L., Tang, L., Cao, W., & Zhu, Y. (2021). Modeling the response of winter wheat phenology to low temperature stress at elongation, booting stages. *Agricultural and Forest Meteorology, 303*, 108376. https://doi.org/10.1016/J.AGRFORMET.2021.108376

Xu, Z., Jiang, Y., & Zhou, G. (2015). Response and adaptation of photosynthesis, respiration, and antioxidant systems to elevated CO_2 with environmental stress in plants. *Frontiers in Plant Science, 6*, 701.

Xu, Y., Chu, C., & Yao, S. (2021). The impact of high-temperature stress on rice: Challenges and solutions. *Crop Journal, 9*(5), 963–976.

Yadav, S. K. (2010). Cold stress tolerance mechanisms in plants. A review. *Agronomy for Sustainable Development, 30*(3), 515–527. https://doi.org/10.1051/agro/2009050

Yadav, A., Khan, Y., & Prasad, M. (2016). Dehydration-responsive miRNAs in foxtail millet: genome-wide identification, characterization and expression profiling. *Planta, 243*, 749–766.

Yamaguchi-Shinozaki, K., & Shinozaki, K. (2006). Transcriptional regulatory networks in cellular responses and tolerance to dehydration and cold stresses. *Annual Review of Plant Biology, 57*(1), 781–803.

Yang, Y., Yu, J., Qian, Q., & Shang, L. (2022). Enhancement of heat and drought stress tolerance in rice by genetic manipulation: A systematic review. *Rice, 15*(1), 1–18.

Yao, H. M., Li, Y. S., Zhang, T. Z., Zhao, J., Wang, C., Wang, H. N., & Fang, Y. F. (2016). Effects of combined drought and salinity stress on germination and physiological characteristics of maize (Zea mays). *Journal of Applied Ecology, 27*(7), 2301–2307.

Yousaf, M. I., Riaz, M. W., Jiang, Y., Aslam, M. Z., Hussain, S., Shehzad, A., Riasat, G., & Manzoor, M. A. (2022). Concurrent effects of drought and heat stresses on physio-chemical attributes, antioxidant status and kernel quality traits in maize (*Zea mays* L.) hybrids. *Frontiers in Plant Science, 13*, 898823.

Yu, W., Ji, R., Wu, J., Feng, R., Mi, N., & Chen, N. (2023). Combined effects of heat and drought stress on the growth process and yield of maize (*Zea mays* L.) in Liaoning Province, China. *Atmosphere, 14*(9), 1397.

Yuping, L., Hongbin, L., Yuanyuan, L., & Suiqi, Z. (2017). Improving water-use efficiency by decreasing stomatal conductance and transpiration rate to maintain higher ear photosynthetic rate in drought-resistant wheat. *Crop Journal, 5*(3), 231–239.

Zhang, J., Li, S., Cai, Q., Wang, Z., Cao, J., Yu, T., & Xie, T. (2020). Exogenous diethyl amino ethyl hexanoate ameliorates low temperature stress by improving nitrogen metabolism in maize seedlings. *PLoS One, 15*(4), e0232294. https://doi.org/10.1371/JOURNAL.PONE.0232294

Zhang, J., Zhang, S., Cheng, M., Jiang, H., Zhang, X., Peng, C., Lu, X., Zhang, M., & Jin, J. (2018a). Effect of drought on agronomic traits of rice, wheat: A meta-analysis. *International Journal of Environmental Research and Public Health, 15*(5), 839. https://doi.org/10.3390/ijerph15050839

Zhang, C. X., Feng, B. H., Chen, T. T., Fu, W. M., Li, H. B., Li, G. Y., Jin, Q. Y., Tao, L. X., & Fu, G. F. (2018b). Heat stress-reduced kernel weight in rice at anthesis is associated with impaired source-sink relationship and sugars allocation. *Environmental and Experimental Botany, 155*, 718–733.

Zhang, W., & Zhao, Y. (2021). The effects of short-term exposure to low temperatures during the booting stage on starch synthesis and yields in wheat grain. *Frontiers in Plant Science, 12*, 684784.

Zhao, C., Liu, B., Piao, S., Wang, X., Lobell, D. B., Huang, Y., Huang, M., Yao, Y., Bassu, S., Ciais, P., & Durand, J. L. (2017). Temperature increase reduces global yields of major crops in four independent estimates. *Proceedings of the National Academy of Sciences, 114*(35), 9326–9331.

Zhao, F. J., Tang, Z., Song, J. J., Huang, X. Y., & Wang, P. (2022). Toxic metals and metalloids: Uptake, transport, detoxification, phytoremediation, and crop improvement for safer food. *Molecular Plant, 15*(1), 27–44.

Zhao, H., Guan, J., Liang, Q., Zhang, X., Hu, H., & Zhang, J. (2021). Effects of cadmium stress on growth and physiological characteristics of sassafras seedlings. *Scientific Reports, 1*, 9913.

Zhao, Q., Zhou, L., Liu, J., Cao, Z., Du, X., Huang, F., Pan, G., & Cheng, F. (2018). Involvement of CAT in the detoxification of HT-induced ROS burst in rice anther and its relation to pollen fertility. *Plant Cell Reports, 37*, 741–757.

Potential Impacts of Heat and Drought on Vegetable Production and Quality

3

Kumari Shubha, Meenu Kumari, Rachana Dubey, Sanjeev Kumar, and Anup Das

Abstract

The increasing global population and climate change present significant challenges to global food security, particularly affecting crop productivity. Drought and heat stress, exacerbated by altered climate conditions, have become critical factors limiting crop yields. Drought events, driven by reduced precipitation and altered rainfall patterns, lead to substantial declines in crop productivity. Similarly, heat stress, characterized by elevated temperatures, adversely affects plant growth and development. Understanding the combined impact of drought and heat stress is crucial, as their effects surpass those observed individually. Vegetable crops, highly sensitive to environmental extremes, face significant yield losses due to high temperatures and limited soil moisture. These stresses affect various physiological and biochemical processes, ultimately impacting productivity and quality. Drought stress, characterized by low rainfall, disrupts essential cellular processes, further compromising growth and productivity. The impacts extend to seed germination and fruit development, affecting crops like tomatoes, potatoes, and leafy vegetables. Additionally, climate change influences insect pests and diseases, altering their ecology and interactions within ecosystems. Elevated temperatures and variable moisture levels directly impact the prevalence and severity of crop diseases, posing additional threats to global food security. Genetic improvement, marker-assisted selection, and biotechnological

K. Shubha (✉) · R. Dubey · S. Kumar
Division of Crop Research, ICAR Research Complex for Eastern Region, Patna, Bihar, India
e-mail: shubha.veg@gmail.com

M. Kumari
ICAR-RCER, Farming System Research Centre for Hill and Plateau Region, Ranchi, Jharkhand, India

A. Das
ICAR Research Complex for Eastern Region, Patna, Bihar, India

R. Dubey et al. (eds.), *Drought and Heat Stress in Agriculture*,
https://doi.org/10.1007/978-981-96-5735-3_3

approaches offer promising avenues for enhancing vegetable crops' resilience to abiotic stresses. Moreover, agronomic practices such as conservation agriculture, mulching, and drip irrigation play crucial roles in mitigating stress impacts and ensuring sustainable production. These strategies highlight the importance of integrating various approaches to address the complex challenges posed by climate change and safeguard global food supplies.

Keywords

Climatic change · Vegetable production · Heat and drought stress · Mitigation strategies

3.1 Introduction

Global food security is currently facing significant challenges due to the rapid increase in population and the drastic changes in climate (Lesk et al., 2016). The consequences of these changes are profound, particularly with regards to crop productivity and overall food security. In the face of shifting climate patterns, drought and heat stress have emerged as critical factors limiting crop yields.

The alteration in climate conditions has led to reduced precipitation and changed rainfall patterns, resulting in more frequent occurrences of droughts worldwide (Lobell et al., 2011). These drought events have severe repercussions, causing substantial declines in crop yields. The negative impacts of drought on plant growth, physiology, and reproduction have been extensively documented (Yordanov et al., 2000; Barnabas et al., 2008), particularly affecting vegetable crops (Ayyogari et al., 2014; Prasad & Chakravorty, 2015; Bisbis et al., 2018). A recent study conducted an analysis on potato productivity, indicating an anticipated decline in all potato-growing states of India (Ayyogari et al., 2014).

Drought stress occurs when there is a decrease in soil and atmospheric humidity coupled with high ambient air temperatures. This condition arises due to an imbalance between the evapotranspiration flux and water intake from the soil (Lipiec et al., 2013). Heat stress, on the other hand, is characterized by elevated soil and air temperatures surpassing a threshold level for a minimum duration, resulting in permanent damage to plant growth and development. A comprehensive multilocation study elucidates the significant impact of temperature variations on crop yields. These adverse environmental conditions are fostering the emergence of drought-prone regions, thereby adversely affecting plant growth and the productivity of major crops (Zhao et al., 2017). It is crucial to consider both drought and heat stress together, as their combined effect surpasses the impact observed when each stressor is considered individually as shown in Fig. 3.1. (Dreesen et al., 2012).

Generally, abiotic stresses such as heat and drought are regulated by multiple genes, and the underlying mechanisms are more intricate compared to traits like biotic stresses, which are typically characterized by monogenic resistance. Moreover, the influence of other abiotic and/or biotic stresses often adds to the

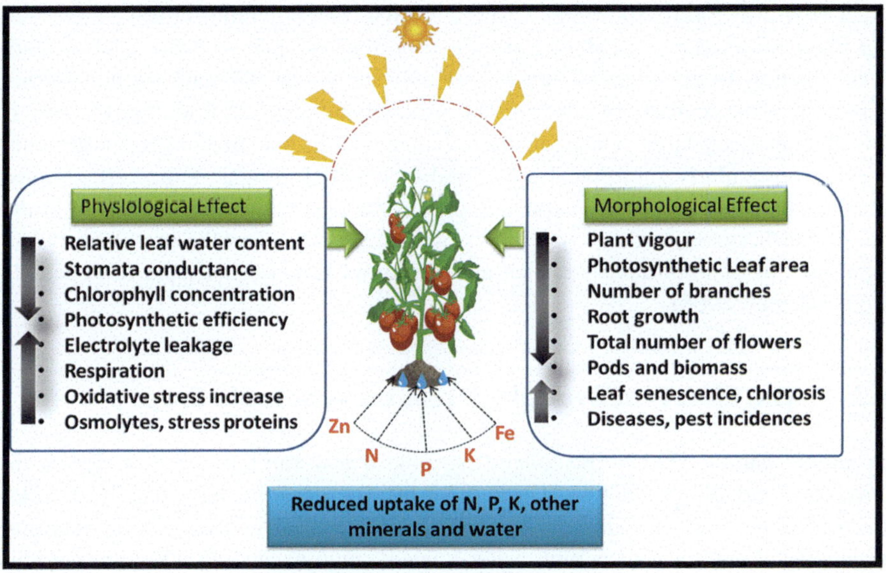

Fig. 3.1 Combined stress of heat and drought in plants

complexity of studying heat and drought responses, presenting additional challenges in research.

3.2 The Effects of High Temperatures on Vegetable Crops

Vegetables are particularly sensitive to environmental extremes, with high temperatures and limited soil moisture being significant contributors to low yields. These factors profoundly impact various physiological and biochemical processes, including reduced photosynthetic activity, altered metabolism and enzymatic activity, thermal injury to tissues, and reduced pollination and fruit set. For instance, in tomatoes, high temperatures can lead to considerable losses in productivity due to reduced fruit set and the production of smaller, lower quality fruits. Pre-anthesis temperature stress in tomatoes is associated with developmental changes in anthers, such as irregularities in epithesium and endothesium, lack of opening of the stromium, and poor pollen formation (Peet et al., 1988; Sato Peet & Thomas, 2002). Similarly, in peppers, exposure to high temperatures during the pre-anthesis stage may not affect pistil or stamen viability, but post-pollination high temperatures can inhibit fruit set, indicating sensitivity of fertilization to temperature stress (Erickson & Markhart, 2002). In Chinese cabbage cultivars, under high temperature stress, there is a slight increase in cell membrane permeability and proline content, which serves as a heat tolerance index (Datta, 2013). Moreover, increased temperatures can also reduce yields by affecting factors such as seed tuber sprouting, growth rates, and tuberization. Each of these factors may be controlled by different genes

(Peet & Wolfe, 2000). Additionally, sudden spikes in temperatures during the growing phase of lettuce and celery can induce bolting (premature seed head production), leading to poor-quality heads and reduced yields. Climate change further complicates the scenario by favoring the survival and reproduction of pests such as the Diamondback Moth (DBM) in warm climates. This can affect brassica-growing regions, particularly subtropical areas, and increasingly impact temperate regions. In arid conditions, thrips are expected to reproduce and survive more easily in the future. Furthermore, higher temperatures will negatively affect the survivability and reproduction of important parasites and predators like Trichoderma in vegetable crops (Deuter, 2008).

3.3 The Effects of Drought Stress on Vegetable Crop

Drought, characterized by prolonged periods of low rainfall, poses a significant threat to vegetable production. Throughout history, droughts have triggered widespread famine and food insecurity, making it a crucial concern for agriculture. Vegetables, known for their high water content-often exceeding 90%, are particularly susceptible to the effects of water scarcity (Allaby & Michael, 1989; CGIAR, 2003). During drought conditions, soil becomes more concentrated with solutes, leading to water moving out of plant cells through osmosis. This imbalance disrupts essential cellular processes like photosynthesis, ultimately stunting growth and reducing productivity. Research has shown that drought stress can result in decreased chlorophyll levels, electrolyte leakage, and compromised leaf health, all of which contribute to diminished yields and quality (Kirnak et al., 2001).

The impacts of drought extend beyond immediate growth to affect seed germination and tuber sprouting. Particularly, sensitive crops like potatoes can experience significant yield reductions even with moderate drought stress (Arora et al., 1987; Jefferies & Mackerron, 1993). Similarly, tomatoes face challenges during critical growth stages such as transplantation, flowering, and fruit development (Nuruddin, 2001). Root and bulb crops, which depend on carbohydrate production and translocation, are also affected by drought. When drought occurs during the enlargement of storage organs, such as potatoes and onions, yields and quality suffer (Anonymous). Additionally, crops like cucumbers, melons, and beans are highly susceptible to drought stress during flowering and fruit development, leading to decreased fruit set and yield (Anonymous). Even leafy vegetables like spinach and lettuce are not immune to drought's effects, experiencing reduced juiciness and flavor. This underscores the importance of addressing the impacts of drought on vegetable production to ensure food security and a consistent supply of nutritious produce (AVRDC, 1990). Impact of high temperature and drought on yield and quality of vegetable is presented in Table 3.1.

The impact of drought stress on the quality of vegetables is multifaceted, encompassing changes in texture, nutritional composition, flavor, aroma, physical appearance, and postharvest characteristics (Shubha et al. 2021a, b). Firstly, water deficiency during growth stages leads to reduced water content in vegetables,

Table 3.1 Impact of high temperature and drought on yield and quality of vegetable

Vegetable	Impact of temperature on quality	References
Tomato	Sunburn disrupts lycopene synthesis, causing yellow areas in affected tissues and preventing fruit setting above 35 °C. Drought induces fruit cracking, which in turn promotes secondary infections	Shubha et al. (2021b)
Potato	Black heart develops in extremely hot weather and under drought stress, typically manifesting as dark-gray to black discoloration at the center of the tuber	Moretti et al. (2012)
Bell pepper	High temperature impairs fertilization and results in deformed fruits. Drought results in fruit cracking leads to secondary infection	Madhavi Reddy et al. (2016)
Cauliflower	High temperatures induce riceyness, with small buds forming on the curd, and cause yellowing and a loose, leafy texture in the curd. Drought causes reduced growth and premature heading	Wu et al. (2012)
Cabbage	Outer leaves display a bleached, papery appearance, and more prone to decay due to heat stress. Drought in head formation reduced yield significantly.	Ackah and Kotei (2021)
Carrot	Carotene synthesis decreases when temperatures fall below 15 °C or rise above 24 °C, resulting in a reduction in color intensity. Drought induces root cracking.	Hazra and Som (2006)
Beet	High temperatures reduce sugar and above 25 °C along with moisture stress cause zoning, with alternate white and colored rings in sliced roots.	Hazra and Som (2006)
Pea	Peas lose tenderness at 30 °C and above due to elevated temperatures and moisture stress hinders enzymatic conversion of starch to sugar, resulting in lower sugar content and reduced quality.	Shubha et al. (2021a)
Palak	Elevated temperatures cause premature bolting, while moisture stress diminishes leaf crispiness and appearance, leading to an overall reduction in quality.	Khosa et al. (2022)

resulting in wilted and less succulent produce. This affects the texture and crispiness, diminishing the overall appeal to consumers. Moreover, drought stress alters the nutrient composition of vegetables, potentially decreasing essential vitamins, minerals, and antioxidants, thereby compromising their nutritional value and health benefits. For example, in tomato (*Solanum lycopersicum*), irrigation at 50% of field capacity resulted in diminished shoot and root biomass, osmotic potential, and stomatal conductance, alongside increases in non-photochemical fluorescence quenching coefficient (NPQ) and accumulation of osmolytes like proline and soluble sugars (Moles et al., 2018). Research by Escalante-Magaña et al. (2019) on pepper varieties (*Capsicum* spp.) revealed a significant reduction in leaf relative water content (RWC) and an increase in electrolyte leakage and proline content as the duration of drought stress increased. This study highlighted the tolerance of pepper plants to water deficit conditions, as indicated by their recovery capacity and maintenance of RWC values similar to the control. Similarly, investigations on cabbage (*Brassica oleracea* var. *capitata*) demonstrated various physiological and biochemical changes under drought stress, including increases in lipid peroxidation, hydrogen peroxide (H_2O_2) content, and electrolyte leakage, coupled with reductions in biometric parameters, photosynthesis, chlorophyll content, and mineral uptake (Sahin et al., 2018). Moreover, controlled drought was employed to enhance the accumulation of cannabinoids and inflorescence dry weight in hemp (*Cannabis sativa*) plants (Caplan et al., 2019). This demonstrated the potential of drought stress to increase the content of beneficial secondary metabolites, such as cannabinoids, in plants subjected to water deficit conditions. In another study on cassava (*Manihot esculenta*) genotypes, certain cultivars showed greater resistance to drought stress, maintaining cell membrane integrity and exhibiting higher levels of antioxidant molecules and enzymes compared to others (Zhu et al., 2020).

Overall, the studies underscore the diverse responses of horticultural species to drought stress and provide valuable insights into the physiological, biochemical, and qualitative changes occurring within plants under water deficit conditions. These findings contribute to a deeper understanding of how different crops cope with and adapt to drought stress, aiding in the development of strategies for mitigating its adverse effects on agricultural productivity.

3.4 Impact of Drought and Heat Stress on Insect Pests and Diseases

Climate change exerts multifaceted impacts on insect pests, profoundly altering their ecology and interactions within ecosystems. As temperatures rise globally, insect pests find expanded territories to colonize, pushing into regions previously uncongenial. Insects, being cold-blooded organisms, usually regulate their body temperatures in line with their surroundings. Consequently, temperature emerges as the foremost environmental factor influencing insect behavior, distribution, development, survival, and reproduction. Predictions of insect life stages often rely on accumulated degree days calculated from a base temperature. Some researchers reported that temperature's impact on insects outweighs that of other environmental factors. For instance, a 2 °C temperature increase might prompt insects to undergo one to five additional life cycles per season (Yamamura & Kiritani, 1998). However, recent studies suggest that variables like high temperature and moisture stress (drought) can also significantly affect insects, particularly in the context of global climate change (Shubha et al., 2021a, b).

Drought and heat stress directly impacted the prevalence and severity of crop diseases, posing significant threats to global food security. With rising temperatures and carbon dioxide levels, as well as variable effects on moisture levels, the consequences for crop health are profound. Elevated temperatures are expected to facilitate the expansion of pathogen and vector distributions, exposing more potential hosts to pathogens and fostering opportunities for pathogen hybridization. The rate of pathogen evolution is closely tied to temperature, as it governs the pace of reproduction for many pathogens. Longer seasons resulting from warmer temperatures provide extended windows for pathogen evolution. Additionally, higher overwintering and over-summering rates due to warmer conditions can accelerate pathogen evolution. Despite potential physiological changes in host plants that may enhance disease resistance under climate change scenarios, pathogens may adapt more swiftly to overcome host resistance, leading to more rapid disease cycles. Furthermore, the efficacy of fungicides and bactericides may be influenced by increased levels of CO_2, moisture, and temperature. Overall, these dynamics highlight the urgent need for comprehensive strategies to mitigate the impact of climate change on crop diseases and safeguard global food supplies.

3.5 Strategies for Heat and Drought Stress Management in Vegetable Crops

3.5.1 Genetic Improvement

Improving crop genetics serves as a vital strategy for enabling plants to adapt to anticipated future climate conditions. This genetic enhancement of crop plants aims to bolster their ability to withstand the adverse impacts of climate change, thereby ensuring sustainable agricultural production and food security. However, genetic improvement encounters challenges due to the complex polygenic nature of abiotic stress tolerance, insufficient selection criteria, and limited understanding of the genetics underlying stress tolerance, making breeding efforts challenging (Ong, 2002). Characterization plays a pivotal role in identifying and describing useful traits present in crop germplasm, facilitating the estimation of inherent variation and diversity among available genetic resources. Such characterization aids in leveraging germplasm for crop improvement initiatives. Genetic improvement relies heavily on the extent of genetic variability within populations. Accessing genetic variability is essential for effective selection processes. In crop breeding, selecting resistant plants from existing populations and developing varieties from their progeny constitute fundamental and effective methods. Notable examples include the development of heat-tolerant hybrids in Chinese cabbage and breeding lines in tomato, as undertaken by the Asian Vegetable Research and Development Centre, Taiwan (Pena & Hughes, 2007b). In India, several institutions have contributed to the development of stress-tolerant vegetable cultivars. For instance, heat and drought-tolerant tomato cultivars were bred at the Indian Agricultural Research Institute, New Delhi, and the Indian Institute of Horticultural Research, Bangalore. Vegetable varieties with various stress tolerance released in India for cultivation are presented in Table 3.2. Harnessing genes from wild relatives presents another avenue for enhancing vegetable crops' resilience to harsh environmental conditions. Wild relatives harbor genes conferring tolerance to various stresses, which can be transferred to cultivated types through conventional breeding or biotechnological methods. Successful introgression of wild genes into cultivated types has been demonstrated in tomato and okra for disease resistance and quality improvement. However, the process of transferring beneficial traits from wild varieties to

Table 3.2 Vegetable varieties with various stress tolerance released in India for cultivation

Crop	Variety	Stress (drought and heat) tolerance
Tomato	Pusa hybrid 1	Fruit set up to 28 °C (high) night temperature
	Pusa Sadabahar	Fruit set at both low (6 °C) and high (30 °C) night temperature
	Arka Vikas	Tolerant to moisture stress
Eggplant	SM-1, SM-19	Drought
	SM-30	
Chilli	Arka Lohit	Drought
Musk melon	Jobner 96–2	High soil pH
Spinach beet	Jobner green	High soil pH (up to 10.5) tolerant
Cucumber	Pusa Barkha	Tolerant to high temperature
	Pusa Uday	Suitable for throughout the year
Carrot	Pusa Kesar, Pusa Vrishti	Tolerant to high temperature
Radish	Pusa Himani	Grown throughout the year
Onion	Arka Kalyan	Drought
Sweet potato	Sree Nandini, Sree Bhadra	Drought
Potato	Kufri Surya	Heat tolerant up to 25 °C night temperature
	Kufri Sheetman and Kufri Sindhuri	Drought
Cassava	H-97	Drought tolerant
	Sree Sahya	

Adapted from Debbarma et al. (2019)

cultivated crops requires meticulous genetic manipulation and breeding efforts (Koundinya et al., 2013).

Biotechnology offers promising tools for enhancing vegetable crops' stress tolerance. Techniques such as tissue culture and genetic engineering enable the screening and development of stress-resistant varieties. For instance, embryo rescue prevents embryo abortion in distant crosses, while somatic hybridization facilitates the fusion of protoplasts from different species. Additionally, somaclonal variation generated during tissue culture allows for the creation of variations in stress tolerance, disease resistance, and herbicide tolerance. Genetic engineering, involving the transfer of genes across species and genera barriers, enables the introduction of novel traits into target species. For example, the introduction of heat shock protein genes from carrot into other vegetable crops demonstrates the potential of genetic engineering for improving stress tolerance in vegetables. Similarly, genes governing drought and salt tolerance, transferred from model plants like *Arabidopsis thaliana* to tomato, have shown promising results. These biotechnological advancements hold significant promise for enhancing stress tolerance in vegetable crops and ensuring food security in the face of climate change (Rai & Rai, 2006).

3.5.2 Marker-Assisted Selection (MAS) for Abiotic Stress Tolerance

MAS plays a crucial role in accelerating conventional breeding by enabling efficient selection based on genetic markers unaffected by external environmental conditions (Collard & Mackill, 2007; Vogel, 2009). It aids in the introgression of superior alleles from wild species into breeding programs and facilitates the pyramiding of resistant genes controlling quantitative traits (Pena & Hughes, 2007a). Anithakumari et al. (2011) identified 23 QTL for recovery after drought stress in potato. Dumont et al. (2009) observed the colocalization of QTL associated with raffinose sugar and RUBISCO activity with frost resistance in pea. AFLP markers were utilized by Muchero et al. (2009) to map ten QTL linked with drought tolerance in cowpea. Additionally, Johnson et al. (2000) identified 13 QTL for taproot length and water extraction ability in lettuce and wild L. serriola using AFLP markers. Despite its success in disease resistance and nutritional quality improvement, its application in enhancing abiotic stress tolerance in vegetables is ongoing (Pena & Hughes, 2007a; Vogel, 2009). However, genomic selection has emerged as a promising approach for improving polygenic traits, surpassing the limitations of MAS by considering all genome-wide markers' effects (Jannink et al., 2010; Wong & Bernardo, 2008). Phenomics, a cutting-edge approach, allows for precise phenotypic characterization of germplasm under controlled environmental conditions (Turyagyenda et al., 2013). It employs high-throughput technologies such as conveyor systems and image-capturing systems to accurately measure plant growth, development, and physiological responses without destructive sampling. By bridging the gap between physiology and plant breeding, phenomics aids in the selection of stress-tolerant plants thus facilitating adaptation to changing environmental conditions (Turyagyenda et al., 2013). This integration of genomics and phenomics provides insights into genotype-environment interactions, enhancing our understanding of stress tolerance mechanisms in vegetable crops.

3.5.3 Agronomic Practices

Agricultural practices such as resource conservation technologies (RCT), conservation agriculture (CA), mulching, cover cropping, organic farming, perennial farming, and agroforestry intervention contribute to climate resilience. CA systems, in particular, demonstrate greater resistance to climate stress during critical growth stages like anthesis compared to conventional tillage methods (Steward et al., 2019). Utilizing large-scale rainout shelters effectively reduces soil moisture, highlighting the need to fill knowledge gaps regarding the long-term climate resilience of these practices. Intercropping baby corn with cowpea, okra, brinjal, and chilli in summer, and with tomato, brinjal, chilli, and pea in autumn-winter, has been shown to be more profitable and productive than sole cropping (Adhikary et al., 2015). This practice optimizes soil moisture and nutrient utilization, particularly in rainfed areas, leading to improved crop productivity. Tree-based intercropping (TBI) systems are effective in mitigating greenhouse gas emissions, with research indicating a reduction in N_2O emissions by 1.2 kg/ha/yr. Intercropping also helps diversify income for farmers, with annual cereals, grain crops, and vegetables grown between rows of perennial tree vegetables like drumstick. Additionally, intercropping can reduce the spread of vector-borne diseases exacerbated by climate change, such as the yellow vein mosaic virus in okra, where baby corn plants act as a barrier to the virus-carrying whitefly (Koundinya & Kumar, 2014; Adhikary et al., 2015). Furthermore, seed priming is a viable technique to counteract the adverse effects of drought stress on crops. This method involves partially hydrating seeds to initiate germination-related metabolic processes without radicle emergence. Primed seeds typically display enhanced germination rates, uniformity, and sometimes overall germination percentages. Various inorganic salts such as NaCl, Na_2SO_4, KCl, KH_2PO_4, $CaCl_2$, and $MgSO_4$, along with organic acids like succinic acid, CCC, and auxins, are commonly used as priming agents. For instance, pre-sowing treatments of okra seeds through hydropriming or osmopriming with substances like Na_2HPO_4, KCl, or $MgSO_4$ have been shown to improve plant growth, seed yield, and quality, particularly under water stress conditions (Menon & Savithri, 2015). The exogenous application of plant growth regulators has emerged as an effective strategy to alleviate the negative impacts of abiotic stresses like drought. Plant hormones such as CCC, salicylic acid, ascorbic acid, NAA, brassinolide, and cytokinins have shown promise in mitigating water stress effects and enhancing crop resilience to extreme water deficit conditions. For instance, both salicylic acid and its derivative, acetylsalicylic acid, when applied via seed soaking or foliar spray, have been shown to protect muskmelon seedlings from drought stress. Similarly, foliar spray of ascorbic acid has been demonstrated to increase transpiration rates in okra under drought stress (Amin et al., 2009). Foliar nutrition application, particularly of potassium and calcium, has been observed to help maintain high tissue water potential during drought conditions, thereby improving drought tolerance through osmotic adjustment. Spraying potassium and calcium during water stress conditions facilitates rapid nutrient absorption, promoting better growth. For example, in bell pepper, foliar application of calcium chloride and potassium chlorides induced drought tolerance, with calcium chloride at 15 mM concentration showing better results under drought conditions (Azam et al., 2016). Vegetable grafting is increasingly recognized as a rapid alternative to traditional breeding methods for enhancing environmental stress tolerance in fruiting vegetables (Rouphael et al., 2008). Grafting, especially in crops like Solanaceous and cucurbits grown in arid and semi-arid regions with long drought periods, has shown promising results. Grafted plants often exhibit improved growth, yield, photosynthesis, water and nutritional status, and reduced accumulation of detrimental elements under adverse soil conditions compared to ungrafted plants. Osmoprotectants, including proline, trehalose, fructan, mannitol, and glycinebetaine, play crucial roles in signaling and regulating plant responses to various stresses, including drought (Sakamoto & Murata, 2002). These compounds aid in osmotic adjustment and protection of subcellular structures in stressed plants. Exogenous application of glycinebetaine has been shown to improve growth and production under stress conditions. Mulching is a valuable technique for mitigating abiotic stresses and improving soil moisture conservation, weed suppression, soil temperature regulation, and soil health. Both plastic and plant residue mulching methods are commonly employed (Kirnak & Demirtas, 2006). Improved methods of seedling production, such as Protray, grown seedlings using coco peat, nylon net protection, and bio-fertilizer/bio-pesticide inoculation at the nursery stage, offer robust, uniform, and healthy seedlings. These seedlings establish better in the main field with less root damage and exhibit resilience to biotic and abiotic stresses, particularly under water stress conditions. Drip irrigation stands out as a superior method of irrigation in horticulture due to its precise and direct application of water to the root zone. It offers significant water savings, increased growth, development, and yields of fruits and vegetables, as well as weed control and labor savings.

Antitranspirants, such as atrazine, can reduce transpiration rates in plants when applied to foliage. Atrazine, at lower concentrations, acts as an antitranspirant by inducing stomatal closure (Moftah & Al-Humaid, 2005). Application of atrazine has been shown to increase potato tuber yield under drought conditions, while kaolin application effectively increases photosynthetic activity and water use efficiency of potato plants under drought stress (Cantore et al., 2009).

3.6 Conclusion

In conclusion, addressing the challenges posed by climate change-induced drought and heat stress on vegetable crops is imperative for ensuring global food security. The multifaceted impacts of these stresses on crop productivity and quality underscore the need for integrated management strategies encompassing genetic improvement, biotechnological interventions, and agronomic practices. By leveraging advancements in genetic enhancement, marker-assisted selection, and biotechnology, vegetable crops can be equipped with enhanced resilience to withstand environmental extremes. Furthermore, the implementation of sustainable agronomic practices such as conservation agriculture, mulching, and drip irrigation can help mitigate stress impacts and optimize resource utilization. However, effective adaptation and mitigation strategies must be accompanied by collaborative efforts among researchers, policymakers, and agricultural stakeholders to facilitate knowledge sharing, technology transfer, and capacity building. By prioritizing resilience-building initiatives and investing in research and development, we can navigate the challenges posed by climate change and safeguard global food supplies for future generations.

References

Ackah, E., & Kotei, R. (2021). Effect of drought length on the performance of cabbage (Brassica oleracea var capitata) in the forest-savannah transition zone, Ghana. *Plant Physiology Reports, 26*, 74–83. https://doi.org/10.1007/s40502-020-00541-5

Adhikary, S., Koundinya, A. V. V., Pandit, M. K., & Bhattacharya, B. (2015). Evaluation of the efficiency of baby corn based vegetable intercropping systems. *International Journal of Plant and Soil Science, 5*(6), 366–374.

Allaby, & Michael. (1989). *Dictionary of the environment* (3rd ed.). New York University Press.

Amin, B. G., Mahleghah, H. M. R., Mahmood, & Hossein, M. (2009). Evaluation of interaction effect of drought stress with ascorbate and salicylic acid on some of physiological and biochemical parameters in okra (*Hibiscus esculentus* L.). *Research Journal of Biological Sciences, 4*, 380–387.

Anithakumari, A. M., Dolstra, O., Vosman, B., Visser, R. G., & van der Linden, C. G. (2011). In vitro screening and QTL analysis for drought tolerance in diploid potato. *Euphytica, 181*, 357–369.

Arora, S. K., Partap, P. S., Pandita, M. L., & Jalal, I. (1987). Production problems and their possible remedies in vegetable crops. *Indian Horticulture, 32*(2), 2–8.

AVRDC. (1990). *Vegetable production training manual* (p. 447). Asian Vegetable Research and Training Center.

Ayyogari, K., Sidhya, P., & Pandit, M. K. (2014). Impact of climate change on vegetable cultivation-a review. *International Journal of Agriculture, Environment and Biotechnology, 7*(1), 145–155.

Azam, M., Noman, M., Abbasi, N., Ramzan, A., Imran, M., Hayat, A., Abbas, H., & Akram, A. (2016). Effect of foliar application of micro-nutrient and soil condition on growth and yield of sweet pepper (*Capsicum annuum* L.). *Science, Technology and Development, 35*(2), 75–81.

Barnabas, B., Jäger, K., & Fehér, A. (2008). The effect of drought and heat stress on reproductive processes in cereals. *Plant, Cell & Environment, 31*, 11–38.

Bisbis, M. B., Gruda, N., & Blanke, M. (2018). Potential impacts of climate change on vegetable production and product quality–A review. *Journal of Cleaner Production, 170*, 1602–1620.

Cantore, V., Pace, B., & Albrizio, R. (2009). Kaolin-based particle film technology affects tomato physiology, yield and quality. *Environmental and Experimental Botany, 66*, 279–288.

Caplan, D., Dixon, M., & Zheng, Y. (2019). Increasing inflorescence dry weight and cannabinoid content in medical cannabis using controlled drought stress. *HortScience, 54*, 964–969.

CGIAR. (2003). *Applications of molecular biology and genomics to genetic enhancement of crop tolerance to abiotic stresses-a discussion document.* Interim Science Council Secretariat.

Collard, B., & Mackill, D. (2007). Marker-assisted selection: An approach for precision plant breeding in the twenty-first century. *Philosophical Transactions*, 557–572.

Datta, S. (2013). Impact of climate change in Indian horticulture – A review. *International Journal of Environmental Science and Technology, 2*(4), 661–671.

Debbarma, S., KumarSingh, A., Daniel, G. R., & Chandra, A. K. (2019). Mitigation of abiotic stresses in vegetable crops. *Plant Stress Bioloy, 137*.

Deuter, P. (2008). Garnaut climate change review. In *Defining the impacts of climate change on horticulture in Australia.*

Dreesen, F. E., De Boeck, H. J., Janssens, I. A., & Nijs, I. (2012). Summer heat and drought extremes trigger unexpected changes in productivity of a temperate annual/biannual plant community. *Environmental and Experimental Botany, 79*, 21–30. https://doi.org/10.1016/j.envexpbot.2012.01.005

Dumont, E., Fontaine, V., Vuylsteker, C., Sellier, H., Bodèle, S., Voedts, N., et al. (2009). Association of sugar content QTL and PQL with physiological traits relevant to frost damage resistance in pea under field and controlled conditions. *Theoretical and Applied Genetics, 118*(8), 1561–1571.

Erickson, A. N., & Markhart, A. H. (2002). Flower developmental stage and organ sensitivity of bell pepper (Capsicum annuum L.) to elevated temperature. *Plant, Cell & Environment, 25*, 123–130.

Escalante-Magaña, C., Aguilar-Caamal, L. F., Echevarría-Machado, I., Medina-Lara, F., Sanchez Cach, L., & Martínez-Estévez, M. (2019). Contribution of glycine betaine and proline to water deficit tolerance in pepper plants. *HortScience, 4*, 1044–1054.

Hazra, P., & Som, M. G. (2006). *Vegetable science.* Kalyani Publishers.

Jannink, J. L., Lorenz, A. J., & Iwata, H. (2010). Genomic selection in plant breeding: From theory to practice. *Briefings in Functional Genomics, 9*(2), 166–177.

Jefferies, R. A., & Mackerron, D. K. L. (1993). Responses of potato genotypes to drought. II. Leaf area index, growth and yield. *Annals of Applied Biology, 122*, 105–112.

Johnson, W. C., Jackson, L. E., Ochoa, O., Van Wijk, R., Peleman, J., St. Clair, D. A., & Michelmore, R. W. (2000). Lettuce, a shallow-rooted crop, and Lactuca serriola, its wild progenitor, differ at QTL determining root architecture and deep soil water exploitation. *Theoretical and Applied Genetics, 101*, 1066–1073.

Khosa, Q., Uz Zaman, Q., An, T., Ashraf, K., Abbasi, A., Nazir, S., et al. (2022). Silicon-mediated improvement of biomass yield and physio-biochemical attributes in heat-stressed spinach (*Spinacia oleracea*). *Crop and Pasture Science, 74*(3), 230–243.

Kirnak, H., & Demirtas, M. N. (2006). Effects of different irrigation regimes and mulches on yield and macro nutrition levels of drip-irrigated cucumber under open field conditions. *Journal of Plant Nutrition, 29*(9), 1675–1690.

Kirnak, H., Kaya, C., Tas, I., & Higgs, D. (2001). The influence of water deficit on vegetative growth, physiology, fruit yield and quality in eggplants. *Bulgarian Journal of Plant Physiology, 27*(3–4), 34–46.

Koundinya, A. V. V., & Kumar, P. P. (2014). Indian vegetable seeds industry: Status and challenges. *International Journal of Plant, Animal and Environmental Sciences, 4*(4), 62–69.

Koundinya, A. V. S., et al. (2013). Horticultural approaches to improve vegetable crop productivity under changing climate: An overview. *Indian Journal of Horticulture, 70*(3), 283–292.

Lesk, C., Rowhani, P., & Ramankutty, N. (2016). Influence of extreme weather disasters on global crop production. *Nature, 529*, 84–87. https://doi.org/10.1038/nature16467

Lipiec, J., Doussan, C., Nosalewicz, A., & Kondracka, K. (2013). Effect of drought and heat stresses on plant growth and yield: A review. *International Agrophysics, 27*, 463–477. https://doi.org/10.2478/intag-2013-0017

Lobell, D. B., Schlenker, W., & Costa-Roberts, J. (2011). Climate trends and global crop production since 1980. *Science, 333*, 616–620. https://doi.org/10.1126/science.1204531

Madhavi Reddy, K., Shivashankara, K. S., Geetha, G. A., & Pavithra, K. C. (2016). Capsicum (hot pepper and bell pepper). In N. Rao, K. Shivashankara, & R. Laxman (Eds.), *Abiotic stress physiology of horticultural crops*. Springer. https://doi.org/10.1007/978-81-322-2725-0_9

Menon, S. S., & Savithri, K. E. (2015). Water stress mitigation in vegetable cowpea through seed hardening and moisture conservation practices. *Journal of Tropical Agriculture, 53*(1), 79–84.

Moftah, A. E., & Al-Humaid, A. I. (2005). Effects of kaolin and pinolene film-forming polymers on water relations and photosynthetic rate of tuberose (*Polianthes tuberosa* L.). *Journal of King Saud University - Agricultural Sciences, 18*(1), 35–49.

Moles, T. M., Mariotti, L., De Pedro, L. F., Guglielminetti, L., Picciarelli, P., & Scartazza, A. (2018). Drought induced changes of leaf-to-root relationships in two tomato genotypes. *Plant Physiology and Biochemistry, 128*, 24–31.

Moretti, C. L., Mattos, L. M., Calbo, A. G., & Sargent, S. A. (2012). Climate changes and potential impacts on postharvest quality of fruit and vegetable crops: A review. *Food Research International, 43*(7), 1824–1832.

Muchero, W., Ehlers, J. D., Close, T. J., & Roberts, P. A. (2009). Mapping QTL for drought stress-induced premature senescence and maturity in cowpea [Vigna unguiculata (L.) Walp.]. *Theoretical and Applied Genetics, 118*, 849–863.

Nuruddin, M. M. (2001). *Effect of water stress on tomato at different growth stages*. M.Sc Thesis.

Ong, C. K. (2002). *Climate change and food security: A framework document*. Institute of Food Security, Environmental Resources, and Agricultural Research, Agricultural University of Norway.

Peet, M. M., & Wolfe, D. (2000). Vegetable crop responses to climatic change. In K. R. Reddy & H. F. Hodges (Eds.), *Climate change and global crop productivity* (pp. 213–243). CABI Publishing.

Peet, M. M., Sato, S., & Gardner, R. (1988). Comparing heat stress on malesterile and male-fertile tomatoes. *Plant, Cell and Environment, 21*, 225–231.

Pena, R., & Hughes, J. (2007a). Improving vegetable productivity in a variable and changing climate. *SATe Journal, 4*(1), 1–22.
Pena, R. J., & Hughes, J. d.' A. (2007b). Breeding for abiotic stress resistance in vegetable crops. In *Advances in breeding techniques for vegetable crops* (pp. 273–313). Springer.
Prasad, B. V. G., & Chakravorty, S. (2015). Effects of climate change on vegetable cultivation-a review. *Nature Environment and Pollution Technology, 14*(4), 923.
Rai, M. K., & Rai, M. (2006). Recent advances in plant biotechnology and its applications in plant tissue culture. *Journal of Plant Biochemistry and Biotechnology, 15*(2), 81–94.
Rouphael, Y., Cardarelli, M., Colla, G., & Rea, E. (2008). Yield, mineral composition, water relations, and water use efficiency of grafted mini-watermelon plants under deficit irrigation. *Horticultural Science, 43*, 730–736.
Sahin, U., Ekinci, M., Ors, S., Turan, M., Yildiz, S., & Yildirim, E. (2018). Effects of individual and combined effects of salinity and drought on physiological, nutritional and biochemical properties of cabbage (Brassica oleracea var. capitata). *Scientia Horticulturae, 240*, 196–204.
Sakamoto, A., & Murata, N. (2002). The role of glycine betaine in the protection of plants from stress: Clues from transgenic plants. *Plant, Cell & Environment, 25*(2), 163–171.
Sato Peet, M. M., & Thomas, J. F. (2002). Determining critical pre-and postanthesis periods and physiological process in *Lycopersicon esculentum* Mill, exposed to moderately elevated temperatures. *Journal of Experimental Botany, 53*, 1187–1195.
Shubha, K., Dhar, S., Maity, A., Raman, R. K., & Mukherjee, A. (2021a). Infuence of leaf architecture on morpho-phenological and yield related traits of garden pea. *Indian Journal of Horticulture, 78*(2), 155–162.
Shubha, K., Mukherjee, A., Singh, N. R., Ray, R. K., & Kumar, U. (2021b). Heat stress tolerance breeding in tomato: A viable option towards climate resilience. *Food and Scientific Reports, 2*(1), 19–25.
Steward, P. R., Thierfelder, C., Dougill, A. J., & Ligowe, I. (2019). Conservation agriculture enhances resistance of maize to climate stress in a Malawian medium-term trial. *Agriculture, Ecosystems & Environment, 277*, 95–104.
Turyagyenda, L. F., Kizito, E. B., Ferguson, M., Baguma, Y., Agaba, M., Harvey, J. J. W., & Osiru, D. S. O. (2013). Physiological and molecular characterization of drought responses and identification of candidate tolerance genes in cassava. *AoB Plants, 5*, 1–27.
Vogel, B. (2009). Smart breeding: Marker-assisted selection: A non-invasive biotechnology alternative to genetic engineering of plant varieties. In N. T. S. Erwood (Ed.), *Greenpeace*.
Wong, C. K., & Bernardo, R. (2008). Genomewide selection in oil palm: Increasing selection gain per unit time and cost with small populations. *Theoretical and Applied Genetics, 116*, 815–824.
Wu, H., Wu, X., Li, Z., et al. (2012). Physiological evaluation of drought stress tolerance and recovery in cauliflower (*Brassica oleracea* L.) seedlings treated with methyl Jasmonate and Coronatine. *Journal of Plant Growth Regulation, 31*, 113–123. https://doi.org/10.1007/s00344-011-9224-x
Yamamura, K., & Kiritani, K. (1998). A simple method to estimate the potential increase in the number of generations under global warming in temperate zones. *Applied Entomology and Zoology, 33*(2), 289–298.
Yordanov, I., Velikova, V., & Tsonev, T. (2000). Plant responses to drought, acclimation, and stress tolerance. *Photosynthetica, 38*, 171–186. https://doi.org/10.1023/A:1007201411474
Zhao, C., Liu, B., Piao, S., Wang, X., Lobell, D. B., Huang, Y., et al. (2017). Temperature increase reduces global yields of major crops in four independent estimates. *Proceedings. National Academy of Sciences. United States of America, 114*, 9326–9331. https://doi.org/10.1073/pnas.1701762114
Zhu, Y., Luo, X., Nawaz, G., Yin, J., & Yang, J. (2020). Physiological and biochemical responses of four cassava cultivars to drought stress. *Scientific Reports, 10*, 6968.

Drought and Heat Stress in Fruit Crops: Implications, Mitigation, and Policy Perspectives

4

Shikha Jain, Anshuman Singh, and Jai Prakash

Abstract

The numerous nutritional benefits associated with fruits make them an essential part of the human diet. They also contribute significantly to crop diversification, employment creation, and better incomes to the growers. However, it is often challenging for fruit crops to reach optimal productivity levels due to abiotic stressors including drought and heat experienced throughout different phenological phases. In addition to hampering harvests, these stressors also endanger food and nutrition security, especially in view of the worsening impacts of climate change. Even though plants have evolved a range of morphological, physiological, and biochemical defense systems to deal with these challenges, even mild-to-moderate intensity of these stresses can severely impair tree growth and productivity; reductions in yield may be as high as 50% under certain conditions. Plant stress response is a multifaceted process mediated, inter alia, by phytohormones, secondary metabolites, transcription factors, and signal transduction. Even though crop improvement through traditional methods still plays a major role in reducing such abiotic risks in fruit crop production, emerging frontiers in biotechnology and nanotechnology would play a crucial role in developing stress-resistant cultivars in the foreseeable future. It is expected that rapid technological advancements in the fields of agro-biotechnology, bioinformatics, and functional and structural genomics will accelerate the development of scion and rootstock cultivars with improved resilience to these and other environmental constraints. In this chapter, we aim to highlight the physiological, biochemical,

S. Jain · J. Prakash (✉)
ICAR-Indian Agricultural Research Institute, Delhi, India
e-mail: jaiprakash@iari.res.in

A. Singh
ICAR-Central Institute for Subtropical Horticulture, Lucknow, Uttar Pradesh, India

R. Dubey et al. (eds.), *Drought and Heat Stress in Agriculture*,
https://doi.org/10.1007/978-981-96-5735-3_4

and genetic components of the recent advances in our understanding of plant drought and heat tolerance. We present an overview of different strategies to adapt to the challenges posed by drought and heat stresses to fruit crop production.

Keywords

Fruits · Bioinformatics · Drought · Heat · Mitigation strategy

4.1 Introduction

Given that a large percentage of the world's population depends on agriculture for their livelihoods and that the population is expected to grow to over 9.7 billion by 2050, the agricultural sector would remain crucial to both food security and economic stability (Castañeda & Doan, 2016). Nonetheless, a number of obstacles such as inadequate irrigation infrastructure, small and dispersed landholdings, poor access to quality planting material and other inputs, limited automation, and excessive dependence on chemical pesticides and fertilizers continue to beset global agricultural production (Dev, 2012). Crop production is additionally hampered by different biotic and abiotic stresses. Abiotic stresses can cause serious damage to plants, resulting in retarded plant development, altered metabolism, and decreased yields. Drought is characterized by extended periods of low freshwater availability, negatively impacting a wide range of ecosystem services and processes that are essential to both ecological balance and human well-being. Droughts can impact large areas over extended periods of time, are difficult to monitor, and are recurrent in nature (Balint et al., 2013). Drought stress occurs when soil and atmospheric humidity are low, and air temperatures are high, leading to an imbalance between evapotranspiration and soil water availability. Conversely, heat stress is defined as elevated soil and air temperatures that remain beyond a threshold level for an extended length of time, permanently harming plant growth and development (Lipiec et al., 2013). Research has demonstrated that heat stress and drought together worsen each other's effects on crop yields, especially in areas where these conditions are common. Given that these stresses might have synergistic effects that can be more harmful than when they occur separately, it is imperative to take both into consideration (Dreesen et al., 2012). Numerous genes and underlying mechanisms are involved in the complicated genetic regulation of responses to abiotic stresses including heat and drought. The genetic underpinnings of abiotic stress responses are more complex than those of biotic stresses, which are frequently regulated by single genes. Examining how plants respond to heat and/or drought stresses often becomes difficult by the impact of other biotic or abiotic stressors. Due to their ability to cause oxidative, osmotic, and thermal alterations, soil salinity and water scarcity are unquestionably two of the biggest obstacles to crop productivity (Landi et al., 2017). Rapid reductions in plant growth and performance can result from simultaneous impacts of high temperatures and water deficits under field conditions, especially in areas with tropical and subtropical climates (Niinemets, 2015).

Increased leaf area and thinning are two common morphological changes brought on by high temperatures (Poorter et al., 2009), which are often accompanied by alterations in leaf anatomy, including smaller cells and higher stomatal density (Król, 2013; Shahinnia et al., 2016). Of India's 159.7 million hectares of agricultural land, 42% are affected by drought (Gogoi & Tripathi, 2019). Thus, a paradigm change is absolutely essential to lessen the negative effects of water scarcity on crop output and livelihoods (Kaushal, 2019). Through an understanding of the diverse expressions and consequences of drought, policymakers, researchers, and farming communities can formulate more focused approaches to alleviate the negative consequences and enhance resilience against subsequent drought incidents.

4.2 Drought and Heat Stresses: A Brief Overview

Understanding the effects of the twin menaces of heat and drought on crop production has gained considerable traction in recent years. Although both stresses alone represent serious risks to the growth and productivity of plants, when combined, they can cause yield reductions that are quite significant (Zhao et al., 2017). Under stressful conditions, both stomatal and non-stomatal factors have a major impact on photosynthesis, a crucial physiological activity in plants. Heat stress and drought can cause non-stomatal limitations such as reduced electron transport capacity and ribulose bisphosphate carboxylase oxygenase (RuBisCO) activity in addition to inhibiting stomatal activity that restricts CO_2 availability. Drought and heat stress combined can also affect Photosystem II, raise leaf temperature, and lower photosynthetic pigment concentrations. Moreover, plants' reproductive processes, such as the growth of pistils and pollen and ovule functions, are also impacted by them, impairing the yields. Reactive oxygen species (ROS) are produced in excess during drought and heat stress, which results in lipid peroxidation, protein denaturation, and damage to the cell membranes. Plant growth and yield are eventually hampered by these physiological and biochemical abnormalities (Prasad & Pisipati, 2011). Drought- and heat-stressed maize plants showed reduced nutrient absorption, decreased root and shoot length, poorer grain and kernel yield, and noticeably higher levels of ROS and malondialdehyde. It was also noted that the activity of antioxidant enzymes including catalase and ascorbate peroxidase were downregulated. Plants frequently prioritize coping strategies according to the degree of the stressors, making their response to the simultaneous presence of heat stress and drought complicated (Rizhsky & Liang, 2004). For example, plants close their stomata early during a drought in order to reduce water loss and maintain water balance. On the other hand, when plants experience heat stress, they raise their stomatal conductance in order to promote transpiration and dissipate heat, which lowers the temperature of their leaves (Rizhsky & Liang, 2004). This varied reaction to co-occurring stresses illustrates the many adaptive mechanisms used by plants to deal with environmental challenges (Rizhsky & Liang, 2004). Rizhsky and Liang (2002) shed light on the intricate relationship between heat stress and drought on plant physiology, with a specific focus on stomatal regulation and leaf temperature. When plants are simultaneously under drought and heat stress, their stomatal closure to minimize water loss takes precedence, which raises the temperature of their leaves relative to those under heat stress alone. Furthermore, the accumulation of osmoprotectants differs remarkably under drought and heat stress conditions.

Proline is a key osmoprotectant that accumulates in plants under drought stress; no proline accumulation has been documented under heat stress. On the other hand, proline buildup can become toxic in the presence of coupled stress, and sucrose becomes the primary osmoprotectant that accumulates in plants faced with drought and heat stresses (Cohen et al., 2020).

4.3 Physiological Responses of Plants to Drought and Heat Stresses

Plants have evolved dynamic responses at various levels to cope with abiotic stresses including drought and heat. These reactions include morphological, physiological, and biochemical adjustments that allow plants to thrive under harsh environmental conditions (Huber & Bauerle, 2016). Extending the root system to extract soil moisture from deeper layers, reducing stomatal number and conductance to minimize water loss through transpiration, decreasing leaf area to reduce surface area exposed to heat and evaporation, increasing leaf thickness to conserve water, and adopting leaf rolling or folding to further reduce water loss are some such adjustments (Goufo et al., 2017). Additionally, deposition of cuticular wax on aerial surfaces contributes to the adaptive response by reducing water loss through transpiration (Lee & Suh, 2013). Conversely, tolerance traits entail maintaining tissue hydrostatic pressure by means of cellular and metabolic alterations, principally through osmotic adjustments. These mechanisms help plants withstand water stress by maintaining cell turgor pressure and preventing cellular damage (Blum, 2017). Plants use a variety of sensors to detect abiotic constraints, and these sensors then start response signaling cascades. These signals are transduced by complex pathways including transcription factors and signaling, each of which has a distinct function in coordinating the plant's stress response. These factors regulate the expression and the activation of stress-responsive genes, which eventually result in adaptive responses that improve plant survival under harsh conditions (Hirayama & Shinozaki, 2010). Furthermore, certain plants are equipped with internal mechanisms that can initiate long-range chemical, electrical, or hydraulic signals that initiate systemic stress reactions. Under drought stress, plants may decrease root hydraulic conductivity to prevent water loss to dry soil, while high temperatures can exacerbate root moisture loss to detrimental levels. Water transport within plants occurs under tension, determined by factors such as soil water availability and atmospheric vapor pressure deficit, which creates turgor pressure within cells (Boursiac et al., 2005). Water transport in roots is influenced by a number of factors, including root anatomy, salinity, and water availability in the soil. Aquaporins, integral membrane proteins that serve as channels for the transfer of small solutes and water, play a significant role in regulating these processes. Their activity can affect water uptake and transport in roots, thereby impacting plant water balance and overall stress response (Vandeleur et al., 2014).

Photosynthesis involves various processes, including the photosystems and photosynthetic pigments, electron transport system, and CO_2 reduction pathways. Any disruption to these components, whether caused by water, salt, or heat stress, can impact photosynthetic efficiency and transpiration rates. Stomatal closure under stressful situations lowers transpiration rates and photosynthetic efficiency by decreasing CO_2 uptake and water loss. However, non-stomatal limitations can also contribute to reduced photosynthetic performance. These limitations may include decreased leaf expansion, leaf senescence, and impaired functioning of the photosynthetic machinery. One prominent explanation for the non-stomatal restrictions seen under stress conditions is the decreased availability of CO_2, which is accompanied by the suppression of important photosynthetic enzymes and ATP synthases. These abnormalities may result in less energy being produced and less carbon assimilated, ultimately affecting the productivity of plants (Wahid et al., 2007). Elevated temperatures have the potential to cause several detrimental effects on plant development and productivity, including a reduction in chlorophyll content, an increase in amylolytic activity, the disintegration of thylakoid grana, and a disruption in assimilate transport (Kozłowska et al., 2007). Decreases in photosynthetic rate are directly associated with Water Use Efficiency (WUE), an important parameter in crop response to osmotic imbalances. Stomatal closure, induced by high temperatures, can increase WUE by reducing transpiration while maintaining CO_2 uptake, although this may come at the expense of reduced photosynthesis. Carbon isotope discrimination is a valuable tool for screening physiological traits related to WUE or "Transpiration Efficiency," allowing for the assessment of phenotypic variation within breeding populations (Ruggiero et al., 2017). Recent advances in photosynthesis research may open up new possibilities for dealing with pertinent problems that arise when crop plants are subjected to severe drought and heat stresses (Kromdijk et al., 2016). Particularly vulnerable to the effects of high temperatures during the reproductive phase of plant development are processes including pollen viability, pollination, growth of the pollen tube, and early embryo development. Male reproductive tissues are generally more sensitive to high temperature stress than female reproductive tissues at all stages of development (Giorno et al., 2013).

4.4 Biochemical Responses of Plants to Drought and Heat Stresses

Signal perception through cell-surface receptors is a fundamental process in all living organisms, crucial for responding to environmental changes and stress conditions. The Wall-Associated Kinases (WAKs) subfamily of the Receptor-Like Kinases (RLK) family, in particular, is essential for plants to perceive signals from their surroundings and nearby cells and to start signaling cascades in response to stress (Shulaev et al., 2008; He et al., 1996). Plants sense stress and use systemic signaling cascades to control different physiological functions. In reaction to environmental cues, aquaporin proteins, which are crucial for controlling hydraulic conductivity and water transport, go through biochemical regulation. Phosphorylation, variations in the pH and calcium levels of the cytoplasm, and relocalization to intracellular compartments are some of these changes (Alleva et al., 2006; Boursiac et al., 2008). Abscisic acid (ABA) is particularly important for stress tolerance (Lata & Prasad, 2011). Hydraulic conductivity and long-distance transmission between roots and shoots are both governed by ABA. It induces responses such as inhibition of leaf expansion and stomatal closure in response to stress, even before discernible changes in leaf water or nutritional status occur (Suzuki et al., 2013). In addition to ABA, osmotic stress triggers the synthesis or catabolism of various other growth regulators, including auxins, cytokinins, ethylene, and gibberellins. These molecules regulate physiological processes to help plants withstand stress by acting as signal molecules in intricate signaling networks (Verma et al., 2016). Stress-related increases in intracellular Ca^{2+} levels are frequently brought on by signal molecules such as diacylglycerol, ROS, and inositol trisphosphate. Calcium binding proteins, acting as Ca^{2+} sensors, detect the elevated Ca^{2+} levels, leading to the activation of calcium-dependent protein kinases. These activated kinases or phosphatases can then modify specific transcription factors (TFs) by phosphorylating or dephosphorylating them, thereby regulating the expression levels of stress-responsive genes (Reddy et al., 2011). In response to early heat stress, there is a surge in Ca^{2+} influx and cytoskeletal restructuring, leading to the upregulation of signaling cascades involving mitogen-activated protein kinases and calcium-dependent protein kinases. Heat stress primarily causes denaturation of proteins, instability of cytoskeleton and nucleic acids, increased fluidity of membranes, and disruption of protein synthesis and degradation. Severe cellular damage or loss may arise from extended exposure to moderately high temperatures or from brief exposure to extremely high temperatures. This may seriously impair plant growth by interfering with ion flux and producing reactive oxygen species (ROS) and other toxic compounds (Wahid et al., 2007). These protective proteins are often associated with stress tolerance, photosynthesis, membrane stability, and maintenance of cellular hydration. Abiotic stresses induce the overproduction of ROS. Plants use two primary mechanisms to combat oxidative stress: first, they prevent or minimize the generation of reactive oxygen species (ROS); second, they detoxify and scavenge ROS using anti-oxidants and enzymes. Additionally, under abiotic stress conditions, plants also produce elevated levels of methylglyoxal, a highly reactive cytotoxic compound that damages cellular function and DNA. Methylglyoxal is detoxified by the glyoxalase system, which converts it to D-lactate via glyoxalase I and glyoxalase II enzymes. (Yadav et al., 2005). New retrograde signals have been discovered recently, such as the metabolite 3′-phosphoadenosine 5′-phosphate, which builds up under drought and strong light conditions. This metabolite moves from the chloroplast to the nucleus, regulating ABA signaling and stomatal closure during oxidative stress, thereby contributing to drought tolerance and activating the high light transcriptome (Pornsiriwong et al., 2017). Plants generate a range of low molecular mass metabolites, such as sugars, soluble carbohydrates, amino acids, and sugar alcohols, to maintain cellular components and restore homeostasis in response to stress. Proline, in particular, plays a crucial role in osmotic adjustment in response to abiotic stress, contributing to stress tolerance. Likewise, while its exact mode of action is unknown, glycine betaine is critical for some plant species' ability to withstand stress. To improve plants' ability to withstand stress, genes linked to the production of glycine betaine have been introduced (Quan et al., 2004).

4.5 Impact of Drought Stress on Fruit Crops

A variety of fruit crops are severely impacted by drought stress, which has an effect on the crops' growth, development, and production. Mango trees under drought stress exhibit a reduction in the quantity of leaves/flushes, length of the flush, and water content of the leaves. Mango trees are unusual in that they can flower despite brief periods of water stress; this is believed to occur because the stress conditions create a stimulus for flowering (Schaffer et al., 1994). Drought stress causes morphological and physiological changes in grapevines. Large xylem channels let grapes tolerate water stress even if they frequently experience it (Serra et al., 2014). When there is a shortage of water, grapevine roots explore the deeper soil layers, resulting in a drop in shoot length and an increase in root length (Hardie & Martin, 2000). Water stress can severely affect bananas, particularly while they are in flowering. According to Murali et al. (2005), the Elakki cultivar is particularly vulnerable to stress during the flower differentiation stage. Rasthali, Robusta, and Karpuravalli cultivars showed a reduction in bunch weight of 42.07%, 25.0%, and 18.83%, respectively, when exposed to water stress (Ravi et al., 2013). Water stress severely impacts the growth and development of papaya fruit. The content of leaves, flowers, and fruit was observed to have decreased by 50%, 86%, and 58%, respectively (Masri et al., 1990). Compared to control and 10 DASI (days after stress imposition)-exposed plants, Gamboa-Tuz et al. (2018) observed that 3-month-old Maradol papaya plants wilted more frequently after 20 DASI.

4.6 Impact of Heat Stress on Fruit Crops

Many fruit crops are susceptible to major adverse effects from heat stress, which can impair their development, growth, and ultimately productivity. Elevated temperatures have the potential to adversely impact the quality of apple fruit by reducing the acid concentration, firmness, and development of water core (Sugiura et al., 2013). Reduced pistil size due to heat stress might produce irregular blooms and less fruit set in apricot plants (Rodrigo & Herrero, 2002). Elevated temperatures (33.5 °C) can impact banana growth and production, with effects on leaf production and relative leaf area growth observed beyond certain temperature thresholds (Turner & Lahav, 1983). High temperatures and low humidity can result in poor fruit set in guava, particularly in regions like North India during summer months (Chadha & Pandey, 1986). Higher temperatures (28/18 °C max/min.) can lead to the development of translucency disorder in pineapple fruits (Paull & Reyes, 1996). Heat stress can reduce pollen viability and fruit set in mango, particularly when temperatures exceed a certain threshold (≥35 °C). In papaya plants, high temperatures can cause female sterility, resulting in nonfunctional female parts and functioning male parts. Increases in temperature exacerbate sterility, especially in bisexual cultivars (Paull & Duarte, 2011). Abnormal pistil development may occur in sweet cherry plants exposed to high temperatures during flower development stages (Beppu et al., 2001). Warm weather can increase the incidence of albinism in strawberries, which causes the fruit to become blotched and have pink or white patches on the surface. The temperature more than 35 °C hindered fruit set in grapes (Fischer & Orduz-Rodriguez, 2012). High diurnal variation has been shown to increase the occurrence of pink berry disease in Thompson Seedless grapes (Sharma et al., 2013). High temperatures reduced the flushing time as well as the flushing interval in litchi. Elevated temperatures have the potential to seriously harm shoot and root development, sunburn leaves, and fruit maturity. They can also produce leaf abscission, dry branches and stems, fruit drop, cracked fruit, and sunburn in litchi (Menzel & Waite, 2005). When the daily air temperature rises above 40 °C, fruit damage from heat is widespread (Singh et al., 2012). The exposed fruit on the south side of the tree is primarily affected by heat damage in litchi fruits.

4.7 Strategies for Mitigating the Effects of Drought and Heat

Over time, numerous agronomic techniques have been developed to improve plant resistance to abiotic stresses like drought, which are becoming more common as a result of climate change (Raza et al., 2019). To mitigate yield losses due to drought stress, it is important to select appropriate tillage systems and provide fertilization and irrigation treatments based on the developmental stage of the plant (Karavidas et al., 2022). Furthermore, a variety of management strategies have been found to maximize the yield of crops under droughts. These include techniques like varying the frequency and timing of sowings, rotating crops, choosing plant species with shorter life cycles, implementing appropriate irrigating techniques, and applying biofertilizers (Abebe et al., 2020). In addition to these management practices, plants themselves have evolved mechanisms to cope with drought stress. The response of plants to drought varies depending on the severity of the stress and the developmental stage of the plant. To adapt to drought conditions, plants can use a variety of strategies, including changing the architecture of their roots, modifying stomatal conductance, building up osmolytes, and turning on stress-responsive genes (Cooper et al., 2006).

The detrimental effects of heat stress on plants can be lessened by applying specific macro- and micronutrients. It has been demonstrated that micronutrients like manganese (Mn), boron (B), and selenium (Se) along with macronutrients like potassium (K) and calcium (Ca) alter stomatal function and initiate physiological and metabolic processes that help maintain high water potential in plant tissues, thereby increasing heat stress tolerance (Waraich et al., 2012). Similarly, the application of silicon (Si) has been found to alleviate the effects of drought stress in some crops (Gong et al., 2005). Higher levels of photosynthetic pigments, enhanced antioxidant activities, and alterations in gene expression linked to stress-related pathways, flavonoid production, and antioxidant responses are all seen in plants treated with silicon (Ma et al., 2016). The potential of plant growth regulators, such as cytokinins, salicylic acid, and ABA, to improve drought tolerance has also been investigated. These substances can increase water potential and chlorophyll content in water-stressed plants (Zhang et al., 2004). Additionally, concentrated ABA or ABA analogs have been investigated to maintain the marketability of horticultural crops by reducing drought stress impacts (Kim & Van Iersel, 2008).

4.8 Management Strategies to Overcome Drought Stress

Drought Tolerant Fruit Crops: Fruit species naturally adapted to arid and semi-arid environments are typically more tolerant to drought stress. Examples of drought-tolerant fruit crops include pomegranate (*Punica granatum*), aonla (*Emblica officinalis*), fig (*Ficus carica*), ker (*Capparis deciduas*), tamarind (*Tamarindus indica*), ber (*Ziziphus species*), karonda (*Carissa carandus*), lasoda (*Cordia myxa*), and khejri (*Prosopis cineraria*) (Table 4.1).

Breeding for Drought Tolerance: Breeding for drought tolerance involves identifying genetic sources of resistance or tolerance in wild varieties, landraces, or existing cultivars. Rootstock breeding is also important, as certain rootstocks can confer drought tolerance to fruit trees. For example, rootstocks such as *Zizyphus rotundifolia* for ber, Saint George and 1103 Paulson for grapes, and Rangpur lime and Marmalade orange for citrus have shown drought tolerance. Jiménez et al. (2013) found that to enhance water-use efficiency (WUE), certain rootstocks when grafted with "Catherina" peach resulted in notable changes in various biochemical parameters under water-deficit conditions. Specifically, the rootstocks "Cadaman," "GF-677," "Rootpac 20" (*Prunus besseyi* x *P. cerasifera*), and "Rootpac R" (*P. cerasifera* x *P. dulcis*) exhibited increased levels of proline in both roots and leaves, sorbitol in leaves, and raffinose in roots.

Table 4.1 Classification of fruits according to relative tolerance to drought (Balerdi et al., 2003)

Tolerant	Moderately tolerant	Sensitive
Ber	Avocado	Citrus
Phalsa	Passion fruit	Papaya
Custard apple	Litchi	Banana
Date Palm	Carambola	Sapota
Pomegranate	Coconut	Rambutan
Grafted Citrus	Guava	Apple
Almond	Longan	
Cashew	Atemoya	

Moisture Conservation Practices: Soil moisture conservation techniques including sod culture, intercropping, cover cropping, pruning, crop load management, and green manuring can minimize the water loss through evaporation and transpiration. These practices help slow water seepage, reduce transpirational losses, improve soil moisture retention, and minimize evaporative losses, ultimately conserving water during periods of drought stress.

4.9 Improved Irrigation Practices

Drip Irrigation: Drip irrigation is highly water-efficient and can significantly reduce water requirements while increasing water use efficiency (WUE) and yield has been implicated in Table 4.2.

Partial Root Zone Drying (PRD): Partial root-zone drying is an irrigation technology that has shown promise for improving water use efficiency without appreciably lowering yields. The method basically entails irrigating about half of a crop's root system while leaving the other half to dry. The wet part of the root system is allowed to dry for a bit before the dry half is watered. This practice helps regulate stomatal aperture, maintain favorable water status, and improve water use efficiency while sustaining fruit crop yield and quality. The re-wetting frequency under PRD should be determined by detection or prediction of completion in extraction of soil water from the drying side. PRD has the potential to maintain or even improve fruit crop yield and quality while increasing WUE by up to 80–92% as in various fruits like grape (Romero et al., 2016) and mango (Dos Santos et al., 2015).

Regulated Deficit Irrigation: In this irrigation technique, full irrigation is applied during drought-sensitive phenological stages and limited irrigation during drought-tolerant phenological stages to save water without compromising the yield. The three main factors that contribute to improved water use efficiency with RDI are: (1) optimized stomatal control that improves the ratio of photosynthesis to transpiration; (2) enhanced guard cell signal transduction network that reduces transpiration water loss; and (3) reduced soil evaporation through partial root-zone watering, which results in smaller evaporative surface areas (Chai et al., 2016). RDI can lead to significant water savings with minimal reduction in yield while improving produce quality. Several researchers have reported water savings from 43% to 65%

Table 4.2 Water use efficiency with drip irrigation (Source: INCID (1994), Drip irrigation in India)

Crops	Yield increase (%)	Water saving (%)	Increase in WUE (%)
Banana	52	45	176
Grapes	23	48	136
Sweet Lime	50	61	289
Pomegranate	45	45	167

under RDI strategy with small reduction in yield, but with higher quality of produce (Mirás-Avalos et al., 2016).

Hydrophilic Polymers (HPs): HPs are a class of polymers that gets dissolved or swollen by water. Utilize hydrophilic polymers, which absorb and retain water, to alleviate stress during drought periods. These polymers swell by absorbing water and releasing it to the plants as needed. HPs improve soil water holding capacity, plant nutritional and water status, soil physical properties, seed germination, and yield while decreasing the negative effects of soil salinity on plants (Finch, 1987). Various HPs used in agriculture like Aquasorb, Agrihope, Broadleaf P4, Pusa Hydrogel, etc. Pusa Hydrogel, a semi-synthetic cross-linked hydrophilic polymer, was developed for improving the water use efficiency of agricultural and horticultural crops under limited irrigation and rainfed conditions. The performance of Pusa Hydrogel has been extensively evaluated across various crops, including groundnut, potato, soybean, mustard, wheat, onion, tomato, cauliflower, carrot, strawberry, opium, maize, sugarcane, paddy, turmeric, chrysanthemum, and cotton. These evaluations have been conducted at institute farms and farmers' fields, spanning multiple locations and in collaboration with different institutes (Mazen et al., 2015; Fidelia & Chris, 2011).

Mulching: Mulching is a technique where a layer of material is spread over the soil surface around plants to conserve soil moisture, suppress weed growth, and maintain soil temperature (Kasirajan & Ngouajio, 2012). Both organic (e.g., straw, leaves, wood chips) and synthetic mulches (e.g., plastic, geotextiles) can be used. Mulching helps create favorable conditions for plant growth and development, thereby enhancing fruit production (Zaen El-deen et al., 2015). Use of mulches to enhance fruit production is depicted in Table 4.3.

Biochar Application: The pyrolysis of biomass produces biochar, a carbon-rich material that has drawn interest for its capacity to improve soil fertility and water retention. By enhancing soil structure, increasing nutrient availability, and improving water-holding capacity, adding biochar to the soil encourages plant establishment and growth during dry spells (Haider & Raza, 2020; Zhang & Ding, 2020). Recent studies have shown a significant improvement in the plant's photosynthetic rates, xylem water potential, chlorophyll content, uptake of essential nutrients like N, Si, P, and K, and growth upon the application of biochar (Semida & Beheiry, 2019). Haider and Raza (2020) observed improvements in various physiological and morphological parameters in water-stressed maize compared to control conditions. This finding suggests that maize plants subjected to water stress may exhibit adaptive responses to cope with the stress, leading to enhanced water use efficiency and stress tolerance. Glaser et al. (2000) revealed that biochar-enriched soil exhibited a notable 18% increase in water retention capacity compared to soil without biochar. Additionally, when biochar was applied to potted *Pyrus ussuriensis* seedlings, it significantly delayed soil moisture loss. Moreover, the use of biochar led to observable changes in chlorophyll fluorescence parameters, ultimately resulting in enhanced photochemical efficiency of photosystem II (Lyu et al., 2016).

Table 4.3 Effect of mulching on fruit crops

Crop	Treatment details	Effects	References
Banana	Wheat Straw and Banana straw (12.5 Kg/plant) in the beginning of summer (Feb)	Increase bunch weight and conservation of soil moisture	–
Citrus	Film mulching, during cell division and early maturing stage	Increase the quality of fruits and sugar content	–
Mango	Plastic mulch (100 μ), during	Enhancement in flowering	CISH,
	FBD (Oct-Nov)	(38–70%) and yield (40–60%)	Lucknow
Guava	Black polythene	Max. no. of fruits as well as highest yield/plant	Das et al. (2010)
	Paddy Straw	Improve fruit quality	Das et al. (2010)

PGPR-Mediated Drought Mitigation: Plant growth-promoting rhizobacteria (PGPR) are bacteria that invade plant roots and stimulate the growth of plants. It is hypothesized that bacterial alterations to the root architecture increase the root's total area, improving its ability to absorb nutrients and water and promoting the root's overall growth. Increased crop productivity may arise from inoculating plants with effective PGPR strains, which may maintain near-average shoot development even under dry conditions (Mohanty et al., 2021). PGPR are beneficial microorganisms found in the rhizosphere that promote plant growth and enhance tolerance to various stresses, including drought. These bacteria colonize the plant roots and provide numerous benefits, such as nitrogen fixation, phosphate solubilization, and production of phytohormones, which help plants cope with drought stress (Xiong & Li, 2020). Some examples of PGPRs include *Pseudomonas, Bacillus, Rhizobium, Azospirillum, and Acetobacter* (Pathania & Rajta, 2020).

Development of Transgenics: Using the transgenic technique, plants are genetically modified to express particular genes that provide resistance to environmental obstacles like drought. Fruit crops can become more resilient through the focused alteration of genes involved in stress response pathways, which is made possible by this strategy (Fujita et al., 2004). However, the use of transgenic techniques in fruit crops for drought tolerance is still limited and requires further research and development. Some important use of this technique in fruits is listed in Table 4.4.

4.10 High Temperature Tolerance Mechanism

Fruit plants use morphological, physiological, and molecular strategies to withstand high temperatures and preserve cellular homeostasis in the face of heat stress.

Morphological Adaptations: Heat and drought tolerance mechanisms in fruit plants include a variety of adaptations that enable them to withstand and survive

Table 4.4 Transgenic approaches for stress tolerance in fruit crops

Fruit	Gene	Remarks	Response
Banana	*MusaDHN-1*	Over expression of dehydrin gene, belonging to a broader class of LEA proteins	Drought
	MusaWRKY71	Encodes a WRKY transcription factor protein	Multiple abiotic stress
	MusaSAP1	Encode zinc finger protein associated proteins (SAP)	Multiple abiotic stress
Citrus	*P5CSF129A*	Osmotic adjustment, protected against ROS by modulating the antioxidant enzymes activity	Water deficit
Mulberry	*hva1*	Encodes a group 3 LEA protein	Drought
	Osmotin	Encoding osmotin and osmotin-like proteins belonging to the plant PR-5 group of proteins	Drought

under adverse environmental conditions. Fruit plants like aonla (Indian gooseberry) and ber (Indian jujube) often develop deep root systems, allowing them to access moisture from deeper soil layers during drought periods. Some fruit plants, such as aonla and ber, may shed their leaves during periods of extreme heat or drought to reduce water loss through transpiration. Fruit plants like aonla, tamarind, and ker (Indian date) may reduce their leaf area to minimize water loss and maintain hydration under drought conditions (Srivastava et al., 2012). Deposition of epicuticular waxes on leaf surfaces can help reduce water loss through transpiration by forming a protective barrier against evaporation. Some fruit plants produce latex in their leaves, which can act as a barrier against water loss and protect against herbivory. Fruit plants may adjust the orientation of their leaves to minimize exposure to direct sunlight and reduce heat stress. This can include vertical orientation or rolling their leaves along the axis to reduce surface area and transpiration. Plants like cactus pear and fig have specialized water-binding mechanisms that allow them to store water efficiently and withstand extended periods of drought. Presence of thorns on fruit plants such as ker and ber can help reduce herbivory pressure and prevent moisture loss by minimizing leaf surface area and maximizing conductive or convective loss of heat (Adams et al., 2001).

Anatomical Adaptations: Anatomical adaptations in fruit crops play a crucial role in their ability to tolerate heat stress. Fruit crops like ber (Indian jujube), ker (Indian date), and cactus pear may have sunken stomata on their leaf surfaces. Sunken stomata are located in pits or depressions on the leaf surface, which can help reduce water loss through transpiration and minimize the risk of desiccation under high temperature conditions. Some fruit crops, such as fig and cactus pear, have thick cuticles covering their leaf surfaces. The thick cuticle acts as a barrier that reduces water loss through transpiration and provides protection against dehydration under heat stress. Fruit crops like ker, lasoda (*Cordia*), and fig may have a wax coating on their leaf surfaces. The wax coating forms a hydrophobic layer that helps reduce water loss by limiting evaporation from the leaf surface and enhances heat tolerance by reflecting sunlight. Some fruit crops, including ber, phalsa (Indian sherbet berry), and tamarind, exhibit pubescence or the presence of fine hairs on their leaf surfaces. Pubescence can create a microclimate around the leaf surface, reducing solar radiation absorption and heat stress. These anatomical adaptations collectively contribute to the heat tolerance of fruit crops by reducing water loss, enhancing heat reflection, and creating a protective barrier against dehydration and excessive solar radiation. By possessing these anatomical features, fruit crops are better equipped to withstand and thrive in environments characterized by high temperatures and water scarcity (Pareek & Sharma, 1991).

Thermo-Tolerance: Thermo-tolerance in plants can be enhanced by the accumulation of certain compounds known as thermo-protectants, as well as phytohormones. These molecules play vital roles in mitigating the adverse effects of heat stress on plant growth and development. Low molecular mass compounds such as proline (Pro), glycine betaine (GB), and trehalose (Tre) act as thermo-protectants in plants. They help maintain cellular integrity and function under heat stress conditions. Exogenous application of these molecules has been shown to benefit plants growing under heat stress by enhancing their ability to withstand high temperatures. Various phytohormones play important roles in regulating plant responses to heat stress. Abscisic acid (ABA) levels increase in response to heat stress (Ding et al., 2010), and it helps induce thermo-tolerance by regulating the expression of heat stress-responsive genes. Salicylic acid (SA) is involved in signaling pathways that confer heat stress tolerance in plants (Wang et al., 2010). Polyamines (PA), such as spermidine and spermine, have been implicated in enhancing thermo-tolerance by modulating various physiological processes. Brassinosteroids (BRs) are involved in regulating plant growth and development, and they also play a role in heat stress responses (El-Bassiony et al., 2012). Ethylene signaling pathways are involved in heat stress responses and thermo-tolerance mechanisms. Nitric oxide (NO) acts as a signaling molecule in plants and is involved in mediating heat stress responses (Hasanuzzaman et al., 2010). Under heat stress conditions, plants induce the production of heat shock proteins (HSPs). These proteins play crucial roles in protecting cellular structures and functions from heat-induced damage. They assist in protein folding, stabilization of membranes, and preventing the aggregation of denatured proteins thus promoting cell survival under high-temperature conditions (Swindell et al., 2007).

4.10.1 Regulation of Gene Expression

When temperatures rise, certain genes are expressed differently in fruit plants. The production of proteins that support stress tolerance and adaptation is triggered by heat stress-responsive genes. Genes that respond to heat, such as those that encode HSPs and other stress-related proteins, are mostly controlled by transcription factors like heat shock factors (HSFs).

4.11 Management Strategies to Overcome Heat Stress

Selection of Fruit Crops: Perennial fruit species adapted to high-temperature conditions such as ber, pomegranate, aonla, fig, ker, tamarind, bordi (*Ziziphus species*), karonda, lasoda, khejri, and Indian fig should be chosen for cultivation in heat-prone areas.

Rootstock Selection: Utilize heat and drought-tolerant rootstock genotypes to improve the response of fruit trees to heat stress and water deficit conditions. Heat and drought-tolerant rootstocks help minimize the immediate effects of dry conditions and promote quicker recovery. Increasing awareness about the use of rootstocks for overcoming the adverse effects of heat and drought has led to their increased adoption in cultivation. (Table 4.5).

Use of Anti-transpirants: Anti-transpirants can be applied to minimize evapotranspiration rates and reduce water loss from plant surfaces. Foliar sprays of 3% kaolin can help reduce heat stress by decreasing evapotranspiration and maintaining plant hydration levels.

Bark Painting: Protect the trunk cambium from high-temperature damage by spraying or painting the bark of exposed trunks and branches with reflective white coating. This helps to reduce heat absorption and prevent damage to the vascular tissues.

4.12 Policy Approaches

Adequate policy support is a prerequisite to foster the use of cutting-edge techniques and technologies for reducing the negative impacts of various stresses on agricultural production. Policies that encourage sustainable agricultural practices should be given the highest priority. Addressing the intricate problems brought on by numerous stressors requires collaboration among different stakeholders, including researchers, legislators, farmers, and industries. By adopting a holistic approach and implementing targeted strategies, it is possible to enhance agricultural resilience and ensure food security in the face of changing environmental conditions.

Some of the emerging policy recommendations include:

A. **For Planners and Line Departments**

It is undoubtedly essential to identify the regions frequently prone to natural disasters and decentralize efforts to address them. An explanation of anticipatory actions and strategies, along with preparation and recovery plans, is provided below:

Drought Monitoring and Early Warning Systems: It is imperative to operationalize early warning systems and drought impact assessment in order to attain cli-

Table 4.5 Heat- and drought-tolerant rootstock in fruit crops (Nimbolkar et al., 2016)

Fruit crops	Rootstocks
Almond	*Prunus amygdaliformis, Prunus elaeagnifolia, Prunus xerophila*
Apple	MM. 101
Ber	Jhar ber (*Ziziphus nummularia*), (*Ziziphus lotus*)
Citrus	Rough Lemon(*Citrus jambhiri*), Rangpur Lime (*Citrus limonia),*Carrizo citrange
Grapes	110 R, Dogridge
Mango	13–1
Peach	Almond, Peach x almond, *Prunus persica x Prunus davidiana*

mate-resilient rainfed agriculture. This involves actively involving remote sensing tools, drought experts, and farming communities to enhance outreach and utilize drought information effectively. Leveraging digital technologies can enhance the implementation of these approaches, facilitating timely responses to drought events.

1. *Custom Hiring Centers for Farm Machinery:* To address labor shortages during extreme weather events, establishing a large number of custom hiring centers can provide farmers with access to farm machinery. This enables them to implement resilient technologies and carry out agricultural activities efficiently, even during labor shortages.
2. *Establishment of Weather Risk Management Centers:* Establishing weather risk management and mitigation centers at the local level, such as at the panchayat or village level, can facilitate anticipatory measures and adaptation to weather-related risks. This includes developing contingency plans, providing seeds of alternate crops, and implementing measures to mitigate the impacts of extreme weather events.
3. *Weather Index-Based Insurance Schemes:* Implementing weather index-based insurance schemes and linking them to credit can provide financial protection to farmers against weather-related risks. Offering various levels and types of policies at affordable premium rates can help ensure widespread adoption and coverage among farmers, thereby enhancing their resilience to extreme weather events.

   By implementing these recommendations, policymakers and stakeholders can strengthen the resilience of rainfed agriculture and improve the ability of farming communities to cope with and adapt to extreme weather events such as droughts.

B. **Farmers**

For farmers, adopting various approaches and strategies can enhance their resilience to extreme weather events. Here are some specific actions they can take:

1. *Shift to Conservation Agriculture:* Implementing conservation agriculture practices, such as minimal soil disturbance, crop rotation, and residue management, can improve soil health and water retention, making farms more resilient to droughts and floods.
2. *Farm-Level Rainwater Harvesting Structures:* Building rainwater harvesting structures on farms, such as ponds, check dams, and rooftop rainwater harvesting systems, can capture and store rainwater for irrigation during dry spells, reducing dependency on erratic rainfall patterns.
3. *Integrated Nutrient Management Modules:* Implementing integrated nutrient management techniques, including the use of organic manures, green manures, and biofertilizers, can improve soil fertility and crop resilience, reducing the impacts of nutrient stress during extreme weather events.
4. *Resilient Crop Varieties:* Planting resilient crop varieties that are adapted to local climatic conditions, such as drought-tolerant or short-duration varieties, can help mitigate the impacts of heat waves, droughts, and floods on crop yields.
5. *Adoption of Intercropping and Novel Farming Systems:* Embracing intercropping and novel farming system combinations, including livestock and fisheries integration, can enhance farm biodiversity and resilience to extreme weather events, while also providing additional sources of income.
6. *Real-Time Contingency Plans:* Developing and implementing real-time contingency plans that have been successfully demonstrated can help farmers respond effectively to sudden weather changes, such as shifting planting dates or adjusting irrigation schedules.

   By adopting these approaches and strategies, farmers can build resilience to extreme weather events, safeguarding their livelihoods and food security in the face of climate change. Additionally, promoting knowledge sharing and capacity building among farmers can enhance their ability to implement these measures effectively.

C. **For Researchers**

For researchers, there are several key actions that can be taken to address the challenges posed by abiotic stresses and extreme weather events:

1. *Focus on Priority Areas:* Researchers should prioritize the focus of their research on the management of various abiotic stresses.
2. *Investment in Abiotic Stress Research:* There should be increased investment in research addressing the management of abiotic stresses. This includes studying the physiological, biochemical, and genetic mechanisms underlying stress responses in crops and developing innovative strategies for stress mitigation and adaptation.
3. *Enhanced Collaboration:* Develop mechanisms to enhance synergies among research institutes focusing on abiotic stresses. Collaboration and knowledge sharing can accelerate progress in developing effective solutions for stress management.

4. *Focus on Degraded Lands:* Research efforts should focus on exploring and documenting mitigation options for abiotic stresses, particularly in degraded lands brought into cultivation. This information can be immediately useful for farmers facing challenges in these environments.
5. *Academic Training and Entrepreneurship:* Augment academic activities to provide training for successful agricultural entrepreneurship in abiotic stressed environments. Equipping researchers with the skills and knowledge to support farmers in managing stressors is crucial for enhancing resilience in agricultural systems.
6. *Promote Public–Private Partnerships (PPP):* Facilitate public–private partnerships to accelerate research and academic activities for abiotic stress management. Collaborations between academia, government agencies, and private sector stakeholders can leverage resources and expertise to address research gaps and translate findings into practical solutions.
7. *Develop Clear Guidelines for Risk Management:* Researchers should work toward developing guidelines for risk management related to extreme weather events. Optimizing data collection for these events can strengthen predictions of occurrence and improve post-hailstorm management of crops, thereby minimizing losses and enhancing resilience.

4.13 Conclusion

The challenges posed by abiotic stresses such as drought and heat for fruit growers are significant, impacting productivity, quality, and overall profitability. The slow pace of adaptation to resistant/tolerant varieties, rootstocks, and cultural practices exacerbates these issues, particularly among small and marginal farmers. The response of plants to stress, including high temperatures and drought, is multifaceted and involves intricate molecular, biochemical, physiological, and morphological mechanisms. Understanding these mechanisms is crucial for developing varieties with high stress tolerance. Given the expected increase in the frequency and intensity of stressors in the future, there is an urgent need to enhance efforts in this area. One approach for improving stress tolerance is to screen wild relatives of crops and related species for traits related to heat tolerance. These species may harbor valuable genetic diversity that can be utilized in breeding programs to develop heat-tolerant varieties. Additionally, conventional breeding methods and biotechnological approaches, such as genetic transformation, can be employed to develop heat-tolerant varieties. Advancements in technologies such as genomics have significantly contributed to our understanding of complex stress tolerance traits in plants. However, the translation of research findings into practical applications requires further validation of products or prototypes in real-world field conditions. Moreover, emerging frontiers in plant research, such as epigenetics, genomic editing (GE), and plant interactions with the soil microbiome, offer new opportunities for discoveries in stress tolerance mechanisms. Exploring these areas can lead to the development of innovative strategies for improving stress resilience in fruit crops. By implementing these strategies and leveraging advancements in technology and scientific understanding, fruit growers can better cope with the challenges posed by abiotic stresses, ultimately improving productivity, quality, and sustainability in fruit production. Additionally, collaboration among researchers, farmers, industry stakeholders, and policymakers is essential for effective implementation and adoption of these solutions.

References

Abebe, A., Tsige, A., Work, M., & Enyew, A. (2020). Optimizing irrigation frequency and amount on yield and water productivity of snap bean (*Phaseolus vulgaris* L.) in NW Amhara, Ethiopia: A case study in Koga and Ribb irrigation scheme. *Cogent Food & Agriculture, 6*, 1773690.

Adams, S. R., Cockshull, K. E., & Cave, C. R. J. (2001). Effect of temperature on the growth and development of tomato fruits. *Annals of Botany, 88*, 869–877.

Alleva, K., Niemietz, C. M., Maurel, C., Parisi, M., Tyerman, S. D., & Amodeo, G. (2006). Plasma membrane of *Beta vulgaris* storage root shows high water channel activity regulated by cytoplasmic pH and a dual range of calcium concentrations. *Journal of Experimental Botany, 57*, 609–621.

Balerdi, C. F., Crane, J. H., & Schaffer, B. (2003). *Managing your tropical fruit grove under changing water table levels*. University of Florida Cooperative Extension Service, Institute of Food and Agricultural Sciences, EDIS.

Balint, Zoltan, & Mutua. (2013). Monitoring drought with the combined drought index in Kenya. *Developments in Earth Surface Processes, 16*, 341–356.

Beppu, K., Ikeda, T., & Kataoka, I. (2001). Effect of high temperature exposure time during flower bud formation on the occurrence of double pistils in 'Satohnishiki'sweet cherry. *Scientia Horticulturae, 87*(1–2), 77–84.

Blum, A. (2017). Osmotic adjustment is a prime drought stress adaptive engine in support of plant production: Osmotic adjustment and plant production. *Plant, Cell & Environment, 40*, 4–10.

Boursiac, Y., Boudet, J., Postaire, O., Luu, D. T., Tournaire-Roux, C., & Maurel, C. (2008). Stimulus-induced downregulation of root water transport involves reactive oxygen species-activated cell signalling and plasma membrane intrinsic protein internalization. *The Plant Journal, 56*, 207–218.

Boursiac, Y., Chen, S., Luu, D. T., Sorieul, M., van den Dries, N., & Maurel, C. (2005). Early effects of salinity on water transport in Arabidopsis roots. Molecular and cellular features of aquaporin expression. *Plant Physiology, 139*, 790–805.

Castañeda, A., & Doan, D. (2016). *Who are the poor in the developing world?*. Poverty and Shared Prosperity Report 2016: Taking on Inequality Background Paper. World Bank Policy Research Working Paper 7844.

Chadha, K. L., & Pandey, R. M. (1986). *Psidium guajava* L. In *Handbook of flowering* (Vol. 5, pp. 43–51).

Chai, Q., Gan, Y., Zhao, C., Xu, H. L., Waskom, R. M., Niu, Y., & Siddique, K. H. (2016). Regulated deficit irrigation for crop production under drought stress. A review. *Agronomy for Sustainable Development, 36*, 1–21.

Cohen, Itay, & Zandalinas. (2020). Meta-analysis of drought and heat stress combination impact on crop yield and yield components. *Physiologia Plantarum*, 171.

Cooper, M., van Eeuwijk, F., Chapman, S. C., Podlich, D. W., & Löffler, C. (2006). Genotype-by environment interactions under water-limited conditions. In *Drought adaptaton in cereals* (pp. 51–96). CRC Press.

Das, B. C., Maji, S., & Mulieh, S. R. (2010). Response of soil covers on guava cv. L-49. *Journal of Crop and Weed, 6*(2), 10–14.

Dev, S. M. (2012). *Small farmers in India: Challenges and opportunities*.

Ding, W., Song, L., Wang, X., & Bi, Y. (2010). Effect of abscisic acid on heat stress tolerance in the calli from two ecotypes of *Phragmites communis*. *Biologia Plantarum, 54*, 607–613.
Dos Santos, M. R., Neves, B. R., Da Silva, B. L., & Donato, S. L. (2015). Yield, water use efficiency and physiological characteristic of "Tommy Atkins" mango under partial rootzone drying irrigation system. *Journal of Water Resource and Protection, 7*(13), 1029.
Dreesen, F. E., De Boeck, H. J., Janssens, I. A., & Nijs, I. (2012). Summer heat and drought extremes trigger unexpected changes in productivity of a temperate annual/biannual plant community. *Environmental and Experimental Botany, 79*, 21–30.
El-Bassiony, A. M., Ghoname, A. A., El-Awadi, M. E., Fawzy, Z. F., & Gruda, N. (2012). Ameliorative effects of brassinosteroids on growth and productivity of snap beans grown under high temperature. *Gesunde Pflanz., 64*, 175–182.
Fidelia, N., & Chris, B. (2011). Environmentally friendly superabsorbent polymers for water conservation in agricultural lands. *Journal of Soil Science and Environmental Management, 2*, 206–211.
Finch, C. A. (1987). Hydrophilic polymers. In *Specialty polymers* (pp. 65–82). Springer.
Fischer, G., & Orduz-Rodriguez, J. O. (2012). Ecofisiologia en frutales. In G. Fischer (Ed.), *Manual para el cultivo de frutales en el trópico* (pp. 54–72). Produmedios.
Fujita, M., Fujita, Y., Maruyama, K., Seki, M., Hiratsu, K., & Ohme-Takagi, M. (2004). A dehydration-induced NAC protein, RD26, is involved in a novel ABA dependent stress-signaling pathway. *The Plant Journal, 39*, 863–876.
Gamboa-Tuz, S. D., Pereira-Santana, A., Zamora-Briseño, J. A., Castano, E., Espadas-Gil, F., & Ayala-Sumuano, J. T. (2018). Transcriptomics and coexpression networks reveal tissue-specific responses and regulatory hubs under mild and severe drought in papaya (*Carica papaya* L.). *Scientific Reports, 1*, 1–6.
Giorno, F., Wolters-Arts, M., Mariani, C., & Rieu, I. (2013). Ensuring reproduction at high temperatures: The heat stress response during anther and pollen development. *Plants, 2*, 489–506.
Glaser, B., Balashov, E., Haumaier, L., Guggenberger, G., & Zech, W. (2000). Black carbon in density fractions of anthropogenic soils of the Brazilian Amazon Region. *Organic Geochemistry, 31*, 669–678.
Gogoi, A., & Tripathi, B. (2019). https://www.indiaspend.com/42-indias-land-area-underdrought-worsening-farm-distress-in-election-year/
Gong, H., Zhu, X., Chen, K., Wang, S., & Zhang, C. (2005). Silicon alleviates oxidative damage of wheat plants in pots under drought. *Plant Science, 169*, 313–321.
Goufo, P., Moutinho-Pereira, J. M., Jorge, T. F., Correia, C. M., Oliveira, M. R., & Rosa, E. A. S. (2017). Cowpea (*Vigna unguiculata* L.Walp.) metabolomics: Osmoprotection as a physiological strategy for drought stress resistance and improved yield. *Frontiers in Plant Science, 8*, 586.
Haider, I., & Raza, M. A. S. (2020). Potential effects of biochar application on mitigating the drought stress implications on wheat (*Triticum aestivum* L.) under various growth stages. *Journal of Saudi Chemical Society, 24*, 974–981.
Hardie, W. J., & Martin, S. R. (2000). Shoot growth on de-fruited grapevines: A physiological indicator for irrigation scheduling. *Australian Journal of Grape and Wine Research, 6*(1), 52–58.
Hasanuzzaman, M., Hossain, M. A., & Fujita, M. (2010). Selenium in higher plants: Physiological role, antioxidant metabolism and abiotic stress tolerance. *Journal of Plant Sciences, 5*, 354–375.
He, Z. H., Fujiki, M., & Kohorn, B. D. (1996). A cell wall-associated, receptor-like protein kinase. *The Journal of Biological Chemistry, 271*, 19789–19793.
Hirayama, T., & Shinozaki, K. (2010). Research on plant abiotic stress responses in the post-genome era: Past, present and future. *The Plant Journal, 61*, 1041–1052.
Huber, A. E., & Bauerle, T. L. (2016). Long-distance plant signaling pathways in response multiple stressors: The gap in knowledge. *Journal of Experimental Botany, 67*, 2063–2079.
Jiménez, S., Dridi, J., Gutiérrez, D., Moret, D., Irigoyen, J. J., Moreno, M. A., & Gogorcena, Y. (2013). Physiological, biochemical and molecular responses in four *Prunus* rootstocks submitted to drought stress. *Tree Physiology, 33*, 1061–1075.

Karavidas, I., Ntatsi, G., Vougeleka, V., Karkanis, A., Ntanasi, T., Saitanis, C., Agathokleous, E., Ropokis, A., Sabatino, L., & Tran, F. (2022). Agronomic practices to increase the yield and quality of common bean (*Phaseolus vulgaris* L.): A systematic review. *Agronomy, 12*, 271.

Kasirajan, S., & Ngouajio, M. (2012). Polyethylene and biodegradable mulches for agricultural applications: A review. *Agronomy for Sustainable Development, 32*(2), 501–529.

Kaushal, M. (2019). Chapter ten – Portraying rhizobacterial mechanisms in drought tolerance: A way forward toward sustainable agriculture. In A. K. Singh, A. Kumar, & P. K. Singh (Eds.), *PGPR amelioration in sustainable agriculture* (pp. 195–216). Woodhead Publishing.

Kim, J., & Van Iersel, M. (2008). ABA drenches induce stomatal closure and prolong shelf life of *Salvia splendens*. *SNA Research Conference, 53*, 107–111.

Kozłowska, M., Rybus-Zajac, M., Stachowiak, J., & Janowska, B. (2007). Changes in carbohydrate contents of *Zantedeschia* leaves under gibberellin-stimulated flowering. *Acta Physiologiae Plantarum, 29*, 27–32.

Król, A. (2013). *The growth and water uptake by yellow seed and black seed rape depending on the State of Soil Compaction*. [dissertation]. Bohdan Dobrzański Institute of Agrophysics PAS.

Kromdijk, J., Głowacka, K., Leonelli, L., Gabilly, S. T., Iwai, M., & Niyogi, K. K. (2016). Improving photosynthesis and crop productivity by accelerating recovery from photoprotection. *Science, 354*, 857–861.

Landi, S., Hausman, J. F., Guerriero, G., & Esposito, S. (2017). Poaceae vs. abiotic stress: Focus on drought and salt stress, recent insights and perspectives. *Frontiers in Plant Science, 8*, 1214.

Lata, C., & Prasad, M. (2011). Role of DREBs in regulation of abiotic stress responses in plants. *Journal of Experimental Botany, 62*, 4731–4748.

Lee, S. B., & Suh, M. C. (2013). Recent advances in cuticular wax biosynthesis and its regulation in Arabidopsis. *Molecular Plant, 6*, 246–249.

Lipiec, J., Doussan, C., Nosalewicz, A., & Kondracka, K. (2013). Effect of drought and heat stresses on plant growth and yield: A review. *International Agrophysics, 27*, 463–477.

Lyu, S., Du, G., Liu, Z., Zhao, L., & Lyu, D. (2016). Effects of biochar on photosystem function and activities of protective enzymes in *Pyrus ussuriensis* Maxim. under drought stress. *Acta Physiologiae Plantarum, 38*, 1–10.

Ma, D., Sun, D., Wang, C., Qin, H., Ding, H., & Li, Y. (2016). Silicon application alleviates drought stress in wheat through transcriptional regulation of multiple antioxidant defense pathways. *Journal of Plant Growth Regulation, 35*, 1–10.

Masri, M., Razak, A. S., & Ghazalli, M. Z. (1990). Response of papaya (*Carica papaya* L.) to limited soil moisture at reproductive stage. *MARDI Research Journal, 18*, 191–196.

Mazen, A. M., Radwan, D. E. M., & Ahmed, A. F. (2015). Growth responses of maize plants cultivated in sandy soil amended by different superabsorbant hydrogels. *Journal of Plant Nutrition, 38*, 325 337.

Menzel, C. M., & Waite, G. K. (2005). Litchi and longan. In *Botany, production and uses* (p. 305). CAB International.

Mirás-Avalos, J. M., Pérez-Sarmiento, F., Alcobendas, R., Alarcón, J. J., Mounzer, O., & Nicolás, E. (2016). Using midday stem water potential for scheduling deficit irrigation in mid–late maturing peach trees under Mediterranean conditions. *Irrigation Science, 34*(2), 161–173.

Mohanty, P., Singh, P. K., Chakraborty, D., Mishra, S., & Pattnaik, R. (2021). Insight into the role of PGPR in sustainable agriculture and environment. *Frontiers in Sustainable Food Systems, 5*, 667150.

Murali, K., Srinivas, K., Shivakumar, H. R., & Kalyanamurthy, K. N. (2005). Effect of soil Moisture stress at different stages on yield and yield parameters on "Elakki" Banana. *Advances in Plant Science, 18*(2), 817.

Niinemets, Ü. (2015). Uncovering the hidden facets of drought stress: Secondary metabolites make the difference. *Tree Physiology, 36*, 129–132.

Nimbolkar, P. K., Shiva, B., & Rai, A. K. (2016). Rootstock breeding for abiotic stress tolerance in fruit crops. *International Journal of Agriculture Environment and Biotechnology, 9*(3), 375–380.

Pareek, O. P., & Sharma, S. (1991). Fruit trees for arid and semiarid lands. *Indian Farming, 41*, 25–33.

Pathania, P., & Rajta, A. (2020). Role of plant growth-promoting bacteria in sustainable agriculture. *Biocatalysis and Agricultural Biotechnology, 30*, 101842.

Paull, R. E., & Duarte, O. (2011). Tropical fruits (Vol. 1). CABI. production. *VI International symposium on irrigation of horticultural crops, 889*, 41–52.

Paull, R. E., & Reyes, M. E. Q. (1996). Pre-harvest weather conditions and pineapple fruit translucency. *Scientia Horticulturae, 66*, 59–67.

Poorter, H., Niinemets, U., Poorter, L., Wright, I. J., & Villar, R. (2009). Causes and consequences of variation in leaf mass per area (LMA): A meta-analysis. *The New Phytologist, 182*, 565–588.

Pornsiriwong, W., Estavillo, G. M., Chan, K. X., Tee, E. E., Ganguly, D., & Crisp, P. A. (2017). A chloroplast retrograde signal, 3″-phosphoadenosine 5″-phosphate, acts as a secondary messenger in abscisic acid signaling in stomatal closure and germination. *eLife, 6*, e23361.

Prasad, P. V. V., & Pisipati, S. (2011). Independent and combined effects of high temperature and drought stress during grain filling on plant yield and chloroplast EF-Tu expression in spring wheat. *Journal of Agronomy and Crop Science, 197*, 430–441.

Quan, R., Shang, M., Zhang, H., Zhao, Y., & Zhang, J. (2004). Engineering of enhanced glycine betaine synthesis improves drought tolerance in maize. *Plant Biotechnology Journal, 2*, 477–486.

Ravi, I., Uma, S., Vaganan, M. M., & Mustaffa, M. M. (2013). Phenotyping bananas for drought resistance. *Frontiers in Physiology, 4*, 9.

Raza, A., Razzaq, A., Mehmood, S. S., Zou, X., Zhang, X., Lv, Y., & Xu, J. (2019). Impact of climate change on crops adaptation and strategies to tackle its outcome: A review. *Plants, 8*, 34.

Reddy, A. S., Ali, G. S., Celesnik, H., & Day, I. S. (2011). Coping with stresses: Roles of calciumand calcium/calmodulin-regulated gene expression. *Plant Cell, 23*, 2010–2032.

Rizhsky, L., & Liang, H. (2002). The combined effect of drought stress and heat shock on gene expression in tobacco. *Plant Physiology, 130*, 1143–1151.

Rizhsky, L., & Liang, H. (2004). When defense pathways collide. The response of arabidopsis to a combination of drought and heat stress. *Plant Physiology, 134*, 1683–1696.

Rodrigo, J., & Herrero, M. (2002). Effects of pre-blossom temperatures on flower development and fruit set in apricot. *Scientia Horticulturae, 92*(2), 125–135.

Romero, P., García, J. G., Fernández-Fernández, J. I., Muñoz, R. G., Del Amor Saavedra, F., & Martínez-Cutillas, A. (2016). Improving berry and wine quality attributes and vineyard economic efficiency by long-term deficit irrigation practices under semi-arid conditions. *Scientia Horticulturae, 203*, 69–85.

Ruggiero, A., Punzo, P., Landi, S., Costa, A., Van Ooosten, M., & Grillo, S. (2017). Improving plant water use efficiency through molecular genetics. *Horticulture, 3*, 31.

Schaffer, B., Whiley, A. W., & Crane, J. H. (1994). Mango. In B. Schaffer & P. C. Andersen (Eds.), *Handbook of environmental physiology of fruit crops* (Sub tropical and tropical crops) (Vol. 2, pp. 165–197). CRC Press.

Semida, W. M., & Beheiry, H. R. (2019). Biochar implications for sustainable agriculture and environment: A review. *South African Journal of Botany, 127*, 333–347.

Serra, I., Strever, A., Myburgh, P. A., & Deloire, A. (2014). The interaction between rootstocks and cultivars (*Vitis vinifera* L.) to enhance drought tolerance in grapevine. *Australian Journal of Grape and Wine Research, 20*(1), 1–14.

Shahinnia, F., Le Roy, J., Laborde, B., Sznajder, B., Kalambettu, P., & Mahjourimajd, S. (2016). Genetic association of stomatal traits and yield in wheat grown in low rainfall environments. *BMC Plant Biology, 16*, 150.

Sharma, J., Upadhyay, A. K., Adsule, P. G., Sawant, S. D., Sharma, A. K., Satisha, J., Yadav, D. S., & Ramteke, S. D. (2013). Effect of climate change on grape and its value-added products. In *Climate-resilient horticulture: Adaptation and mitigation strategies* (pp. 67–80).

Shulaev, V., Cortes, D., Miller, G., & Mittler, R. (2008). Metabolomics for plant stress response. *Physiologia Plantarum, 132*, 199–208.

Singh, G., Natt, V., Pandey, S. D., Ray, P. K., & Singh, H. S. (2012). *The Litchi*. FAO.

Srivastava, S., Pathak, A. D., Gupta, P. S., Shrivastava, A. K., & Srivastava, A. K. (2012). Hydrogen peroxide-scavenging enzymes impart tolerance to high temperature induced oxidative stress in sugarcane. *Journal of Environmental Biology, 33*, 657–661.

Sugiura, T., Ogawa, H., Fukuda, N., & Moriguchi, T. (2013). Changes in the taste and textural attributes of apples in response to climate change. *Scientific Reports, 3*, 1–7.

Suzuki, N., Miller, G., Salazar, C., Mondal, H. A., Shulaev, E., & Cortes, D. F. (2013). Temporal-spatial interaction between reactive oxygen species and abscisic acid regulates rapid systemic acclimation in plants. *Plant Cell, 25*, 3553–3569.

Swindell, W. R., Huebner, M., & Weber, A. P. (2007). Transcriptional profiling of Arabidopsis heat shock proteins and transcription factors reveals extensive overlap between heat and non-heat stress response pathways. *BMC Genomics, 8*, 125.

Turner, D. W., & Lahav, E. (1983). The growth of banana plants in relation to temperature. *Functional Plant Biology, 10*(1), 43–53.

Vandeleur, R. K., Sullivan, W., Athman, A., Jordans, C., Gilliham, M., & Kaiser, B. N. (2014). Rapid shoot-to-root signalling regulates root hydraulic conductance *via* aquaporins. *Plant, Cell & Environment, 37*, 520–538.

Verma, V., Ravindran, P., & Kumar, P. P. (2016). Plant hormone mediated regulation of stress responses. *BMC Plant Biology, 16*, 86.

Wahid, A., Gelani, S., Ashraf, M., & Foolad, M. R. (2007). Heat tolerance in plants: An overview. *Environmental and Experimental Botany, 61*, 199–223.

Wang, L. J., Fan, L., Loescher, W., Duan, W., Liu, G. J., Cheng, J. S., Luo, H. B., & Li, S. H. (2010). Salicylic acid alleviates decreases in photosynthesis under heat stress and accelerates recovery in grapevine leaves. *BMC Plant Biology, 10*, 34.

Waraich, E. A., Ahmad, R., Halim, A., & Aziz, T. (2012). Alleviation of temperature stress by nutrient management in crop plants: A review. *Journal of Soil Science and Plant Nutrition, 12*, 221–244.

Xiong, Y. W., & Li, X. W. (2020). Root exudates-driven rhizosphere recruitment of the plant growth-promoting rhizobacterium *Bacillus flexus* KLBMP 4941 and its growth-promoting effect on the coastal halophyte *Limonium sinense* under salt stress. *Ecotoxicology and Environmental Safety, 28*(194), 110374.

Yadav, S. K., Singla-Pareek, S. L., Ray, M., Reddy, M. K., & Sopory, S. K. (2005). Methylglyoxal levels in plants under salinity stress are dependent on glyoxalase I and glutathione. *Biochemical and Biophysical Research Communications, 337*, 61–67.

Zaen El-deen, E. M. A., Attia, M. F., Laila, F. H., Shahin, M. F. M., Genaidy, E. A., & Merward, M. A. (2015). Soil mulching and foliar anti-transpirations effect on soil, growth and nutrients status of young mango trees cultivated in Toshka. *Journal of Agricultural Technologies, 11*, 1013–1032.

Zhang, M., Duan, L., Zhai, Z., Li, J., Tian, X., & Wang, B. (2004). Effects of plant growth regulators on water deficit-induced yield loss in soybean. In *Proceedings of the 4th international crop science congress (Brisbane, QLD)*.

Zhang, Y., & Ding, J. (2020). Biochar addition alleviates the negative effects of drought and salinity stress on soybean productivity and water use efficiency. *BMC Plant Biology, 20*(288).

Zhao, C., Liu, B., Piao, S., Wang, X., Lobell, D. B., & Huang, Y. (2017). Temperature increase reduces global yields of major crops in four independent estimates. *Proceedings. National Academy of Sciences. United States of America, 114*, 9326–9331.

Livestock Production Under Drought and Heat Stress

5

Sudhanshu Shekhar, R. R. K. Sinha, Kaushalendra Kumar, and Ravikant Nirala

Abstract

Climate change is manifesting globally in the form of altered weather patterns, significantly impacting livestock production. Scientific consensus confirms that climate change, driven by increased human activities, has severe social and economic repercussions. Greenhouse gases (GHGs) trap solar heat, leading to observed climatic shift, particularly heat stress and drought, affecting both humans and animals. These changes have resulted in substantial economic losses for poor and marginal farmers due to decreased livestock productivity. Droughts, characterized by prolonged low precipitation and high temperatures, lead to reduced forage availability, compromised animal health, and decreased production efficiency. The Indian subcontinent, highly susceptible to droughts, experience frequent agricultural water scarcity, impacting livestock growth, reproduction, and survival. Effecting drought mitigation strategies focuses on optimizing resource use, ensuring livestock survival, and minimizing productivity losses. This includes improved housing, feeding, and health management practices tailored to sustain livestock in drought prone regions. Addressing both immediate relief and long-term prevention measures is essential for mitigating the adverse impacts of drought on livestock and farming communities.

S. Shekhar (✉)
Krishi Vigyan Kendra Ramgarh (ICAR-RCER, Patna), Mandu, Jharkhand, India

R. R. K. Sinha · R. Nirala
Department of Livestock Production Management, Bihar Animal Sciences University, Bihar Veterinary College Campus, Patna, Bihar, India

K. Kumar
Department of Animal Nutrition, BiharAnimal Sciences University, Bihar Veterinary College Campus, Patna, Bihar, India

R. Dubey et al. (eds.), *Drought and Heat Stress in Agriculture*,
https://doi.org/10.1007/978-981-96-5735-3_5

Keywords

Climate changes · Greenhouse gases · Livestock production · Economic loss

5.1 Introduction

Globally, climate change is now evident through shifting weather patterns that have both direct and indirect effects on livestock production. Ample scientific evidence confirms that our climate is indeed changing, and, increasingly, this shift is linked to human activities or actions that carry serious consequences for social and economic development. Greenhouse gases (GHGs) absorb heat from the sun within the atmosphere, limiting the escape of this heat into space. This trapped heat has been identified as a key driver of climate changes observed throughout the twentieth century. The effects of these changes are particularly visible in the form of increased heat stress on animals and prolonged drought conditions, both of which impact humans and animals alike. Persistent heatwaves and droughts due to climate change have significantly reduced livestock productivity, leading to severe economic losses, especially for poor and marginal farmers. Droughts and extreme heat bring numerous challenges, affecting livestock growth rates, milk and egg production, reproductive performance, health, and even survival, as well as the availability of feed. The Indian subcontinent is especially susceptible to these climate impacts, with droughts occurring frequently and with varying intensity and impact across different geographic regions. Addressing these challenges is essential to secure livestock production and protect the livelihoods of vulnerable communities. Developing resilient agricultural practices that can adapt to these environmental changes is crucial for sustaining food security and supporting rural economies in the face of a changing climate.

5.2 Drought

Drought is defined as a condition in which precipitation and other water resources fall below expected levels, while demand for water remains unchanged. More specifically, it is characterized by prolonged dry weather due to low rainfall and high temperatures. In agriculture, drought leads to crop failures resulting from insufficient soil moisture, which can cause food shortages and, in severe cases, famine. For livestock and poultry production, drought reduces forage availability, slows animal growth rates, and disrupts production efficiency. The biological systems of animals are also affected, with drought leading to poor growth, reproductive issues, weakened immunity, increased susceptibility to disease, and lower quality feed. These factors collectively reduce animal health and productivity. Effective drought mitigation strategies focus on preserving livestock health, as severe drought can cause up to 20% body weight loss in animals. These strategies aim to minimize productivity losses and safeguard animal lives by making the most of limited resources.

5.2.1 Causes of Drought

Droughts occur when there is prolonged low precipitation, often influenced by atmospheric conditions such as climate change, ocean temperature shifts, alterations in the jet stream, and changes in the local landscape. According to the International Rescue Committee, drought can be triggered by both natural and human factors. Human-induced causes include climate change, global warming, deforestation, intensive agriculture, and high water demands, such as heavy upstream water usage that can lead to drought conditions in downstream regions.

5.2.2 Impact of Drought

Drought significantly affects both livestock and farming communities. Consequences include vast tracts of uncultivated land, water supply via tankers, socioeconomic challenges for farmers, migration of people and animals, distressed livestock sales, and high mortality rates in drought-affected areas. In regions with irregular monsoons, persistent water scarcity has severely impacted agriculture. In India, as noted by Patil et al. (2006), drought is endemic in some areas. However, drought management often takes the form of crisis response rather than proactive, long-term planning to address natural disasters. While immediate relief is essential during scarcity, there is a pressing need for preventive measures to mitigate such situations. Beyond the immediate lack of water, drought conditions lead to reduced nutritional resources and increased exposure of animals to harsh environmental conditions, particularly heat stress. During droughts, forage scarcity raises the risks of animal poisonings and nutritional imbalances. Poppenga and Puschner (2014) highlight additional drought-related risks, such as impaired water quality, diminished feed quality, nutrient deficiencies, and increased cases of plant poisonings, all of which threaten livestock health and productivity. Water and nutrient shortages, coupled with environmental stressors, lead to numerous issues in livestock, including anorexia, lethargy, anemia, stunted growth, poor productivity, reproductive issues like repeat breeding and anoestrus, skin and skeletal disorders, secondary infections due to weakened immunity, and even mortality. With limited fodder availability, animals often turn to unconventional feed sources, increasing exposure to anti-nutritional factors. Drought also leads to shifts in livestock production systems. Many livestock farmers are compelled to migrate to more fertile areas, sell animals for slaughter, or transition from large ruminants to small ruminants to cope with the scarcity.

5.2.3 Mitigation Strategy for Drought Management

In drought-prone areas of the country, sustaining livestock populations amid recurring drought or delayed monsoons presents numerous challenges. Effective drought management involves practices that enhance sustainability and minimize livestock

Table 5.1 Mitigation strategies for drought management in livestock production

Mitigation strategy	Description
Housing optimization	Ensure proper ventilation in sheds to mitigate heat stress. Optimize stocking density to prevent overcrowding, which can exacerbate heat stress and increase competition for resources.
Feeding management	Increase feeding frequency to compensate for reduced forage availability. Provide feed during cooler periods, such as early mornings and late evenings, to encourage higher dry matter intake.
Grazing practices	Schedule grazing activities during early hours and limit duration to minimize exposure to intense sunlight. Opt for grazing locations nearer to water sources and with partial shade to reduce heat stress.
Water management	Ensure access to clean and ample water sources at all times. Implement water conservation measures such as rainwater harvesting and efficient irrigation techniques.
Health care	Prioritize preventive health measures such as vaccinations and deworming to bolster immune response. Monitor animal health closely and promptly address any signs of distress or illness.

mortality. Key management practices—such as housing, feeding, breeding, and health—must be optimized to support livestock survival. For housing, proper ventilation in animal sheds and maintaining optimal stocking density are essential. Feeding should occur more frequently, especially in the early morning and late evening, to encourage higher dry matter intake. Open and partially shaded housing may be preferable to keeping animals tied in enclosed sheds. Additionally, grazing should ideally be limited to early hours and nearby areas to reduce exposure to prolonged sunlight. These practices collectively help improve the resilience and adaptability of livestock in challenging conditions (Table 5.1).

5.2.4 Nutritional Strategies for Drought Stress Management

Efficient feeding practices are essential to make the most of limited resources, especially in drought-prone areas. Green fodder conservation methods become crucial in anticipation of drought. However, when unpreparedness is a factor, feed enrichment and supplementation strategies are even more critical. The choice of supplements, particularly those rich in energy, protein, and essential minerals, plays a vital role. Research into suitable fodder species, alternative fodder sources, and ways to optimize storage, transport, and nutrient availability for different livestock species is needed to ensure sustainable livestock production. This focus on optimized feeding and resource use is key to supporting livestock in areas where drought is a persistent threat.

5.2.5 Feeding Management

The performance of livestock is closely linked to their nutrient needs and intake, with the quantity and quality of available forage being the primary factors that regulate nutrient intake in grazing cattle. When forage drops below a minimum threshold, animal performance declines. A shortage of green fodder often leads to vitamin deficiencies, especially in vitamins A and E, making supplementation essential. Supplementary feeding with grains or oil cakes twice a week can help, though grains are often difficult to justify due to scarcity and storage challenges. Alternative concentrate sources for draft animals, such as fodder trees and urea-treated straw, can be valuable. Where feasible, by-products like oil seed cakes should be used, provided they are reasonably priced. Locally available tree leaves can also serve as animal feed, depending on the livestock species. Due to the scarcity of quality fodder, livestock are more likely to graze on poisonous plants, raising toxicity risks. Farmer awareness is crucial, particularly regarding anti-nutritional and toxic elements like nitrate/nitrite and hydrogen cyanide (HCN) poisoning. In available fallow lands, farmers are encouraged to grow short-duration, drought-tolerant crops and grasses as fodder with any rainfall. Partially damaged or discarded grains, such as wheat, can also be directed toward feeding productive animals, supporting their nutrition in challenging conditions.

5.2.6 Hay and Silage

Hay and silage making are primary options for farmers aiming to conserve large quantities of forage. In drier climates, haymaking remains significant; however, over the past 30 years, the proportion of forage preserved as silage has increased while hay usage has declined (Wilkinson et al., 1996). Ensiling has several advantages over haymaking: it enables quick conservation of large amounts of forage, is less dependent on weather, and is highly compatible with mechanization. However, a key drawback of silage is its reduced feeding value compared to the original crop (Charmley, 2000). Silage typically consists of preserved forages, crop residues, or agricultural byproducts, maintained through natural or artificial acidification, and used as animal feed during feed shortages (Mannetje, 1999). Charmley (2000) also suggests that future advancements could yield silages with feeding values surpassing those of the original crops. Physical treatments to enhance intake and digestibility offer promising improvements in silage quality.

5.2.7 Complete Feed Blocks

Crop residues are often low in nutritional value, bulky, and fibrous, making them less effective as feed. Furthermore, their availability is seasonal and regional, and they are often underutilized or wasted. In emergency situations, complete feed block technology has been employed to prevent hunger and death in livestock. Depending on the productivity levels of animals, Densified Total Mixed Ration Blocks (DTMRBs) or Densified Total Mixed Ration Pellets (DTMRPs) can be formulated using various ingredients, including minerals, vitamins, and feed additives, to ensure adequate nutrition.

5.2.8 Urea Molasses Treatment

Urea treatment of poor-quality fodder can be highly beneficial during drought periods when applied judiciously and under controlled conditions. Urea serves as a nonprotein nitrogen source, which rumen microbes utilize to synthesize microbial protein. This process enhances the availability of valuable protein for ruminants, helping to improve their nutrition and overall health.

5.2.9 Drought-Tolerant Fodder Varieties

The strategy should focus on the selection and breeding of high-yielding, stress-tolerant, and short-duration fodder crops and varieties. Emphasis should be placed on improving yields through sustainable production practices, efficient conservation techniques, and strengthening the value chain for dairy and meat producers by offering essential services to enhance productivity and sustain livelihoods. Research efforts should aim to increase fodder yields from both cultivated crops on agricultural lands and from wastelands and community pastures (Hegde, 2010).

5.2.10 Hydroponics

Hydroponic green fodder refers to fodder grown from seeds without soil, using water or nutrient-rich solutions. The yield and quality of the sprouts produced are influenced by various factors, including soaking time, grain quality, variety, and treatments, as well as temperature, humidity, nutrient supply, the depth and density of grains in troughs, and the presence of mold (Sneath & McIntosh, 2003). Bradley and Marulanda (2000) reported that hydroponic green fodder production uses only 10–20% of the water required to produce the same amount of crop in traditional soil-based cultivation.

5.2.11 Alternate Feeds Resources

In various agro-ecological systems, there are numerous alternative feed resources that should be identified and analyzed for future use in livestock feeding. Several new feed resources have been evaluated and proven beneficial. In India, protein cakes extracted from seeds of neem, castor, karanj, palm, and mahua have been found suitable for livestock feed after detoxification. The fruit and vegetable processing, packaging, distribution, and consumption industries generate significant quantities of fruit and vegetable waste (FVW). According to Wadhawa and Bakshi (2013), FVW such as baby corn husks, cauliflower and cabbage leaves, pea pods, sarsonsaag waste, culled snow peas, tomato pomace, citrus, carrot and bottle gourd pulp, as well as banana and mango peels, are rich in nutrients. These by-products can be fed to livestock in their natural form, after drying or ensiling with cereal straws to improve palatability, nutrient utilization, and overall livestock health and performance.

5.2.12 Improvement of Nutrient Utilization by Animals

Considerable research is being conducted to enhance nutrient utilization in animals, particularly in terms of dry matter intake, digestibility, and assimilation. Feeding chaffed feed and ensuring timely feeding practices contribute to more efficient feed utilization by reducing waste and improving digestibility. The use of feed additives and supplements made from locally available materials, along with protein and fat protection strategies, is becoming increasingly common to enhance nutrient utilization. Genetic manipulation of rumen microbes is also being explored to improve the efficiency of feed and fodder utilization. During periods of feed scarcity, it is crucial to manage the nutrition of individual animals to sustain their health, productivity, and even survival. Providing concentrate mixtures, along with appropriate supplementation of minerals, vitamins, and antioxidants, based on the specific needs of the animals, should be done under the guidance of experts.

5.2.13 Fodder Banks and Fodder Grids

In drought-prone regions, a traditional coping mechanism involves storing fodder during surplus periods for use during times of scarcity. The establishment of fodder banks through Goshalas, farmers' groups, cooperative dairies, etc., at the Gram Panchayat level can be promoted to ensure fodder security during droughts. These fodder banks should focus on "bailing intensification," storage, and the transport of fodder from surplus areas to meet the needs of deficit areas. Creating fodder banks in villages within drought-prone areas could serve as an effective solution and contingency plan to address fodder shortages during lean periods, particularly in times of drought and floods. Under the "fodder grid" concept, these banks should be linked across states so that if one area faces a shortage, supplies can be redirected from other regions (Table 5.2).

Table 5.2 Nutritional strategies for drought stress management

Nutritional strategies	Description
Feeding management	Adjust feeding practices to meet nutrient requirements. Supplement green fodder with grains/oil cakes twice weekly. Utilize fodder trees and urea-treated straw.
Hay and silage	Conserve forage through haymaking or ensiling. Silage offers advantages in mechanization and weather resilience, though may reduce feeding value.
Complete feed blocks	Utilize densified total mixed ration blocks/pellets to provide balanced nutrition, particularly during emergencies.
Urea molasses treatment	Treat poor-quality fodder with urea to improve microbial protein synthesis and enhance protein availability for ruminants.
Drought-tolerant varieties	Select and breed high-yielding, stress-tolerant fodder crops. Focus on sustainable production practices and efficient value chain strengthening.
Hydroponics	Produce green fodder in water/nutrient-rich solutions. Offers high yields with minimal water usage.
Alternate feed resources	Evaluate and utilize various alternate feed resources, including neem, castor, and karanj protein cakes, fruit and vegetable waste, and agricultural by-products.
Nutrient utilization	Improve nutrient intake, digestibility, and assimilation through chaffed feed, feed additives/supplements, and genetic manipulation of rumen microbes.
Fodder banks and grids	Establish fodder banks at local and regional levels for storage and distribution of fodder during drought periods. Foster collaboration across states for supply resilience.

5.3 Livestock Management Strategies

Maintaining good health and performance records of livestock is essential for dairy and livestock producers to make informed management decisions, particularly when it comes to culling and herd reduction. Livestock should be grouped based on their nutritional requirements, ensuring each group receives proper feeding and facilitating easier assessment of body condition scores. Additionally, young stock should be weaned as early as possible to allow mature females to maintain optimal body condition. While selling animals is sometimes necessary, it is important to retain the most desirable younger stock to preserve and enhance the herd's genetics. Culling decisions should primarily be based on the productivity of the animals.

5.3.1 Reproduction Management

Animals that are maintaining normal or acceptable growth, milk production, and cyclicity should be bred at the appropriate time. Repeat breeders should be treated based on veterinary advice. Infertility treatments should be postponed until drought conditions are resolved; however, cyclic animals should still be inseminated through artificial insemination (AI). During hot summer months, insemination should be conducted during the cooler hours of the day, such as in the morning or evening. To prevent periparturient problems, such as premature births, retention of placenta, dystocia, and mastitis, it is beneficial to supplement vitamin A, D, and E along with selenium orally or via intramuscular injection during the last trimester of pregnancy, particularly in areas where forage is scarce.

5.3.2 Health Management

During drought conditions, stress can lead to a reduced immune response in animals, making them more susceptible to a range of pathogens. Global warming and drought can also increase the likelihood of the emergence or re-emergence of pathogens, vectors, and changes in the epidemiological patterns of diseases. Extreme drought can drive livestock migration toward less affected areas, which presents a significant risk for the spread of diseases, especially those more prevalent in dry regions. Farmers often prioritize securing food for their families before addressing the needs of their livestock, diverting limited resources to mobilize feed and fodder. This often results in reduced spending on animal health, increasing vulnerability to various pathogens. The rapid spread of diseases can occur due to the differing susceptibility of migrating animals, potentially leading to the emergence of new pathogens that livestock have not adapted to, which can alter the epidemiology of tropical enzootic diseases, such as bluetongue and theileriosis. The increased population of disease-carrying vectors during these times exacerbates the risks. Additionally, poor-quality dry fodder, sudden changes in diet, and reduced water intake can lead to digestive issues in livestock, including poor salivation, low feed intake, constipation, and anorexia.

5.3.3 Development of Early Warning System

Implementing early warning systems for impending droughts can help mitigate the severity of their impact on livestock farmers. Additionally, research into locally available alternative feeds and fodder resources, along with clear guidelines for their storage and use, can be highly beneficial for farmers in drought-prone areas. During times of surplus availability, hay and silage, as well as feed ingredients unfit for human consumption, can be preserved. Grasses from common lands, the periphery of forest areas, wastelands, and farmlands can be stored as briquettes or high-density stacks, while crop residues, coarse cereals, and legumes can also be stored in fodder banks for future use. Climate change is emerging as one of the most significant long-term challenges for farmers and livestock owners worldwide. It is increasingly recognized as one of the most critical environmental issues facing the global community today.

5.3.4 Climate Change and Heat Stress

Climatic factors are among the primary and most critical limitations to animal production in warm regions, and global warming is expected to exacerbate heat stress-related issues in livestock. The most significant stress livestock face due to climate change is heat stress, which is one of the key environmental factors that challenges animal production in many parts of the world. While animals can adapt to hot climates, the physiological mechanisms that help them survive can often negatively affect their overall performance. The susceptibility of livestock to heat stress varies based on species, genetic traits, life stage, and nutritional status. Heat stress, in particular, has become one of the most impactful factors on ruminant livestock production in the face of changing climate conditions, severely affecting both production and reproduction.

5.4 Heat Stress

Heat stress is a form of hyperthermia, where the body's physiological systems fail to regulate its temperature within a normal range, leading to an elevated body temperature. Humidity plays a crucial role in exacerbating heat stress. The most commonly used indicator of heat stress, the temperature–humidity index (THI), is derived from the temperature and relative humidity levels. When ambient temperatures are high, animals expend energy on panting and sweating to cool down, which reduces their feed intake and overall production. Animals require an environment that ensures thermal comfort over a wide range of temperatures. The temperature at which they perform most efficiently is closely linked to their metabolic activity and corresponds to a specific range of core body temperatures. A significant increase or decrease in core temperature beyond acceptable physiological limits can disrupt cell integrity. Temperature preferences vary among species and can change throughout the day, depending on their need for heat production. The thermos-neutral zone is the range of environmental temperatures in which an animal's body temperature is maintained with minimal effort from its thermoregulatory mechanisms, allowing it to perform optimally. This zone varies depending on factors such as age, species, breed, insulation, nutritional status, previous acclimatization to temperature, production levels, housing conditions, behavioral responses, and the time of day.

5.4.1 Reasons and Signs of Heat Stress

Animals are more susceptible to heat stress during the summer due to factors such as higher temperature-humidity index (THI), increased solar radiation, and lower wind speed. Recent rainfall followed by a rise in humidity can reduce sweating and breathing, making it harder for animals to dissipate heat. Prolonged exposure to high maximum and minimum temperatures or excessive environmental humidity also contributes significantly to heat stress. The absence of natural cloud cover can intensify solar radiation, increasing the risk of heatstroke. Similarly, warm, cloudy nights hinder heat dissipation, further raising the risk of heat stress. Sudden shifts in climate conditions may also predispose animals to heat stress. Husbandry practices and farm layout can sometimes exacerbate the onset of heat stress. Certain animal characteristics make individuals more vulnerable to heat stroke. For example, indigenous breeds (such as Bos indicus, e.g., Sahiwal) are more heat-tolerant than exotic breeds (such as Bos Taurus, e.g., Ayrshire). Animals with lighter coat colors (e.g., cattle compared to buffalo) generally tolerate heat better. Animals with coarse hair coats, like sheep, are more prone to heat stress compared to those with finer coats, like goats. Obese and emaciated animals are particularly susceptible to heat stroke, as are recently weaned or older animals. Indigenous breeds are more likely to adapt to local climate conditions if temperature changes occur gradually. Lactating or peri-parturient animals may face greater risks due to hormonal changes, which make them more sensitive to increases in temperature or humidity. Additionally, pests like flies and mosquitoes can create discomfort, leading animals to crowd together in search of vector-free areas. Common signs of heat stress in animals include seeking shade, increased water intake and reduced feed intake, preference for standing over lying down, rapid respiration, higher body temperature, excessive salivation, and open-mouth panting.

5.4.2 Measurement of Heat Stress in Animals

The temperature humidity index (THI) is a simple method to determine the level of heat stress in animals. THI is a combination of ambient temperature (Ta) and relative humidity (RH) into a single value to estimate the degree of stress. According to Thom (1959) $THI = 9/5 \times [(Ta \times 17.778) - (0.55 - (0.55 \times RH/100) \times (T\text{-}14.444)]$, value of $THI < 72$ indicates thermo-neutral conditions and THI 76–78.5 represents mild to moderate heat stress. Another index, THI, derived from a combination of wet and dry bulb air temperatures for a particular day, is expressed using a specific formula as per Mcdowell (1976), $THI = 0.72\ (W°C + D°C) + 40.6$, where $W°C =$ wet bulb and $D°C =$ dry bulb. The THI values of 70 or less comfortable, 75–78 stressful, and values above 78 extreme distress. The THI in combination with the temperature and humidity is expressed using the following formula: THI = (Dry-Bulb Temperature °C) + (0.36 Dew Point Temperature °C) + 41.2). THI exceeding 72 indicates mild stress, 80 medium stress, and above 90 severe heat stress in cattle.

5.4.3 Impact of Heat Stress on Livestock Growth, Production, and Reproduction

High environmental temperatures, solar radiation, and humidity are significant stress factors for animals, particularly in hot regions of the world. Heat stress triggers various physiological and behavioral responses in livestock, which can vary in intensity and duration depending on the animal's genetic makeup and environmental conditions (Freeman, 1987). In humid tropical regions, extended periods of high temperatures and humidity diminish the effectiveness of non-evaporative cooling mechanisms (conduction, convection, and radiation). As a result, animals increasingly rely on evaporative cooling, such as sweating and panting, to manage heat stress (Kimothi & Ghosh, 2005). Heat stress reduces feed intake, body weight, and milk production, while also affecting cellular and molecular stress responses, which in turn influence mammary gland metabolism, energy partitioning, and immune function. In extreme conditions, heat stress may also lead to increased mortality rates. It can impair reproductive processes, reducing libido, fertility, and embryonic survival in animals, and contribute to electrolyte loss due to increased heat dissipation (Coppock et al., 1982). Heat stress particularly affects reproductive functions, with males experiencing decreased sperm production and females suffering from reduced fertility. For example, buffaloes are more vulnerable to heat stress during the summer, as their dark skin and sparse body hair absorb more heat, and their poor heat dissipation mechanisms (due to fewer sweat glands) exacerbate the problem (Marai & Haeeb, 2010). In lactating dairy cows, heat stress leads to the loss of sodium and potassium in the serum (West, 1999) and can also reduce the birth weights of calves. Additionally, high ambient temperatures can negatively impact cellular structure and function, including immune system cells (Iwagami, 1996). Heat stress in lactating animals significantly reduces roughage intake, gut motility, and rumination, which in turn affects dietary protein utilization and body protein metabolism (Ames et al., 1980). Temperature extremes also reduce disease resistance in dairy calves (Stott et al., 1976). In poultry, heat stress affects comfort levels, productivity, growth rate, feed conversion, and live weight gain, altering behavioral, physiological, and immunological responses (Etches et al., 1995; Yalcin et al., 1997). As ambient temperatures rise, poultry must balance heat production and loss, which leads to a 5% reduction in feed consumption for every 1 °C increase in temperature between 32 and 38 °C (Sohail et al., 2012). Heat stress also reduces dietary digestibility and decreases plasma protein and calcium levels (Bonnet et al., 1997; Zhou et al., 1998). In laying hens, it results in decreased egg production and egg quality, as the bird diverts energy to regulate body temperature, which also affects egg shell thickness due to hyperventilation and reduced CO_2 levels in the blood (Campos, 2000). Heat stress impairs broiler chicken growth by affecting plasma triiodothyronine and thyroxine levels (Sahin et al., 2001), which are critical growth promoters. Heat exposure can negatively affect organ and muscle metabolism, persisting even after slaughter (Gregory, 2010), and chronic heat stress reduces fat deposition and meat quality (Imik et al., 2012). It also affects meat chemical composition, leading to higher meat pH, less expressed juice, and increased cooking and drip loss. In broilers, chronic heat stress reduces breast muscle proportion and increases thigh muscle proportion, with lower protein content and higher fat deposition (Lara & Rostagno, 2013; Zhang et al., 2012). Additionally, heat stress results in reduced egg shell thickness, increased egg breakage, and lower egg weight and quality (Lin et al., 2004; Lara & Rostagno, 2013), as well as adversely affecting semen production in breeder cocks (Banks et al., 2005). The immune-suppressing effects of heat stress on broilers and laying hens in hot climates may also alter global disease distribution (Ghazi et al., 2012; Guis et al., 2012).

5.5 Adaptive Mechanisms of Livestock to Heat Stress

5.5.1 Behavioral and Physiological Adaptability

When animals are exposed to heat stress, they attempt to modify their behavior in order to cope with the situation. These behavioral changes include reduced standing, increased lying, increased water intake, decreased defecation and urination, and less rumination. Dark-colored coats in animals provide an advantage over lighter colored coats in cooler environments, as they help conserve solar energy and reduce the need for food energy to maintain homeothermy. To regulate body temperature, respiration rate, pulse rate, rectal temperature, and sweating rate all fluctuate with the seasons in response to changes in environmental temperature. An increase in respiration rate is the first reaction when animals encounter temperatures exceeding the thermoneutral zone. This response triggers heat receptors in the peripheral skin, which send nerve impulses to the heat center in the hypothalamus. The intensity of physiological responses to thermal stress and the time it takes for the animal to return to normal after the stress is removed are important indicators of their ability to handle heat. Sweating is considered more effective for heat dissipation than respiratory evaporative cooling in livestock. Native animals typically rely more on cutaneous evaporative cooling than exotic breeds during thermal stress at high ambient temperatures.

5.5.2 Neuron-Endocrine Response to Stress in Animals

The stress-regulating systems in animals vary depending on individual factors such as previous experiences, physiological condition, genetic traits, and the duration and intensity of stress. Stress triggers several neuroendocrine responses, activating various hormonal axes and the release of hormones that support adaptive and behavioral reactions. These stress hormones prioritize energy distribution to maintain muscle and neural functions, enhance environmental perception, increase glucose levels in the brain, adjust cardiovascular and respiratory functions, and modulate immune responses. However, they ultimately lead to reduced productivity and reproductive performance. The hypothalamic-pituitary-adrenal (HPA) axis is activated by stressors such as drought and nutritional challenges, which increase the secretion of glucocorticoids. The sympathetic-adrenal-medullary (SAM) axis is also activated, triggering the autonomic nervous system to release catecholamines like adrenaline and noradrenaline. Adrenal and thyroid hormones play key roles in thermoregulation and metabolic responses during stress. Activation of the HPA axis raises adrenocorticotropic hormone levels, which stimulates the production of glucocorticoids, especially cortisol, a critical stress-relieving hormone. Cortisol helps regulate behavioral and neuroendocrine responses during stress by promoting hepatic gluconeogenesis, which produces glucose from non-carbohydrate sources to maintain energy balance and support vital functions.

5.6 Mitigation of Heat Stress

Heat stress has a profound impact on production, reproduction, and overall health, leading to substantial economic losses. To mitigate its effects, a comprehensive approach should be developed, focusing on strategies such as shelter management, microclimatic adjustments, proper feeding and nutrition, and the genetic selection of animals for enhanced resilience to heat stress.

5.6.1 Animal Housing and Ventilation

During heat stress, animals can be cooled in two primary ways: by enhancing their heat dissipation mechanisms or by modifying their immediate environment through micro-climatic adjustments. Providing shelters or shades effectively shields animals from radiant heat and helps regulate their body temperature. The addition of trees or other vegetative cover around the area further mitigates the radiative heat load. Roofing materials should be poor conductors of heat, with optimal housing featuring roofs painted white and insulated with materials such as puff or straw. Additionally, seasonal vegetable plants like bottle gourd, pumpkin, and ridge gourd grown on earth can further reduce the shed's internal temperature. Evaporative cooling techniques, such as fogging, misting, or sprinkling (with or without fans, depending on the season), are also effective methods of cooling (Atrian & Shahryar, 2012). The orientation of the shed should align with the north-south direction in the northern hemisphere and east-west in the southern hemisphere to minimize direct solar radiation exposure. To ensure proper heat dissipation for cattle, the shed must have adequate ventilation, which can be achieved by increasing airflow through open walls, using fans, or raising the building's height. Providing shade alone can reduce a cow's respiration rate by 30%, and, when combined with sprinklers, it can lower the respiration rate by up to 67%, while also reducing rectal temperatures.

5.6.2 Cooling Systems in the Farm

To effectively manage heat stress, it is crucial to provide access to fresh, cool, and clean drinking water under shaded areas. The combination of fans and water sprinkling offers one of the most effective cooling methods. However, excessive sprinkling should be avoided, as it can wet the bedding, making the animals more susceptible to mastitis and other diseases. Ensuring a constant supply of cool, clean water is one of the best practices for reducing heat stress. Other cooling methods include providing shade, using commercial coolers, tunnel ventilation, shower/fanning stations, fans, cooling ponds, and center pivots. Fans work by removing body heat through convection, cooling the animal's surface, while sprinklers soak the animal's coat to the skin, allowing heat loss through conduction. The combination of fans and sprinklers provides both conduction and evaporative cooling, as the fans help evaporate the water that absorbs body heat. Significant relief has been observed in cows using this combined approach, with respiration rates reduced by 50%, bringing them down to around 50 breaths per minute.

5.6.3 Animal Handling

Reducing stocking density during hot weather allows animals to dissipate body heat more efficiently. Studies on animal handling show that moving animals through working facilities requires energy expenditure, which raises their average body temperature by 0.5–1.0 °C (0.9–1.8 °F), depending on ambient conditions. Therefore, to promote animal comfort, it is recommended to minimize handling on hot days. Additionally, animal transportation should be carried out with minimal stock and during the cooler hours of the day.

5.6.4 Nutritional Strategies

Heat-stressed animals often experience reduced reproductive and productive performance. Feeding high-quality forages and balanced rations can help mitigate some of the effects of heat stress and enhance animal performance. During hot, dry summers, animals tend to decrease their feed intake, which contributes to reduced productivity. In such situations, strategies like frequent feeding, improving forage quality, using palatable feeds, and ensuring balanced and nutrient-dense nutrition are necessary. West (1999) reported that lower dry matter intake (DMI) and reduced nutrient availability during hot weather lead to decreased nutrient utilization efficiency. Therefore, supplementing dairy animals with nutrient-enriched feed, such as good-quality green fodder, concentrates, bypass fat, and protein, can improve both production and reproduction during hot weather. Sweating leads to mineral losses, especially potassium, and hyperventilation alters blood acid-base balance, reducing bicarbonate and buffering capacity while increasing urinary excretion of electrolytes. Thus, electrolyte supplementation becomes essential. Among the three main rumen-produced volatile fatty acids, propionate is primarily converted into glucose by the liver. Highly fermentable starches, like grains, increase propionate production in the rumen, but excessive grain feeding can risk ruminal acidosis in heat-stressed cows. Improved dietary energy density, coupled with limited fat feeding to prevent ruminal and metabolic disorders, is essential. Supplementing saturated fatty acids at 1.5–3.0% of the diet dry matter has been shown to increase milk yield, milk fat content, and reduce peak rectal temperatures in heat-stressed cows. Feeding during the cooler hours of the day, along with increasing the number of daily feedings, proves beneficial for combating heat stress. Feed additives such as niacin, sodium bicarbonate buffer, antioxidants, fungal yeast culture, lipoic acid, and thiazolidinediones are effective practices for alleviating heat stress. El-Tarabany and El-Bayoumi (2015) found that niacin and antioxidant supplementation improved fertility and reduced heat stress in buffaloes. Additionally, vitamins C and E are known to reduce heat stress-induced oxidative stress (Seyrek et al., 2004). Nutritional strategies like antioxidant supplementation (e.g., vitamin A, selenium, and zinc) and ruminant-specific live yeast can also provide significant benefits.

5.6.5 Breeding Management

It is recommended to continue artificial insemination (AI) breeding rather than using bulls, as natural breeding can lead to infertility in both bulls and cows due to summer heat stress. Cows tend to show fewer heat symptoms during periods of heat stress compared to when they are in thermal comfort, making it essential to implement an effective heat detection program to identify cows with subtle heat signs.

Breeding should take place during the cooler hours of the day, preferably in the morning or late evening. Likewise, heat detection should also be performed during these cooler times.

5.6.6 Genetic Selection

In light of the changing climate, it is crucial to harness the genetic potential of indigenous livestock breeds across various species. Identifying and selecting heat-tolerant, high-producing breeds for breeding programs can help livestock cope with heat stress, even in crossbreeding scenarios (Kimothi & Ghosh, 2005). Productive traits should be assessed to evaluate the performance of these indigenous breeds. Following a thorough evaluation, suitable breeds should be developed to ensure their survival in local environmental conditions. Under hot conditions, cattle with darker coat colors and long hair tend to be less adaptable than those with lighter coats, smaller hair, thinner skin, fewer hair follicles, and larger hair coat diameters (Bernabucci et al., 2010; Hoffmann, 2010). McManus et al. (2009) found that animals with light or white coats reflect 50–60% of direct solar radiation compared to their dark-colored counterparts. A good example of an adaptable breed is the Tharparkar cattle, with highly pigmented skin that blocks direct UV radiation penetration (Uttarani et al., 2014). Additionally, genetic selection based on specific molecular markers for heat tolerance will be instrumental in alleviating heat stress in cattle and buffaloes by identifying heat-resistant animals.

5.7 Conclusion

Climate change poses a significant threat to the sustainability of livestock production systems, with mitigation and adaptation strategies being essential to reduce greenhouse gas emissions from livestock. Scientific research plays a vital role in supporting the livestock sector in addressing the challenges posed by climate change. Droughts lead to reduced forage availability, decreased animal growth rates, and decreased efficiency in livestock and poultry production. These conditions negatively impact the biological systems of animals, causing poor thriftiness, compromised reproduction, weakened immunity, increased disease susceptibility, and poor-quality feed, ultimately resulting in a decline in animal health and productivity. Effective management practices, such as optimizing housing, feeding, breeding, and health protocols, are crucial for ensuring livestock survival during drought conditions. Similarly, heat stress alters rumen motility and microflora, impacting feed digestibility and fermentation processes. It shortens and weakens estrus duration, inhibits follicular development through the regulation of reproductive hormones, and suppresses embryonic growth due to impaired placental function, ultimately affecting offspring growth and immunity. The impact of heat stress can be mitigated through appropriate farm design, close monitoring, proactive preventive measures, and proper planning before temperatures and humidity rise.

Standardizing mineral supplements to maintain acid-base balance should also be considered for animals under thermal stress. Physical modifications to the environment and nutritional management are key strategies for alleviating heat stress.

References

Ames, D. R., Brink, D. R., & Willms, C. L. (1980). Adjusting protein in feed lot diet during thermalstress. *Journal of Animal Science, 50*(1), 1–6.

Atrian, P., & Shahryar, H. A. (2012). Heat stress in dairy cows. *Research in Zoology, 2*(4), 31–37.

Banks, S., King, S. A., Irvine, D. S., & Saunders, P. T. K. (2005). Impact of a mild scrotal heat stress on DNA integrity in murine spermatozoa. *Reproduction, 129*(4), 505–514.

Bernabucci, U., Lacetera, N., Baumgard, L. H., Rhoads, R. P., Ronchi, B., & Nardone, A. (2010). Metabolic and hormonal acclimation to heat stress in domesticated ruminants. *Journal of Animal Science, 4*(7), 1167–1183.

Bonnet, S., Geraert, P. A., Lessire, M., Carre, B., & Guillaumin, S. (1997). Effect of high ambient-temperature on feed digestibility in broilers. *Poultry Science, 76*(6), 857–863.

Bradley, P., & Marulanda, C. (2000). Simplified hydroponics to reduce global hunger. *Acta Horticulture, 554*, 289–295.

Campos, E. J. (2000). *Avicultura: razoes, fatos e divergencias*. FEP-MVZ Escola de Veterinaria da UFMG. 311 p.

Charmley, E. (2000). *Towards improved silage quality–A review*. Water resources SystemsDivision. http://www.nih.ernet.in/rbis/india_information/draught.htm

Coppock, C. E., Grant, P. A., & Portzer, S. J. (1982). Lactating dairy cow responses to dietary sodium, chloride, bicarbonate during hot weather. *Journal of Dairy Science, 65*(4), 566–576.

El-Tarabany, M. S., & El-Bayoumi, K. M. (2015). Reproductive performance of backcross Holstein xBrown Swiss and their Holstein contemporaries under subtropical environmentalconditions. *Theriogenology, 83*, 444–448.

Etches, R. J., John, T. M., & Verrinder Gibbins, A. M. (1995). Behavioural, physiological, neuroendocrine and molecular responses to heat stress. In N. J. Daghir (Ed.), *Poultryproduction in hot climates* (pp. 31–65). CAB International.

Freeman, B. M. (1987). The stress syndrome. *World's Poultry Science Journal, 43*(1), 15–19.

Ghazi, S. H., Habibian, M., Moeini, M. M., & AbdolMohammadi, A. R. (2012). Effects of different levelsof organic and inorganic chromium on growth performance and immune competence ofbroilers under heat stress. *Biological Trace Element Research, 146*(3), 309–317,

Gregory, N. G. (2010). How climatic changes could affect meat quality. *Food Research International, 43*, 1866–1873.

Guis, H., Caminade, C., Calvete, C., Morse, A. P., Tran, A., & Baylis, M. (2012). Modelling the effectsofpast and future climate on the risk of bluetounge emergence in Europe. *Journal of the Royal Society Interface, 9*(67), 339–350.

Hegde, N. G. (2010). *Forage resource development in India*. Souvenir of IGFRI Foundation Day. November 2010.

Hoffmann, I. (2010). Climate change and the characterization, breeding and conservation of animalgenetic resources. *Animal Genetics, 41*, 32–46.

Imik, H., Ozlu, H., Gumus, R., Atasever, M. A., Urgar, S., & Atasever, M. (2012). Effects of ascorbic acidand alpha-lipoic acid on performance and meat quality of broilers subjected to heat stress. *British Poultry Science, 53*(6), 800–808.

Iwagami, Y. (1996). Changes in the ultrasonic of human cells related to certain biologicalresponses under hyper-thermic culture conditions. *Human Cell, 9*(4), 353–366.

Kimothi, S. P., & Ghosh, C. P. (2005). *Strategies for ameliorating heat stress in dairy animals* (pp. 371–377). DairyYear book.

Lara, L. J., & Rostagno, M. H. (2013). Impact of heat stress on poultry production. *Animals, 3*, 356–369.

Lin, H., Mertens, K., Kemps, B., Govaerts, T., De Ketelaere, B., Baerdemaeker, D., Decuypere, J., & Buyse, J. (2004). New approach of testing the effect of heat stress on eggshell quality: Mechanicaland material properties of eggshell and membrane. *British Poultry Science, 45*(4), 476–482.

Mannetje, L. (1999). Introduction to the conference on silage making in the tropics. In L. Mannetje (Ed.), *Proc. FAO e-Conf. on Trop. Silage. FAO Plant Prod. and Protect Paper161. 1 Sept. – 15 Dec. 1999. Paper 1.0* (pp. 1–3).

Marai, I. F. M., & Haeeb AAM. (2010). Buffalo's biological functions as affected by heat stress—A review. *Livestock Science, 127*, 89–109.

Mcdowell, L. R. (1976). *Mineral deficiencies and toxicities and their effect on beef production in developing countries* (pp. 216–241). Symposium: Beef cattle production in developing countries, University of Edinburgh, Scotland.

McManus, C. M., Paludo, G. R., Louvandini, H., Gugel, R., Sasaki, L. C. B., & Paiva, S. R. (2009). Heattolerance in Brazilian sheep: Physiological and blood parameters. *Tropical Animal Health Production, 41*, 95–101.

Patil, N. V., Mathur, B. K., Patel, A. K., Patidar, M., & Mathur, A. C. (2006). ICAR short courseon "Feeding of livestock during drought and scarcity" during 1st November to 10th November, 2006 at central arid zone research institute, Jodhpur.

Poppenga, R. H., & Puschner, D. V. M. (2014). *Drought related poisonings and nutritional risksto cattle* (p. 10). School of Veterinary Medicine University of California.

Sahin, N., Sahin, K., & Kucuk, O. (2001). Effects of vitamin E and vitamin A supplementation onperformance, thyroid status and serum concentrations of some metabolites and mineralsin broilers reared under heat stress (32°C). *Veterinary Medicine, 46*(11–12), 286–292.

Seyrek, K., KarginKiral, F., & Bildik, A. (2004). Chronic ethanol induced oxidative alterations inthe rat tissues and protective effect of vitamin E. *The Indian Veterinary Journal, 81*, 1102–1104.

Sneath, R., & McIntosh, F. (2003). *On farm review of hydroponic fodder production for beefcattle* (pp. 1–54). Meat and Livestock Australia Limited.

Sohail, M. U., Hume, M. E., Byrd, J. A., Nisbet, D. J., Ijaz, A., Sohail, A., Shabbir, M. Z., & Rehman, H. (2012). Effect of supplementation of prebiotic mannanoligo saccharides and probiotic mixtureon growth performance of broilers subjected to chronic heat stress. *Poultry Science, 91*(9), 2235–2240.

Stott, G. H., Wiersma, F., Mevefec, B. E., & Radwamki, F. R. (1976). Influence of environment on passiveimmunity in calves. *Journal of Dairy Science, 59*(7), 1306–1311.

Thom, E. C. (1959). The discomfort index. *Weatherwise, 12*, 57–59.

Uttarani, M., Singh, S. V., Upadhyay, R. C., Kumar, S., Beenam, & Singh, A. K. (2014). Expression ofskin colour genes in Tharparkar cattle during summer and winter season. *Journal of Environmental Research and Development, 9*(01), 113–119.

Wadhawa, M., & Bakshi, M. P. S. (2013). *Utilization of fruit and vegetable wastes as livestock feedand as substrate for generation of value added products* (Vol. 2013). FAO.

West, J. W. (1999). Nutritional strategies for managing the heat-stressed dairy cow. *American Society ofAnimal Science and American Dairy Science Association, 2*, 21–35.

Wilkinson, J. M., Wadephul, F., & Hill, J. (1996). *Silage in Europe - A survey of 33 countries*. Chalcombe Publications.

Yalcin, S., Settar, P., Ozkan, S., & Cahaner, A. (1997). Comparative evaluation of three commercialbroiler stocks in hot versus temperate climates. *Poultry Science, 76*(7), 921–929.

Zhang, Z. Y., Jia, G. Q., Zuo, J. J., Zhang, Y., Lei, J., Ren, L., & Feng, D. Y. (2012). Effects of constant andcyclic heat stress on muscle metabolism and meat quality of broiler breast fillet and thighmeat. *Poultry Science, 91*(11), 2931–2937.

Zhou, W. T., Fijita, M., Yamamoto, S., Iwasaki, K., Ikawa, R., Oyama, H., & Horikawa, H. (1998). Effectsof glucose in drinking water on the changes in whole blood viscosity and plasmaosmolality of broiler chickens during high temperature exposure. *Poultry Science, 77*(5), 644–647.

Climate Change and Its Impact on Fisheries and Aquaculture

6

Jaspreet Singh and Kamal Sarma

Abstract

Fish and fisheries play an important role in achieving food and nutrition security, apart from providing a source of income and livelihood to millions of people across the globe. In the last few decades, aquaculture has continued to significantly expand in terms of area and production, making it one of the fastest growing production sectors. However, the sustainability of this sector is at stake due to climate change and related issues. Climate change is anticipated to significantly affect several aspects, including ecosystems, overall productivity, society, and economies, increasing pressure on livelihoods and food supplies. In 2020, an estimated 58.5 million people were engaged in capture fisheries (65%) and aquaculture (35%) (FAO, The State of World Fisheries and Aquaculture 2022. Rome, FAO, Towards Blue Transformation, 2022). Furthermore, most people directly and indirectly associated with the fisheries sector belong to the small and marginal category; consequently, the impact will be even more severe on them. Several observable changes have already been evident in recent times, which can be directly attributed to climate change, like drastic changes in weather conditions, increasing frequency of cloud bursts, changes in hydrological regimes, heavy wind and storms, increased incidence of flood and droughts, etc. These effects can consequently lead to changes in fish productivity, disease infestation, decline in production from inland and aquaculture systems, etc. The impact of climate change, especially on developing nations, will be predominantly more severe due to their constrained capabilities and restricted resources to address

J. Singh
Germplasm Exploration Evaluation and Conservation Division, ICAR National Bureau of Fish Genetic Resources, Lucknow, India

K. Sarma (✉)
DLFM, ICAR- Research Complex for Eastern Region, Patna, Bihar, India

R. Dubey et al. (eds.), *Drought and Heat Stress in Agriculture*,
https://doi.org/10.1007/978-981-96-5735-3_6

this pressing issue effectively. In the context of India, the nation exhibits a significant susceptibility to the impacts of climate change due to its pronounced physical exposure to climate-related disasters. Specifically, approximately 65% of the country is prone to drought, 12% is susceptible to flooding, and 8% is at risk of cyclones. Moreover, the Indian economy and population rely heavily on climate-sensitive sectors such as agriculture, forests, tourism, and fisheries. Hence, it is imperative for long-term management strategies to include the possible ramifications of climate change and catastrophes on the livelihoods and companies associated with the economy. In this chapter, extreme events related to climate change, like, the occurrence of cyclones, rising sea levels, prolonged periods of drought, intrusion of saline water, flooding, alterations in precipitation patterns, and changes in ocean acidity as well as changes on aquaculture production and its implications on the sector's sustainability, shall be discussed.

6.1 Introduction

Fisheries and aquaculture play an important role in achieving food and nutrition security and providing employment and livelihood to millions of people across the globe. During 2022, the total fisheries and aquaculture production of the world was 222.3 million tons (MT) (185.4 MT of aquatic animals and 37.8 MT of algae), which was mainly due to the growth of aquaculture, particularly in Asia (FAO, 2024). The fish consumption in 2022 was 20.7 kg per capita, double the average of 9.9 kg per capita in 1960s. In 2022, the global fisheries and aquaculture goods trade yielded approximately USD 195 billion (FAO, 2024). Around 850 million people globally live within a 100 km range of tropical coastal habitats, including coral reefs and mangroves (Seggel & De Young, 2016). These habitats offer a range of benefits, such as guaranteeing food security, safeguarding coastlines, providing cultural services, and producing income through activities like fishing, aquaculture, and tourism. Fish production in India has also shown an unmatched surge from 0.75 million tonnes in 1950–1951 to 16.25 million tonnes during 2021–2022, where inland fish production recorded around 75% of the total fish production, with an annual growth of 7% over the last 5 years. Similarly, the export earnings recorded an impressive average annual growth rate of about 10% in recent years (56.6 thousand crores in 2021–2022) (HBFS, 2022). During the last two and a half decades, there has been a marked shift in the inland fisheries sector. Freshwater aquaculture, with a share of 34% in Inland Fisheries in the mid-1980s, has increased to about 76% in recent years (AR, 2023). Though the present major share of fish production comes from aquaculture, about 70 and 80% of aquaculture ventures globally are still considered small-scale, often family-based activities which are integrated at the farm level with crop and livestock farming and where part of the production may be retained for household consumption and nearly 90% of whom live in Asia (De Silva & Soto, 2009; Subasinghe et al., 2012). Hence, for sustainable production and incessant aquaculture growth, special attention must be given to the small and marginal farmers.

This sun-rising sector also provides the livelihood of millions of people directly and indirectly. It is estimated that around 58.5 million people globally are employed in the primary fisheries sector. Including subsistence and secondary sector workers and their dependents, about 600 million individual's livelihood heavily depends on fisheries and aquaculture (FAO, 2022). In India also this sector provides livelihood to about 16 million fishers and fish farmers at the primary level and several lakhs along the value chain (AR, 2023). With this large population, directly and indirectly, depending on fisheries and aquaculture, any drastic change in the production structure shall not only create a significant impact on their income and livelihood security but will also hamper their nutritional security.

Climate change is becoming a global issue and has already been showing its impact in all sectors of agriculture and its allied disciplines, natural ecosystems, livestock production systems, etc., thereby increasing pressure on livelihoods and food supplies. As per an estimate, approximately 65% of the country is prone to drought, 12% is susceptible to flooding, and 8% is at risk of cyclones (Salim & James, 2018). It is also now making a noticeable impact on the fisheries and aquaculture sector. Globally, aquatic food provides about 17% of animal proteins and 7% of all proteins (2019) and for 3.3 billion people, aquatic food provides at least 20% of the average per capita intake of animal protein (Cochrane et al., 2009; FAO, 2022). Due to its challenge to social welfare, equity, and the sustainability of future development, developing countries are generally believed to be more vulnerable to climate change impacts due to limited resources and low adaptive capacity (IPCC, 2014; Schneider et al., 2001). Hence, making agriculture and allied sectors climate-smart is very important to ensure the food and nutrition security of the growing global population at a time when there are problems of small and fragmented land holding, land degradation, resource crunch, carbon emissions, pollution, etc. (Kochewad et al., 2022).

Since 200 years of excessive greenhouse gas (GHG) emissions from the burning of fossil fuels for energy generation, transportation, and industry, as well as deforestation and intensive agriculture, it is now widely acknowledged that climate change is not just a potential threat but an inevitable reality (IPCC, 2023). Substantial evidence from indirect and direct observations, historical reconstructions, and more recent instrumental observations and conceptual and numerical models support recent understanding of climate change. Barange et al. (2018) show that human activities, including burning gas and oil, deforestation, and intensive agriculture, are the primary cause of the observed warming since the mid-twentieth century. Only 1% of the additional heat energy caused by anthropogenic climate change activities remains in the atmosphere, with the Earth's oceans absorbing 93%. Approximately 3–4% of the remaining heat is acquired through ice and snow melting (Reid & Hill, 2016). The ocean possesses a vast thermal capacity, and even a minor change in the transfer of heat between the ocean and the atmosphere would substantially impact the Earth's overall air temperature (Reid & Hill, 2016). Moreover, the ocean has assimilated around 25% of the carbon dioxide released due to human activities (Barange et al., 2018), thereby playing a crucial role in regulating the Earth's climate.

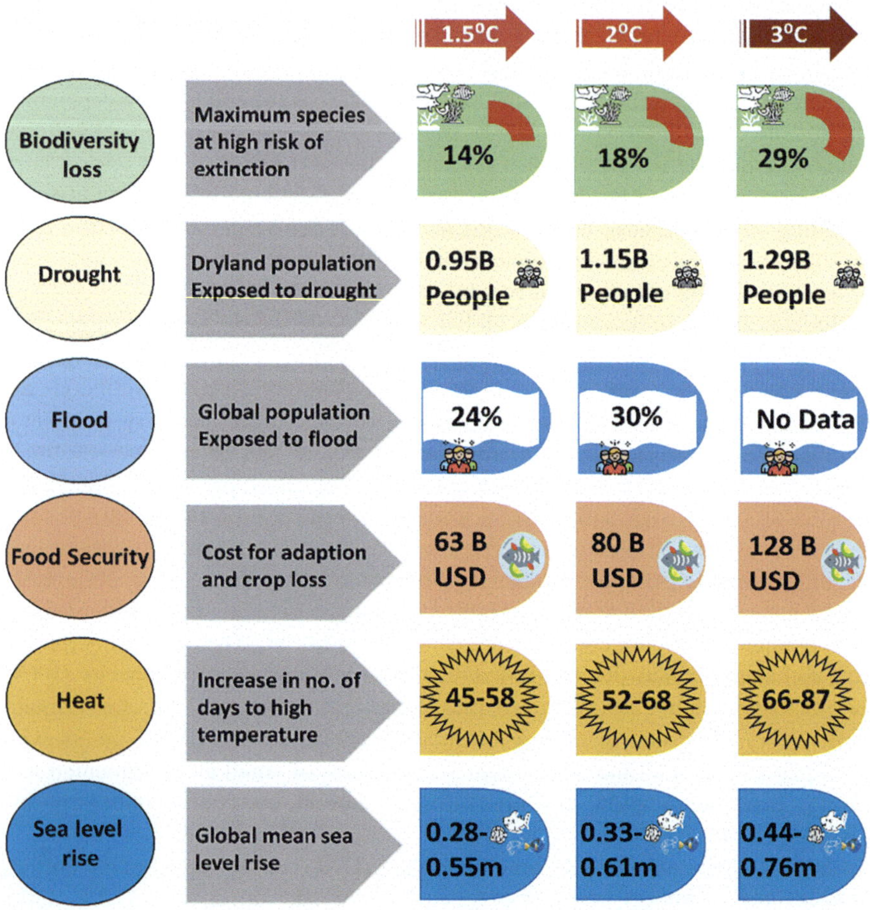

Fig. 6.1 Climate change and its risk at different temperature levels. (IPCC, 2023)

The IPCC Sixth Assessment Report (AR6) disclosed that the climate system has undeniably been warming and that numerous changes observed since 1950 are incomparable compared to previous periods. Globally, the Earth's average surface temperature has risen by over 1.1 °C between 2011 and 2020, with more gain over land (1.59 °C) than the ocean (0.88 °C) (IPCC, 2023). Global risk at different levels of increasing temperature is given in Fig. 6.1. Global warming has been shown by human activity, mostly through greenhouse gases (GHGs) (carbon dioxide [CO_2], methane [CH_4], and nitrogen dioxide [NO_2]) emissions. The global emissions of greenhouse gases have consistently risen, primarily due to unsustainable energy usage, land use and changes, consumption and production patterns, and human lifestyles. Historically and presently, these contributions have been unequal across areas, countries, and individuals (IPCC, 2023). As per the IPCC (AR 6) report, atmospheric CO_2 concentration reached 410 parts per million in 2019, surpassing any recorded levels in the past two million years. Additionally, methane and nitrous oxide concentrations reached 1866 parts per billion and 332 parts per billion, respectively, beyond any levels observed in the past 800,000 years. IFAD and other development organizations have recognized climate change as a major threat to mankind and the poorest and most vulnerable individuals will be disproportionately affected by its impacts (IFAD, 2022).

Climate change has been widely recognized as a threat to global food production and poses a major threat to both the quality and quantity of production (Beach & Viator, 2008; Myers et al., 2017; Maulu et al., 2021). Fisheries and aquaculture productivity is predicted to decline significantly in the tropical regions of Asia and Africa due to rising temperatures and increased greenhouse gas emissions (Ruby & Ahilan, 2018; FAO, 2020). These changes can cause significant damage to important ecosystems such as coral reefs, mangroves, wetlands, and other aquatic ecosystems, resulting in declined fish populations and production. Thus, to minimize consequences, it is urgently necessary to create vocational alternatives to lessen fishing pressure and increase fishermen's adaptive capacity based on the development of resilient infrastructure in the aquaculture and fisheries sectors. The IPCC, 2023 report presented the projected alteration in the maximum fisheries catch from 2081 to 2099, considering the expected levels of global warming ranging from 3.4 to 5.2 °C (mean 4.3 °C).

Due to their importance in global food security, nutrition, and livelihoods, climate change impacts on fisheries and aquaculture sustainability have recently received attention (Barange et al., 2018; Muhala et al., 2021; Maulu et al., 2021). Climate change threatens the aquaculture value chain, notwithstanding its instability (Cochrane et al., 2009; Brander et al., 2017; Barange et al., 2018; Muhala et al., 2021). Extensive research and analysis have been conducted to examine the impact of climate change on aquaculture, both on a regional and worldwide level, and most of the studies have shown a negative impact on production (De Silva & Soto, 2009; Soheila & Bahram, 2010; Klinger et al., 2017; Handisyde et al., 2017; Barange et al., 2018; Maulu et al., 2021). Climate change is expected to have direct and indirect effects on aquaculture productivity. The physiological characteristics of shellfish and finfish stocks in production systems are altered due to climate change. Biological transformations refer to changes in the breeding, migration, and feeding patterns, which could impact production and fish populations. On the other hand, the indirect effects manifest through changes in the primary and secondary productivity and the overall structure of ecosystems. Additionally, it can impact the availability and cost of input supplies, such as fishmeal-based feed, medicines, fish oil, and other goods and services required by fishers and aquaculture producers. Fish prices and the costs related to fishing and fish farming could also increase. The possible ramifications of this could greatly impact both food security and employment opportunities in regions that rely on fisheries. This chapter will provide a comprehensive overview of global climate change's causes, general concerns, and impacts on fisheries and aquaculture. It will also discuss potential mitigation strategies and the development of suitable monitoring technologies.

6.2 Climate Change Impacts on Fisheries and Aquaculture

6.2.1 Marine Fisheries

Marine fisheries play a crucial role in prosperity and ensuring many coastal nations' food security, livelihoods, and environmental health (Giron-Nava et al., 2021; FAO, 2020; Heck et al., 2023). Global marine fisheries contribute 17% of the total animal protein for direct human consumption, and world fisheries generate around US$ 150 billion in trade (FAO, 2022). Furthermore, fisheries offer employment opportunities to around 260 million individuals, both full-time and part-time, and 22 million are engaged in small-scale fisheries (Teh & Sumaila, 2013). Despite this, marine fisheries face several challenges due to anthropogenic activities, such as overexploitation, habitat degradation, and pollution. Besides, it is also vulnerable to the impacts of climate change, which pose a risk to both marine fisheries and the communities that rely on them (Belhabib et al., 2018; Free et al., 2019; Heck et al., 2023).

The catch potential of marine capture fisheries is expected to be affected by changes in the chemical and physical conditions of the ocean such as variations in sea surface temperature (SST), salinity, coastal erosion, and ocean acidification that alter aquatic animal growth and metabolism and biology. These changes impact the food webs, productivity, fish immunity balance, and spatial distribution of fish species responsive to these effects (Barange & Perry, 2009; Doney et al., 2020; Heck et al., 2023). In their study, Cheung et al. (2019) projected that the impact of climate change on the maximum catch potential in the exclusive economic zones (EEZs) of the world will result in a decrease in production ranging from 2.8% to 5.3% by the years 2050 and 2095. The decline in production has to be faced by tropical countries such as Nauru, Kiribati, Tuvalu, Ecuador, Palau, Micronesia, and Tokelau. The foreseen ecological changes could indirectly impact fisher communities, risking food security and economic conditions due to more expenses to mitigate these changes and, ultimately, a decline in earnings from the fisheries (Eurich et al., 2024). Furthermore, certain coastal regions are susceptible to rising sea levels and a rise in both the frequency and intensity of extreme weather events. These events can result in the disruption and destruction of coastal infrastructure, as well as the loss of vital habitats for fish, including intertidal wetlands such as marshes and mangroves, coral reefs, and the diversity of fish species that inhabit them (Heck et al., 2023). Additionally, these changes pose an increased risk to fishermen while at sea.

The productivity of marine fisheries is anticipated to be affected by climate change, mostly due to alterations in temperature, acidity, deoxygenation, the consequences of sea level rise, and other factors. These changes are predicted to surpass natural variability in tropical regions initially. Worldwide, Kelp, seagrass, coral reefs, mangrove forests, salt marshes, and other organisms that form their environment are susceptible to climate change impacts (Trégarot et al., 2024). The increase in temperatures and the ocean's acidity are endangering coral reefs on a global scale, and Klein et al. (2024) projected that in future scenarios characterized by >4 °C warming, >93% of global reef cells will be at risk by the end of the century. Furthermore, ocean acidification is modifying the chemical composition of ocean carbonate, impeding coral growth and deteriorating the physical integrity of reefs (Kay et al., 2023). The deterioration of coral reefs in Southeast Asia poses a significant threat to the interconnected system of organisms that rely on them for survival. This endangers the diverse range of species and the fishing industry and negatively affects the region's food security, coastal defense, and tourism (Kay et al., 2023). The potential economic consequences of this degradation could result in the region losing billions of dollars in revenue.

Substantial data indicates that climate change has caused significant alterations in the world's oceans. The rate at which rising ocean temperature was measured at 0.88 °C between 1850 and 2020. Furthermore, projections suggest that the oxygen levels in the ocean may decrease by 3–4% by the year 2100. Ocean acidification has already caused a 30% reduction in pH (Ding et al., 2017). These changes may be occurring at a rate that surpasses the capacity of species to adapt. The stressors encompassing global warming, ocean acidification, and increasing deoxygenation are generally known as climate change's "deadly trio" (Bijma et al., 2013). The simultaneous occurrence of these factors leads to synergistic effects that harm marine life and disrupt the structure and functions of the ecosystem. Climate change is predicted to significantly impact the marine environment by altering the physical and chemical characteristics of seawater, such as temperature, salinity, currents, vertical stratification, and oxygen levels. Rising SST and eutrophication levels promote pelagic (floating) algae blooming, increasing the frequency of "green tides" and hazardous algal blooms, which may be the reason for mass mortality of important seagrass beds and corals through shading, anoxia, and excessive nutrient loading.

Climate change is anticipated to impact target species directly and indirectly. The direct impacts encompass alterations in fish stocks' reproduction, distribution, and population dynamics. Indirect impacts encompass alterations in the production and composition of the ecosystems that serve as crucial sources of sustenance and refuge for fish. The production of marine capture fisheries has remained unchanged over the previous decade due to overfishing, unregulated fishing practices, habitat damage, and pollution. The situation may worsen with the impact of climate change (Ding et al., 2017). The increase in water temperature can affect the variety, distribution, number, and timing of fish species. Severe weather events like storms, floods, and droughts will significantly hinder fishing. The rise in sea level will decrease fish productivity and harm the livelihoods of communities. Some tropical fish populations may be at risk of regional extinction. Other individuals may migrate to regions closer to the Earth's poles. Coastal ecosystems and resources are susceptible to rising sea levels, increasing sea temperatures, excessive nutrient enrichment (eutrophication), and introduction of invasive species. Many fish species have a limited range of ideal temperatures that are optimal for their basic metabolic functions and the availability of feeding organisms. Species with a short life span and quick turnover, such as plankton and small pelagic fishes, are the most prone to undergo such alterations. Over a few years to a decade, there can be significant and sudden changes in the distribution, recruitment, and abundance of numerous species. Fluctuations in population size will modify the makeup of the species. Possible alterations in net primary production and transfer to higher trophic levels may occur over several decades. According to research by the Indian Council of Agricultural Research, several marine species in India will react to climate change in the following ways: (i) Higher temperatures can lead to alterations in the species composition of phytoplankton; (ii) Small pelagic species may expand their geographical boundaries; (iii) Certain species may also be discovered in deeper seas; and (iv) Phenological changes may also occur (Vivekanandan, 2010).

Prior to 1985, small pelagic species such as oil sardine and Indian mackerel were restricted to the Malabar upwelling zone. However, in the past two decades, their catch distribution has expanded between 14°N and 20°N. In 1985, the Indian mackerel primarily inhabited surface and subsurface seas. However, since 2003, 15% of this fish has been caught by bottom trawlers around the Indian coast. In their study, Singh et al. (2023a) observed a rise in the occurrence of unintentional capture of pelagic fish in trawl fisheries along the south Konkan coast over the years. Singh et al. (2023b) revealed that catch per hour of trawl fisheries declined from 198 to 33.7 kg h^{-1} along the west coast of India. Vivekanandan (2010) reported a change in the spawning season of the threadfin bream *Nemipterus japonicus* and *N. mesoprione* throughout the Indian coast. The spawning time shifted from the warmer months of April–September to the cooler months of October–March. Unfortunately, the *Lacterius lacterius* population has been reduced in the Gulf of Mannar, India. Prowse et al. (2009) conducted a study which found that the movement of species toward the north and the disturbance and competition caused by invasive species are currently happening and will continue to impact marine, terrestrial, and freshwater communities of Northern Canada.

6.2.2 Inland Fisheries

Freshwater fish resources are located and exploited from various freshwater habitats, such as rivers, lakes, reservoirs, ponds, streams, and wetlands, on all continents except Antarctica. Globally, inland capture fish production was estimated at 11.5 million tons in 2022, at a historically high level (FAO, 2022). Asia accounted for about 66% of the overall inland fish production, with Africa being the second largest contributor. China, the leading inland fisheries producer since the mid-1980s, was surpassed by India, producing 1.8 million tons (FAO, 2022). Freshwater ecosystems are sustained by inland fisheries, which can cut across multiple climatic zones and be transboundary. Inland fisheries exhibit wide association from small-scale to medium-large-scale commercial fisheries operating in inland waters for their survival and livelihood with specific economic considerations. Recreational fishing is a substantial industry widespread in developed countries and is increasingly important. Human interference has caused substantial global changes in the biosphere, encompassing the Earth's atmosphere, soil, and oceans. Human actions have threatened around 65% of the world's inland waters' ability to sustain human populations, including inland fisheries. Human actions include overexploiting natural resources, altering and degrading habitats, pollution, industrial and thermal contaminations,

GHG emissions, and allowing non-native species to travel outside their natural habitats. Interactions between other anthropogenic stresses and climate change have had, and will continue to have, a significant and frequently unpredictable impact on inland fisheries. Due to the great diversity of inland terrain and human activity, the effects of climate change on inland fisheries differ over time and from place to place (Barange et al., 2018).

Freshwater ecosystems possess a high degree of sensitivity in reflecting climate change. The global average temperature increased from 0.85 to 1.1 °C from 1880 to 2020; these changes have been happening for many decades and may have affected our understanding toward the environment (IPCC, 2023; Barange et al., 2018). Inland fisheries are expected to be affected by climate change due to changes in temperature, groundwater levels, thermal contamination, water quality, and unexpected or minimal precipitation, which will alter stream and river flows. All these occurrences may affect inland fish diversity, directly affecting subsistence and recreational fishers' livelihoods, human communities, and economies. According to Das et al. (2019), climate change will have a range of consequences on the physical, chemical, and biological features of freshwater ecosystems, such as unexpected weather events, i.e., severe droughts, flooding, and heat waves, higher water temperature, increasing sea levels would result in the intrusion of saltwater into coastal aquifers, leading to the displacement of freshwater by saltwater, which all have an impact on inland as well marine fisheries. These factors will affect the growth, reproduction, and survival of fish and their predator–prey relationship, as indicated by Hershkovitz et al. (2015) and Kanno et al. (2016). Furthermore, climate change is predicted to alter important life history events of fish, such as migration, reproductive changes, spawning, species interaction, disease risk, immune function, and the geographical distribution of fish populations (Fig. 6.2) (Lynch et al., 2016). Climate change directly impacts the chemical properties of water bodies, including dissolved oxygen, salinity, evaporation, contamination, and nutrient content levels

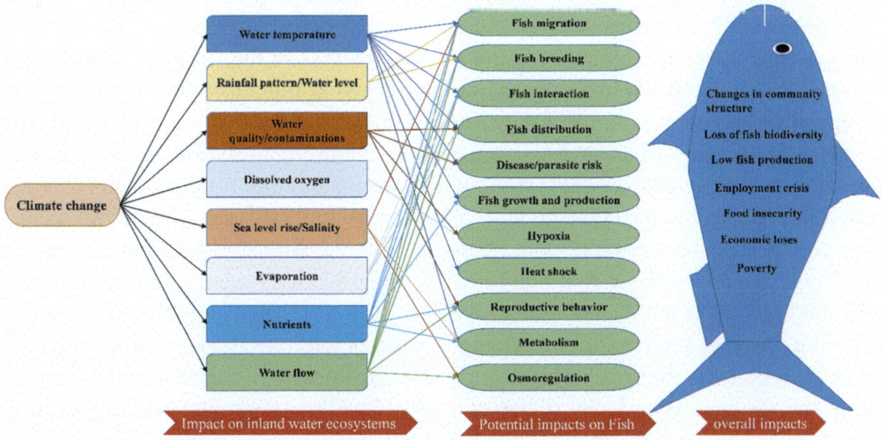

Fig. 6.2 Climate change impact on inland fisheries and related livelihood

(Miller et al., 2014; Das et al., 2019). In the future, water demands are expected to rise due to population expansion, land use changes, food demand, changes in food production systems, and increased industrial needs. This is anticipated to result in further degradation and challenges for inland fisheries. In this regard, Barange et al. (2018) estimated a detailed projected future of risk posed by climate change, population expansion, and water stress on the main contributors of inland fisheries which will be of great use.

In addition to alterations in air and ocean temperatures, climate change has significantly impacted the global hydrological cycle, resulting in modifications to the timing, quantity, and nature of precipitation. These changes vary regionally in terms of their magnitude and direction. The ongoing changes will persist, with the magnitude of alterations in precipitation (and evaporation) intensifying. Alterations in precipitation have consequential consequences on various aspects of the hydrological cycle, such as runoff, groundwater replenishment, and river discharge, subsequently affecting the accessibility, quality, and source of water. These factors, in turn, impact the ability of freshwater systems to deliver ecosystem services, such as supporting fisheries, diluting trash and pollutants, and providing water for extraction. This can lead to conflicts between different user groups. One of the most significant (non-climate) anthropogenic stresses on inland fisheries is the regulation of rivers. These factors lead to habitat degradation and fragmentation, noticeable changes in community structure, the extinction of sensitive species, and a loss of population connectivity. Döll and Zhang (2010) proposed that climate change could have a more significant effect on ecologically relevant river flow characteristics by the middle of the twenty-first century compared to the impact of river regulation and water abstraction up to the present time. This suggests that climate change has the potential to impact inland fisheries significantly. Riverine fisheries are likely to be disproportionately affected by water abstraction and climate change, particularly in large, intensively fished rivers like the Yangtze, Mekong, Zambezi, and Ganges, as noted by McIntyre et al. (2016), also pointed out that global riverine fish catch scaled steeply (and positively) with discharge.

At present, 24 countries are undergoing a state of medium stress. Nevertheless, these countries are anticipated to encounter high stress in the future. Furthermore, 60 countries currently experience medium stress levels and are expected to continue facing comparable stress in the future. Additionally, it was observed that numerous countries in this classification have undertaken extensive alterations to river basins and have erected extensive water storage, irrigation, and damming systems to meet their water requirements, consequently affecting inland fisheries. The inland fisheries in major river basins such as the Yangtze, Ganga, Amazon, Mekong, La Plata, and Congo are currently at risk due to habitat degradation, overexploitation, pollution, the introduction of non-native species, and the effects of climate change, including a rise in temperature, flooding, erratic precipitation, etc. Climate change has been linked to changes in hydrological conditions in significant water bodies within the Yangtze River basin (Chen et al., 2017). Climate projections suggest that in the twenty-first century, the Yangtze River basin will experience a rise in temperature, high precipitation, and low discharge. It has had a direct impact on the physicochemical conditions in the region, which could cause species relocation in their suitable habitat and may enhance the population of non-native species in a particular region (Davidson et al., 2011; Rahel & Olden, 2008).

The Ganges River basin is one of the largest and most densely populated river basins, with a population density of 550 individuals/km2, and it covers an area of around 1.7 million km2 and comprises portions of China (3.1%), India (79.1%), Nepal (13.5%), and Bangladesh (4.3%) (Sanghi & Kaushal, 2014). The basin plays a vital role in the region's economic activities, serving as the primary source for more than 90% of wheat and 60% of rice production in India. However, a substantial decline in rice and wheat yields is expected from 2011 to 2040 due to the scarcity of water resources, impeding its ability to keep up with increasing needs (Dulal, 2014). Several studies have forecasted that the air temperature in the Ganga River basin will rise by 2–5 by the end of the twenty-first century (Jeuland et al., 2016; Nepal and Shrestha, 2015). Additionally, the temperature is anticipated to rise more during winter than in summer. The warming in the Ganga basin will result in a 10% impact in ice melting from upland areas. It will contribute to water flow during seasons of low precipitation in the Ganga basin. According to Palmer et al. (2008), the Ganges River's discharge is projected to increase by 17% by the 2050s. However, future water flows are expected to decrease by 10–24%. From 2046 to 2065, the Ganges basin is expected to experience a decline of 18% in the average amount of water flowing into it (Nepal & Shrestha, 2015). This loss in water supply is primarily due to a reduction in melt runoff, but it is partially offset by increased rainfall upstream.

Xing (2013) highlighted major land-use changes such as dam construction, water diversion, extensive rubber plantations, and the development of transnational transportation infrastructure that could impact the lower Mekong River basin (and its fisheries) by 2050. The lower Mekong River basin is one of the most climate-vulnerable regions in the world (IRG, 2010), mainly due to its significance as a rice-growing region and the likelihood of extreme weather events like drought and flooding. Climate change will make the area more vulnerable to poverty and food insecurity. According to the *MRC* Council assessment, fish output in the lower Mekong basin will drop by at least 15% in dry conditions, and modelling shows the impact of tributary dams will further lower the biomass of migrating fishes by 19% (Barlow et al., 2008; Ziv et al., 2012).

According to the combined stress projection, the Congo Basin appears to experience comparatively less risks than the rest of the world (Barange et al. (2018). Even while overall flooding of wetlands enhanced the potential output of fish (More breeding), there may still be problems with greater precipitation because higher river flows may paradoxically limit the catchability of fish. By the end of the twenty-first century, the human population in the region is expected to have doubled, which implies that rising demands for fish as food and increased habitat degradation from deforestation for agriculture would be the main threats to fisheries Barange et al. (2018). Major producers of inland fisheries, i.e., China and India, are predicted to face more future stresses due to climate change, which will affect their inland fisheries. However, many countries responsible for producing about 60% of the world's inland fisheries are predicted to face medium- or relatively low-level future stress and will not be subject to the most severe effects of climate change (Barange et al., 2018). Even though these countries are currently under low-stress conditions, anthropogenic drivers such as the introduction of non-native fish and other taxa, excessive fishing, excessive water extraction, and the alteration, deterioration, and destruction of important habitats will impact the future of inland fisheries.

6.2.3 Aquaculture

Aquaculture has emerged as the only practical means of supplying aquatic food to meet the global demand for nutrition and livelihood following a period of stagnation or decrease in capture fish production. Out of the predicted 178 million tons of world fish production, aquaculture's share has increased steadily from 0.7 MMT to 87.5 MMT (49%) (FAO, 2022). Additionally, it is anticipated that by 2030, aquaculture production will account for 53% of all fish produced worldwide, up from its current 49% (FAO, 2020). Climate change's effects on aquaculture have garnered much attention recently because it plays a major role in global food security, nutrition, and livelihoods (Blanchard et al., 2017; Dabbadie et al., 2019). In recent decades, aquaculture enterprises have experienced the most significant growth rate in the food production industry and still possess substantial potential for additional expansion. However, the expansion of aquaculture has generated worries regarding possible environmental problems, such as eutrophication, water contamination, decline in biodiversity, land use alterations, and increased greenhouse gas emissions (climate change). Most recent research on aquaculture indicates that certain effects of climate change, such as rising temperatures, altered precipitation patterns, and more frequent extreme events, have already been observed in water resources. However, there are still other consequences that are currently being experienced (Blanchard et al., 2017; Troell et al., 2014; Zolnikov, 2019; Maulu et al., 2021). The entire aquaculture value chain is projected to be vulnerable to the effects of climate change, notwithstanding significant areas of uncertainty (Cochrane et al., 2009; Barange et al., 2018; Dabbadie et al., 2019). It took more than 15 years for researchers to focus seriously on climate change and aquaculture after it was first mentioned in publications around the end of the 1980s. The need for more clarity regarding the links between aquaculture and climate change was a major obstacle to its previous prioritizing. Climate change research has reviewed various topics, including global and marine/fishery models, the effects on fish and shellfish biology, disease, changes in environmental factors like temperature, acidity, and salinity, and the assessment of livelihoods, resilience, and sustainability.

Climate change has both direct and indirect impact on aquaculture and that could be for a short or long period of time. Climate change's direct or short-term impacts involve altering finfish and shellfish's physical and physiological characteristics in production systems due to extreme weather conditions, disease circumstances, toxic algae and parasite invasions. On the other hand, indirect or long-term impacts can manifest by altering the primary and secondary productivity and structure of ecosystems (eutrophication), as well as affecting the availability and cost of input supplies (seed, feed, fishmeal, fish oil), product prices, and other goods and services required by aquaculture producers (Dabbadie et al., 2019; Maulu et al., 2021). Environmental shifts may have an impact on aquaculture system productivity and raise the risk of disease outbreaks in farmed fish. Fish metabolism and physiology are altered by relatively slight temperature changes, which have an impact on growth, fertility, eating behavior, distribution, migration, and abundance. Warmer temperatures generally have the following effects on parasites: faster growth and maturation, an earlier start to spring maturation, higher parasite mortality, more generations annually, higher rates of parasitism and disease, earlier and prolonged transmission, and the potential for continuous, year-round transmission (Marcogliese, 2008). Aquaculture activities like input supplies (power/electricity), feed production, and transportation are recognized mainly as the primary means by which aquaculture contributes to greenhouse gas emissions and the overall contribution of this sector is relatively small in comparison to other food production sectors (Cochrane et al., 2009; Adhikari et al., 2018; Barange et al., 2018).

On a global scale, the fishing and aquaculture sectors released 172.3 million metric tons (mmt) and 385 mmt of carbon dioxide (CO_2) in 2012 and 2010, respectively. Compared to other forms of food production, these emissions were relatively low, at 0.5% of the total CO_2 emissions worldwide during the same year (Hall et al., 2011; IMO, 2015; Barange et al., 2018). Similarly, MacLeod et al. (2020) estimated 263Mt CO_2 emissions from total aquaculture production in 2017, representing around 0.5% of the total CO_2 emissions in the same year. They stated that the regional distribution of emissions approximately corresponds to productivity. East Asia and South Asia represent the majority of production, contributing to 88% of the total, and are responsible for nearly 90% of the overall emissions. For instance, cyprinids contribute to 31% of both emissions and production. Nevertheless, certain cases deviate from the general trend. Shrimp, for instance, contribute to 21% of emissions, accounting for only 10% of output. On the other hand, bivalves are responsible for 7% of emissions while representing 21% of production.

Handisyde et al. (2017) predicted the vulnerability of aquaculture (freshwater, brackish water, and mariculture) to global climate change across several levels. According to their analysis, Asia is expected to have the most vulnerability in freshwater aquaculture, with Viet Nam being the most vulnerable country, followed by Bangladesh, Myanmar, and China. Similarly, the countries Ecuador, Vietnam, Belize, Egypt, and Taiwan have shown the most vulnerable in brackish water aquaculture. Likewise, Norway, Chile, China, and Vietnam have showed the highest vulnerability in mariculture aquaculture. Figure 6.3 depicts countries vulnerable to climate change regarding freshwater, brackish, and marine aquaculture.

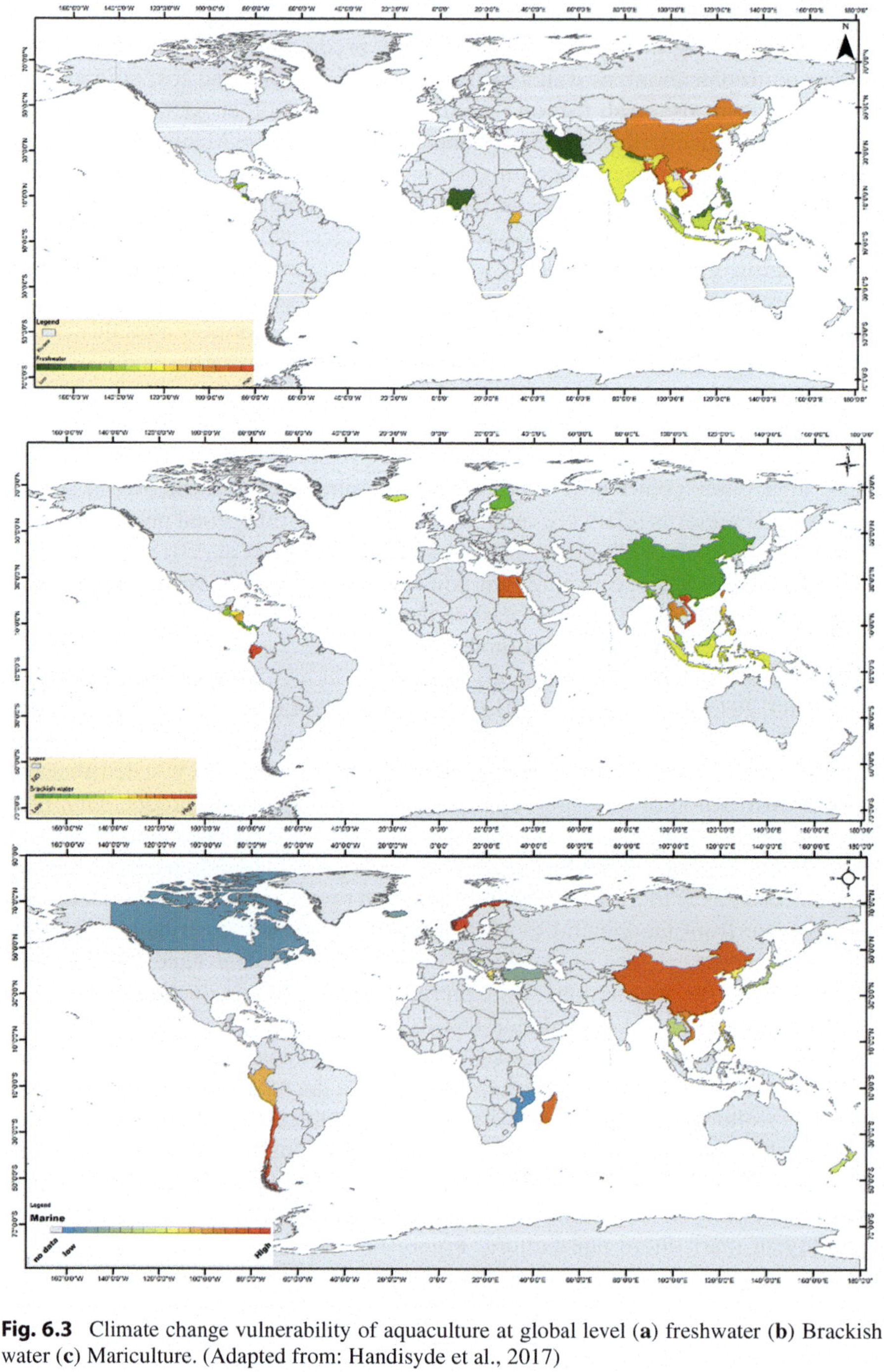

Fig. 6.3 Climate change vulnerability of aquaculture at global level (**a**) freshwater (**b**) Brackish water (**c**) Mariculture. (Adapted from: Handisyde et al., 2017)

6.3 Conclusion

The potential effects of climate change on the productivity of aquaculture and fisheries, as well as the implications for the industry's long-term survival, have been highlighted in this chapter. Globally, fisheries and aquaculture have made substantial contributions to both the assurance of adequate food supply and the sustenance of the livelihoods of millions of people. By the year 2032, the production of aquatic animals is expected to increase by 10%, with 90% of the additional output intended for human use. The significant increase in aquaculture production has mostly contributed to the average yearly growth in global food fish consumption, which stood at 3.2% from 1961 to 2016. The growth rate of this phenomenon is twice as high as the rate of human population increase, which was at 1.6% throughout the same era. The average fish consumption increased from 9.0 kg in 1961 to 20.2 kg in 2015.

The worldwide occurrence of climate change and its impacts on the fishing industry has been widely noted in many places worldwide. The sector is adversely affected by climate change, either directly or indirectly. Most small-scale fishers are expected to experience a more significant influence from climate change in the future. Climate phenomena, such as rising sea levels, cyclones, droughts, floods, and increased sea surface temperatures (SST), adversely affect various natural ecosystems, including wetlands, estuary, coral reefs, seagrass beds, mangroves, and riverine ecosystem. These variables directly influence fish's physiology, biology, and breeding behaviors. Developing countries located in the southern and southeaster regions of the Mediterranean and Black Sea are deemed to have greater susceptibility to climate change and possess limited ability to adapt, making them more sensitive to its impacts. All of these instances emphasize the significance of adaptive capacity, whether it is limited or not, as a primary factor contributing to vulnerability. An additional significant finding, derived from several chapters and illustrated by the tuna examples, is that anticipated shifts in species distribution are likely to generate new conflicts or intensify existing ones among users. These conflicts may arise within countries or when the distribution of key species changes across borders between neighboring countries or between a country's Exclusive Economic Zone (EEZ) and international waters. When fish resources are shared across international boundaries, any changes in their distribution can cause disagreements over how they should be divided. This was the case when Atlantic mackerel started moving northwards and westwards, resulting in a decrease in their abundance in Norwegian waters and an increase in the waters of Iceland and the Faroe Islands. Despite being viewed as the sole solution to meet the increasing worldwide demand for aquatic products, the aquaculture industry is increasingly vulnerable to the tangible and impending hazards posed by the impacts of human-caused climate change. Furthermore, climate change presents a global menace to food production, the particular hazards to fisheries and aquaculture are expected to differ based on factors such as geographic location, climate, national economy, water conditions, production methods, production scale, and the species being cultivated by aquaculture producers. To ensure the productivity of aquaculture and mitigate the effects of climate change, aquaculture producers need to quickly adapt to existing alternatives and concurrently develop long-term changes to their production processes to reduce the repercussions.

However, it has noted numerous limitations that need to be considered in future research. Specifically, the evaluation focuses just on the production and input supply stages of the aquaculture value chain. It does not cover other significant stages, like trade, processing, and consumption of the products. Additionally, the majority of research focused only on the environmental components of sustainability, disregarding the social and economic dimensions. With the ongoing expansion of the aquaculture industry and the increasing clarity of climate change, it is imperative to adopt a comprehensive approach in predicting the effects of climate change on aquaculture and effectively mitigating these impacts. As a result, the implementation of mitigation and adaptation techniques would yield greater effectiveness. Nevertheless, this will necessitate advancements in research, particularly in locations that are more susceptible due to their limited ability to adapt. Hence, underprivileged economies stand to gain more advantages from global collaboration.

References

Adhikari, S., Keshav, C. A., Barlaya, G., Rathod, R., Mandal, R. N., Ikmail, S., Saha, G. S., De, H. K., Sivaraman, I., Mahapatra, A. S., & Sarkar, S. (2018). Adaptation and mitigation strategies of climate change impact in freshwater aquaculture in some states of India. *Journal of Fisheries Sciences.com, 12*(1), 16–21.

Adopted, I .P. C. C., (2014). *Climate change 2014 synthesis report*. IPCC: Geneva, Szwitzerland, pp.1059–1072

AR (2023) Department of Fisheries, Ministry of Fisheries, Animal Husbandry and Dairying Government of India. https://dof.gov.in/sites/default/files/2023-04/Final_Annual_Report_2022-23_English.pdf

Barange, M., & Perry, R. I. (2009). Physical and ecological impacts of climate change relevant to marine and inland capture fisheries and aquaculture. In *Climate change implications for fisheries and aquaculture: Overview of current scientific knowledge* (pp. 7–106). FAO Fisheries and Aquaculture Technical Paper, 530.

Barange, M., Bahri, T., Beveridge, M. C. M., Cochrane, K. L., Funge-Smith, S., & Poulain, F. (Eds.). (2018). *Impacts of climate change on fisheries and aquaculture: Synthesis of current knowledge, adaptation and mitigation options*. FAO Fisheries and Aquaculture Technical Paper No. 627 (p. 628). FAO.

Barlow, C., Baran, E., Halls, A. S., & Kshatriya, M. (2008). How much of the Mekong fish catch is at risk from mainstream dam development? *Catch and Culture, 14*(3), 16–21.

Beach, R. H., & Viator, C. L. (2008). The economics of aquaculture insurance: An overview of the US pilot insurance program for cultivated clams. *Aquaculture Economics & Management, 12*(1), 25–38.

Belhabib, D., Dridi, R., Padilla, A., Ang, M., & Le Billon, P. (2018). Impacts of anthropogenic and natural "extreme events" on global fisheries. *Fish and Fisheries, 19*(6), 1092–1109.

Bijma, J., Pörtner, H. O., Yesson, C., & Rogers, A. D. (2013). Climate change and the oceans–What does the future hold? *Marine Pollution Bulletin, 74*(2), 495–505.

Blanchard, J. L., Watson, R. A., Fulton, E. A., Cottrell, R. S., Nash, K. L., Bryndum-Buchholz, A., Büchner, M., Carozza, D. A., Cheung, W. W., Elliott, J., & Davidson, L. N. (2017). Linked sustainability challenges and trade-offs among fisheries, aquaculture and agriculture. *Nature Ecology & Evolution, 1*(9), 1240–1249.

Brander, K., Cochrane, K., Barange, M., & Soto, D. (2017). Climate change implications for fisheries and aquaculture. In *Climate change impacts on fisheries and aquaculture: A global analysis* (Vol. 1, pp. 45–62).

Chen, Y., Zhang, S., Huang, D., Li, B. L., Liu, J., Liu, W., Ma, J., Wang, F., Wang, Y., Wu, S., & Wu, Y. (2017). The development of China's Yangtze River Economic Belt: How to make it in a green way. *Science Bulletin, 62*(9), 648–651.

Cheung, W. W., Bruggeman, J., & Butenschön, M. (2019). Projected changes in global and national potential marine fisheries catch under climate change scenarios in the twenty-first century. In M. Barange, T. Bahri, M. Beveridge, K. Cochrane, S. FungeSmith, & F. Poulain (Eds.), *Impacts of climate change on fisheries and aquaculture* (pp. 63–85). FAO.

Cochrane, K., De Young, C., Soto, D., & Bahri, T. (2009). *Climate change implications for fisheries and aquaculture* (p. 212). FAO Fisheries and aquaculture technical paper, 530.

Dabbadie, L., Aguilar-Manjarrez, J., Beveridge, M. C., Bueno, P. B., Ross, L. G., & Soto, D. (2019). Effects of climate change on aquaculture: Drivers, impacts and policies. In *Impacts of climate change on fisheries and aquaculture* (p. 449).

Das, B. K., Sarkar, U. K., & Roy, K. (2019). Global climate change and inland open water fisheries in India: Impact and adaptations. In *Climate change and agriculture in India: Impact and adaptation* (pp. 79–95).

Davidson, A. M., Jennions, M., & Nicotra, A. B. (2011). Do invasive species show higher phenotypic plasticity than native species and, if so, is it adaptive? A meta-analysis. *Ecology Letters, 14*(4), 419–431.

De Silva, S. S., & Soto, D. (2009). Climate change and aquaculture: Potential impacts, adaptation and mitigation. In *Climate change implications for fisheries and aquaculture: Overview of current scientific knowledge* (pp. 151–212). FAO Fisheries and Aquaculture Technical Paper, 530.

Ding, Q., Chen, X., Hilborn, R., & Chen, Y. (2017). Vulnerability to impacts of climate change on marine fisheries and food security. *Marine Policy, 83*, 55–61.

Döll, P., & Zhang, J. (2010). Impact of climate change on freshwater ecosystems: A global-scale analysis of ecologically relevant river flow alterations. *Hydrology and Earth System Sciences, 14*(5), 783–799.

Doney, S. C., Busch, D. S., Cooley, S. R., & Kroeker, K. J. (2020). The impacts of ocean acidification on marine ecosystems and reliant human communities. *Annual Review of Environment and Resources, 45*, 83–112.

Dulal, H. B. (2014). Governing climate change adaptation in the Ganges basin: assessing needs and capacities. *International Journal of Sustainable Development & World Ecology, 21*(1), 1–14.

Eurich, J. G., Friedman, W. R., Kleisner, K. M., Zhao, L. Z., Free, C. M., Fletcher, M., Mason, J. G., Tokunaga, K., Aguion, A., Dell'Apa, A., & Dickey-Collas, M. (2024). Diverse pathways for climate resilience in marine fishery systems. *Fish and Fisheries, 25*(1), 38–59.

FAO. (2020). *The State of World Fisheries and Aquaculture 2022. Sustainability in Action*. FAO.

FAO. (2022). *The State of World Fisheries and Aquaculture 2022. Towards Blue Transformation*. FAO. https://doi.org/10.4060/cc0461en

FAO. (2024). *The State of World Fisheries and Aquaculture 2024*. Blue Transformation in action. https://doi.org/10.4060/cd0683en

Free, C. M., Thorson, J. T., Pinsky, M. L., Oken, K. L., Wiedenmann, J., & Jensen, O. P. (2019). Impacts of historical warming on marine fisheries production. *Science, 363*(6430), 979–983.

Giron-Nava, A., Lam, V. W., Aburto-Oropeza, O., Cheung, W. W., Halpern, B. S., Sumaila, U. R., & Cisneros-Montemayor, A. M. (2021). Sustainable fisheries are essential but not enough to ensure well-being for the world's fishers. *Fish and Fisheries, 22*(4), 812–821.

Hall, S. J., Delaporte, A., Phillips, M. J., Beveridge, M. C. M., & O'Keefe, M. (2011). *Blue frontiers: Managing the environmental costs of aquaculture*. WorldFish Centre. 93 pp. Also available at http://pubs.iclarm.net/resource_centre/WF_2818.pdf

Handisyde, N., Telfer, T. C., & Ross, L. G. (2017). Vulnerability of aquaculture-related livelihoods to changing climate at the global scale. *Fish and Fisheries, 18*(3), 466–488.

HBFS. (2022). *Handbook on fisheries statistics 2018*. Department of Fisheries, Ministry of Fisheries, Animal Husbandry & Dairying, Government of India, Krishi Bhawan, New Delhi. 198 pp.

Heck, N., Beck, M. W., Reguero, B., Pfliegner, K., Ricker, M., & Prütz, R. (2023). Global climate change risk to fisheries–A multi-risk assessment. *Marine Policy, 148*, 105404.

Hershkovitz, Y., Dahm, V., Lorenz, A. W., & Hering, D. (2015). A multi-trait approach for the identification and protection of European freshwater species that are potentially vulnerable to the impacts of climate change. *Ecological Indicators, 50*, 150–160.

IMO (International Maritime Organization). (2015). *The third IMO greenhouse gas study 2014*. International Maritime Organisation. 295 pp.

IFAD. (2022). International Fund for Agricultural Development—IFAD. In *The Europa directory of international organizations* (pp. 340–343). Routledge

IPCC. (2023). Summary for policymakers. In Core Writing Team, H. Lee, & J. Romero (Eds.), *Climate Change 2023: Synthesis Report. Contribution of Working Groups I, II and III to the Sixth Assessment Report of the Intergovernmental Panel on Climate Change* (pp. 1–34). IPCC. https://doi.org/10.59327/IPCC/AR6-9789291691647.001

IRG (International Resources Group). (2010). *USAID Asia-Pacific regional climate change adaptation assessment*. Final report: Findings and recommendations (p. 135). International Resources Group.

Jeuland, M. A. (2016). Climate change and the Ganges Basin. In *The Ganges River Basin* (pp. 86–104). Routledge

Kanno, Y., Pregler, K. C., Hitt, N. P., Letcher, B. H., Hocking, D. J., & Wofford, J. E. (2016). Seasonal temperature and precipitation regulate brook trout young-of-the-year abundance and population dynamics. *Freshwater Biology, 61*(1), 88–99.

Kay, S., Avillanosa, A. L., Cheung, V. V., Dao, H. N., Gonzales, B. J., Palla, H. P., Praptiwi, R. A., Queiros, A. M., Sailley, S. F., Sumeldan, J. D., & Syazwan, W. M. (2023). Projected effects of climate change on marine ecosystems in Southeast Asian seas. *Frontiers in Marine Science, 10*, 495.

Klein, S. G., Roch, C., & Duarte, C. M. (2024). Systematic review of the uncertainty of coral reef futures under climate change. *Nature Communications, 15*(1), 2224.

Klinger, D. H., Levin, S. A., & Watson, J. R. (2017). The growth of finfish in global open-ocean aquaculture under climate change. *Proceedings of the Royal Society B: Biological Sciences, 284*(1864), 20170834.

Kochewad, S. A., Saroj, P. L., Sejjian, V., Sarma, K., Chavan, S. B., Pradhan, A., & Pathak, H. (2022). *Climate-smart integrated farming system. Technical Bulletin* (p. 65). ICAR-National Institute of Abiotic Stress Management.

Lynch, A. J., Myers, B. J., Chu, C., Eby, L. A., Falke, J. A., Kovach, R. P., Krabbenhoft, T. J., Kwak, T. J., Lyons, J., Paukert, C. P., & Whitney, J .E. (2016). Climate change effects on North American inland fish populations and assemblages. *Fisheries, 41*(7), 346–361.

MacLeod, M. J., Hasan, M. R., Robb, D. H., & Mamun-Ur-Rashid, M. (2020). Quantifying greenhouse gas emissions from global aquaculture. *Scientific Reports, 10*(1), 11679.

Marcogliese, D. J. (2008). The impact of climate change on the parasites and infectious diseases of aquatic animals. *Revue Scientifique et Technique, 27*(2), 467–484.

Maulu, S., Hasimuna, O. J., Haambiya, L. H., Monde, C., Musuka, C. G., Makorwa, T. H., Munganga, B. P., Phiri, K. J., & Nsekanabo, J. D. (2021). Climate change effects on aquaculture production: Sustainability implications, mitigation, and adaptations. *Frontiers in Sustainable Food Systems, 5*, 609097.

McIntyre, P. B., Reidy Liermann, C. A., & Revenga, C. (2016). Linking freshwater fishery management to global food security and biodiversity conservation. *Proceedings of the National Academy of Sciences, 113*(45), 12880–12885.

Miller, K. M., Teffer, A., Tucker, S., Li, S., Schulze, A. D., Trudel, M., Juanes, F., Tabata, A., Kaukinen, K. H., Ginther, N. G., & Ming, T. J. (2014). Infectious disease, shifting climates, and opportunistic predators: Cumulative factors potentially impacting wild salmon declines. *Evolutionary Applications, 7*(7), 812–855.

Muhala, V., Chicombo, T. F., Macate, I. E., Guimarães-Costa, A., Gundana, H., Malichocho, C., Hasimuna, O. J., Remédio, A., Maulu, S., Cuamba, L., & Bessa-Silva, A. R. (2021). Climate change in fisheries and aquaculture: Analysis of the impact caused by Idai and Kenneth cyclones in Mozambique. *Frontiers in Sustainable Food Systems, 5*, 714187.

Myers, S. S., Smith, M. R., Guth, S., Golden, C. D., Vaitla, B., Mueller, N. D., Dangour, A. D., & Huybers, P. (2017). Climate change and global food systems: Potential impacts on food security and undernutrition. *Annual Review of Public Health, 38*, 259–277.

Nepal, S., & Shrestha, A. B. (2015). Impact of climate change on the hydrological regime of the Indus, Ganges and Brahmaputra river basins: a review of the literature. *International Journal of Water Resources Development, 31*(2), 201–218.

Palmer, M. A., Reidy Liermann, C. A., Nilsson, C., Flörke, M., Alcamo, J., Lake, P. S., & Bond, N. (2008). Climate change and the world's river basins: anticipating management options. *Frontiers in Ecology and the Environment, 6*(2), 81–89.

Prowse, T. D., Furgal, C., Wrona, F. J., & Reist, J. D. (2009). Implications of climate change for northern Canada: Freshwater, marine, and terrestrial ecosystems. *AMBIO: A Journal of the Human Environment, 38*(5), 282–289.

Rahel, F. J., & Olden, J. D. (2008). Assessing the effects of climate change on aquatic invasive species. *Conservation Biology, 22*(3), 521–533.

Reid, P. C., & Hill, C. (2016). Ocean warming: Setting the scene. In *Explaining ocean warming: Causes, scale, effects and consequences* (p. 17).

Ruby, P., & Ahilan, B. (2018). An overview of climate change impact in fisheries and aquaculture. *Climate Change, 4*(13), 87–94.

Salim, S. S., & James, H. E., (2018). *3" International Symposium on Aquaculture and Fisheries Education (ISAFE3)*. Theme: Fisheries Education for Sustainable Blue Economy Integrating climate change and Blue Economy in fisheries research and education in India (pp. 124–138). ICAR-CIFE.

Sanghi, R., & Kaushal, N. (2014). Introduction to our national river Ganga via cmaps. In *Our national River Ganga: Lifeline of millions* (pp. 3–44). Springer.

Schneider, S., Sarukhan, J., Adejuwon, J., Azar, C., Baethgen, W., Hope, C., Moss, R., Leary, N., Richels, R., & Van Ypersele, J. P. (2001). Overview of impacts, adaptation, and vulnerability to climate change. *Climate Change*, 75–103.

Seggel, A., & De Young, C. (2016). *Climate change implications for fisheries and aquaculture: Summary of the findings of the Intergovernmental Panel on Climate Change Fifth Assessment Report* (p. I). FAO Fisheries and Aquaculture Circular, (C1122).

Singh, J., Jaiswar, A. K., Ahirwal, S. K., Gogoi, P., Gurjar, U. R., Samanta, R., Sarma, K., & Shenoy, L. (2023a). Geo-spatial distribution and trends of trawl catch and bycatch off the south Konkan coast, Maharashtra, India. *Regional Studies in Marine Science, 63*, 103022.

Singh, J., Sarma, K., Jaiswar, A. K., Mohite, A. S., Ahirwal, S. K., Samanta, R., & Shenoy, L. (2023b). Comparative footprint studies of single and multiday trawl fishing along Ratnagiri coast, Maharashtra, India. *Indian Journal of Fisheries, 70*(2).

Soheila, K. Y., & Bahram, S. (2010). The effects of climate change on aquaculture. *International Journal of Environment Science and Development, 11*(5), 2010–2264.

Subasinghe, R. P., Ahmad, I., Kassam, L., Krishnan, S., Nyandat, B., Padiyar, A., Phillips, M., Reantaso, M., Miao, W., & Yamamoto, K. (2012). Protecting small-scale farmers: a reality within a globalized economy?

Teh, L. C., & Sumaila, U. R. (2013). Contribution of marine fisheries to worldwide employment. *Fish and Fisheries, 14*(1), 77–88.

Trégarot, E., D'Olivo, J. P., Botelho, A. Z., Cabrito, A., Cardoso, G. O., Casal, G., Cornet, C. C., Cragg, S. M., Degia, A. K., Fredriksen, S., & Furlan, E. (2024). Effects of climate change on marine coastal ecosystems–A review to guide research and management. *Biological Conservation, 289*, 110394.

Troell, M., Naylor, R. L., Metian, M., Beveridge, M., Tyedmers, P. H., Folke, C., Arrow, K. J., Barrett, S., Crépin, A. S., Ehrlich, P. R., & Gren, Å. (2014). Does aquaculture add resilience to the global food system? *Proceedings of the National Academy of Sciences, 111*(37), 13257–13263.

Vivekanandan, E. (2010). *Impact of climate change in the Indian marine fisheries and the potential adaptation options*.

Xing, L. (2013). Land-use change in the Mekong region. In *The water-food-energy Nexus in the Mekong Region: Assessing development strategies considering cross-sectoral and transboundary impacts* (pp. 179–189).

Ziv, G., Baran, E., Nam, S., Rodríguez-Iturbe, I., & Levin, S. A. (2012). Trading-off fish biodiversity, food security, and hydropower in the Mekong River Basin. *Proceedings of the National Academy of Sciences, 109*(15), 5609–5614.

Zolnikov, T. R. ed. (2019). *Global adaptation and resilience to climate change*. Springer International Publishing. ISBN 978-3-030-01213-7. https://doi.org/10.1007/978-3-030-01213-7

Agroforestry and Climate Adaptation: Overcoming Drought and Heat Challenges

7

Abhishek Kumar, Sanjeev Kumar, Rachana Dubey, Anup Das, and Sangram Chavan

Abstract

Heat and drought stress significantly impact crop productivity, leading to potential dramatic yield losses. The gradual global temperature increase, coupled with prolonged durations of heat stress, poses a considerable threat to agricultural sustainability and productivity worldwide. To address these challenges, agroforestry emerges as a promising solution to enhance the resilience of agriculture, particularly in drought or water-deficit situations. In arid and semiarid areas, farmers have traditionally adopted agroforestry to mitigate the impact of climate-related challenges. Agroforestry systems incorporating *Prosopis cineraria* and *Acacia senegal* have traditionally proven effective in enhancing crop yields under heat stress conditions in arid regions of India and Africa. The diversification of conventional mono-crop-based agriculture with woody perennials is a key strategy. This practice not only enhances soil health but also improves soil moisture retention, reduces crops' evaporative demand and soil temperature, provides shade and windbreaks, and buffers extreme temperature. The numerous benefits of integrating trees into agricultural systems contribute significantly during drought years by boosting food grain production, providing fodder for livestock, offering alternative food sources for farmers, and generating timber for sale. Despite the evident advantages of agroforestry in managing heat and draught hazards, its widespread adoption faces constraints. These constraints need to be explored and addressed to polarize agroforestry among farmers. This chapter emphasizes the significance of employing tree-based systems to enhance the resilience of fragile agriculture in the face of these challenges.

A. Kumar (✉) · S. Kumar · R. Dubey · A. Das
Division of Crop Research, ICAR Research Complex for Eastern Region, Patna, Bihar, India

S. Chavan
ICAR-National Institute of Abiotic Stress Management, Baramati, Maharashtra, India

R. Dubey et al. (eds.), *Drought and Heat Stress in Agriculture*,
https://doi.org/10.1007/978-981-96-5735-3_7

Keywords

Heat stress · Drought · Agroforestry, management

7.1 Introduction

Drought and heat stress are the two most devastating natural hazards that significantly affect agricultural productivity and ultimately food security worldwide (Fahad et al., 2017). These environmental stressors have profound implications for agricultural sustainability, which can severely impact crop yields, water availability, soil health, livestock health, ecosystem stability, and rural livelihood. The primary factor limiting plant productivity globally is water scarcity, and drought plays a significant role in the development of eco-physiological adaptations in plants (Flexas et al., 2004). Drought disrupts water availability, leading to soil moisture deficits and hindered plant growth. Concurrently, elevated temperatures exacerbate these effects, accelerating evaporation rates and intensifying water stress. Reduced water availability coupled with heat stress can diminish crop quality, decrease yields, and compromise food security. Additionally, these conditions exacerbate soil degradation, exacerbating erosion and nutrient depletion. The impacts of drought and heat stresses are amplifying and becoming more frequent and intense due to climate change. Many climate projections indicate that climate change poses a threat to humankind through a rise in average temperature, increasing the intensity and frequency of drought, change in the distribution of annual precipitation, etc. (Sterl et al., 2008). According to the Intergovernmental Panel on Climate Change (IPCC), the frequency of droughts has increased in many regions, and extreme heat events have become more frequent and severe.

Restrictions in the amount or frequency of precipitation are usually the cause of drought. According to the IPCC (2001), drought intensity is the decrease in precipitation amount below a specific threshold, such as the negative departure from a multi-annual average, whereas, drought interval is the decrease in precipitation frequency. These two factors work together to determine how the drought would impact the agricultural productivity in the affected area. In India, more than 60% of agricultural area is under rainfed conditions and it is fragile to be affected by many stressors including drought and heat. A recent report by the UN Convention to Combat Desertification titled "Drought in Numbers in 2022" includes India as one of the countries badly affected by drought and enrolled in the UN's Global Drought Vulnerability Index. The vulnerable people rising out of poverty in the affected regions are suffering from high levels of debt and face a wide range of risks. The worst-hit states include Andhra Pradesh, Bihar, Uttar Pradesh, Jharkhand, Gujarat, Karnataka, Maharashtra, Rajasthan, Telangana, and West Bengal which is the home to nearly 40% of the country's population. Addressing the multifaceted impacts of drought and heat stress on agriculture requires comprehensive strategies integrating sustainable land management practices, alternate land use systems and technologies, crop diversification, and soil and moisture conservation techniques to mitigate risks and enhance the resilience of agricultural systems in the face of climatic challenges.

Agroforestry, the integrated management of trees with crops or livestock, has been viewed as one of the potential solutions to enhance the resiliency of agriculture under drought and heat stress. Agroforestry is the practice of cultivating trees with agricultural crops on the same piece of land with the aim to maximize productivity through greater photosynthetic efficiency, obtaining water and nutrients from deeper layers, and improving the microclimate. It is one of the most effective land-use systems that can enhance resilience to drought and heat stress while providing multiple environmental, economic, and social benefits. It enhances farm income, reclaims degraded lands and wastelands, lessens the negative effects of climate factors, promotes industrial growth, helps improve the environment for better health, and supplements the 6Fs (food, fodder, fertilizers, fibers, fruits, and fuelwood) (Chavan et al., 2016). It necessarily involves perennial components, such as forest trees, fruit trees, shrubs, bamboos, and palms, which are more resilient to harsh environmental circumstances due to their hardiness and complexity.

Trees are known for their hardy character, which includes their ability to withstand water shortage, waterlogging, and salt-affected soils. Through their deep roots, trees explore a huge soil volume of water and nutrients and replenish the soil with leaf litter that helps to maintain production during stressful situations. By incorporating woody perennials into conventional mono-cropping agriculture, it is possible to diversify the cropping system while improving soil health, moisture retention, crop evaporative demand, soil temperature, shadow production, and windbreak properties. Because of their multiple benefits, trees will contribute to increased food grain output, cattle feed, alternative food sources for farmers, and timber sales during drought years. In this way, Agroforestry provides a wide range of outputs, including tangible benefits, viz., food, fuelwood, timber, fertilizer, and nonfood products (NTFPs), as well as intangible benefits, viz., soil and water conservation, carbon sequestration and several other ecosystem services that provide climate resilience. Hence, it has been considered a crucial component of climate-smart agriculture to adapt and mitigate the effects of climate change (Newaj et al., 2016).

7.2 Understanding Drought and Heat Stress

Drought and Heat stress are complex phenomena characterized by prolonged periods of low rainfall and elevated temperatures, respectively. Drought can be taken as a more or less short period of abnormally dry weather, where the lack of precipitation causes serious hydrological imbalance (World Meteorological Organization, 1992). It can occur in various forms, including meteorological drought (lack of precipitation), hydrological drought (low water availability in rivers and reservoirs), and agricultural drought (insufficient soil moisture for crops). The possibilities of occurrence of drought in India vary from once in two years in the Western part of Rajasthan to once in 15 years in north-east Indian states (NRAA, 2009). The long-term trend shows that drought is experienced at least once in five years in almost every state of India except in the northeast (Chand & Raju, 2009). Heat stress refers to prolonged exposure to high temperatures often combined with high humidity, leading to physiological strain and adverse health effects in plants, animals, and humans. Heat stress can cause heat-related illnesses, reduced productivity in livestock, decreased crop yields, and even mortality in vulnerable populations.

Drought and heat stress are influenced by a combination of natural variability and human activities. Natural factors such as changes in atmospheric circulation patterns, including shifts in the jet stream and variations in oceanic temperature gradients, can lead to prolonged periods of reduced rainfall and elevated temperatures, creating conditions conducive to drought and heat stress. Human activities exacerbate these natural processes. Deforestation, urbanization, and land use changes alter surface albedo and disrupt local climate patterns, leading to decreased moisture availability and increased temperatures. Additionally, unsustainable water management practices, including over-extraction of groundwater and inefficient irrigation techniques, contribute to water scarcity, amplifying the impacts of drought. Furthermore, anthropogenic climate change, driven primarily by the emission of greenhouse gases (GHGs) from burning fossil fuels and deforestation, exacerbates heat stress by raising global temperatures and intensifying heat waves.

Both drought and heat stress are interconnected with each other, as high temperatures can exacerbate drought conditions by increasing evaporation rates and drying out soils. Together, they present complex challenges for adaptation and resilience efforts, requiring integrated strategies at local, regional, and global levels to mitigate their impacts and build more climate-resilient societies. Therefore, understanding the drought and heat stress-related mortality in plants and finding out innovative and sustainable strategies involving integrated approaches that combine climate adaptation and sustainable land management to mitigate their impacts on agriculture under the changing climatic scenario is of utmost importance.

7.3 Impacts of Drought and Heat Stress in a Changing Climate Scenario

In a changing climate scenario, drought and heat stress exert profound ecological impacts, reshaping ecosystems and endangering biodiversity. The rapid changes in precipitation patterns and temperatures are causing a decline in water sources, leading to habitat loss and species decline. Terrestrial ecosystems are also suffering due to vegetation wilting, forests becoming susceptible to wildfires, and grasslands degrading. Native flora and fauna struggle to adapt, causing species migrations and local extinctions. These disruptions affect pollination, nutrient cycling, and soil fertility, and vulnerable species face increased risks of extinction. Human health is also affected due to drought and heat stress which is noticeable in the form of heat-related illnesses, waterborne diseases, and mental health issues. These conditions are more frequent and intense, causing heatstroke, dehydration, and respiratory problems, especially among vulnerable populations. Dwindling water supplies and compromised sanitation increase the risk of waterborne diseases like diarrheal illnesses. Prolonged drought and heat also contribute to stress, anxiety, and depression, especially in communities reliant on agriculture or facing economic hardship. These consequences amplify social inequalities, trigger migration, and exacerbate socioeconomic disparities. Displacement becomes common, with marginalized populations, including low-income communities and Indigenous groups, facing disproportionate hardships due to limited access to resources and adaptive capacity. Conflicts over dwindling water and food supplies escalate, further exacerbating social tensions and community cohesion.

The agricultural sector faces severe consequences from drought and heat stress, which present formidable challenges to global food security and economic stability. Erratic precipitation patterns and rising temperatures disrupt traditional farming practices, diminishing crop yields and quality. Staple crops such as maize, wheat, and rice are particularly vulnerable, facing stunted growth, reduced photosynthesis, and increased susceptibility to pests and diseases. Reduced soil moisture and increased evaporation rates intensify water stress, compelling farmers to implement costly irrigation measures or face crop failures. Livestock farming suffers as pastures dry up, water sources dwindle, and animals experience heat-related stress, leading to decreased milk production and higher mortality rates. Smallholder farmers, already marginalized, bear the brunt of these impacts, facing diminished incomes and heightened food insecurity. Moreover, the economic ramifications ripple through supply chains, with rising food prices and increased costs for agricultural inputs exacerbating the strain on vulnerable communities.

Conservation efforts are challenged, necessitating adaptive management strategies and protected areas to protect biodiversity and ecosystem resilience. Adaptation measures, including the adoption of drought-resistant crop varieties, improved water management practices, and diversification of livelihoods, are crucial for bolstering resilience and mitigating the agricultural impacts of drought and heat stress. However, concerted global efforts to mitigate climate change are essential to address the root causes of these challenges and ensure the long term sustainability of agricultural systems and food security for future generations in the face of a changing climate.

Plant growth and biomass yield are negatively impacted by drought and heat stresses. Drought-related plant mortality is primarily caused by carbon starvation, hydraulic failure, and biotic agent amplification (McDowell et al., 2008). When a drought lasts longer than the amount of carbon stored in the atmosphere, photosynthesis is suppressed, leading to carbon starvation. Hydraulic failure occurs when drought intensity is high enough to cause xylem cavitation and embolism, leading to the formation of air bubbles and the breakage of the water column. While plants that forgo hydraulic conductivity can carry out photosynthesis across a wider range of water potentials, they may experience hydraulic failure in the event of an extended drought (McDowell et al., 2008). Xylem embolism lowers photosynthetic activity in plants and affects the productivity of tree plantations by decreasing the hydraulic efficiency and slowing down the movement of water and nutrients required for plant growth, ultimately leading to hydraulic failure and thereby to the death of plants

(Brodribb & Herve, 2009). Recent research indicates that the mass plant mortality crisis is mostly caused by hydraulic failure rather than the depletion of nonstructural carbohydrate reserves (Rowland et al., 2015). Plant mortality resulting from both carbon starvation and hydraulic failure can be exacerbated or enhanced by amplification of biotic agents, such as insects and viruses.

7.4 Principles of Agroforestry

Agroforestry is a collective name for land-use systems and technologies where woody perennials are deliberately used on the same land-management units as agricultural crops and/or animals, in some form of spatial arrangement or temporal sequence. In agroforestry systems, there are both ecological and economical interactions between the different components (Lundgren & Raintree, 1983). It serves as a bridge between agriculture and forestry, promoting mixed land-use practices that maximize production while ensuring sustainable management of farmers' land within the system (Nair, 1993). It offers resilience against extreme weather conditions and demonstrates better adaptability within today's changing climate scenario (Van Noordwijk et al., 2021). Being an intentional, intensive, interactive, and integrated system, it can be distinguished from other farming practices, toward creating favorable conditions for biodiversity conservation and soil health improvement apart from increasing land productivity (Newaj et al., 2013).

Agroforestry systems ideally encompass three essential attributes, viz., productivity, sustainability, and adaptability. These systems typically aim to maintain or enhance the production of desired agricultural commodities while improving overall land productivity, which is achieved through increased yields of tree products, improved outputs from associated crops, reduced inputs into cropping systems, and greater labor efficiency (Nair, 1993). Being practiced on 13.75 million ha area in India under diversified types of systems (Chavan et al., 2015), agroforestry practices contribute to conserve the natural resource base by leveraging the beneficial impacts of perennial woody plants on soil quality. While some farming communities have already embraced agroforestry, the successful adoption of improved technologies in new areas should also conform to local farming practices and acceptance by these communities (Sanou et al., 2017).

7.5 Agroforestry and Sustainable Developmental Goals (SDGs)

Agroforestry has been acknowledged by numerous international organizations and conventions, such as the Kyoto Protocol of 2001, the REDD & REDD+ mechanism, the Sustainable Development Goals (SDGs), and the Paris Agreement, for its potential to address climate change adaptation and mitigation. Agroforestry plays a vital role in contributing to several Sustainable Development Goals (SDGs) outlined by the United Nations. Agroforestry plays a crucial role in achieving at least 9 out of

17 sustainable development goals (SDG) (Chavan et al., 2020). Here is how agroforestry intersects with various SDGs:

1. SDG #1: *No Poverty*: Agroforestry provides sustainable livelihoods for rural communities by diversifying income sources through the production of timber, fruits, nuts, and other non-timber forest products.
2. SDG #2: *Zero Hunger*: Agroforestry enhances food security by increasing agricultural productivity, improving soil fertility, and diversifying crop yields, thereby reducing dependency on single crops vulnerable to climate variability.
3. SDG #3: *Good Health and Well-being*: Agroforestry promotes healthier environments by reducing air pollution, enhancing water quality, and providing access to nutritious foods, contributing to improved health outcomes for communities.
4. SDG #5: *Gender Equality*: Agroforestry empowers women by providing opportunities for income generation, improving access to resources, and promoting gender-inclusive decision-making processes within communities.
5. SDG #6: *Clean Water and Sanitation*: Agroforestry helps to conserve water resources by reducing soil erosion, enhancing groundwater recharge, and maintaining water quality through the protection of riparian zones and watersheds.
6. SDG #7: *Affordable and Clean Energy*: Agroforestry can contribute to renewable energy production through the cultivation of biomass for bioenergy, such as wood pellets or biogas from organic waste.
7. SDG #12: *Responsible consumption and production:* Agroforestry is a sustainable method of managing land that combines the wise use of natural resources with production diversification. The production of agricultural and wood-based commodities on a sustainable basis without depleting natural resources and as low as external inputs
8. SDG #13: *Climate Action*: Agroforestry mitigates climate change by sequestering carbon dioxide, reducing greenhouse gas emissions, and enhancing ecosystem resilience to extreme weather events, such as droughts and floods.
9. SDG #15: *Life on Land*: Agroforestry conserves biodiversity by providing habitat for wildlife, preserving the genetic diversity of tree species, and restoring degraded landscapes thus contributing to the conservation and sustainable use of terrestrial ecosystems.

Overall, agroforestry offers a holistic approach to sustainable development, addressing environmental, social, and economic challenges while contributing to the achievement of the SDGs on a global scale.

7.6 Benefits of Agroforestry under Drought and Heat Stress

Agroforestry involves complex interactions among biophysical resources, *viz.*, light, space, water, and nutrients through the integration of trees and crops, which are beneficial under the adverse condition of drought and heat stresses for three major reasons. Firstly, trees have a beneficial effect on soil fertility through greater organic matter production and recycling of nutrients (Young, 1986). Secondly, a combination of annual crops and trees raises biomass production because differences in rooting depth enable uptake of more water and nutrients (Huxley, 1983). Finally, the presence of trees acts as a protective barrier against soil erosion or as windbreaks (Wiersum, 1984). These interactions include both aboveground and below-ground interactions. Aboveground interactions that account for productivity enhancement during drought include amelioration of microclimate, viz., shading, change in temperature, buffering of humidity, reduction in windspeed and evapotranspiration (Huxley, 1983; Monteith et al., 1991), and the temporal sharing and complementary utilization of light and space. Possible below-ground interactions in agroforestry systems which often resulted in sustained productivity under drought and heat stresses, may include judicious utilization of below-ground resources, *viz.*, water and nutrient through spatial sharing and vertical partitioning of uptake between deep rooting trees and shallow rooting crops, and nitrogen transfer from nitrogen-fixing trees. The net effect of these interactions is expressed in terms of such quantifiable responses as higher system productivity, microclimate amelioration, and soil improvement (Rao, 1998).

The multifaceted advantages offered by agroforestry for sustaining production level under drought and heat stress within agricultural systems are given hereunder:

(a) *Productivity Enhancement*: To enhance productivity in the face of drought and heat stress, well-crafted agroforestry models play a pivotal role. These models, characterized by their diverse array of species and crops, not only improve productivity but also adhere to sustainable development principles. Compared to traditional single-focus agricultural or forestry methods, agroforestry systems offer superior economic returns. This economic advantage is further bolstered by ongoing ecological and climatic shifts, ensuring their long-term relevance and effectiveness (Sahoo et al., 2021). Agroforestry is especially significant for smallholder farmers and rural communities as it enhances food security, income, and overall health through a wide range of benefits, both tangible and intangible (Dhyani et al., 2013). By integrating food, fuel, and fiber production, agroforestry enhances overall biomass and yield, thereby supporting nutritional security and sustaining livelihoods (Montagnini & Metzel, 2017).
(b) *Water Conservation*: Tree components can provide better stress tolerance due to extensive root systems that explore more groundwater. They can help regulate water availability by enhancing water infiltration and moisture retention in the soil through their deep root systems. Tree roots also create channels for water to penetrate deeper into the soil, reducing surface runoff and increasing groundwater recharge. Tree canopy helps to reduce water loss through evapotranspiration by providing shade and shelter to crops thus minimizing soil moisture depletion during periods of drought. (Anglaaere et al., 2011).
(c) *Hydraulic Lift and Redistribution*: In agroforestry systems, trees and annual crops have contrasting abilities to access water resources due to their differing root structures. Trees, with their deep and extensive root networks, can tap into water stored in deeper soil layers, which is crucial during dry periods when

surface soils become arid and lack moisture. In contrast, annual crops generally have shallower roots that limit their access to deep soil moisture. However, in agroforestry settings where trees and crops are grown together, a process called hydraulic lift can occur. During times of low transpiration, typically at night, trees draw water from deep soil layers. This water is then transported upward through the tree's vascular system and released near the soil surface through root exudates or leakage. This phenomenon enhances the availability of moisture for nearby crops, thereby contributing to more efficient water use and resilience in the agroecosystem (Burgess et al., 1998).

(d) *Soil Protection:* Agroforestry generates adaptation benefits through its impact on reducing soil and water erosion, and reducing variability in crop output (Verchot et al., 2007). It can increase soil porosity, reduce runoff, and increase soil cover, leading to increased water infiltration and retention in the soil profile, which can reduce moisture stress during low rainfall years (Pramova et al., 2012). Trees in agroforestry systems play a crucial role in mitigating soil erosion and conserving soil and water resources by intercepting rainfall, thereby reducing soil erosion caused by drought-induced water runoff. The extensive root systems of trees stabilize soils by binding soil particles, enhancing soil structure and fertility, improves water infiltration, and retention capacity, which are crucial for mitigating drought impacts.

(e) *Soil Health Improvement*: The organic matter produced by trees, such as leaf litter and root exudates, enriches the soil and improves its water-holding capacity. This organic matter also supports the growth of beneficial soil microorganisms, further enhancing soil structure and fertility. A constant addition of organic matter to the soil through decaying dead roots of trees leads to increases in the carbon status of the soil (Dhyani & Tripathi, 2000). Furthermore, leguminous tree components in agroforestry also help in fixing atmospheric nitrogen and make it available to companion crop.

(f) *Restoration of Degraded Lands*: Agroforestry has also been proven the best scientific method for restoring degraded lands to their natural state and managing resources sustainably (Tripathi et al., 2009). In agroforestry system, cycling of nutrients is an important aspect as a considerable amount of nutrients are returned through litterfall in the form of leaves, twigs, branches, etc., and are made available for reabsorption. Litterfall, litter decomposition, and subsequent nutrient release are primary biological pathways in agroforestry for nutrient transfer from vegetation to soils. These processes are crucial for controlling nutrient cycling and preserving soil fertility in agroecosystems (Triadiati et al., 2011).

(g) *Microclimate Modification*: Trees in agroforestry systems enhance microclimates by creating cooler and more humid environments compared to open areas. This moderates temperature extremes and mitigates heat stress on crops, livestock, and soil, thereby promoting favorable conditions for agricultural productivity during periods of high temperatures. The adjustment of microclimates by trees is valued for providing diverse ecosystem services, notably through shade provision. This shading effect significantly lowers temperatures under tree canopies, offering relief from heat stress, which is particularly advantageous for plants and animals, especially during the hottest parts of the day. Furthermore, trees and shrubs in agroforestry systems serve as natural windbreaks and shelterbelts, reducing wind speed and evaporation rates. This helps to minimize water loss from soil and crops, maintaining soil moisture levels and reducing the risk of drought stress.

(h) *Biodiversity Conservation*: Agroforestry promotes biodiversity by integrating multiple tree, crop, and livestock species within the same landscape. (Montagnini & Metzel, 2017). Agroforestry systems diversify available resources, including water, nutrients, and sunlight, among different layers and components. This diversification reduces competition for resources among plants and enhances the resilience of agricultural systems to drought and heat stress by spreading risks across different species and increasing overall ecosystem stability.

(i) *Carbon Sequestration*: Trees in agroforestry systems play a major role in the regulation of atmospheric accumulation of GHGs, especially CO_2 through carbon sequestration, and thereby helping to mitigate climate change and reduce the intensity of heat stress by lowering greenhouse gas concentrations (IPCC, 2000). Agroforestry systems have been recognized as having the highest potential for carbon sequestration by the year 2040 among all the land use types (as per the Land-use, Land-use Change and Forestry Report of the IPCC, 2000). The carbon sequestration potential of agroforestry systems depends on the plant characteristics (tree species, age, planting density, growth rate, intercrops, etc.), system characteristics (structure and stability), and management factors (tillage, fertilization, residues, and harvesting regime) (Feliciano et al., 2018). Globally, agroforestry systems can have a wide range of carbon stocks ranging from 29 to 228 tons Carbon ha^{-1} with a median value of 95 tons Carbon ha^{-1} (Albrecht & Kandji, 2003).

Through harnessing these benefits, agroforestry practices contribute to sustainable land management and climate adaptation efforts, mitigating the impacts of drought and heat stress while promoting biodiversity and ecosystem resilience. The multifaceted benefits of trees will help during drought years to boost food grain production, fodder for livestock, alternate sources of food for farmers, and timber for sale. But at the same time, drought also hinders tree growth, biomass production, and in extreme cases also causes the mortality of trees. Therefore, understanding the impact of drought and heat stress on tree growth and physiology and, in turn, responses of trees to drought, their physiological mechanisms to cope with drought with anatomical aspects, and land-use management techniques is important to enhance tree performance under drought conditions.

7.7 Impact of Drought and Heat Stress on Tree Growth and Physiology

The negative impact of drought and heat stress on tree growth can be clearly observed through several factors such as reduced seed germination, seedling recruitment, stunted growth and productivity, increased mortality among mature trees, heightened susceptibility to pests and diseases, and greater sensitivity to fire damage. (Reichstein et al., 2013). Sapeta et al. (2013) reported a reduction in growth (elongation of stem, emergence of leaves, and total leaf area) to optimize water loss in response to drought stress in *Jatropha curcas.* Ogaya and Peñuelas (2007) observed a reduction in diameter increment, an increase in mortality, and a decrease in aboveground biomass increment as responses to drought in *Arbutus unedo, Phillyrea latifolia,* and *Quercus ilex.*

The adverse effects of drought and heat stress on tree physiology are well documented by several researchers. Abobatta (2019) reported a number of significant adverse effects of the drought on trees, including lower photosynthetic active radiation absorption, decreased radiation-use efficiency, and a decreased tree harvest index. Hassan et al. (2021) reported decreased chlorophyll levels and relative water content (RWC) in leaves of *Tectona grandis* due to drought. They also observed a 46.4% reduction in photosynthesis, an 89.5% decrease in stomatal conductance, and an 86.4% decline in transpiration rates. Wang et al. (2014) studied drought stress in *Hevea brasiliensis* and found a reduction in leaf relative water content (RWC), along with elevated activities of peroxidase (POD) and superoxide dismutase (SOD), as well as an increase in proline content. The drought also caused damage to cell membranes and led to an accumulation of soluble sugars.

The survival of plants during drought and the effects of drought on carbon dynamics are primarily determined by various plant hydraulic traits. According to McDowell et al. (2008), plants that sacrifice hydraulic conductivity can maintain photosynthesis over a broader range of water potentials, which in turn, may render them susceptible to hydraulic failure during prolonged drought conditions. Xylem embolism decreases photosynthetic activity in plants and impacts the productivity of tree plantations by reducing hydraulic efficiency, which slows the transport of water and nutrients essential for growth (Brodribb & Hervé, 2009). Ultimately, failure to restore water in the xylem conduits can lead to hydraulic failure and plant mortality during drought stress.

7.7.1 Response Mechanism of Trees to Drought and Heat Stress:

Drought can provoke tree responses from the molecular to the forest stand level (Hamanishi & Campbell, 2011). Trees employ a variety of interconnected and coordinated morphological, anatomical, and physiological responses to mitigate water loss during drought. These responses encompass tissue-specific adaptations that operate within the distinct microenvironments of roots, stems, and leaves (Fig. 7.1, Choat et al., 2018).

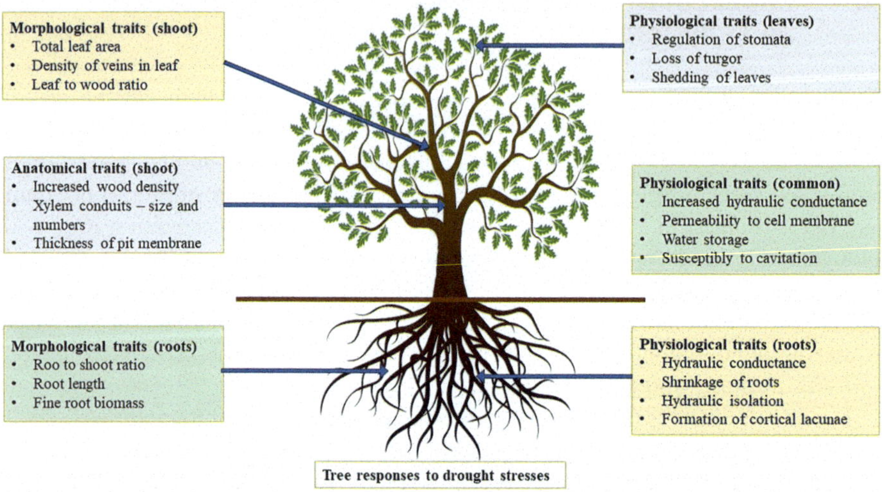

Fig. 7.1 Tree hydraulic traits commonly associated with drought-induced mortality. (Modified from Choat et al., 2018)

Morphological responses to drought are reduction in plant growth attributes, total leaf area, and increased root-to-shoot ratios through the reallocation of more biomass in roots than the shoots. However, the biomass and life span of fine roots are often reduced to minimize transpiration and respiration losses under drought conditions (Arend et al., 2011). The decision to maintain or shed old roots and produce new ones is influenced by water uptake and carbon expenditure.

Anatomical responses are gradual tapering of xylem conduits from roots to the aboveground parts, increasing their cell wall thickness and formation of smaller vessels and ellipsoid tracheids in stems, helping to achieve an optimal structure for the functioning of the vascular elements, and providing resistance against xylem cavitation (embolism). Cell wall lignification and suberization help to minimize water loss and cell dehydration, whereas, rhizodeposits enhance ecosystem resilience under drought by stimulating microbial activity.

Physiological responses to drought include leaf shedding, decreased transpiration, photosynthesis, regulation of water balance, and root biomass partitioning. Tree roots are responsible for water and nutrient uptake and act as sensors for water-deficit conditions and send signals to the shoots. These signals are mediated by abscisic acid (stress hormone), which regulates water loss by promoting closure of stomata enhancement of the biosynthesis of osmolytes (osmoregulation). Water uptake by trees under drought condition is mainly transcellular, controlled by the amount and activity of water channels (Aquaporins) in the cell membranes (Chaumont & Tyerman, 2014).

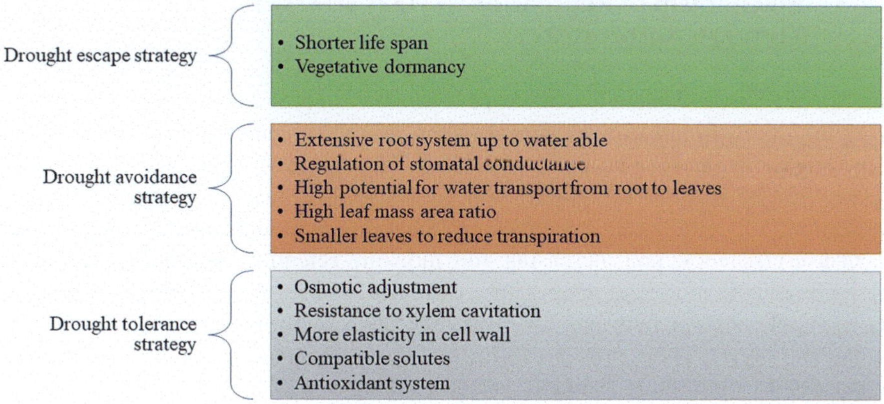

Fig. 7.2 Strategies and mechanisms of drought resistance in trees. (Modified from Vilagrosa et al., 2012)

7.7.2 Drought Adaptation Strategies in Trees

Traditionally, plant resistance to drought has been categorized into escape, avoidance, and tolerance strategies (Fig. 7.2, Vilagrosa et al., 2012). Plants with short lifespans often survive drought by completing their life cycle before its onset. Perennial plants such as trees and shrubs employ either avoidance or tolerance mechanisms to cope with drought stress.

7.7.2.1 Drought Avoidance

Drought avoidance refers to a plant's ability to maintain higher tissue water content by regulating the balance between water loss and uptake rates during water-deficient conditions. This strategy involves minimizing water loss through early stomatal closure, limiting shoot growth, enhancing water uptake via deep taproot systems, and accumulating solutes in root tissues (Levitt, 1980).

7.7.2.2 Drought Tolerance

During severe drought conditions, when avoidance mechanisms are no longer effective, plants protect their tissues from cellular damage through tolerance mechanisms. Drought tolerance refers to the ability of plants to withstand low tissue water content by maintaining cell turgor through osmoregulation with protective solutes and proteins, and by enhancing resistance against hydraulic conductivity in the protoplasm (Morgan, 1984). Physiological traits that facilitate gas exchange, constant water transport, and improved cell survival under low moisture and water potential conditions support this strategy. This includes adjustments in osmotic pressure, reinforcement of structural wood, and reduction of xylem susceptibility to embolism (Hacke et al., 2006).

7.7.3 Mechanism of Regulation of Heat Stresses Through Agroforestry

Agroforestry trees shield adjacent crops from hot winds and heat stresses by modifying the microclimate around tree canopies, thereby enhancing crop yields while providing fodder and shelter for livestock. Under extreme heat stresses, agroforestry trees bolster agricultural climate resilience through three main mechanisms. Firstly, conserving soil and water resources by reducing evapotranspiration, improving water-use efficiency, and increasing soil moisture retention; secondly, regulating microclimates to moderate temperature extremes; and finally, utilizing tree residuals as supplementary nutrients for crops, thereby promoting nutrient recycling in deeper soil layers. These processes illustrate the significant impact of trees on crop production by optimizing available resources (Ong & Kho, 2015).

The impacts of heat stresses are mitigated by agroforestry through one of the following processes:

- Presence of trees in agroforestry provides cooling effects under heat stresses through two main mechanisms: firstly, by shading, which intercepts solar radiation and prevents it from warming the surfaces beneath the canopy; and secondly, by cooling through transpiration, where water vapor released through leaves' stomata converts energy into latent heat flux. These dual processes demonstrate trees' effectiveness in mitigating heat and enhancing environmental comfort (Gao et al., 2020; Rahman et al., 2020).
- The shade provided by trees effectively reduces daytime surface temperatures by limiting the amount of sunlight reaching the ground (Scott et al., 1999). In contrast, while grass has a noticeable impact on surface temperatures, its influence on local and global air temperatures is relatively limited, potentially diminishing its role in enhancing human comfort. Tree shade, on the other hand, plays a crucial role in delivering localized cooling effects, significantly enhancing human comfort in specific regions (Armson, 2012).
- Trees with higher evapotranspiration rates play a crucial role in moderating local microclimates by releasing water vapor through their leaves, which cools the surrounding air (Pataki et al., 2011). This process enhances heat dissipation from the Earth's surface to the atmosphere, contributing to a cooler environment in their immediate vicinity. Species with dense foliage and a high rate of transpiration are particularly effective in this regard, as they continuously release moisture into the air. This produces a cooling effect which helps in mitigating the intensity of heat during hot periods, thereby improving environmental comfort, and potentially benefiting nearby vegetation and surroundings.
- The size of tree stands significantly influences air temperatures and relative humidity, both within individual stands and across larger stands. Smaller stands typically experience higher temperatures, lower humidity levels, and greater variations in temperature and humidity compared to larger stands, highlighting the importance of stand size in moderating microclimatic conditions (Jiao et al., 2017).

- The architecture of trees, along with leaf shape, canopy shape, canopy density, and evapotranspiration patterns, all influence temperatures beneath their canopies. The tree canopy serves as a barrier to incoming solar radiation, reducing the amount of energy absorbed by the surface during the day and thereby lowering surface temperatures (Rahman et al., 2019).
- Windbreaks in arid regions can cool the surface up to ten times the height of the trees, illustrating their effective cooling capacity. In temperate areas, windbreaks create distinct temperature patterns; temperatures on the leeward side are higher during the day and lower at night compared to adjacent open fields (Tamang et al., 2010).
- During the day, the trees with dense canopies effectively reduce near-surface air temperatures, providing significant cooling benefits and acting as efficient temperature regulators that mitigate heat stress. However, at night, these same trees with their thick canopies impede radiative cooling, leading to heat retention beneath their leafy cover (Wujeska-Klause & Pfautsch, 2020).

7.7.4 Strategies for Sustainable Agroforestry Productivity Under Drought and Heat Stress

It is critical for researchers and farmers to comprehend how crops and trees react to drought and heat stress in safeguarding food security and addressing climate change challenges. Agricultural crops are significantly more susceptible to drought and heat as compared to trees. The trees exhibit higher efficiency in absorbing water and nutrients from the soil during extreme weather conditions, rendering them more resilient to drought than agricultural crops (Kumar et al., 2020).

Different moisture conservation techniques efficiently collect rainwater, resulting in consistent in situ soil moisture levels and substantial improvements in plant survival and growth in semiarid regions. Strategies such as staggered contour trenches (Ranjan et al., 2020), double trenches (Kumar et al., 2019), and V-shaped micro-catchments have been shown to enhance the growth and yield of agroforestry tree species (Kumar et al., 2018). The implementation of *Prosopis cineraria*-based agroforestry in water-stressed areas could enhance the growth and yield of understory crops by modifying the local microclimate conditions (Kumar et al., 2018). Additionally, the vegetative measures improve the bio-physicochemical characteristics of the soil and preserve soil and moisture by lowering erosion losses (Uthappa et al., 2015). Therefore, to increase plant productivity in places with limited water resources, these methods should be used in situ.

Enhancing drought resilience through agroforestry needs thoughtful implementation of additional strategies, such as selecting appropriate provenances and breeding for optimal phenology in various crops. For instance, greater rooting depths, increased root hair density, or smaller root structures can all enhance a plant's ability to absorb moisture during periods of water scarcity. Furthermore, it is crucial for agroforesters, silviculturists, agronomists, breeders, ecologists, physiologists, and plant experts to comprehend the principles of hydraulic functioning and devise strategies to improve plant productivity, thereby maximizing the performance of trees and crops in drought-prone environments.

7.8 Case Studies: Relevance of *P. cineraria*-based Agroforestry Systems in Providing Resilience Against Drought and Heat Stress in the Arid Region of India

In India, arid ecosystems cover approximately 18.8% of the world's area, with Rajasthan alone accounting for 61%, followed by Gujarat, Punjab, and Haryana. These regions are characterized by challenging conditions such as scanty rainfall (100–400 mm), extreme temperatures (2–48 °C), high wind velocities (up to 40 km/h), and high evaporation rates (1500–2000 mm/year), compounded by poor soil fertility and low water retention. Despite these adversities, traditional agroforestry systems have evolved to sustain livelihoods and environmental security.

Farmers in these regions traditionally retain scattered shrubs and trees, such as *P. cineraria* (Khejri), *Acacia nilotica, Tecomella undulata, Azadirachta indica, Ziziphus mauritiana,* and *Ailanthus excelsa*, for their numerous benefits. *P. cineraria*, known as the "king of the desert" or "lifeline of the desert," covers more than 40% of the area in Rajasthan. This resilient tree species, growing up to 10 meters in height, exhibits high tolerance to extreme temperatures (up to 50 °C) and low rainfall (less than 100 mm). Khejri trees provide a wide array of ecosystem services, such as food, fodder, shade, shelter, fuelwood, esthetic value, and soil fertility improvement, and almost all parts of the tree are utilized. Additionally, this tree also contributes to fixing nitrogen biologically, improving soil health through leaf litter, extracting water from deep soil layers with its root system, and fostering microflora and phytomass under its shade.

Research has documented that the microclimate created under Khejri trees regulates extreme heat, temperatures, humidity, and soil moisture, leading to improved nutrient content in the soil (e.g., available N, P, K). Studies have shown a significant reduction (16.07%) in temperature under Khejri canopies compared to open conditions during extreme summer months. This microenvironmental enhancement contributes to better seedling establishment and has been linked to increased crop yields (10–15%) in various crops like cowpea, cluster bean, pearl millet, mustard, taramira, wheat, and mungbean. Therefore, *P. cineraria*-based agroforestry systems are pivotal in sustaining productivity and alleviating the challenges posed by drought and heat stress.

7.9 Policy Implications and Recommendations

To harness the full potential of agroforestry in addressing drought and heat stress under changing climate conditions, policymakers and stakeholders should consider the following recommendations:

1. Policy Support: Governments should provide policy support and incentives for agroforestry adoption, including financial incentives, technical assistance, and extension services. Policy measures such as subsidies, tax incentives, and land tenure reforms can facilitate the transition to agroforestry and encourage investment in sustainable land management practices.
2. Research and Innovation: Increased investment in research and innovation is needed to develop and promote climate-resilient agroforestry practices tailored to local environmental conditions and socioeconomic contexts. Research priorities should include the development of drought-tolerant tree species, agroforestry-based adaptation strategies, and climate-smart agricultural technologies.
3. Capacity Building: Capacity building and training programs should be implemented to enhance farmers' knowledge and skills in agroforestry management. Farmer field schools, demonstration plots, and knowledge exchange networks can facilitate peer-to-peer learning and knowledge sharing among farmers, extension agents, and researchers.
4. Market Access: Efforts should be made to improve market access for agroforestry products, including fruits, nuts, timber, and non-timber forest products. Market infrastructure development, value chain integration, and certification schemes can help farmers access premium markets and realize the economic benefits of agroforestry.

7.10 Conclusion

Agroforestry emerges as a multifaceted solution to combat the challenges posed by drought and heat stress in agriculture. By integrating trees with traditional farming practices, agroforestry enhances soil health and water retention while promoting biodiversity and providing supplementary income streams for farmers. These systems exemplify sustainable land management practices that not only mitigate climate risks but also bolster the resilience of agricultural landscapes against fluctuating environmental conditions. As global climate patterns continue to shift, embracing agroforestry offers a viable pathway toward ensuring food security, promoting environmental sustainability, securing the long-term prosperity of farming communities, and safeguarding livelihoods for present and future generations. This approach not only adapts to changing climatic conditions but also contributes positively to the overall health and resilience of ecosystems and rural economies worldwide.

References

Abobatta, W. F. (2019). Influence of climate change on citrus growth and productivity (effect of temperature). *Advances in Agricultural Technology & Plant Sciences, 2*(3), 180036.

Albrecht, A., & Kandji, S. T. (2003). Carbon sequestration in tropical agroforestry systems. *Agriculture, Ecosystems and Environment, 99*, 15–27.

Anglaaere, L. C. N., Cobbina, J., Sinclair, F. L., & McDonald, M. A. (2011). The effect of land use systems on tree diversity: Farmer preference and species composition of cocoa-based agroecosystems in Ghana. *Agroforestry Systems, 81*, 249–265.

Arend, M., Kuster, T., Gunthardt-Goerg, M., & Dobbertin, M. (2011). Provenance-specific growth responses to drought and air warming in three European oak species (*Quercus robur, Q. petraea* and *Q. pubescens*). *Tree Physiology, 31*, 287–297. https://doi.org/10.1093/treephys/tpr004

Armson, D. (2012). *The effect of trees and grass on the thermal and hydrological performance of an urban area*. The University of Manchester.

Brodribb, T. J., & Herve, C. (2009). Hydraulic failure defines the recovery and point of death in water-stressed conifers. *Plant Physiology, 149*(1), 575–584.

Burgess, S. S., Adams, M. A., Turner, N. C., & Ong, C. K. (1998). The redistribution of soil water by tree root systems. *Oecologia, 115*, 306–311.

Chand, R., & Raju, S. S. (2009). Instability in Indian agriculture during different phases of technology and policy. *Indian Journal Of Agricultural Economics, 64*(2), 187–207.

Chaumont, F., & Tyerman, S. D. (2014). Aquaporins: Highly regulated channels controlling plant water relations. *Plant Physiology, 164*(4), 1600–1618.

Chavan, S. B., Keerthika, A., Dhyani, S. K., Handa, A. K., Newaj, R., & Rajarajan, K. (2015). National Agroforestry Policy in India: A low hanging fruit. *Current Science, 108*, 1826–1834.

Chavan, S. B., Uthappa, A. R., Sridhar, K. B., Keerthika, A., Handa, A. K., Ram, N., Naresh, K., Dhiraj, K., & Chaturvedi, O. P. (2016). Trees for life: Creating sustainable livelihood in Bundelkhand region of Central India. *Current Science, 11*(6), 994–1002.

Chavan, S. B., Keerthika, A., Bhat, S. S., Handa, A. K., Rajarajan, K., & Ahmad, S. (2020). Poplar (*Populus deltoides*) in Jammu and Kashmir, India: Facts and fiction. *Current Science, 119*(6), 910–911.

Choat, B., Brodribb, T. J., Brodersen, C. R., et al. (2018). Triggers of tree mortality under drought. *Nature, 558*, 531–539. https://doi.org/10.1038/s41586-018-0240-x

Dhyani, S. K., & Tripathi, R. S. (2000). Biomass and production of fine and coarse roots of trees under agrisilvicultural practices in north-east India. *Agroforestry Systems, 50*(2), 107–121.

Dhyani, S. K., Handa, A. K., & Uma. (2013). Area under agroforestry in India: An assessment for present status and future perspective. *Indian Journal of Agroforestry, 15*(1), 1–11.

Fahad, S., Bajwa, A. A., Nazir, U., Anjum, S. A., et al. (2017). Crop production under drought and heat stress: Plant responses and management options. *Frontiers in Plant Science, 8*, 1147. https://doi.org/10.3389/fpls.2017.01147

Feliciano, D., Ledo, A., Hillier, J., & Nayak, D. R. (2018). Which agroforestry options give the greatest 27 soil and above ground carbon benefits in different world regions? *Agriculture, Ecosystems and Environment, 28*. https://doi.org/10.1016/j.agee2017.11032

Flexas, J., Bota, J., Loreto, F., Cornic, G., & Sharkey, T. D. (2004). Diffusive and metabolic limitations to photosynthesis under drought and salinity in C3 plants. *Plant Biology, 6*, 269–279. https://doi.org/10.1055/s-2004-820867

Gao, K., Santamouris, M., & Feng, J. (2020). On the efficiency of using transpiration cooling to mitigate urban heat. *Climate, 8*(6), 69.

Hacke, U. G., Sperry, J. S., Wheeler, J. K., & Castro, L. (2006). Scaling of angiosperm xylem structure with safety and efficiency. *Tree Physiology., 26*, 689–701.

Hamanishi, E. T., & Campbell, M. M. (2011). Genome-wide responses to drought in forest trees. *Forestry: An International Journal of Forest Research, 84*(3), 273–283. https://doi.org/10.1093/forestry/cpr012

Hassan, M. A. E., Santhoshkumar, A. V., Hrideek, T. K., Jijeesh, C. M., & Joseph, J. (2021). Variability in drought response among the plus tree accessions of *Tectona grandis* (Linn f.) from the provenances of Kerala, South India. *Acta Physiologiae Plantarum., 43*(3), 1–12.

Huxley, P. A. (1983). Some characteristics of trees to be considered in agroforestry. In P. A. Huxley (Ed.), *Plant research and agroforestry* (pp. 3–12). ICRAF.

IPCC. (2000). In R. T. Watson, I. Noble, B. Bolin, N. H. Ravindranath, D. J. Verardo, & D. J. Dokken (Eds.), *Special report on land use, land use change and forestry* (p. 377). Cambridge University Press.

Jiao, M., Zhou, W., Zheng, Z., Wang, J., & Qian, Y. (2017). Patch size of trees affects its cooling effectiveness: A perspective from shading and transpiration processes. *Agricultural and Forest Meteorology, 247*, 293–299.

Kumar, M., Suresh Kumar, A. R., Uthappa, & Kalappurakkal, S. (2018). Influence of *Prosopis cineraria* L. on microclimate conditions under agroforestry system in arid region, India. *Indian Journal of Agroforestry, 20*(1), 35–39.

Kumar, M., Pramanik, M., Ranjan, R., Ramesha, M. N., Kumar, P., & Raj, K. (2019). Effect of in-situ moisture conservation on establishment of tree borne oilseeds (TBOs) plantation in red soils of Bundelkhand region. In P. Mishra, N. K. Sharma, D. Mandal, G. Kumar, G. Kaushal, & A.'d. R. J. Singh (Eds.), *Sustainable land management: Issues, problems and prospects*. Satish Serial Publishing House.

Kumar, R., Bhatnagar, P. R., Kakade, V., & Dobhal, S. (2020). Tree plantation and soil water conservation enhances climate resilience and carbon sequestration of agro ecosystem in semi-arid degraded ravine lands. *Agricultural and Forest Meteorology, 282-283*, 107857.

Levitt, J. (1980). *Responses of plant to environmental stress: Water, radiation, salt and other stresses* (p. 365). Academic Press.

Lundgren, B. O., & Raintree, J. B. (1983). Sustained agroforestry. In B. Nestel (Ed.), *Agricultural Research for Development: Potentials and challenges in Asia* (pp. 37–49). ISNAR, The Hague.

McDowell, N. W., Pockman, W. T., Craig, D. A., David, D. B., Neil, C., Thomas, K., & Jennifer, P. (2008). Mechanisms of plant survival and mortality during drought: Why do some plants survive while others succumb to drought? *New Phytologist, 178*(4), 719–739. https://doi.org/10.1111/j.1469-8137.2008.02436.x

Montagnini, F., & Metzel, R. (2017). The contribution of agroforestry to sustainable development goal 2: End hunger, achieve food security and improved nutrition, and promote sustainable agriculture. In F. Montagnini (Ed.), *Integrating landscapes: Agroforestry for biodiversity conservation and food sovereignty, advances in agroforestry, 12*. Springer International Publishing Agency.

Monteith, J. L., Ong, C. K., & Corlett, J. E. (1991). Microclimatic interactions in agroforestry systems. *Forest Ecology and Management, 45*, 31–44.

Morgan, J. M. (1984). Osmoregulation and drought stress in higher plants. *Annual Review of Plant Physiology, 35*, 299–319.

Nair, P. R. (1993). *An introduction to agroforestry*. Springer Science and Business Media.

Newaj, R., Chavan, S., & Prasad, R. (2013). Agroforestry as a strategy for climate change adaptation and mitigation. *Indian Journal of Agroforestry, 15*(2), 41–48.

Newaj, R., Chavan, S. B., Alam, B., & Dhyani, S. K. (2016). Biomass and carbon storage in trees grown under different agroforestry systems in semi-arid region of Central India. *Indian Forester, 142*(7), 642–648.

NRAA. (2009). *Drought Management Strategies—2009* (p. 70). Draft paper of National Rainfed Area Authority, Ministry of Agriculture, Government of India.

Ogaya, R., & Peñuelas, J. (2007). Tree growth, mortality, and above-ground biomass accumulation in a holm oak forest under a five-year experimental field drought. *Plant Ecology, 189*(2), 291–299.

Ong, C. K., & Kho, R. M. (2015). A framework for quantifying the various effects of tree-crop interactions. In *Tree-crop interactions: Agroforestry in a changing climate* (pp. 1–23). CABI.

Pataki, D., McCarthy, H., & Litvak, E. (2011). Transpiration of urban forests in the Los Angeles metropolitan area. *Ecological, 21*(3), 661–677.

Pramova, E., Locatelli, B., Djoudi, H., & Somorin, O. A. (2012). Forests and trees for social adaptation to climate variability and change. *WIREs Climate Change, 3*(6), 581–596.

Rahman, M. A., Moser, A., Rötzer, T., & Pauleit, S. (2019). Comparing the transpirational and shading effects of two contrasting urban tree species. *Urban Ecosystems, 22*(4), 683–697.

Rahman, M. A., Stratopoulos, L. M. F., Moser-Reischl, A., Zölch, T., Häberle, K. H., Rötzer, T., Pretzsch, H., & Pauleit, S. (2020). Traits of trees for cooling urban heat islands: A meta-analysis. *Building and Environment, 170*, 106606. https://doi.org/10.1016/j.buildenv.2019.106606

Ranjan, R., Pramanik, M., Tiwari, S. P., Kumar, M., & Yadav, R. S. (2020). Rain water harvesting in non-arable land using staggered trenching in semi-arid climate of Bundelkhand. *Journal of Agricultural Physics, 20*(1), 69–74.

Rao, M. R., Nair, P. K. R., & Ong, C. K. (1998). Biophysical interactions in tropical agroforestry systems. *Agroforestry Systems, 38*, 3–50.

Reichstein, M., Bahn, M., & Ciais, P. (2013). Climate extremes and the carbon cycle. *Nature, 500*, 287–295.

Rowland, L., da Costa, A. C. L., Galbraith, D. R., Oliveira, S., Binks, O. J., Oliveira, A. R., & Pullen, A. M. (2015). Death from drought in tropical forests is triggered by hydraulics not carbon starvation. *Nature, 528*(7580), 119–122. https://doi.org/10.1038/nature15539

Sahoo, S., Lenka, B., Raj, D., & Jhariya, M. (2021). *Climate change impacts and mitigation through sustainable agroforestry practices. In advances in Sustainable Development and Management of Environmental and Natural Resources; Economic Outlook and Opinions; Apple* (Vol. 1, pp. 265–308). Academic Press Inc.

Sanou, L., Savadogo, P., Eugene, E. E., & Thiombiano, A. (2017). Drivers of farmers' decisions to adopt agroforestry: Evidence from the Sudanian savanna zone, Burkina Faso. *Renewable Agriculture and Food Systems*, 1–18.

Sapeta, H., Costa, J. M., Lourenco, T., Maroco, J., Van der Linde, P., & Oliveira, M. M. (2013). Drought stress response in *Jatropha curcas*: Growth and physiology. *Environmental and Experimental Botany, 85*, 76–84.

Scott, K. I., Simpson, J. R., & Mc pherson, E.G. (1999). Effects of tree cover on parking lot microclimate and vehicle emissions. *Journal of Arboriculture, 25*(May), 129–142.

Sterl, A., Severijns, C., Dijkstra, H., Hazeleger, W., et al. (2008). When can we expect extremely high surface temperatures? *Geophysical Research Letters, 35*, L14703. https://doi.org/10.1029/2008GL034071

Tamang, B., Andreu, M. G., & Rockwood, D. L. (2010). Microclimate patterns on the leeside of single-row tree windbreaks during different weather conditions in Florida farms: Implications for improved crop production. *Agroforestry Systems, 79*(1), 111–122.

Triadiati, Tjitrosemito, S., Guhardja, E., Sudarsono, Qayim, I., & Leuschner, C. (2011). Litterfall production and leaf-litter decomposition at natural forest and cacao agroforestry in Central Sulawesi, Indonesia. *Asian Journal of Biological Sciences, 4*(3), 221–234.

Tripathi, O. P., Pandey, H. N., & Tripathi, R. S. (2009). Litter production, decomposition and physico-chemical properties of soil in 3 developed agroforestry systems of Meghalaya, Northeast India. *African Journal of Plant Science, 3*(8), 160–167.

Uthappa, A. R., Chavan, S. B., Dhyani, S. K., Handa, A. K., & Ram, N. (2015). Trees for soil health and sustainable agriculture. *Indian Farming, 65*(3), 2–5.

Van Noordwijk, M., Coe, R., Sinclair, F. L., et al. (2021). Climate change adaptation in and through agroforestry: Four decades of research initiated by Peter Huxley. *Mitigation and Adaptation Strategies for Global Change, 26*, 18.

Verchot, L. V., van Noordwijk, M., Kandji, S., Tomich, T. P., Ong, C. K., & Albrecht. (2007). Climate change: Linking adaptation and mitigation through agroforestry. *Mitigation and Adaptation Strategies for Global Change, 12*, 901–918.

Vilagrosa, A. E., Chirino, J. J., Peguero-Pina, T. S., Barigah, H., Cochard, E., & Gil-Pelegrin, E. (2012). Xylem cavitation and embolism in plants living in water-limited ecosystems. In R. Aroca (Ed.), *Plant responses to drought stress* (pp. 63–109). Springer Berlin Heidelberg. https://doi.org/10.1007/978-3-642-32653-0_3

Wang, G., & Dillon, M. E. (2014). Recent geographic convergence in diurnal and annual temperature cycling flattens global thermal profiles. *Nature Climate Change, 4*, 988–992.

Wiersum, K. F. (1984). Surface erosion under various tropical agroforestry systems. In C. L. O'Loughlin & A. J. Pearce (Eds.), *Symposium on effects of Forest land use on Erosion and slope stability* (pp. 231–239). Environment and Policy Institute East-West Center.

Wujeska-Klause, A., & Pfautsch, S. (2020). The best urban trees for daytime cooling leave nights slightly warmer. *Forests, 11*(9), 945.

Young, A. (1986). *The potential of agroforestry for soil conservation and erosion control* (p. 68). ICRAF Working Paper 42.

Part II

Mitigation and Adaptation Options for Building Resilience Towards Drought and Heat Stress

Mitigating the Impact of Drought and Heat Stress on Crop Productivity and Environmental Sustainability

8

Kanu Murmu, Arup Sarkar, Sramana Sen Sarma, Kalyan Jana, and Sidhu Murmu

Abstract

The impacts of elevated temperatures and prolonged periods of water scarcity provide substantial obstacles to the cultivation of plants and the implementation of sustainable agricultural methods on a global scale. The objective of this study is to provide a comprehensive analysis of the various ways in which plants respond and adapt to drought and elevated temperature. This includes examining the effects on different plant components such as roots, shoots, and grain yield. Additionally, we will explore potential management strategies that might help alleviate the detrimental consequences of these stressors. The findings presented in this review are primarily derived from recent scientific research published in reputable scholarly journals. The publication's sections focus on various plant responses, including root growth, transpiration, photosynthesis, water use efficiency (WUE), phenotypic flexibility, accumulation of low molecular mass compounds (such as proline and gibberellins), and the expression of specific genes and proteins to enhance tolerance to abiotic stresses. Plants use a variety of adaptive and avoidance mechanisms to endure unfavorable conditions. Understanding the physiological and biochemical responses of crops to harsh conditions is of paramount importance to devise strategies and methods for enhancing plant tolerance. This chapter also addresses management practices about soil and crop cultivation that can aid in alleviating the adverse effects of heat stress and drought. Conducting experiments that encompass the assessment of plant assimilate partitioning, phenotypic plasticity, and the discovery of stress-tolerant plant

K. Murmu (✉) · A. Sarkar · S. S. Sarma · K. Jana
Department of Agronomy, Bidhan Chandra Krishi Viswavidyalaya,
Mohanpur, Nadia, West Bengal, India

S. Murmu
Agricultural Chemistry and Soil Science, Bidhan Chandra Krishi Viswavidyalaya,
Mohanpur, Nadia, West Bengal, India

R. Dubey et al. (eds.), *Drought and Heat Stress in Agriculture*,
https://doi.org/10.1007/978-981-96-5735-3_8

genotypes is crucial for advancing our understanding of response complexity and the future of plant breeding. Agronomic strategies such as modifying the timing of planting, employing seed priming techniques, utilizing foliar application of diverse macro- and micronutrients, and implementing the use of rhizobacteria have the potential to partially alleviate the negative impact of heat and drought stress. Breeding methodologies such as trait-based selection, inheritance research including marker-based selection, and genetic techniques utilizing transcriptome and metabolome analysis have the potential to facilitate the selection and development of crops with enhanced heat and drought stress adaptability and mitigation capabilities.

Keywords

Drought · Heat · Stress · Sustainability · Seed priming · Foliar application

8.1 Introduction

There is a speculation that the global population will exceed 9.7 billion by the year 2050, with a projected increase in the proportion of humans relying exclusively on agriculture for their sustenance which is expected to rise from approximately 65% to as high as 90% in emerging nations (Castañeda et al., 2016). Consequently, both the economy and the food supply of a nation will heavily rely on the agricultural sector. But the agricultural operations face numerous challenges, including limited access to sufficient irrigation infrastructure, the presence of tiny and fragmented landholdings, the unavailability of high-quality seeds, inadequate automation, and excessive reliance on chemical fertilizers and pesticides. Because of this, crop quality and yield were drastically reduced after harmful abiotic environmental pressures were introduced to the soil (Shrivastava & Kumar, 2015) and hence crop production and yields, as well as food security, are negatively impacted by abiotic conditions including drought and heat (Garai et al., 2020). As the climate continues to shift, the severity of abiotic pressures is expected to rise, which could affect global food security. Increased crop variability is predicted to make plants more vulnerable to heat and drought (Mamrutha et al., 2020). However, crop plants are unable to reach their full genetic potential due to natural abiotic stressors (Tuteja & Sopory, 2008). Insects, viruses, bacteria, fungi, and spiders are all examples of living things that can induce biotic stress. Even after repeated exposures, the plant is unable to acquire an adaptive tolerance to biotic stress. This is why biotic stress plays such a significant role in both pre- and postharvest failures (Singla & Krattinger, 2016). Hotter weather has posed serious threats to both food production and the safety of our diets. Heat stress on crops has a devastating effect on small farmers. Drought and heat stress are notoriously hard to track because of their recurrent, widespread, and unpredictable nature (Kukal & Irmak, 2018) and is associated with decreased soil water levels and consequent causes crop failures, severely affecting food production all over the world; and the socioeconomic drought that relates to failure of supply and demand of various commodities due to drought (Florez-Sarasa et al., 2020). Drought has a lasting impact, and the losses it causes are costly. Reduced annual rainfall in rain-fed agricultural areas, the occurrence of inter-seasonal dry spells, and moisture scarcity all contribute to the inevitable development of drought-like conditions, while the cost of groundwater pumping continues to rise (Waterland et al., 2010). Increasing or decreasing temperature thresholds, even by one degree, will have a significant impact on crop yields. Changes in a plant's physiology, morphology, anatomy, and biochemistry due to heat stress can affect the crop's yield. While stresses are detrimental throughout a plant's life cycle, their impact is more felt during the reproductive phase due to the greater impact on yield. Drought, extreme temperatures, excessive soil salinity, metal toxicity, and oxidative stressors are all examples of abiotic stress. Damage to a plant's growth, metabolism, and production, as well as a shift in genetic behavior that might induce mutations in offspring, can be irreversible when exposed to these conditions (Balla et al., 2011; Dreesen et al., 2012; Joshi et al., 2020). Therefore, it becomes vital to make a paradigm shift toward sustainable agriculture and find solutions to issues related to water scarcity and its impact on food security (Kyei-boahen et al., 2017). It is imperative to devise methods that can enhance the endurance levels of a plant to drought stress, allowing improved growth and production of crops to satisfy the ever-increasing food requirement in conditions of limited water availability.

8.2 Impact of Drought and Heat Stress on Crop Plants

The rapid growth of the world population and the significant shifts in climate patterns pose substantial challenges to global food security. In general, climate is one of the main determinants of agricultural production and climate alteration might cause variability in agricultural production (Murmu et al., 2022). Climate change poses a significant threat to the sustainability of agricultural systems. The decrease in yearly precipitation, coupled with the unpredictable nature of monsoon precipitation, presents a significant risk of recurrent droughts on a global scale. As a result of the negative effects of climate variability (temperature and rainfall) and climate-driven extremes (flood, drought, heat stress, cold waves, and storms) on crop and livestock production, food security has become an issue of paramount importance (Sarkar et al., 2023). The co-occurrence of drought and heat stress is commonly observed in semiarid regions or areas experiencing drought conditions, and has been the subject of substantial scholarly investigation. However, there remains a limited understanding regarding the collective impact of these factors on plant physiology and growth (Nadeem et al., 2019). The rate of photosynthesis is a physiological reaction exhibited by plants when subjected to stressful situations. This response is affected by limitations in stomatal conductance, such as inadequate availability of CO_2, as well as non-stomatal limitations, including reduced electron transfer capacity and ribulose bisphosphate carboxylase/oxygenase (RuBisCO) activity. Nevertheless, the simultaneous occurrence of drought and heat stress might result in the deterioration of Photosystem II, decreased activity of RuBisCO, elevated leaf temperature, reduced stomatal conductance, and decreasing levels of photosynthetic pigments. According to the findings of Raja et al. (2020), the tomato plants exhibited a reduction of 80% in chlorophyll concentrations and a loss of 57% in carotenoid concentrations when subjected to the combined effects of drought and heat stress. The combined occurrence of drought and heat stress has been observed to have an impact on various aspects of plant reproduction, including the growth of pistils and pollen, the functioning of ovules, and the weight of grains. Notably, the levels of free proline have been observed to significantly increase in crops and other plant species. Proline has the ability to accumulate to large concentrations in plant cells as an osmoprotectant, without causing disruption to the cellular structure. The buildup of proline has been found to play a significant role in facilitating osmotic adjustment, detoxification of reactive oxygen species (ROS), and maintenance of membrane integrity in plants under conditions of environmental stress (Zhang et al., 2011). The formation of hydrogen peroxide (H_2O_2) in response to drought stress is influenced by factors such as genotype variation, duration of water deficit, and plant age, all of which have significant implications. Hydrogen peroxide (H_2O_2) serves as a significant regulator of various physiological processes, including senescence, photorespiration, photosynthesis, stomatal movement, cell cycle, growth, and development. Lipid peroxidation is widely regarded as one of the most dependable indicators for distinguishing between genotypes that are tolerant or vulnerable to stress conditions and is observed in both cellular and organelle membranes when the level of reactive oxygen species (ROS) beyond a certain threshold (Bailly et al., 2019). Reactive oxygen species (ROS) have been found to elicit cellular damage by several mechanisms, including the oxidation of proteins, the inactivation of enzymes, alterations in gene expression patterns, and the breakdown of biomembranes and hence playing a crucial role in the adaptation of plants to abiotic stressors (Mahalingam 2015). The stability can be influenced by a decrease in antioxidants or an increase in the production of reactive oxygen species (ROS), leading to oxidative stress. This oxidative stress can cause damage to cellular macromolecules and membranes, as well as an increase in lipid peroxidation. Additionally, these plants demonstrated diminished activities of antioxidant enzymes such as catalase and ascorbate peroxidase, reduced nutrient absorption, shortened root and shoot length, and decreased grain and kernel yield. In instances of drought, plants exhibit premature closure of stomata as a mechanism to mitigate water loss. Conversely, under heat stress, stomatal conductance is heightened to facilitate leaf temperature reduction via transpiration. This increase in temperature was attributed to the prioritization of stomatal closure, which aimed to minimize water loss, over the cooling of leaves through open stomata (Karthikeyan et al., 2012). The susceptibility of plants to drought and pathogen infection may be attributed to their inability to effectively adjust tolerance mechanisms hence exacerbating the detrimental effects of either stressor. In a study conducted by Lastochkina et al. (2020), it was observed that subjecting wheat plants to both drought stress and the fungal pathogen *Fusarium culmorum*, which is known to cause common root rot and seedling blight, led to elevated levels of malondialdehyde, enhanced leaf yellowing, reduced growth of both roots and shoots, and a decline in biomass accumulation. These effects were found to be more pronounced

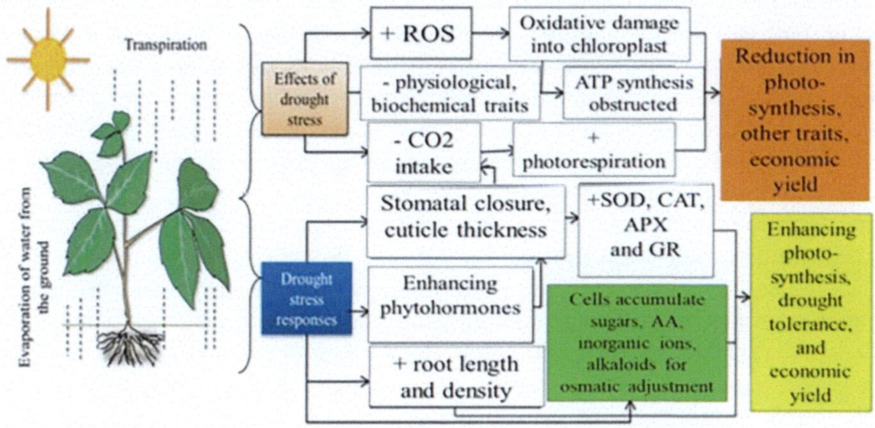

Fig. 8.1 Adverse effects and adaptations of plants to drought stress. (Modified from Ullah et al. and https://doi.org/10.3390/plants10020259). – means reduce; + means increase

when compared to the impact of each stressor individually. Under certain instances, plants experiencing drought-induced stress exhibited resilience against specific diseases that typically thrive under persistently moist or humid climatic circumstances. In the presence of drought stress, *N. benthamiana* plants exhibited a reduced number of disease signs when infected with the fungal pathogen *Sclerotinia sclerotiorum*, which is known to cause white mold. This contrasted with the well-watered plants that were also infected with the pathogen (Ramegowda et al., 2013). Nevertheless, the interplay of stress responses, such as those caused by drought and pathogen infection, is a multifaceted phenomenon (Fig. 8.1).

Consequently, additional research endeavors, such as simulation modeling and transcriptomic investigations, are necessary to generate detailed forecasts regarding the initiation of pathogen infection under drought stress conditions. Moreover, these studies aim to comprehend the repercussions of both drought and pathogen stress across various soil types and pathogens, with the ultimate goal of devising strategies to alleviate their detrimental effects (Jones et al., 2019). The occurrence of such concurrences is far more detrimental than the occurrence of drought alone, potentially resulting in substantial agricultural harm. In contrast, drought alone caused agricultural damage up to 50 billion dollars in the United States during the same time frame (Suzuki et al., 2014). Plants employ many strategies to endure drought conditions, including the creation of reactive oxygen species, the generation of stress hormones such as ethylene and abscisic acid (ABA), and alterations in root and shoot morphology (Chiappero et al., 2019; Bhat et al., 2021). Several immediate reactions can be observed, including reduction in carbon assimilation, closing of stomata, correction of osmotic balance, limitation of growth, alterations in hydraulic processes, transport of signals, and activation of cell-drought signaling mechanisms. Typically, these responses do not inflict significant harm upon plants, given that normal conditions are promptly restored. Extended durations of stress have been found to result in heightened susceptibility to severe and enduring damage, and in some cases, even mortality, among plant organisms. According to Kaur and Asthir (2017), long-term reactions to certain conditions can manifest as inhibited shoot growth, metabolic adaptations, reduced transpiration area, kernel abortion, and senescence. Drought stress is widely recognized as a significant constraint on plant growth and establishment, impacting the elongation and expansion growth of plants as a result of reduced turgor pressure. The process of osmotic control plays a crucial role in maintaining cell turgor and aiding the survival of plants, particularly in pearl millet, even when subjected to extreme drought stress conditions (Shao et al., 2008). In order to mitigate water usage during periods of drought, plants employ a strategy of reducing shoot growth through a decrease in the number of tillers (Soda et al., 2010). In general, short-duration varieties tend to exhibit superior performance under stressful situations when compared to long-duration varieties. The diverse root systems of these genotypes are the contributing factor in this regard. The reduction in plant size and stomatal conductance is believed to be a significant element in protecting plants against stress, perhaps leading to decreased productivity (Deikman et al., 2012). Drought stress has been observed to have a negative impact on leaf growth in various plant species, such as Populus, soybean, and several other species (Farooq et al., 2009). The study conducted by Manivannan et al. (2007) found a correlation between the level of abscisic acid (ABA) in the roots and shoots and an improvement in the root-to-shoot proportion in a drought environment (Farooq et al., 2009). Typically, crop species with shallow root systems have lower levels of drought tolerance compared to plants with deep root systems. This adaptive mechanism serves to safeguard plants against drought conditions by mitigating water loss from the roots (Demirevska et al., 2009). These communication pathways facilitate the transmission of both positive and negative signals between the roots and shoots hence regulating the growth rate and behavior of the plant. The prevailing framework for understanding plant reactions to arid soil conditions relies on hydraulic signaling, which encompasses a sequence of events including reduced water intake by the roots, subsequent declines in water potential and turgor within the leaves, closure of stomata, diminished leaf elongation, and osmotic adjustment (Rollins et al., 2013). Researchers have successfully identified a range of physiological characteristics that play a crucial part in facilitating the continuous growth of plants under environmental challenges. These parameters include variation in relative water content (RWC), electrolyte leakage, water use efficiency (WUE), stomatal conductance, photosynthesis, transpiration, chlorophyll leaf fluorescence, and chlorophyll color index. When plants encounter a water deficit under field conditions, their primary reaction is to regulate the opening and closing of stomata as a means to minimize water loss through transpiration (Gupta et al., 2020). Plants that are sensitive to water deficits and those that are resilient to drought have discernible differences in their stomatal movements. The presence of drought stress in plants induces the closure of stomata hence restricting the flow of gases between the internal environment of the leaf and the surrounding atmosphere. The closure of stomata in response to drought stress is recognized as a constraint on the photosynthetic process; yet, it is universally regarded as an effective approach for reducing water loss. The relative water content (RWC) is a measurement that quantifies the maximal water capacity of a tissue. The relative water content is considered a very suitable indicator for assessing the water status of plants due to its direct correlation with the physiological implications of cellular water deficiency. The relative water content (RWC), which is adversely impacted by many pressures, particularly drought stress, is regarded as a more reliable measure of water status compared to the water potential attained by plants while mitigating drought stress (Rai et al., 2015).

8.3 Methods to Overcome Drought and Heat Stress

Drought and heat are significant abiotic stressors that have a detrimental influence on agricultural yield and pose a threat to global food security. This is particularly concerning in light of the ongoing and escalating effects of climate change, which contribute to the heightened occurrence and intensity of these stress factors. Plants have evolved intricate mechanisms at various levels, including morphology, physiology, and biochemistry, enabling them to evade or adjust to adverse environmental circumstances. However, it is important to note that even the most moderate levels of heat and drought stress have a detrimental impact on crop productivity (Meena et al., 2017). Moreover, numerous autonomous investigations have demonstrated that elevated temperature and prolonged periods of drought have the potential to significantly diminish agricultural productivity, leading to a potential decline in crop yields of up to 50%. The physiological reaction to stress is a multifaceted process that encompasses various elements such as signaling pathways, transcription factors, hormonal responses, and secondary metabolite production. The impact of heat stress on the reproductive phase of growth, which ultimately affects grain yield, has been seen to be more pronounced in certain crop species (Kumar & Verma, 2018). The application of biotechnology, particularly in the fields of genomics and information technology, holds promise for alleviating the adverse impacts of heat and drought. This can be achieved by the implementation of agronomic management strategies and the creation of crop varieties that exhibit enhanced productivity in stressful environments.

8.4 Application of Plant Nutrients

In the setting of heat stress, the maintenance of proper hydration can be achieved by the utilization of diverse macronutrients and micronutrients. The physiological and stomatal activities in plants are significantly impacted by macronutrients and micronutrients, such as potassium, calcium, boron, selenium, and manganese. Moreover, it has been discovered that the application of micronutrients through foliar spray not only aids in mitigating stress but also enhances the nutritional composition of crops (Sarkar et al., 2021). Foliar application is linked with the advantage of fast and effective utilization of nutrients, reduction of losses through leaching, fixation, and regulating the nutrient uptakes of plants (Saikh et al., 2022). Nitrogen (N) is a crucial macronutrient that is commonly lacking in the majority of soils, particularly in dry and semiarid locations. It has a significant role in various structural, genetic, and metabolic processes within plants. The absorption and utilization of nitrogen by plants during periods of water stress play a crucial role in plant development and production. This process has a direct impact on carbon partitioning and facilitates the buildup of soluble carbohydrates, particularly starch, which subsequently enhances leaf growth. The inhibition of growth caused by water deficit is mostly attributed to a shortfall of nitrogen resulting from drought conditions. Consequently, this phenomenon has a direct impact on the capacity of the roots to efficiently uptake water and nutrients, deviating from the expected levels during regular transpiration. The utilization of phosphorus (P) fertilizer has been shown to significantly enhance plant growth in the presence of drought circumstances. The augmentation of stomatal conductance, photosynthesis, cell-membrane stability, water relations, and drought tolerance has been ascribed as the underlying mechanisms for the beneficial impacts of P on plant growth in drought conditions. Phosphorus is another important primary nutrient that enhances root growth, nodule, and seed development, also provides vital metabolic functions like photosynthesis, carbon partitioning, energy storage, and transfer that improve colonization of rhizobium and nitrogen fixation (Mahato et al., 2022). One crucial strategy for enhancing phosphorus (P) intake is to leverage the symbiotic relationship between plant roots and mycorrhiza. This association not only improves plant growth and tolerance to drought but also facilitates the uptake of many essential nutrients such as phosphorus, zinc, copper, manganese, and iron. Moreover, it is crucial to note that under drought conditions, chloroplasts experience significant potassium (K) loss, which leads to a further decline in photosynthesis and the initiation of additional reactive oxygen species (ROS) production. This discussion provides substantial evidence in favor of the notion that heightened levels of drought stress lead to proportional increases in the requirement for potassium (K) to sustain photosynthesis and safeguard chloroplasts from oxidative damage. There are several potential strategies that might be employed to mitigate the negative impacts of drought on crop plants by enhancing water usage efficiency, particularly in relation to potassium (K) nutrition. In instances of water shortage, the presence of potassium (K) in plants enhances their ability to withstand water stress by optimizing the utilization of soil moisture, particularly when compared to plants that are poor in potassium (Singh et al., 2010; Sarkar et al., 2022). Zinc (Zn) plays a significant role in modulating auxin levels and has been recognized for its function as a coenzyme in the biosynthesis of tryptophan, a key precursor in auxin creation. The administration of zinc (Zn) leads to an elevation in auxin levels hence promoting root growth and subsequently enhancing drought tolerance in plants (Karmakar et al., 2022). The inclusion of boron in plant systems has been shown to enhance their ability to withstand drought conditions through many mechanisms, such as the facilitation of sugar transfer, promotion of flower retention, facilitation of pollen synthesis, and enhancement of seed germination. Copper is an essential element in the process of lignin production, which plays a crucial role in enhancing cell wall integrity and preventing wilting. Drought stress has a negative impact on all of these physiological processes in plants. Adequate copper (Cu) nutrition mitigates the negative consequences of drought, including the reduction of stem and twig dieback, leaf yellowing, stunted development, and susceptibility to withering of pale green leaves, and enhances carbohydrate (CHO) and nitrogen metabolism thus promoting plant growth. The use of silicon has been shown to enhance plant development and increase resistance to both biotic and abiotic stressors. The inclusion of silicon (Si) in maize plants resulted in an enhancement of water use efficiency. This was achieved by a reduction in leaf transpiration and a decrease in the water flow rate within the xylem vessel. Sorghum bicolor (L.) may experience enhanced water intake and transport under drought conditions through the involvement of Si can effectively mitigated oxidative stress in potted wheat subjected to drought conditions by modulating the activities of antioxidant enzymes.

8.5 Plant Growth Regulators

Plant growth regulators, also known as phytohormones, have been acknowledged for their capacity to mitigate stress in plants. The plant growth-promoting rhizobacteria (PGPR) are microorganisms that inhabit the rhizosphere, a region surrounding plant roots, and derive their metabolic energy from root exudates (Wozniak & Galazka, 2019). The abundance and variety of plant growth-promoting rhizobacteria (PGPRs) present in the rhizosphere are influenced by the physiological characteristics of the soil and the availability of nutrients. The primary roles of plant growth-promoting rhizobacteria (PGPRs) encompass various essential functions, including nitrogen fixation, phosphate solubilization, micronutrient adaptation, phytohormone release, soil composition maintenance, bioremediation of contaminated soil, induction of systemic resistance against pests, insects, and fungal pathogens, as well as alleviation of abiotic stresses such as salinity and drought. According to the nitrogenase, the enzyme makes it easier for PGPRs to transform atmospheric nitrogen into nitrate and ammonia, which plants may use. It has been observed that plant growth-promoting rhizobacteria (PGPRs) facilitate the solubilization of phosphate by synthesizing organic acids that form complexes with phosphate, so augmenting its accessibility to the plant. Plant growth-promoting rhizobacteria (PGPRs) have been found to enhance the accessibility of micronutrients such as iron through the synthesis of siderophores. Siderophores are small molecules with a strong affinity for iron, allowing them to effectively chelate and make it available for plant uptake (Singh et al., 2020). The presence of plant growth hormones, specifically auxins and gibberellins, the ability to regulate pathogens through the synthesis of antibiotics, hydrogen cyanide, and lytic enzymes, as well as the capacity to remediate soil through the production of diverse degradative enzymes, are additional notable features of PGPRs. Therefore, through the synthesis of phytohormones, plant growth-promoting rhizobacteria (PGPRs) can enhance the structure of plant root systems, thereby facilitating the plant's ability to endure periods of drought-induced stress. However, when the severity of drought circumstances intensifies, there is a steady loss of water resources, leading to a decline in shoot growth. While the decrease in shoot growth serves to limit transpiration rates and mitigate the risk of dehydration, it concurrently diminishes the plant's overall productivity and growth (Ngumbi & Kloepper, 2016). Plant growth-promoting rhizobacteria (PGPRs) have the ability to secrete many hormones and enzymes, including cytokines, auxins, gibberellins, and ACC deaminase. These substances play a crucial role in facilitating near-normal shoot growth rates in plants by promoting tissue and cell elongation and counteracting the negative impacts of ethylene. According to Zheng et al. (2015), plant growth-promoting rhizobacteria (PGPRs) possess the capacity to modulate water content through the modification of hydraulic conductivity and stomatal apertures. The precise mechanism underlying this process remains elusive. Nevertheless, there is speculation that plant growth-promoting rhizobacteria (PGPRs) may contribute to the augmentation of the plant's relative water content by exogenous production of osmoprotectants such as proline and phytohormones like indole acetic acid. The application of PGPR (plant growth-promoting rhizobacteria) to plants experiencing drought stress circumstances led to a significant augmentation in the levels of carotenoids. Carotenoid production plays a crucial role in the functioning of the photosynthetic apparatus in plants, and its levels are significantly influenced under conditions of drought-induced stress. Nevertheless, the use of plant growth-promoting rhizobacteria (PGPRs) can result in a significant tenfold augmentation in carotenoid levels in plants. The production of ACC deaminase is a prominent approach utilized by plant growth-promoting rhizobacteria (PGPRs) to alleviate the impact of drought stress on plants, thereby promoting plant survival, as well as enhancing their overall growth and development (Katiyar et al., 2016). In general, the utilization of plant growth-promoting rhizobacteria (PGPRs) presents a viable strategy for mitigating both biotic and abiotic challenges, while concurrently fostering sustainable agricultural practices (Kumar et al., 2020). The effective commercialization of this product will be contingent upon the collaboration and partnership of professionals in the fields of agriculture, microbiology, biotechnology, industry, and farming. Each action undertaken will contribute to the advancement of sustainable agriculture (Fig. 8.2).

8.6 Seed Priming

Seed priming is a refined agricultural practice that entails controlled hydration of seeds before planting, to exert a lasting impact on the crop's performance (Muhie et al., 2021). The process of priming has been found to induce a heightened tolerance to stress in seedlings, as well as throughout the entire growth period of a crop (Jana et al., 2022). Priming has been found to elicit accelerated germination and consistent emergence thus enabling farmers to mitigate the time lost due to drought conditions (Dey et al., 2021). These features facilitate the establishment of a crop by enhancing the availability of soil water and nutrients hence promoting healthier growth. In contrast, seed priming has been found to facilitate the expansion of leaf area, allowing for increased absorption of solar radiation which in turn, leads to

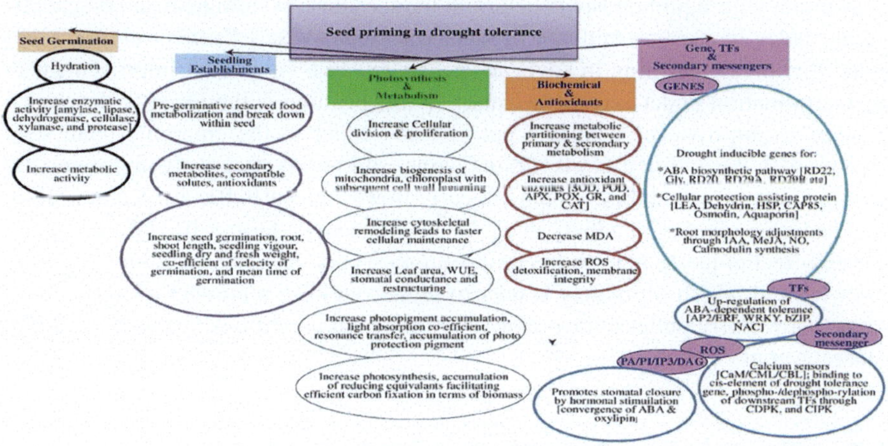

Fig. 8.2 Systematic representation of seed priming mechanism in drought stress tolerance. (Source, Saha et al., 2022) (*ROS reactive oxygen species, TFs transcription factors, NO nitric oxide, CAT catalase, POX peroxidase, JA jasmonic acid, SOD superoxide dismutase, GR glutathione reductase, APX ascorbate peroxidase, HSP heat shock protein, LEA late embryogenic protein, ABA abscisic acid)

enhanced photosynthetic activity and ultimately results in higher crop yields. The aforementioned benefits suggest that seed priming may be a highly effective strategy for mitigating the negative impacts of terminal heat and drought stress (Singhal & Bose, 2020). In chemical priming, different types of nitrates, nitric oxide, hydrogen sulfide, melatonin, and polyamines are in use which ameliorate the effects of abiotic stresses in agricultural crops (Savvides et al., 2016; Singhal et al., 2021a). The use of hormones, like GA3, kinetin, salicylic acid (SA), and ABA for priming purposes, is referred to as hormonal priming; during abiotic and biotic stress hormonal priming has been found effective to improve crop stand (Sytar et al., 2019; Rhaman et al., 2021). Additionally, the treatment of selenium (Se) and silicon (Si) during priming has been found to provide protection against drought stress (Ramirez-Villegas et al., 2018). The application of priming agents such as KH_2PO_4, Na_2HPO_4, and water has been observed to enhance the drought tolerance of crops. The observed phenomenon could perhaps be attributed to hormonal action, specifically involving salicylic acid, abscisic acid, and ethylene. Priming can improve germination by enhancing the physiological metabolism like activity of alpha amylase, increase in soluble and proline contents during germination without involving any lag phase or activation period, which increases seedling vigor in normal as well as in stress conditions (Guan et al., 2009). It ameliorates the adverse effects of various stresses and increases the yield in critical environmental conditions by manipulating the defense mechanism-related enzymes and metabolites during the vegetative growth phase. Seed priming with mannose has improved drought tolerance by increasing the antioxidant levels, reducing oxidative injuries, and accumulating higher amounts of reducing sugars for osmotic regulation (Conrath et al., 2011).

Biopriming with liquid phosphobacterium induces the physiological and biochemical attributes in okra to mitigate drought tolerance. Vaseii Kashani et al. (2015) found that matrix priming in soybean seeds showed a significant improvement in seed germination under limited irrigation. Nutri-priming is the recently developed strategy, to improve the available nutrient and water to the emerging plant seed priming with the addition of magnesium, zinc, and boron effectively improving the germination, growth and development, early flowering, early maturity, grain filling rate, and crop yield (Choudhary et al., 2021). It seems a farmer-friendly, practical, and easy technique to help in the uniform emergence of seeds with high seedling vigor in field which ultimately results in higher yield in a number of agricultural crops, particularly under unfavorable environmental conditions (Jisha et al., 2013; Paparella et al., 2015).

8.7 Drought Regulation by the Application of Biochar

Biochar is a carbonaceous material that is produced by the process of pyrolysis, which involves the thermal decomposition of biomass under oxygen-limited conditions. In recent years, there has been a notable increase in the recognition of biochar's significance as a means of alleviating plant stress. This recognition stems from its capability to augment soil fertility, boost nutrient absorption and growth, and improve the water retention capabilities of plants experiencing stress (Haider et al., 2020). The utilization of biochar as a method for improving soil quality is contingent upon various factors related to its physical and chemical properties. These factors encompass the type of material used as feedstock, the conditions under which pyrolysis occurs, as well as the specific surface area and porosity of the biochar (Afshar et al., 2016; Zhang et al., 2020). Biochar, when introduced into the soil, exerts a substantial influence on various soil characteristics, including surface area, porosity, pore-size distribution, water-holding capacity, and numerous chemical and biological factors. This can be attributed to the negatively charged surface, high surface area, and carbon sequestration ability of biochar (Durukan et al., 2020). Enhancing soil fertility significantly improves the growth and establishment of plants in drought circumstances. Biochar has the ability to effectively absorb and retain nutrients within the soil, allowing plant roots to subsequently acquire these nutrients without the need to extend further into the ground for water absorption. Several studies have documented a significant enhancement in various physiological and growth parameters of plants when biochar is applied. These parameters include photosynthetic rates, xylem water potential, chlorophyll content, phytohormone production, uptake of essential nutrients such as nitrogen (N), silicon (Si), phosphorus (P), and potassium (K), as well as overall biomass and growth (Ali et al., 2017; Semida et al., 2019). According to Abideen et al. (2019), plants subjected to biochar treatment exhibited reduced oxidative stress as a result of enhanced synthesis of scavenging enzymes and improved effectiveness of photosystem II. In a study conducted by Haider et al. (2020), it was observed that water-stressed maize exhibited enhanced transpiration rates, relative water content, root and shoot length,

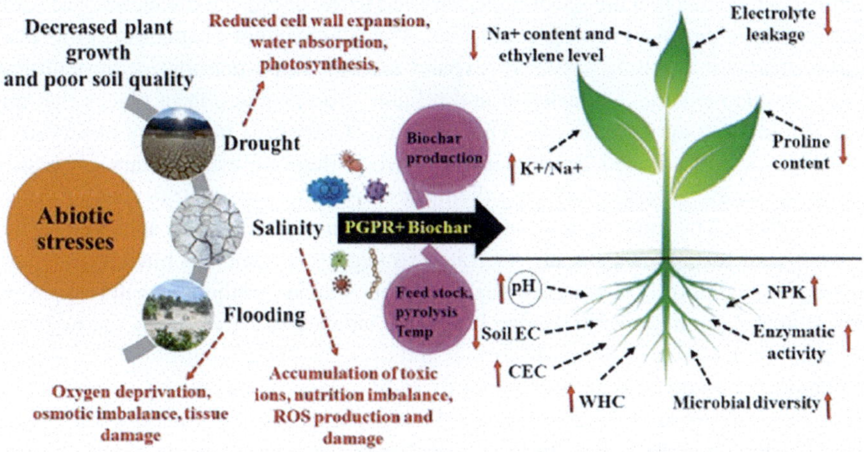

Fig. 8.3 Effects of plant growth-promoting bacteria (PGPB) and biochar on plant resilience to drought, salinity, and flooding stress. (Source, https://doi.org/10.1016/j.ecoenv.2022.114408)

plant biomass, and a better osmotic defense system when compared to the control group. While the utilization of biochar as a means to mitigate the effects of drought stress may appear promising, there remains a significant lack of knowledge regarding the potential dangers and risks associated with this technology. Several significant drawbacks have been identified in relation to biochar. These include the limited availability of feedstock, the inefficient utilization of energy during the production process, the binding and deactivation of nutrients, the presence of herbicides and insecticides, the elevation of soil pH, the release of toxic substances and heavy metals into the soil, and the disruption of the natural microbial flora of the soil (Oni et al., 2019). Moreover, it should be noted that the effects of biochar on plant growth and development may vary, with certain species exhibiting limited or negligible benefits compared to others. In the study conducted by Afshar et al. (2016), it was observed that the application of biochar did not provide any noteworthy enhancements in leaf weight, photosynthetic rate, leaf area, and plant height of milk thistle (Silybum marianum L. Gaertn) under conditions of drought stress. The synthesis of biochar using crop remains as feedstock is a labor-intensive and costly procedure for farmers. Extensive research is necessary to gain a comprehensive understanding of the environmental impacts associated with the technology of biochar (Fig. 8.3).

8.8 Future Response of Drought

Enhancing drought resistance (DR) in crop plants is a formidable task for plant breeders and crop physiologists due to its intricate genetic nature, including several pathways. The successful cultivation of crop plants that are resistant to drought necessitates the strategic combination and interaction of several mechanisms, features, and genes that are specifically tailored to the unique characteristics of each crop and its respective growing conditions. The achievement of success in this particular endeavor not only serves to expand the cultivated regions for agricultural plants but also ensures a consistent and reliable crop production in places that are susceptible to drought conditions. The presence of variety is frequently observed in wild species and adapted genotypes that have undergone evolutionary processes driven by natural selection, making them the most reliable source of disease resistance features. The improvement of drought resistance (DR) can be achieved by the evaluation of resources using an integrated strategy that combines phenotyping and genotyping techniques. Drought-resistant plants should combine a better root system, stomatal regulation, WUE, and hormonal balance while avoiding the negative effects on grain yield under both normal and drought stress conditions. The timing of planting is a crucial determinant for drought escape in regions characterized by subtropical, tropical, and rainy climates. One potential solution to address the issue of early planting in dry conditions is to enhance water use efficiency (WUE) and adjust the timing of sowing to mitigate drought stress, particularly during critical growth phases. The implementation of a cropping system that achieves equilibrium between water availability, nutrient content, and balance will result in improved canopy growth and yield through the enhancement of biomass production. Therefore, the manipulation of planting dates to modify biomass production per unit of transpiration can serve as a beneficial approach for enhancing drought resistance or drought escape. Topsoil alterations have an impact on soil water through processes such as infiltration, evaporation, and the transmission of heat between the environment and the soil. Various methods can be employed to save water in soil, and one such method is mulching. Mulching has the potential to significantly decrease the amount of water lost by evaporation, while simultaneously improving the absorption and retention of water in the root zone of plants. The present irrigation systems, including drip, film hole, and sprinkler irrigation, have demonstrated enhanced efficiency and water conservation capabilities, resulting in improved water use efficiency (WUE) and grain yield as compared to surface irrigation. In addition, it has been shown that the application of additional irrigation and the implementation of soil management practices during the reproductive growth stage can lead to a significant increase in crop production, with potential enhancements of up to 70%. In today's world, as the demand for higher outputs, improved production of crops, soil fertility, and sustainable agriculture is increasing, research is shifting toward rhizoengineering to fabricate a unique environment for the interaction of plants and microorganisms. Plant growth-promoting rhizobacteria are a promising component of rhizoengineering and their benefits are numerous and evident. PGPRs are effective in the mitigation of many abiotic and biotic stresses. Overall, the application of PGPRs is a promising approach to combat biotic and abiotic stresses and promote sustainable agriculture. Every step taken will be a step toward sustainable agriculture. Therefore, a holistic crop improvement strategy should involve the deployment of high crop yield potential and the utilization of a combination of morpho-physiological, biochemical, and anatomical adaptive responses to drought stress.

8.9 Conclusion

The impact of drought on plant productivity is significant, leading to a substantial decrease in agricultural economic viability. Numerous approaches have been devised to address the issue of drought, each possessing distinct advantages and limits. Nevertheless, the implementation of appropriate mitigation methods is crucial in addressing drought circumstances, as it not only offers direct and indirect advantages but also improves overall plant performance. The utilization of plant growth-promoting rhizobacteria (PGPRs), seed priming, and biochar technology exhibits notable economic advantages, since it may be implemented on modest landholdings without necessitating substantial energy inputs. However, it is imperative to acknowledge the potential good effects of these strategies for mitigating drought. Consequently, greater attention should be devoted to ensuring their widespread accessibility to farmers of average means globally.

References

Abideen, Z., Koyro, H., Huchzermeyer, B., Ansari, R., Zulfiqar, F., & Gul, B. (2019). Ameliorating effects of biochar on photosynthetic efficiency and antioxidant defense of phragmites karka under drought stress. *Plant Biology, 22*, 259–266.

Afshar, R. K., Hashemi, M., DaCosta, M., Spargo, J., & Sadeghpou, A. (2016). Biochar application and drought stress effects on physiological characteristics of silybum marianum. *Communications in Soil Science and Plant Analysis, 47*, 743–752.

Ali, S., Rizwan, M., Qayyum, M. F., Ok, Y. S., Ibrahim, M., Riaz, M., Arif, M. S., Hafeez, F., Mohammad, I., & Al-Wabel, A. N. S. (2017). Biochar soil amendment on alleviation of drought and salt stress in plants: A critical review. *Environmental Science and Pollution Research, 24*, 12700–12712.

Bailly, C. (2019). The signaling role of ROS in the regulation of seed germination and dormancy. *The Biochemical Journal, 476*, 3019–3032.

Balla, K., Rakszegi, M., Li, Z. G., Bekes, F., Bencze, S., & Veisz, O. (2011). Quality of winter wheat in relation to heat and drought shock after anthesis. *Czech Journal of Food Sciences, 29*, 117–128.

Bhat, M. A., Mir, R. A., Shah, A. S., Zagar, S. M., Rahman, S., & Jan, A. T. (2021). Mechanistic insights of CRISPR/Cas-mediated genome editing towards enhancing abiotic stress tolerance in plants. *Physiologia Plantarum, 172*, 1255–1268.

Castaneda, R., Doan, D., Newhouse, D. L., Nguyen, M., Uematsu, H., & Wagner de Azevedo, J. P. (2016). *Who are the poor in the developing world?*. World Bank Policy Research Working Paper No. 7844, Available at SSRN: https://ssrn.com/abstract=2848472

Chiappero, J., del Rosario Cappellari, L., Lucas, G., Alderete, S., Palermo, T. B., & Erika, B. (2019). Plant growth-promoting rhizobac- teria improve the antioxidant status in Mentha piperita grown under drought stress leading to an enhancement of plant growth and total phenolic content. *Industrial Crops and Products, 139*, 111553.

Choudhary, S.K., Kumar, V., Singhal, R.K., Bose, B., Chauhan, J., Alamri, S., Siddiqui, M. H., Javed, T., Shabbir, R., Rajendran, K. & Iqbal, M.A. (2021). Seed priming with Mg $(NO_3)_2$ and $ZnSO_4$ salts triggers the germination and growth attributes synergistically in wheat varieties. *Agronomy, 11*, 2110.

Conrath, U. (2011). Molecular aspects of defense priming. *Trends in Plant Science, 16*, 524–531.

Deikman, J., Petracek, M., & Heard, J. E. (2012). Drought tolerance through biotechnology: improving translation from the laboratory to farmers fields. *Current Opinion in Biotechnology, 23*, 243–250.

Demirevska, K., Zasheva, D., Dimitrov, R., Simova-Stoilova, L., Stamenova, M., & Feller, U. (2009). Drought stress effects on Rubisco in wheat: Changes in the Rubisco large subunit. *Acta Physiologiae Plantarum, 31*(6), 1129–1138.

Dreesen, P. E., De Boeck, H. J., Janssens, I. A., & Nijs, I. (2012). Summer heat and drought extremes trigger unexpected changes in productivity of a temperate annual/biannual plant community. *Environmental and Experimental Botany, 79*, 21–30.

Durukan, H., Demirbas, A., & Türkekul, İ. (2020). Effects of biochar rates on yield and nutrient uptake of sugar beet plants grown under drought stress. *Communications in Soil Science and Plant Analysis, 51*(21), 2735–2745.

Farooq, M., Wahid, A., Kobayashi, N., Fujita, D., & Basra, S. M. (2009). Plant drought stress: Effects, mechanisms and management. *Sustainable Agriculture, 29*, 153–188.

Florez-Sarasa, I., Fernie, A. R., & Gupta, K. J. (2020). Does the alternative respiratory pathway offer protection against the adverse effects resulting from climate change? *Journal of Experimental Botany, 71*, 465–469.

Garai, S., Bhowal, B., Pareek, A., Singla-Pareek, S. L., Kaur, C., & Sopory, S. K. (2020). Expression dynamics of glyoxalase genes under high temperature stress in plants. *Plant Physiology Reports*. https://doi.org/10.1007/s40502-020-00545-1

Guan, Y. J., Hu, J., Wang, X. J., & Shao, C. X. (2009). Seed priming with chitosan improves maize germination and seedling growth in relation to physiological changes under low temperature stress. *Journal of Zhejiang University. Science. B, 10*, 427–433. https://doi.org/10.1631/jzus.B0820373

Gupta, P. K., Balyan, H. S., Sharma, S., & Kumar, R. (2020). Genetics of yield, abiotic stress tolerance and biofortification in wheat (*Triticum aestivum* L.). *Theoretical and Applied Genetics, 133*(5), 1569–1602.

Haider, I., Raza, M. A. S., Iqbal, R., Usman Aslam, M., Habib-ur-Rahman, M., Shameem, R., Khan, M. T., Mahran Aslam, M., Waqas, M., & Ahmad, S. (2020). Potential effects of biochar application on mitigating the drought stress implications on wheat (*Triticum aestivum* L.) under various growth stages. *Journal of Saudi Chemical Society, 24*, 974–981.

Jana, K., Sarkar, A., Mondal, R., & Agrawal, R. K. (2022). Quality and green fodder productivity of forage maize (*Zea mays* L.) as influenced by different seed priming techniques under rainfed situation. *Journal of Crop and Weed, 18*(3), 71–77. https://doi.org/10.22271/09746315.2022.v18.i3.1619

Jisha, K. C., Vijaya kumari, K. & Puthur, J. T. (2013). Seed priming for abiotic stress tolerance: an overview. *Acta Physiologiae Plantarum, 35*, 1381–1396. https://doi.org/10.1007/s11738-012-1186-5

Jones, P., Garcia, B. J., Furches, A., Tuskan, G. A., & Jacobson, D. (2019). Plant host-associated mechanisms for microbial selection. *Frontiers in Plant Science, 10*, 862.

Joshi, R., Sahoo, K. K., Singh, A. K., Anwar, K., Gautam, R. K., Sopory, S. K., Pareek, A., & Singla-Pareek, S. L. (2020). Engineering trehalose biosynthesis pathway improves yield potential in rice under drought, saline and sodic conditions. *Journal of Experimental Botany, 71*, 653–668.

Karmakar, A., Goswami, S. B., Sarkar, A., Jana, K., & Murmu, K. (2022). Growth, grain yield and soil nutrient status of Kharif Rice (Oryza sativa L.) as influenced by integrated nutrient management in lower gangetic plains. *International Journal of Environment and Climate Change, 12*(11), 2177–2184.

Katiyar, D., Hemantaranjan, A., & Singh, B. (2016). Plant growth promoting Rhizobacteria-an efficient tool for agriculture promotion. *Advance in Plants Agricultural Research, 4*(6), 426–434.

Karthikeyan, R., Venkatesh, P., & Chandrasekhar, N. (2012). Morphoanatomical studies of leaves of Abutilon indicum (linn.) sweet. *Asian Pacific Journal of Tropical Biomedicine, 10*, 464–469.

Kaur, G., & Asthir, B. (2017). Molecular responses to drought stress in plants. *Biological Plant, 61*, 201–209. https://doi.org/10.1007/s10535-016-0700-9

Kukal, M. S., & Irmak, S. (2018). Climate-driven crop yield and yield variability and climate change impacts on the U.S. great plains agricultural production. *Scientific Reports, 8*, 3450.

Kumar, A., & Verma, J. P. (2018). Does plant-microbe interaction confer stress tolerance in plants: A review? *Microbiological Research, 207*, 41–52.

Kumar, P., Yadav, S., & Singh, M. P. (2020). Bioregulators application improved heat tolerance and yield in chickpea (Cicer arietinum L.) by modulating zeaxanthin cycle. *Plant Physiology Reports*. https://doi.org/10.1007/s40502-020-00555-z

Kyei-boahen, S., Savala, C. E. N., Chikoye, D., Abaidoo, R., & Kyei-Boahen, S. (2017). Growth and yield responses of cowpea to inoculation and phosphorus fertilization in different environments. *Frontiers in Plant Science, 8*, 646.

Lastochkina, O., Garshina, D., Allagulova, C., Fedorova, K., Koryakov, I., & Vladimirova, A. (2020). Application of endophytic Bacillus subtilis and salicylic acid to improve wheat growth and tolerance under combined drought and Fusarium root rot stresses. *Agronomy, 10*(9), 1343.

Mahalingam, R. (2015). Consideration of combined stress: a crucial paradigm for improving multiple stress tolerance in plants. Combined stresses in plants. *Physiological, Molecular, and Biochemical Aspects*, 1–25.

Mahato, T., Murmu, K., Sarkar, A., Maji, S., & Das, P. (2022). Yield intensification of spring mungbean (Vigna radiata L.) through exogenous foliar nutrition. *Journal of Crop and Weed, 18*(3), 38–43.

Mamrutha, H. M., Rinki Khobra, Sendhil, R., Renu Munjal, Sai Prasad, S. V., Suma Biradar, Mavi, Tapamay Dhar, G. S., Raj Bahadur, Bhagwan, J. H., Surya Prakash, Hoshiyar Singh, Shukla, R. S., Meera Srivastava, Chandrakant Singh, Gosavi, A. B., Salunke, V. D., Vipin Chandra Dhyani & Singh, G. P. (2020). Developing stress intensity index and prioritizing hotspot locations for screening wheat genotypes under climate change scenario. *Ecological Indicators, 118*, 106714.

Manivannan, P., Abdul Jaleel, C., Sankar, B., Kishorekumar, A. Somasundaram, R., Lakshmanan, G.M.A. & Panneerselvam, R. (2007). Growth, biochemical modifications and proline metabolism in *Helianthus annuus* L. as induced by drought stress. *Colloids and Surfaces B: Biointerfaces, 59*(2), 141–149.

Meena, K. K., Sorty, A. M., Bitla, U. M., Choudhary, K., Gupta, P., Pareek, A., Singh, D. P., Prabha, R., Sahu, P. K., Gupta, V. K., Singh, H. B., Krishanani, K. K., & Minhas, P. S. (2017). Abiotic stress responses and microbe-mediated mitigation in plants: The omics strategies. *Frontiers in Plant Science, 8*, 172.

Muhie, S., Memis, N., Özdamar, C., Gökdas, Z., & Demir, I. (2021). Biostimulant priming for germination and seedling quality of carrot seeds under drought, salt and high temperature stress conditions. *International Journal of Agriculture Environment and Food Sciences, 5*, 342–349.

Murmu, K., Das, P., Sarkar, A., & Bandopadhyay, P. (2022). Organic agriculture: As a climate change adaptation and mitigation strategy. *Zeichen Journal, 8*(03), 171–187. ISSN No: 0932-4747.

Nadeem, M., Li, J., Yahya, M., Sher, A., Ma, C., Wang, X., & Qiu, L. (2019). Research progress and perspective on drought stress in legumes: A review. *International Journal of Molecular Sciences, 20*, 2541.

Ngumbi, E., & Kloepper, J. (2016). Bacterial-mediated drought tolerance: Current and future prospects. *Applied Soil Ecology, 105*, 109–125.

Oni, B. A., Oziegbe, O., & Obembe, O. O. (2019). Significance of biochar application to the environment and economy. *Annals of Agricultural Science, 64*, 222–236.

Paparella, S., Araújo, S.S., Rossi, G., Wijayasinghe, M., Carbonera, D. & Balestrazzi, A. (2015). Seed priming: state of the art and new perspectives. *Plant Cell Reports, 34*, 1281–1293.

Rai, N., Rai, K. K., Tiwari, G. & Singh, P. K. (2015). Changes in free radical generation, metabolites and antioxidant defense machinery in hyacinth bean (*Lablab purpureus*. L) in response to high temperature stress. *Acta Physiologiae Plantarum, 37*, 37–46.

Raja, V., Qadir, S. U., Mohammed Nasser, A., & Parvaiz, A. (2020). Impact of drought and heat stress individually and in combination on physio-biochemical parameters, antioxidant responses, and gene expression in Solanum lycopersicum. *3 Biotech, 10*, 208.

Ramegowda, V., Senthil-Kumar, M., Ishiga, Y., Kaundal, A., Udayakumar, M., & Kirankumar, S. M. (2013). Drought stress acclimation imparts tolerance to sclerotinia sclerotiorum and pseudomonas syringae in nicotiana benthamiana. *International Journal of Molecular Sciences, 14*, 9497–9513.

Ramirez-Villegas, J., Heinemann, A. B., Pereira de Castro, A., Breseghello, F., Navarro-Racines, C., Li, T., Rebolledo, M. C., & Challinor, A. J. (2018). Breeding implications of drought stress under future climate for upland rice in Brazil. *Global Change Biology, 24*, 2035–2050.

Rhaman, M. S., Imran, S., Rauf, F., Khatun, M., Baskin, C. C., Murata, Y., & Hasanuzzaman, M. (2021). Seed priming with phytohormones: An effective approach for the mitigation of abiotic stress. *Plants, 10*, 37.

Rollins, J. A., Habte, E., Templer, S. E., Colby, T., Schmidt, J., & von Korff, M. (2013). Leaf proteome alterations in the context of physiological and morphological responses to drought and heat stress in barley (Hordeum vulgare L.). *Journal of Experimental Botany, 64*(11), 3201–3212.

Saikh, R., Murmu, K., Sarkar, A., Mondal, R., & Jana, K. (2022). Effect of foliar zinc application on growth and yield of rice (Oryza sativa) in the Indo-Gangetic Plains of India. *Nusantara Bioscience, 14*(2), 182–187.

Sarkar, A., Jana, K., & Mondal, R. (2021). Growth and yield of hybrid mustard (Brassica juncea L.) as influenced by foliar nutrition in Gangetic plains of West Bengal. *Journal of Crop and Weed, 17*(3), 35–40.

Sarkar, A., Jana, K., Mondal, R., Mondal, K., Banerjee, S., & Murmu, K. (2022). Effect of foliar nutrition on growth, oil yield, production economics of hybrid mustard (Brassica juncea L.) and soil microbial properties. *The Pharma Innovation Journal, 11*(6), 1456–1460.

Sarkar, A., Murmu, K., Murmu, S., Jana, K., & Sen, S. S. (2023). Legume forages: A multidisciplinary approach for ensuring agricultural sustainability and global food security under altered climate conditions. In *Sustainable solutions for a changing world*. ISBN: 978-81-19149-40-7.

Savvides, A., Ali, S., Tester, M., & Fotopoulos, V. (2016). Chemical priming of plants against multiple abiotic stresses: Mission possible? *Trends in Plant Science, 21*, 329–340.

Semida, W. M., Beheiry, H. R., Sétamou, M., Simpson, C. R., El-Mageed, T. A. A., Rady, M. M., & Nelson S. D. (2019). Biochar implications for sustainable agriculture and environment: A review. *South African Journal of Botany, 127*, 333–347.

Shao, H. B., Chu, L. Y., Jaleel, C. A. & Zhao, C. X. (2008). Water-deficit stress-induced anatomical changes in higher plants. *Comptes Rendus Biologies, 331*, 215–225.

Shrivastava, P., & Kumar, R. (2015). Soil salinity: A serious environmental issue and plant growth promoting bacteria as one of the tools for its alleviation. *Saudi Journal of Biological Sciences, 22*, 123–131. https://doi.org/10.1016/j.sjbs.2014.12.001

Singh, R. (2010). Effect of fertilizers on yield characteristics if mustard (Brassica juncea L. czern& cross). *Journal of Phytology, 2*(10), 20–24. ISSN: 2075-6240.

Singh, R. K., Singh, P., Li, H. B., Guo, D. J., Song, Q. Q. & Yang, L. T. (2020). Plant-PGPR interaction study of plant growthpromoting diazotrophs *Kosakonia radicincitans* BA1 and *Stenotrophomonas maltophilia* COA2 to enhance growth and stress-related gene expression in Saccharum spp. *Journal of Plant Interaction, 15*, 427–445. https://doi.org/10.1080/17429145.2020.18.57857

Singhal, R. K., & Bose, B. (2020). Wheat seedlings as affected by Mg (NO3)2 and ZnSO4 priming treatments. *World Scientific News, 144*, 13–29.

Singla, J. & Krattinger, S. G. (2016). Biotic stress resistance genes in wheat. In: Wrigley, C., Corke, H., Seetharaman, K., Faubion, J. (Eds.), *Encyclopedia of Food Grains*, (Vol. 2, pp. 388–392).

Soda, M. E., Nadakuduti, S. S., Pillen, K., & Uptmoor, R. (2010). Stability parameter and genotype mean estimates for drought stress effects on root and shoot growth of wild barley pre- introgression lines. *Molecular Breeding, 26*, 583–593.

Suzuki, N., Rivero, R. M., et al. (2014). Abiotic and biotic stress combinations. *New Phytologist, 203*, 32–43.

Sytar, O., Kumari, P., Yadav, S., Brestic, M., & Rastogi, A. (2019). Phytohormone priming: Regulator for heavy metal stress in plants. *Journal of Plant Growth Regulation, 38*, 739–752.
Tuteja, N., & Sopory, S. K. (2008). Chemical signaling under abiotic stress environment in plants. *Plant Signaling & Behavior, 3*, 525–536. https://doi.org/10.4161/psb.3.8.6186
Waterland, N. L., Campbell, C. A., Finer, J. J., & Jones, M. L. (2010). Abscisic acid application enhances drought stress tolerance in bedding plants. *HortScience, 45*, 409–413.
Wozniak, M., & Galazka, A. (2019). Endophytic bacteria potentially promote plant growth by synthesizing different metabolites and their phenotypic/physiological profiles in the Biolog GEN III microplate TM Test. *International Journal of Molecular Sciences, 20*, 5283.
Yaojun, Z., Jiaqi, D., Hong, W., Su, L., & Zhao, C. (2020). Biochar addition alleviates the negative effects of drought and salinity stress on soybean productivity and water use efficiency. *BMC Plant Biology, 20*, 288.
Zhang, M., Chen, Q., & Shihua Shen, S. (2011). Physiological responses of two Jerusalem artichoke cultivars to drought stress induced by polyethylene glycol. *Acta Physiologiae Plantarum, 33*, 313–318.
Zhang, C., Li, J., Wu, X., Long, Y., An, H., Pan, X., Li, M., Dong, F. & Zheng, Y. (2020). Rapid degradation of dimethomorph in polluted water and soil by Bacillus cereus WL08 immobilized on bamboo charcoal–sodium alginate. *Journal of Hazardous Materials, 398*, 122806.
Zheng, M., Tao, Y., Hussain, S., Jiang, Q., Peng, S., Huang, J., et al. (2015). Seed priming in dry direct-seeded rice: consequences for emergence, seedling growth and associated metabolic events under drought stress. *Plant Growth Regulation*. https://doi.org/10.1007/s10725-015-0083-5

Potential Mitigation of Drought and Heat Stress in Common Food Crops Using Indigenous Rhizobacterial Populations

9

B. N. Aloo, O. J. Were, and T. Vishal

Abstract

Drought and heat stress present formidable challenges to global food production systems and food security, especially in the wake of climate change. Cognizant of this, there is a need to explore various options for their mitigation to reduce abiotic plant stress. Rhizobacteria are plant-root-residing bacteria that establish intimate relations with their host plants and provide them with several benefits, such as abiotic stress tolerance. Thus, rhizobacteria can be explored as environmentally friendly option for drought and heat stress mitigation in plants. This chapter presents the potential of rhizobacteria for drought and heat stress mitigation in common food crops from a global perspective. This chapter first unravels the impacts of drought and heat stress on crop plants and provides insights into rhizobacteria-mediated stress mitigation. This chapter also encompasses the multifarious mechanisms through which rhizobacteria enhance plant resistance to drought and heat stress, from the alleviation of oxidative stress to the regulation of phytohormones, and the challenges and future directions regarding this potential. This chapter includes recent research reports to shed light on the potential of rhizobacteria-mediated drought and heat stress mitigation in agriculture. This chapter also highlights the urgency of harnessing these beneficial microbes to ensure sustainable crop production and enhance global food security. By exploring the potential of rhizobacteria for drought and heat stress mitigation in

B. N. Aloo (✉)
Department of Biological Sciences, University of Eldoret, Eldoret, Kenya
e-mail: baloo@uoeld.ac.ke

O. J. Were
Department of Seed, Crop and Horticultural Sciences, University of Eldoret, Eldoret, Kenya

T. Vishal
Department of Biotechnology, Graphic Era (Deemed to be University), Dehradun, Uttarakhand, India

R. Dubey et al. (eds.), *Drought and Heat Stress in Agriculture*,
https://doi.org/10.1007/978-981-96-5735-3_9

plants, this chapter provides a comprehensive understanding of their potential in promoting the resilience of agricultural ecosystems, to ensure food production in a constantly changing climate which is anticipated to become harsher for crop production.

Keywords

Drought stress · Heat stress · Rhizobacteria · Mitigation · Abiotic stress · Climate change

9.1 Introduction

The global population has increased multifold in the last few decades, and so has agricultural intensification to meet the growing demand for food. It is estimated that the demand for food will further escalate by more than 50% of the current production by the year 2050 to meet the rising demand (FAO, 2017).

This infers that new and more strategies must be explored for sustainable agricultural intensification. However, global agricultural production systems are frequently threatened by a multitude of abiotic and biotic stresses. Biotic stresses emanate from pests, weeds, and diseases that destroy crops and reduce yields, thereby threatening food security (Lima et al., 2019; Umar et al., 2021). Similarly, abiotic stresses such as heat, drought, salinity, and acidity also intensify the vulnerability of crops in agricultural systems (Suzuki et al., 2014). The impending increase in demand for food, together with the continued threats of abiotic and biotic stresses, presents a daunting challenge to global food production systems.

Drought and heat stress are the most common types of plant abiotic stresses in different parts of the world. The severity of drought and heat stress in plants depends on the occurrence and distribution of rainfall, temperatures, and the moisture-holding capacity of soils. Numerous research findings have shown that drought and heat stress affect the yield of many crop plants like legumes such as lentils (*Lens culinaris*) and soybeans (*Cicer arietinum*) (Awasthi et al., 2014; Sehgal et al., 2017), and cereals like maize, wheat, and rice (Chukwudi et al., 2021; Lawas et al., 2018; Qi et al., 2022), which are essential for global food and security. The impacts of drought and heat stress on plants are thus immense. In bread wheat, for instance, the physiological and biochemical attributes were recently shown to be tremendously altered by combined or individual drought and heat stresses during the terminal growth phase (Sattar et al., 2020).

Drought and heat stress to plants are anticipated to increase multifold because of climate change; hence, the imperative for innovative and sustainable solutions in agriculture has never been more pressing. It is in this context that this chapter aims to present the potential of indigenous rhizobacterial populations for the mitigation of drought and heat stress in common food crops. This chapter starts by examining the various implications of drought and heat stress on crops and shifts the focus to plant-root-residing microbes or rhizobacteria with the remarkable capacity to establish intimate relations with host plants and eliciting numerous benefits, including resistance to various abiotic stresses. This chapter includes a systematic examination of the multifarious mechanisms through which the rhizobacteria act in plants to mitigate drought and heat stress. This chapter also addresses the challenges and future directions in harnessing the full potential of rhizobacteria for the resilience of crop production systems to abiotic stresses like drought and heat. This chapter aims to provide a comprehensive overview and lay the foundation for application in agricultural systems to promote resilience to abiotic stresses, and ensure food production in an ever-changing climate. The urgency to harness these beneficial microorganisms is becoming more apparent sustainable food production and global food security.

9.2 Effects of Drought and Heat Stress on Common Crop Plants

Drought and heat stress present formidable challenges and impacts in agricultural systems. The impacts of drought and heat stress on crops are huge and manifest in reduced yields, growth, and quality (Aneja et al., 2022) (Fig. 9.1). These abiotic stresses interfere with critical physiological processes in plants, hinder water uptake and photosynthesis, and inhibit nutrient absorption (Du et al., 2020a, b; Joseph et al., 2021; Shahriari, 2021). Drought also affects the antioxidant defense system in plants that result in oxidative stress owing to the formation of reactive oxygen species (ROS), such as hydrogen peroxide (Rai et al., 2019). Since the excessive accumulation of ROS inhibits the formation of photosynthetic pigments, drought-affected plants may experience impacts on their photosynthetic machinery (Sekmen et al., 2014; Silva et al., 2010). Photosystem II is particularly more vulnerable to oxidative damage. Reactive oxygen species, especially singlet oxygen, can damage the reaction center of Photosystem II, causing a decline in the efficiency of water-splitting and electron transport (Chen et al., 2016). The inhibition of photosynthesis by ROS

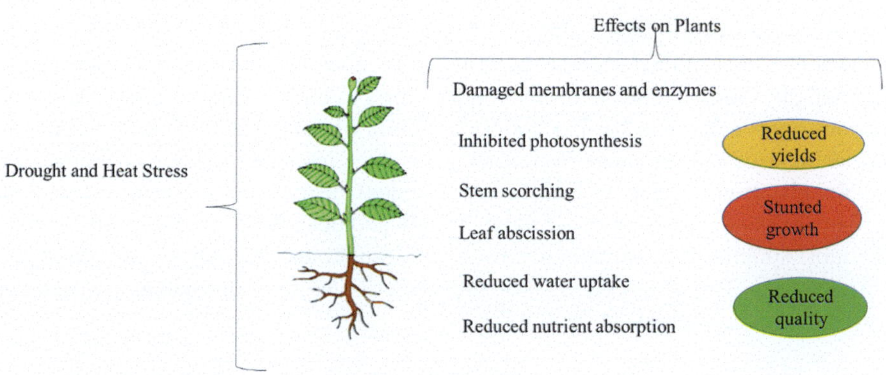

Fig. 9.1 Effects of drought and heat stress on crop plants

has cascading impacts on plant physiology, resulting in reduced biomass accumulation, stunted growth, lower yields, and impaired nutrient uptake. In wheat, for instance, drought can considerably reduce the grain-filling period, peduncle length, plant height, thousand grains weight, the number of spikes per plot, the number of grains per spike, grain yield, biomass, and harvest index (Pour-Aboughadareh et al., 2020). Similar results have also been obtained from lentils (Sehgal et al., 2017).

Generally, drought or heat-affected crops face increased susceptibility to diseases, pest infestations, and diminished resilience to fluctuations in other environmental conditions (Fatima et al., 2022). Drought stress is especially challenging because its severity depends on the occurrence and distribution of rainfall, evaporative demands, and moisture-storing capacity of soils, all of which are unpredictable. At the moment, the climate has changed all around the globe by continuously increasing temperature levels. Under extreme drought conditions, the amount of soil water available to plants is steadily decreasing, causing plants to die prematurely.

Drought causes plants to lower their shoot growth, as well as their metabolic demands as has recently been established in cotton (He et al., 2022) and soybeans (Du et al., 2020b). On the other hand, when plants are subjected to heat stress, their seed germination rate, photosynthetic efficiency, and yield all suffer. Under heat stress, during the reproductive growth period, the function of a petal cell is lost, and the anther is dysplastic. For example, maize yields decrease sharply when the plants are exposed to temperatures greater than approximately 29–30 °C (Liu et al., 2018). The consequences of diminished crop productivity due to drought and heat stress reverberate across nations (Table 9.1). The increased vulnerability of agricultural systems to drought and heat shocks has caused the escalation of food prices, thereby affecting accessibility to staple foods for many people worldwide.

Existing options for mitigating the impacts of drought and heat stress on crops encompass a range of approaches, ranging from the genetic modification of plants to agronomic practices like irrigation management and soil moisture conservation. While these strategies have yielded some success, they are disadvantaged by various inherent limitations. For instance, genetic modification and breeding programs often necessitate prolonged timelines and encounter regulatory hurdles (Bapela et al., 2022). Similarly, agronomic practices might require substantial resources and may not be universally applicable due to varying agroecological contexts. Moreover, most approaches often discourse the symptoms rather than the underlying biochemical and physiological mechanisms of plant stress tolerance. Besides, the efficacy of various approaches can be dependent on factors like the availability of resources, infrastructure, and adoption by farmers, thereby presenting further challenges in their application and impact. This has necessitated the quest for innovative and more sustainable strategies to improve plant drought and heat tolerance. Presently, among the most sought-after technologies for plant drought and heat stress mitigation are rhizobacteria. This chapter discusses rhizobacteria-mediated drought and heat stress mitigation in crop plants in Sect. 9.3.

Table 9.1 Effects of drought and heat stress in plants

Stress	Plant	Effects	Test conditions	Country of study	References
Drought	Cassava (*Manihot esculenta*)	Reduced plant height, stem diameter, leaf number, leaf water content, net photosynthetic rate	Greenhouse	China	Shan et al. (2018)
	Chrysanthemum indicum	Reduced shoot and root lengths, number of flowers, photosynthetic activity, stomatal conductance	Potted	India	Sahithi et al. (2021)
	Cowpeas (*Vigna unguiculata*)	Reduced fresh leaf weight, oxidized soluble sugars and lipids in leaves and roots	In vitro	Canada	Jayawardhane et al. (2022)
	Rapeseed (*Brassica naous*)	Alteration of physiological and anatomical parameters	Potted/growth chamber	China	Zhu et al. (2021)
	Sesame (*Sesamum indicum*)	Reduced germination percentage, coefficient of variation in germination time, germination index, and seedling vigor index	Growth chamber	Pakistan	Ahmed et al. (2022)
	Tomatoes (*Solanum lycopersicum*)	Oxidative stress and the production of antioxidants	Potted	China	Zhou et al. (2019)
	Wheat (*Triticum aestivum*)	Decreased grain weight	Potted/growth chamber	China	Li et al. (2018)

(continued)

Table 9.1 (continued)

Stress	Plant	Effects	Test conditions	Country of study	References
		Altered physiological and biochemical functions	Pot experiment	Pakistan	Sattar et al. (2020)
		Reduced grain filling period, plant height, peduncle length, number of spikes, number of grains per spike, grains weight, grain yield, biomass, and harvest index	Field	Iran	Pour-Aboughadareh et al. (2020)
	Maize (*Zea mays*)	The ratio of gross primary production to chlorophyll fluorescence decreased with increasing drought stress levels, and the canopy stomata conductance both declined	Field	China	Chen et al. (2021)
	Soybean (*Glycine max*)	Reduced evapotranspiration and aboveground biomass accumulation	Pot experiment	China	Cui et al. (2021)
Heat	Mung bean (*Vigna radiate)*	Leaf damage, membrane damage, reduced chlorophyll, chlorophyll fluorescence, leaf water content, seed-filling rate and duration	Pots in an outdoor natural environment	India	Priya et al. (2023)
	Rice (*O. sativa*)	Delayed germination, and reduced shoot/root dry and fresh weight, length	In vitro	India	Vibhuti et al. (2015)
	Tomato (*S. lycopersicum*)	Production of reactive oxygen species like superoxide and hydrogen peroxide	Greenhouse	Korea	Muneer et al. (2016)

9.3 Rhizobacteria-Mediated Drought and Heat Stress Mitigation in Crop Plants

Rhizobacteria, which are natural inhabitants of plant root environments, possess a remarkable capacity to mitigate stress tolerance in crops and offer a promising approach to sustainable agricultural practices amid escalating environmental challenges and climate change. Sections 9.3.1, 9.3.2, 9.3.3 and 9.3.4 present some of the rhizobacteria mediated drought and heat stress mitigation mechanisms in various crop plants. A graphical presentation of these mechanisms is provided in Fig. 9.2 while a summary of rhizobacteria-mediated drought and heat stress mitigation in major crops is provided in Table 9.2.

9.3.1 Physiological and Molecular Alterations

The symbiotic interaction between rhizobacteria and crops fosters numerous mechanisms that collectively enhance plant resilience to drought and heat stress. The beneficial microbes help in augmenting drought and heat stress tolerance by priming the plant's defense mechanisms thus enabling them to withstand and adapt to adverse environmental conditions, and are often referred to as plant growth-promoting rhizobacteria (PGPR). Studies have shown that most PGPR contribute to drought and heat stress mitigation in plants through the alteration of various molecular and physiological responses, for example, in maize (Ahmad et al., 2019, 2023; Notununu et al., 2022).

From the molecular perspective, drought-tolerant PGPR can modulate or regulate the expression of stress-responsive genes and induce drought tolerance in plants as has recently been shown in *Arabidopsis thaliana* by Yasmin et al. (2022). Similar

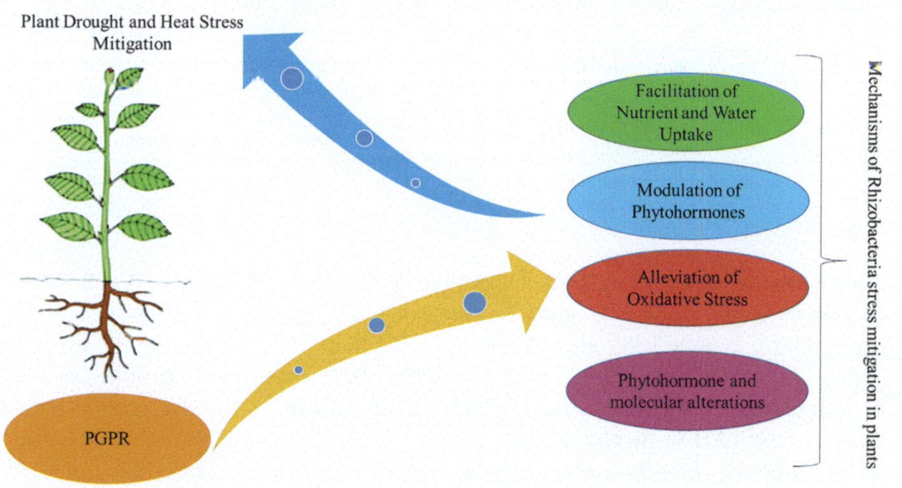

Fig. 9.2 Mechanisms of Rhizobacteria-mediated stress mitigation in plants

Table 9.2 Rhizobacteria-mediated drought and heat stress mitigation in plants

Plant	Rhizobacteria	Test conditions	References
Barley	*Consortia*	Controlled culture room	Slimani et al. (2023)
Chickpea	*Stenotrophomonas* sp.	Greenhouse pot experiment	Sharma et al. (2023)
Common bean (*Phaseolus vulgaris*)	*Bacillus subtilis*	Controlled growth chamber	de Lima et al. (2019)
Faba bean (*Vicia faba*)	*Rhizobium leguminosarum biovar viciae, Pseudomonas putida*	Field	Mansour et al. (2021)
Lentils	*Rhizobium laguerreae, Enterobacter aerogenes, Bacillus* sp.	Greenhouse	Taha et al. (2022)
Maize (*Zea mays*)	*Bacillus, Pseudomonas*	Controlled	Ahmad et al. (2023)
	Pseudomonas sp.	Greenhouse pot experiment	Pereira et al. (2020)
	Pseudomonas spp., *Bacillus* sp.	Pot experiment; greenhouse	Ojuederie and Babalola (2023)
	Bacillus megaterium, Pseudomonas fluorescens, Bacillus pumilus, Pseudomonas putida	Pot and field	Kálmán et al. (2023)
Pearl millet	*Shewanella putrefaciens, Cronobacter dublinensis*	Pot experiment	Manjunatha et al. (2022)
Rice (*Oryza sativa*)	*Bacillus licheniformis, Bacillus haynesii, Bacillus paralicheniformis*	Pots under field conditions	Joshi et al. (2020)
Ryegrass (*Lolium perenne*)	*Bacillus, Pseudomonas*	Germination bags; Greenhouse	He et al. (2021)
Tomato (*Solanum lycopersicum*)	*Not mentioned*	Field	Andryei et al. (2021)
	Bacillus safensis	Greenhouse	Mukhtar et al. (2023)
Wheat (*Triticum aestivum*)	*Bacillus paramycoides, Bacillus paranthracis*	Controlled	Yadav et al. (2022)
	Enterobacter cloacae, Bacillus cereus, Bacillus megaterium	Pot experiment; greenhouse	Shankar and Prasad (2023)

studies include Liu et al. (2020). These studies demonstrate that PGPR can initiate alterations at the molecular level to improve plant response to drought and heat stress. These genetic level alterations by PGPR help plants withstand the negative impact of drought stress. For instance, the expression profile of the 9-cis-epoxycartenoid dioxygenase gene, which is crucial for the production of abscisic acid (ABA) in plants, was shown to increase under drought (Ali et al., 2020). Similarly, *Bacillus amyloliquefaciens* and *B. subtilis* increased the transcription rate of auxin and triggered the production of osmolyte glycine betaine and choline, exerting drought resistance in *Arabidopsis* by altering the expression of cell wall structure governing genes (Zhang et al., 2007).

In *Pseudomonas aeruginosa*, water stress triggers transcription of genes in the alginate biosynthesis gene cluster, which are prominent in eliminating drought stress in *Arabidopsis* (Yasmin et al., 2022). Similar alginate-triggered drought stress endurance is reported in wheat inoculated with *Bacillus thuringiensis*. According to Sen et al. (2018), *Pseudomonas putida* regulates the expression of essential polyamine biosynthesis genes and boost the levels of putrescine and spermidine in plant cells in water-deficient conditions. A recent real-time PCR study showed that inoculating *Moringa oleifera* with *Bacillus pumilus* also upregulated phytoene synthase, γ-tocopherol methyl-transferase, phytoene desaturase, lycopene β cyclase, and thymidylate synthase genes which in turn increased the levels of tocopherols, carotenoids, and folate in the leaves (Sonbarse et al., 2020).

9.3.2 Alleviation of Oxidative Stress

Under drought and heat stress, plants are presented with heightened oxidative stress due to imbalanced production of ROS and the antioxidant defense system. Cognizant of this, drought and heat stress can be mitigated through the alleviation of oxidative stress. In wheat seedlings in water-deficient conditions, inoculation with *B. thuringiensis* enhanced the activity of the ROS-scavenging enzymes such as superoxide dismutase and catalases, which enhanced their survival (Timmusk et al., 2014). Still in wheat, *Bacillus safensis* or *Ochrobactrum pseudogregnonense* applied to seeds improved the activity of antioxidative enzymes and triggered the accumulation of nonenzymatic antioxidants (Chakraborty et al., 2013). Also, the enhanced activity of catalases, peroxidase, and superoxide dismutase diminished the oxidative stress in the plants and resulted in higher shoot and root biomass.

In lemongrass, PGPR have also been shown to improve antioxidant properties and growth (Mirzaei et al., 2020). Recently, drought tolerance in *Pennisetum glaucum* by ACC-deaminase-producing *B. amyloliquefaciens* was shown to correlate with an increased antioxidant defense system (Murali et al., 2021). Generally, bacterial priming leads to less oxidative stress, better survival, and higher growth parameters, such as fresh and dry weights and water content in all aforementioned studies. However, it is not yet clear whether bacteria trigger the increased production of antioxidants in plants or if the increased antioxidant activity and amount are due to bacteria's own production of antioxidants. It is also likely that PGPR has other modes of action that alleviate plant drought and heat stress, such as the production and modulation of several phytohormones as discussed in Sect. 9.3.3.

9.3.3 Modulation of Phytohormones

Phytohormones play a crucial role in the normal development and growth of plants, but they also have an importance in stress response, including drought (Vurukonda et al., 2016). Some PGPR can produce or activate the production of phytohormones in plants, and, thus, affecting plant hormonal balances. Bacterial production of phytohormones is mainly limited to auxins such as indole-3-acetic acid (IAA), indole-3-carboxylic acid, or indole-3-lactic acid. Rhizobacteria also coordinate the modulation of other stress-related phytohormones like abscisic acid (ABA), cytokinins, and gibberellins which are very important in regulating plant stress responses. This is through microbial stimulation of their synthesis, thereby modulating the plant's physiological responses to drought and heat stress. This regulatory role not only helps plants to mitigate water loss through the regulation of stomatal closure, but also promotes adaptive mechanisms in plants confronted with unfavorable environmental conditions.

According to Tanveer and Ali (2022), different strains of *Bacillus* and *Rhizobium* produce auxin and enhance the growth of mung beans (*Vigna radiata* L.) under drought stress. Inoculation of wheat seedlings with IAA-producing *Klebsiella* sp. leads to improved root length and number during drought stress. It triggered higher fresh and dry weights of roots and shoots (Gontia-Mishra et al., 2016). These changes in root morphology are associated with drought tolerance improvement due to improved nutrient and water uptake as discussed in Sect. 9.3.4. The content of ABA may be modified in plants inoculated with PGPR. Indeed, wheat seedlings inoculated with *B. subtilis* showed a lower ABA concentration under drought stress compared to non-inoculated seedlings, most likely from the production of 1-amino cyclopropane-1-carboxylate (ACC)-deaminase (Barnawal et al., 2017). In inoculated plants, ABA decreased by 30% compared to non-primed stressed control, while the shoot dry weight increased by 28% and the root dry weight by 17% (Barnawal et al., 2017).

Plant growth-promoting rhizobacteria that produce ACC-deaminase cause the degradation of ACC which is a precursor of ethylene into ammonium and α-ketobutyrate (Singh et al., 2015). Bacteria that produce ACC-deaminase thus reduce ethylene quantities in plants and contribute to drought and heat stress mitigation. This is because the upsurge of ethylene can cause inhibitory consequences such as inhibition of seed germination and root development. The ACC-deaminase is produced by many PGPRs such as *Bacillus, Rhizobium, Sinorhizobium, Pseudomonas, Variovorax, Azospirillum, or Burkholderia* (Duan et al., 2021; Gowtham et al., 2020; Murali et al., 2021). PGPR that produces ACC-deaminase increases plant growth by modulating ethylene production in response to several stresses, most importantly drought. Wheat seedlings inoculated with ACC-deaminase-producing strain *B. subtilis* showed a diminished content in ACC and better photosynthetic efficiency under drought stress (Barnawal et al., 2017). Similarly, priming wheat seedlings with *Klebsiella* sp. caused higher root length and number, enhanced fresh and dry weight of shoots and roots, and better relative water content (Gontia-Mishra et al., 2016). These studies show that PGPR reduce the negative effects of the stress hormone—ethylene.

9.3.4 Facilitation of Nutrient and Water Uptake

Drought and heat stress often interfere with nutrient uptake and assimilation in plants that exacerbate their negative effects on crop productivity. Rhizobacteria use various strategies, like the production of siderophores, exopolysaccharides, and organic acids, to enhance nutrient availability and uptake by plants under drought and heat stress. The sustained nutrient acquisition under drought and heat stress thus ensures the resilience of crops facing these stresses.

The production of auxins by PGPR may modify the root system architecture and facilitate nutrient and water uptake. In wheat, for example, auxins increase root surface area and allow a better assimilation of water and nutrients, improving global plant growth and health during drought stress (Raheem et al., 2018). Similarly, in maize, PGPR improve growth and nutrient use efficiency under water-deficit conditions (Pereira et al., 2020). The intricate interplay between rhizobacteria and plants that mitigate the impacts of drought and heat stress involves a wide range of mechanisms. By exploiting these mechanisms, agricultural systems can be made sustainable through the mitigation of the effects of climate-induced stresses on crops. As research progresses, a deeper comprehension of these interactions and their potential to ensure food security and sustainability in agriculture is becoming more urgent.

9.4 Challenges and Future Directions for Application

Several limitations and hurdles still hamper the development and utilization of rhizobacteria in drought and heat stress in common food crops. Challenges encompass the variability in efficacy among different rhizobacterial strains and their performance in diverse environmental conditions. Most studies have focused on the amelioration of the effects of drought or heat on yield as single stress factors without emphasizing the roles of rhizobacteria in alleviating combined abiotic stresses. The scalability and commercial viability of rhizobacterial products, logistical constraints in application methods, and limited awareness among farmers also impede widespread adoption. Furthermore, regulatory frameworks, intellectual property rights, and economic constraints pose hurdles in translating research findings into tangible on-field applications.

Experiments carried out in the greenhouse and in vitro conditions are not often sustainable under field conditions and consistency is not maintained because in the field they are exposed to natural environmental conditions which are not present in the greenhouse or in vitro (Aloo et al., 2022). Moreover, other factors like soil pH, soil structure and texture, nutrient availability, and mineral composition also vary under field conditions which alters the efficacy of microbes. Navigating this challenge demands concerted efforts from multidisciplinary stakeholders, including researchers, policymakers, farmers, and industry players.

Current research studies spotlight the exploration of novel rhizobacterial strains with enhanced stress-alleviating capabilities. Advancements in omics technologies enable a deeper exploration of microbial communities and their functional roles in promoting plant resilience. Emerging innovations encompass the development of precision microbiome engineering tools for a targeted manipulation of plant–microbe interactions for tailored stress tolerance. In this front, metagenomics is to be one of the greatest tools in discovering new cultivable flora from different environments (Fadiji et al., 2022). Similarly, genetic manipulation to enhance the expression of stress-responsive genes or engineer plants to foster symbiotic relationships with specific rhizobacterial strains opens new frontiers in augmenting stress tolerance. However, ethical considerations, public acceptance, and regulatory frameworks remain key issues in navigating the path toward deploying genetically modified organisms in agriculture.

Currently, efforts are geared toward the use of rhizobacterial consortia and mixtures of arbuscular mycorrhizal fungi (AMF) and PGPR to boost drought and heat resistance in plants (Aalipour et al., 2020; Azizi et al., 2021; Ikan et al., 2023). These studies have reported better plant performance and superior yields in plants inoculated with a consortium of PGPR or together with AMF. In a recent study by Laranjeira et al. (2022), inoculation with a mixture of PGPR improved plant growth parameters and grain yield; however, the consortium of PGPR and AMF had a synergistic effect that resulted in significant increases in grain yield. Indeed, combining different species in the same microbial formulation broadens the spectrum of action by targeting various biological mechanisms. Similar results have also been shown by Ikan et al. (2023) in durum wheat. Thus, such consortia or mixtures of AMF and PGPR should be explored for better alleviation of drought and heat stress in plants.

9.5 Conclusion

The investigation of rhizobacteria-mediated strategies as an approach to tackling drought and heat stress in agricultural systems is quickly gathering momentum. Rhizobacteria have multifarious mechanism of enhancing stress tolerance, for example, by alleviating oxidative damage, modulating stress-responsive hormones, and facilitating nutrient uptake. Harnessing beneficial microbes for plant abiotic stress mitigation can result in more sustainable agricultural practices that extend beyond mere productivity; but resonates in securing livelihoods and ensuring access to food for the increasing global population. However, harnessing rhizobacteria for abiotic plant stress mitigation is accompanied by various bottlenecks and opportunities that shape research and application. The continued exploration of rhizobacteria in agricultural systems herald a future of sustainable, climate-resilient agriculture and global food production in the face of escalating environmental challenges.

References

Aalipour, H., Nikbakht, A., Etemadi, N., Rejali, F., & Soleimani, M. (2020). Biochemical response and interactions between arbuscular mycorrhizal fungi and plant growth promoting rhizobacteria during establishment and stimulating growth of Arizona cypress (*Cupressus arizonica G.*) under drought stress. *Scientia Horticulturae, 261*, 108923.

Ahmad, I., Zaib, S., Alves, P., Luthe, D., Bano, A., & Shakeel, S. (2019). Molecular and physiological analysis of drought stress responses in Zea mays treated with plant growth promoting rhizobacteria. *Biologia Plantarum, 63*, 536–547.

Ahmad, M., Imtiaz, M., Nawaz, M. S., Mubeen, F., Sarwar, Y., Hayat, M., Asif, M., Naqvi, R. Z., Ahmad, M., & Imran, A. (2023). Thermotolerant PGPR consortium B3P modulates physio-biochemical and molecular machinery for enhanced heat tolerance in maize during early vegetative growth. *Annals of Microbiology, 73*(1), 34.

Ahmed, M., Kheir, A. M., Mehmood, M. Z., Ahmad, S., & Hasanuzzaman, M. (2022). Changes in germination and seedling traits of sesame under simulated drought. *Phyton, 91*(4), 713–726.

Ali, S., Hayat, K., Iqbal, A., & Xie, L. (2020). Implications of abscisic acid in the drought stress tolerance of plants. *Agronomy, 10*(9), 1323. https://doi.org/10.3390/agronomy10091323

Aloo, B. N., Tripathi, V., Makumba, B. A., & Mbega, E. R. (2022). Plant growth-promoting rhizobacterial biofertilizers for crop production: The past, present, and future. *Frontiers in Plant Science, 13*, 1002448.

Andryei, B., Horváth, K. Z., Agyemang Duah, S., Takács, S., Égei, M., Szuvandzsiev, P., & Neményi, A. (2021). Use of plant growth promoting rhizobacteria (PGPRs) in the mitigation of water deficiency of tomato plants (*Solanum lycopersicum* L.). *Journal of Central European Agriculture, 22*(1), 167–177.

Aneja, P., Dwivedi, A., & Ranjan, A. (2022). Physiology of crop yield under heat stress. In R. R. Kumar, S. Praveen, & G. K. Rai (Eds.), *Thermotolerance in crop plants* (pp. 45–79). Springer.

Awasthi, R., Kaushal, N., Vadez, V., Turner, N. C., Berger, J., Siddique, K. H., & Nayyar, H. (2014). Individual and combined effects of transient drought and heat stress on carbon assimilation and seed filling in chickpea. *Functional Plant Biology, 41*(11), 1148–1167.

Azizi, S., Kouchaksaraei, M. T., Hadian, J., Abad, A. R. F. N., Sanavi, S. A. M. M., Ammer, C., & Bader, M. K.-F. (2021). Dual inoculations of arbuscular mycorrhizal fungi and plant growth-promoting rhizobacteria boost drought resistance and essential oil yield of common myrtle. *Forest Ecology and Management, 497*, 119478.

Bapela, T., Shimelis, H., Tsilo, T. J., & Mathew, I. (2022). Genetic improvement of wheat for drought tolerance: Progress, challenges and opportunities. *Plants, 11*(10), 1331.

Barnawal, D., Bharti, N., Pandey, S. S., Pandey, A., Chanotiya, C. S., & Kalra, A. (2017). Plant growth-promoting rhizobacteria enhance wheat salt and drought stress tolerance by altering endogenous phytohormone levels and TaCTR1/TaDREB2 expression. *Physiologia Plantarum, 161*(4), 502–514.

Chakraborty, U., Chakraborty, B. N., Chakraborty, A. P., & Dey, P. L. (2013). Water stress amelioration and plant growth promotion in wheat plants by osmotic stress tolerant bacteria. *World Journal of Microbiology and Biotechnology, 29*, 789–803.

Chen, Y.-E., Liu, W.-J., Su, Y.-Q., Cui, J.-M., Zhang, Z.-W., Yuan, M., Zhang, H.-Y., & Yuan, S. (2016). Different response of photosystem II to short and long-term drought stress in Arabidopsis thaliana. *Physiologia Plantarum, 158*(2), 225–235. https://doi.org/10.1111/ppl.12438

Chen, J., Liu, X., Du, S., Ma, Y., & Liu, L. (2021). Effects of drought on the relationship between photosynthesis and chlorophyll fluorescence for maize. *IEEE Journal of Selected Topics in Applied Earth Observations and Remote Sensing, 14*, 11148–11161.

Chukwudi, U. P., Kutu, F. R., & Mavengahama, S. (2021). Heat stress effect on the grain yield of three drought-tolerant maize varieties under varying growth conditions. *Plants, 10*(8), 1532.

Cui, Y., Ning, S., Jin, J., Jiang, S., Zhou, Y., & Wu, C. (2021). Quantitative lasting effects of drought stress at a growth stage on soybean evapotranspiration and aboveground Biomass. *Water, 13*(1), 18.

de Lima, B. C., Moro, A. L., Santos, A. C. P., Bonifacio, A., Araujo, A. S. F., & de Araujo, F. F. (2019). *Bacillus subtilis* ameliorates water stress tolerance in maize and common bean. *Journal of Plant Interactions, 14*(1), 432–439. https://doi.org/10.1080/17429145.2019.1645896

Du, Y., Zhao, Q., Chen, L., Yao, X., & Xie, F. (2020a). Effect of drought stress at reproductive stages on growth and nitrogen metabolism in soybean. *Agronomy, 10*(2), 302.

Du, Y., Zhao, Q., Chen, L., Yao, X., Zhang, W., Zhang, B., & Xie, F. (2020b). Effect of drought stress on sugar metabolism in leaves and roots of soybean seedlings. *Plant Physiology and Biochemistry, 146*, 1–12.

Duan, B., Li, L., Chen, G., Su-Zhou, C., Li, Y., Merkeryan, H., Liu, W., & Liu, X. (2021). 1-Aminocyclopropane-1-Carboxylate deaminase-producing plant growth-promoting rhizobacteria improve drought stress tolerance in grapevine (Vitis vinifera l.). *Frontiers in Plant Science, 12*, 706990.

Fadiji, A. E., del Orozco-Mosqueda, M. C., de los Santos-Villalobos, S., Santoyo, G., & Babalola, O. O. (2022). Recent developments in the application of plant growth-promoting drought adaptive rhizobacteria for drought mitigation. *Plants, 11*(22), 3090.

FAO. (2017). *The future of food and agriculture—trends and challenges* (pp. 1–180). Food and Agriculture Organization of the United Nations. https://www.fao.org/3/i6583e/i6583e.pdf

Fatima, Z., Naz, S., Iqbal, P., Khan, A., Ullah, H., Abbas, G., Ahmed, M., Mubeen, M., & Ahmad, S. (2022). Field crops and climate change. In W. N. Jatoi, M. Mubeen, A. Ahmad, M. A. Cheema, Z. Lin, & M. Z. Hashmi (Eds.), *Building climate resilience in agriculture: Theory, practice and future perspective* (pp. 83–94). Springer.

Gontia-Mishra, I., Sapre, S., Sharma, A., & Tiwari, S. (2016). Amelioration of drought tolerance in wheat by the interaction of plant growth-promoting rhizobacteria. *Plant Biology, 18*(6), 992–1000. https://doi.org/10.1111/plb.12505

Gowtham, H., Singh, B., Murali, M., Shilpa, N., Prasad, M., Aiyaz, M., Amruthesh, K., & Niranjana, S. (2020). Induction of drought tolerance in tomato upon the application of ACC deaminase producing plant growth promoting rhizobacterium *Bacillus subtilis* Rhizo SF 48. *Microbiological Research, 234*, 126422.

He, A., Niu, S., Yang, D., Ren, W., Zhao, L., Sun, Y., Meng, L., Zhao, Q., Paré, P. W., & Zhang, J. (2021). Two PGPR strains from the rhizosphere of Haloxylon ammodendron promoted growth and enhanced drought tolerance of ryegrass. *Plant Physiology and Biochemistry, 161*, 74–85.

He, J., Hu, W., Li, Y., Zhu, H., Zou, J., Wang, Y., Meng, Y., Chen, B., Zhao, W., & Wang, S. (2022). Prolonged drought affects the interaction of carbon and nitrogen metabolism in root and shoot of cotton. *Environmental and Experimental Botany, 197*, 104839.

Ikan, C., Ben-Laouane, R., Ouhaddou, R., Ghoulam, C., & Meddich, A. (2023). Co-inoculation of arbuscular mycorrhizal fungi and plant growth-promoting rhizobacteria can mitigate the effects of drought in wheat plants (*Triticum durum). Plant Biosystems, 157*(4), 1–20.

Jayawardhane, J., Goyali, J. C., Zafari, S., & Igamberdiev, A. U. (2022). The response of Cowpea (Vigna unguiculata) plants to three abiotic stresses applied with increasing intensity: Hypoxia, salinity, and water deficit. *Metabolites, 12*(1), 38. https://doi.org/10.3390/metabo12010038

Joseph, J., Luster, J., Bottero, A., Buser, N., Baechli, L., Sever, K., & Gessler, A. (2021). Effects of drought on nitrogen uptake and carbon dynamics in trees. *Tree Physiology, 41*(6), 927–943.

Joshi, B., Chaudhary, A., Singh, H., & Kumar, P. A. (2020). Prospective evaluation of individual and consortia plant growth promoting rhizobacteria for drought stress amelioration in rice (*Oryza sativa* L.). *Plant and Soil, 457*, 225–240.

Kálmán, C. D., Nagy, Z., Berényi, A., Kiss, E., & Posta, K. (2023). Investigating PGPR bacteria for their competence to protect hybrid maize from the factor drought stress. *Cereal Research Communications, 52*, 1–22.

Laranjeira, S., Reis, S., Torcato, C., Raimundo, F., Ferreira, L., Carnide, V., Fernandes-Silva, A., & Marques, G. (2022). Use of plant-growth promoting rhizobacteria and mycorrhizal fungi consortium as a strategy to improve chickpea (*Cicer arietinum* L.) productivity under different irrigation regimes. *Agronomy, 12*(6), 1383. https://doi.org/10.3390/agronomy12061383

Lawas, L. M. F., Zuther, E., Jagadish, S. K., & Hincha, D. K. (2018). Molecular mechanisms of combined heat and drought stress resilience in cereals. *Current Opinion in Plant Biology, 45*, 212–217.

Li, X., Lv, X., Wang, X., Wang, L., Zhang, M., & Ren, M. (2018). Effects of abiotic stress on anthocyanin accumulation and grain weight in purple wheat. *Crop and Pasture Science, 69*(12), 1208–1214.

Lima, L. K. S., de Jesus, O. N., Soares, T. L., de Oliveira, S. A. S., Haddad, F., & Girardi, E. A. (2019). Water deficit increases the susceptibility of yellow passion fruit seedlings to *Fusarium* wilt in controlled conditions. *Scientia Horticulturae, 243*, 609–621.

Liu, X., Zhou, Y., Xiao, J., & Bao, F. (2018). Effects of chilling on the structure, function and development of chloroplasts. *Frontiers in Plant Science, 9*, 1715.

Liu, W., Sikora, E., & Park, S.-W. (2020). Plant growth-promoting rhizobacterium, *Paenibacillus polymyxa* CR1, upregulates dehydration-responsive genes, RD29A and RD29B, during priming drought tolerance in arabidopsis. *Plant Physiology and Biochemistry, 156*, 146–154.

Manjunatha, B. S., Nivetha, N., Krishna, G. K., Elangovan, A., Pushkar, S., Chandrashekar, N., Aggarwal, C., Asha, A. D., Chinnusamy, V., Raipuria, R. K., Watts, A., Bandeppa, S., Dukare, A. S., & Paul, S. (2022). Plant growth-promoting rhizobacteria *Shewanella putrefaciens* and *Cronobacter dublinensis* enhance drought tolerance of pearl millet by modulating hormones and stress-responsive genes. *Physiologia Plantarum, 174*(2), e13676. https://doi.org/10.1111/ppl.13676

Mansour, E., Mahgoub, H. A., Mahgoub, S. A., El-Sobky, E.-S. E., Abdul-Hamid, M. I., Kamara, M. M., AbuQamar, S. F., El-Tarabily, K. A., & Desoky, E.-S. M. (2021). Enhancement of drought tolerance in diverse Vicia faba cultivars by inoculation with plant growth-promoting rhizobacteria under newly reclaimed soil conditions. *Scientific Reports, 11*(1), 24142.

Mirzaei, M., Ladan Moghadam, A., Hakimi, L., & Danaee, E. (2020). Plant growth promoting rhizobacteria (PGPR) improve plant growth, antioxidant capacity, and essential oil properties of lemongrass (*Cymbopogon citratus*) under water stress. *Iranian Journal of Plant Physiology, 10*(2), 3155–3166.

Mukhtar, T., Ali, F., Rafique, M., Ali, J., Afridi, M. S., Smith, D., Mehmood, S., Amna, Souleimanov, A., & Jellani, G. (2023). Biochemical characterization and potential of *Bacillus safensis* strain SCAL1 to mitigate heat stress in *Solanum lycopersicum* L. *Journal of Plant Growth Regulation, 42*(1), 523–538.

Muneer, S., Ko, C. H., Wei, H., Chen, Y., & Jeong, B. R. (2016). Physiological and proteomic investigations to study the response of tomato graft unions under temperature stress. *PLoS One, 11*(6), e0157439. https://doi.org/10.1371/journal.pone.0157439

Murali, M., Singh, S. B., Gowtham, H., Shilpa, N., Prasad, M., Aiyaz, M., & Amruthesh, K. (2021). Induction of drought tolerance in *Pennisetum glaucum* by ACC deaminase producing PGPR-*Bacillus amyloliquefaciens* through antioxidant defense system. *Microbiological Research, 253*, 126891.

Notununu, I., Moleleki, L., Roopnarain, A., & Adeleke, R. (2022). Effects of plant growth-promoting rhizobacteria on the molecular responses of maize under drought and heat stresses: A review. *Pedosphere, 32*(1), 90–106.

Ojuederie, O. B., & Babalola, O. O. (2023). Growth enhancement and extenuation of drought stress in maize inoculated with multifaceted ACC deaminase producing rhizobacteria. *Frontiers in Sustainable Food Systems, 6*, 1076844.

Pereira, S., Abreu, D., Moreira, H., Vega, A., & Castro, P. (2020). Plant growth-promoting rhizobacteria (PGPR) improve the growth and nutrient use efficiency in maize (*Zea mays* L.) under water deficit conditions. *Heliyon, 6*(10), e05106.

Pour-Aboughadareh, A., Mohammadi, R., Etminan, A., Shooshtari, L., Maleki-Tabrizi, N., & Poczai, P. (2020). Effects of drought stress on some agronomic and morpho-physiological traits in durum wheat genotypes. *Sustainability, 12*(14), 5610.

Priya, M., Bhardwaj, A., Jha, U. C., HanumanthaRao, B., Prasad, P. V. V., Sharma, K. D., Siddique, K. H. M., & Nayyar, H. (2023). Investigating the influence of elevated temperature on nutritional and yield characteristics of mung bean (*Vigna radiata* L.) genotypes during seed filling in a controlled environment. *Frontiers in Plant Science, 14*, 1233954.

Qi, Y., Zhang, Q., Hu, S., Wang, R., Wang, H., Zhang, K., Zhao, H., Ren, S., Yang, Y., & Zhao, F. (2022). Effects of high temperature and drought stresses on growth and yield of summer maize during grain filling in North China. *Agriculture, 12*(11), 1948.

Raheem, A., Shaposhnikov, A., Belimov, A. A., Dodd, I. C., & Ali, B. (2018). Auxin production by rhizobacteria was associated with improved yield of wheat (*Triticum aestivum* L.) under drought stress. *Archives of Agronomy and Soil Science, 64*(4), 574–587.

Rai, R., Singh, S., Rai, S., Shankar, A., Chatterjee, A., & Rai, L. (2019). Abiotic stress-mediated oxidative damage in plants: An overview. In A. Chatterjee, A. Shankar, S. Singh, V. Kesari, R. Rai, A. K. Patel, & L. Rai (Eds.), *Approaches for enhancing abiotic stress tolerance in plants* (pp. 227–252). CRC Press.

Sahithi, B. M., Razi, K., Al Murad, M., Vinothkumar, A., Jagadeesan, S., Benjamin, L. K., Jeong, B. R., & Muneer, S. (2021). Comparative physiological and proteomic analysis deciphering tolerance and homeostatic signaling pathways in Chrysanthemum under drought stress. *Physiologia Plantarum, 172*(2), 289–303.

Sattar, A., Sher, A., Ijaz, M., Ul-Allah, S., Rizwan, M. S., Hussain, M., Jabran, K., & Cheema, M. A. (2020). Terminal drought and heat stress alter physiological and biochemical attributes in flag leaf of bread wheat. *PLoS One, 15*(5), e0232974.

Sehgal, A., Sita, K., Kumar, J., Kumar, S., Singh, S., Siddique, K. H., & Nayyar, H. (2017). Effects of drought, heat and their interaction on the growth, yield and photosynthetic function of lentil (*Lens culinaris* Medikus) genotypes varying in heat and drought sensitivity. *Frontiers in Plant Science, 8*, 1776.

Sekmen, A. H., Ozgur, R., Uzilday, B., & Turkan, I. (2014). Reactive oxygen species scavenging capacities of cotton (*Gossypium hirsutum*) cultivars under combined drought and heat induced oxidative stress. *Environmental and Experimental Botany, 99*, 141–149.

Sen, S., Ghosh, D., & Mohapatra, S. (2018). Modulation of polyamine biosynthesis in Arabidopsis thaliana by a drought mitigating *Pseudomonas putida* strain. *Plant Physiology and Biochemistry, 129*, 180–188. https://doi.org/10.1016/j.plaphy.2018.05.034

Shahriari, Y. (2021). Plant drought stress: From impact of drought stress on plant morphological, biochemical and physiological features to role of nutrients in drought stress alleviation. *Nveo-Natural Volatiles & Essential Oils Journal, 8*(4), 5344–5369.

Shan, Z., Luo, X., Wei, M., Huang, T., Khan, A., & Zhu, Y. (2018). Physiological and proteomic analysis on long-term drought resistance of cassava (*Manihot esculenta* Crantz). *Scientific Reports, 8*(1), 1–12.

Shankar, A., & Prasad, V. (2023). Potential of desiccation-tolerant plant growth-promoting rhizobacteria in growth augmentation of wheat (*Triticum aestivum* L.) under drought stress. *Frontiers in Microbiology, 14*, 1017167.

Sharma, A., Vaishnav, A., Jamali, H., Keswani, C., Srivastava, A. K., Kaushik, R., Gupta, S., & Bansal, Y. K. (2023). Unraveling the plant growth-promoting mechanisms of Stenotrophomonas sp. CV83 for drought stress tolerance and growth enhancement in chickpea. *Journal of Plant Growth Regulation, 42*(10), 6760–6775. https://doi.org/10.1007/s00344-023-11010-2

Silva, E. N., Ferreira-Silva, S. L., de Vasconcelos Fontenele, A., Ribeiro, R. V., Viégas, R. A., & Silveira, J. A. G. (2010). Photosynthetic changes and protective mechanisms against oxidative damage subjected to isolated and combined drought and heat stresses in Jatropha curcas plants. *Journal of Plant Physiology, 167*(14), 1157–1164.

Singh, R., Shelke, G., Kumar, A., & Jha, P. (2015). Biochemistry and genetics of ACC deaminase: A weapon to "stress ethylene" produced in plants. *Frontiers in Microbiology, 6*, 00937. https://www.frontiersin.org/articles/10.3389/fmicb.2015.00937

Slimani, A., Raklami, A., Oufdou, K., & Meddich, A. (2023). Isolation and characterization of PGPR and their potenzial for drought alleviation in barley plants. *Gesunde Pflanzen, 75*(2), 377–391.

Sonbarse, P. P., Kiran, K., Sharma, P., & Parvatam, G. (2020). Biochemical and molecular insights of PGPR application for the augmentation of carotenoids, tocopherols, and folate in the foliage of Moringa oleifera. *Phytochemistry, 179*, 112506. https://doi.org/10.1016/j.phytochem.2020.112506

Suzuki, N., Rivero, R. M., Shulaev, V., Blumwald, E., & Mittler, R. (2014). Abiotic and biotic stress combinations. *New Phytologist, 203*(1), 32–43.

Taha, K., El Attar, I., Hnini, M., Raif, A., Béna, G., & Aurag, J. (2022). Beneficial effect of Rhizobium laguerreae co-inoculated with native *Bacillus sp*. and *Enterobacter aerogenes* on lentil growth under drought stress. *Rhizosphere, 22*, 100523.

Tanveer, S., & Ali, B. (2022). Evaluation of *Bacillus* and *Rhizobium* strains to enhance the growth of Vigna radiata (L.) under drought stress. *Pak-Euro Journal of Medical and Life Sciences, 5*(1), 101–112.

Timmusk, S., Abd El-Daim, I. A., Copolovici, L., Tanilas, T., Kännaste, A., Behers, L., Nevo, E., Seisenbaeva, G., Stenström, E., & Niinemets, Ü. (2014). Drought-tolerance of wheat improved by rhizosphere bacteria from harsh environments: Enhanced biomass production and reduced emissions of stress volatiles. *PLoS One, 9*(5), e96086.

Umar, O. B., Ranti, L. A., Abdulbaki, A. S., Bola, A., Abdulhamid, A., Biola, M., & Victor, K. (2021). Stresses in plants: Biotic and abiotic. In M. Ansari (Ed.), *Current trends in wheat research*. Intechopen.

Vibhuti, C. S., Bargali, K., & Bargali, S. (2015). Seed germination and seedling growth parameters of rice (*Oryza sativa* L.) varieties as affected by salt and water stress. *Indian Journal of Agricultural Sciences, 85*(1), 102–108.

Vurukonda, S. S. K. P., Vardharajula, S., Shrivastava, M., & SkZ, A. (2016). Enhancement of drought stress tolerance in crops by plant growth promoting rhizobacteria. *Microbiological Research, 184*, 13–24. https://doi.org/10.1016/j.micres.2015.12.003

Yadav, V. K., Yadav, R. C., Choudhary, P., Sharma, S. K., & Bhagat, N. (2022). Mitigation of drought stress in wheat (*Triticum aestivum* L.) by inoculation of drought tolerant Bacillus paramycoides DT-85 and Bacillus paranthracis DT-97. *Journal of Applied Biology and Biotechnology, 10*(1), 59–69.

Yasmin, H., Bano, A., Wilson, N. L., Nosheen, A., Naz, R., Hassan, M. N., Ilyas, N., Saleem, M. H., Noureldeen, A., & Ahmad, P. (2022). Drought-tolerant *Pseudomonas* sp. showed differential expression of stress-responsive genes and induced drought tolerance in Arabidopsis thaliana. *Physiologia Plantarum, 174*(1), e13497.

Zhang, H., Kim, M.-S., Krishnamachari, V., Payton, P., Sun, Y., Grimson, M., Farag, M. A., Ryu, C.-M., Allen, R., Melo, I. S., & Paré, P. W. (2007). Rhizobacterial volatile emissions regulate auxin homeostasis and cell expansion in Arabidopsis. *Planta, 226*(4), 839–851. https://doi.org/10.1007/s00425-007-0530-2

Zhou, R., Kong, L., Yu, X., Ottosen, C.-O., Zhao, T., Jiang, F., & Wu, Z. (2019). Oxidative damage and antioxidant mechanism in tomatoes responding to drought and heat stress. *Acta Physiologiae Plantarum, 41*(2), 20. https://doi.org/10.1007/s11738-019-2805-1

Zhu, J., Cai, D., Wang, J., Cao, J., Wen, Y., He, J., Zhao, L., Wang, D., & Zhang, S. (2021). Physiological and anatomical changes in two rapeseed (*Brassica napus* L.) genotypes under drought stress conditions. *Oil Crop Science, 6*(2), 97–104. https://doi.org/10.1016/j.ocsci.2021.04.003

Influencing and Managing Abiotic Stresses in Cereals

10

K. Sammi Reddy, M. G. Vinay, E. Senthamil, Hanamant M. Halli, P. S. Basavaraj, C. B. Harisha, and K. M. Boraiah

Abstract

Cereals are crucial for global food security, providing essential calories and proteins to a large segment of the global population and serving as key livestock feed. Additionally, cereals are integral to various industrial applications, including biofuels, paper, and pharmaceuticals, playing important role in economic and societal importance across the globe. Abiotic stresses such as drought, heat, salinity, cold, and waterlogging severely impact cereal crops, leading to substantial yield reductions and also grain quality including nutrient composition. These stresses disrupt vital physiological processes like photosynthesis and nutrient uptake, directly affecting plant growth and productivity. Effectively managing these stresses is crucial for sustaining cereal crop yields and ensuring global food security. This chapter is vital for developing effective strategies to manage abiotic stresses in cereal crops, offering a dual perspective highlighting the impacts of these stresses and agronomic and genetic interventions to reduce impacts posed by different stresses. Importantly, the chapter provides a detailed exploration of adaptive mechanisms and resilience traits in cereals, informing breeding programs and agronomic practices designed to enhance stress tolerance. By synthesizing cutting-edge research and practical management techniques, the

K. Sammi Reddy · H. M. Halli · P. S. Basavaraj · C. B. Harisha · K. M. Boraiah (✉)
ICAR-National Institute of Abiotic Stress Management,
Malegaon, Baramati, Pune, Maharashtra, India
e-mail: boraiah.m@icar.gov.in

M. G. Vinay
Department of Agronomy, GKVK, University of Agricultural Sciences,
Bangalore, Karnataka, India

E. Senthamil
Department of Agronomy, University of Agricultural Sciences, Dharwad, Karnataka, India

R. Dubey et al. (eds.), *Drought and Heat Stress in Agriculture*,
https://doi.org/10.1007/978-981-96-5735-3_10

chapter equips researchers, farmers, and policymakers with the knowledge needed to implement sustainable solutions that can mitigate the effects of abiotic stresses, ultimately contributing to global food security.

Keywords

Abiotic stresses · Cereal crop · Crop yields · Food security

10.1 Introduction

Cereals hold a paramount position in global food production, directly contributing as major components of food grain production and indirectly supporting livestock production. Hence, cereals serve as vital sources of sustenance for both human beings and livestock across the world. The cereals, belonging to the Gramineae family, are cultivated extensively for their digestible constituents and unique botanical structure, encompassing the bran, endosperm, and embryo. The prominent grain species within the cereal family include wheat (*Triticum*), barley (*Hordeum*), oat (*Avena*), rye (*Secale*), maize (*Zea*), rice (*Oryza*), pearl millet (*Pennisetum*), and sorghum (*Sorghum*), with wheat and rice standing out as particularly dominant. Originating from diverse locations, cereal crops have adapted to thrive in a broad range of climatic conditions. As the most abundantly grown and consumed commodities worldwide, cereals and their derivatives constitute staple diets in many countries, providing unparalleled dietary energy and serving as principal sources of essential macro- and micronutrients. According to the Food and Agriculture Organization (FAO, 2023), cereals contribute an estimated 51% of calories and 47% of protein in an average human diet. Cereal crops not only serve as a feed and dietary foundation but also play a crucial role in various industrial applications such as starch, biofuels, enzymes, and biopesticides, as well as pharmaceutical, paper, and textile industries.

The total annual cereal production globally is about 2500 million tons. In order to feed the growing population and meet their specific dietary and nutritional demands, it is also essential to gear up cereal production with desired quality. However, in the next decade of 2023–2032, the growth of these crops demand is expected to be slower than the past decade due to weaker growth in feed demand, biofuels and other industrial uses with witnessing the saturation in the consumption of most cereals (OECD/FAO, 2023). Population-driven increases in the consumption of wheat and rice are expected in Asia, and millet, sorghum, and maize are expected in Africa. In contrast, the growing role of rice in African diets is projected to translate into continued increases in its per capita food use in that continent. The region-specific cereal consumption pattern depends on production factors determined by their cultivation, which varies widely in different countries and depends on policies and economy besides favorable growing climatic conditions.

Further, it projected that in the next decade, global production of cereals will increase from its current level by about 320 Mt. to 3.1 bln t by 2032, mainly contribution from maize, rice, and other coarse grains mostly millets expected to originate primarily in Asian and African countries. The expected growth will be due to higher yields attributed to the broader availability and adoption of new and improved varieties, and improved agricultural practices in addition efficient and intensive use of inputs including land. Additionally, global market imbalance could revive policies aimed at increasing domestic production. On the other hand, climate change, lack of access to new technologies in a few countries, as well as insufficient funding and poor support from the government and new environmental policies, may limit the growth of cereal production (OECD/FAO, 2023).

Globally, the major cereal crops like maize, sorghum, and millets are grown mainly as rainfed crops, and, thus, these crops are more prone to the uncertainties of rains and extreme weather conditions and hence frequently witnessing abiotic stresses like drought, waterlogging, heat, and salinity. Similarly, winter cereals such as wheat, barley, oats, and triticale are often affected by salinity, drought, and extreme thermal stresses. The reports of varying levels of yield of penalty in different cereals indicate the crop's specific vulnerability to various abiotic stresses (Pathak et al., 2022), and losses depend on the duration and severity of stress/es besides crop stage and its inherent tolerance capacity. For instance, severe drought combined with heat at the flowering stage may cause complete abortion of ears and lead to 100% yield loss in maize (Boraiah et al., 2022).

Despite increasing demand for wheat by about 60% in 2050 to feed an estimated 9.7 billion people in the world, an increment in productivity was acceptable to meet demand, as it increased by only 1.1% (Anonymous, 2018). Recently, several international organizations have forecasted that the extreme events of abiotic stresses, such as drought, high temperature, and salinity, will reduce wheat yield by 20–30%, especially in developing countries. The projected impact of global warming indicates that for every 1 °C temperature rise, there is an estimated 4.0–6.5% reduction in wheat production (Trnka et al., 2019). As a piece of evidence, wheat crops are witnessing severe yield loss in Australia and Pakistan, followed by India and China, due to heat stress (Zulkiffal et al., 2021). Besides global warming, the deficit in irrigation water forces wheat crops to face the negative impacts of drought. Despite the fact that approximately 80% of the wheat-growing area in India is under irrigation, only two-thirds of these areas receive partial irrigation out of the six recommended irrigations (Singh et al., 2022).

The consequence of abiotic stresses accelerated by global climate change, along with extreme weather in local conditions, forced some countries/territories to seek external assistance for food. The countries facing unfavorable crop production prospects are countries where current conditions indicate a high likelihood that cereal production would fall below the 5-year average as a result of a reduction of the area planted and yields due to diverse abiotic stresses influenced by adverse weather conditions (FAO, 2023). Additionally, abiotic stresses also have repercussions on the quality of cereals, affecting their grain compositions, including phytochemicals, vitamins, fiber, protein, free amino acids, sugars, and oils.

Overall, abiotic stresses, exacerbated by climate change and compounded by non-climate stressors like population growth and increased demand for animal-sourced products, pose a significant threat to the global food production system, including cereals cultivation, leading to regional food and nutritional insecurity in several countries (FAO, 2023). Further, recent trends in crop research and forecasting studies highlight the importance of abiotic stress in crops, especially cereals (Morris et al., 2024). The majority of these studies have identified various approaches, such as predicting climate abnormalities that impact cereal-dominated food supplies, developing yield loss models, investigating the effects of abiotic stresses, and understanding plant responses through physiological and molecular genetic studies.

Recognizing that abiotic stresses pose significant challenges to cereal productivity and quality, it is imperative to address and mitigate these challenges through a combination of agronomic and genetic interventions. Keeping abreast of recent advancements in these fields and disseminating this knowledge is crucial for alleviating and minimizing the impact of abiotic stresses on cereal production. In this chapter, we discussed the effect of abiotic stresses on cereal productivity and quality. Further, we explore potential management strategies to alleviate these stresses and conclude by proposing a way forward for achieving sustainable cereal production.

10.2 Abiotic Stresses Affecting Cereals Production

10.2.1 Water Deficit (Drought)

Among different abiotic stresses, drought significantly affects both the yield and quality of cereal crops worldwide. The Global Report on Food Crises (FSIN and Global Network Against Food Crises, 2023) highlighted that out of 258 million people, 56.8 million faced both severe food and nutritional insecurity, primarily resulting from sustained drought in the Horn of Africa, devastating flooding in Pakistan, and tropical storms, cyclones, and drought in Southern Africa. This underscores the detrimental impact of abiotic stresses on agriculture, particularly the crop production system. Additionally, crop-based rainfed agriculture, constituting 80% of global farming, is a critical factor influencing global food security due to its susceptibility to fluctuations in food production. With over 65% of global cereal croplands depending on rainfed conditions, it contributes to approximately 60% of the world's cereal production, including 90% of millet. The countries relying on rainfed cereal crop production often experience food insecurity due to crop failures attributed to water deficit drought events.

The drought is a regular climatic event in the Guinea Savannah ecological zone of Nigeria. It varies in time of occurrence and severity from year to year and thus often affects cereal farming (Ojebisi & Kolade-Oje, 2021). Durowoju et al. (2023) noticed that cereal crops in the Kaduna River Basin of Nigeria were affected by drought and found positive correlations between crop yields and drought indices, with rainfall and soil moisture as the most significant influencing factors. While four major abiotic stresses, namely droughts, heatwaves, cold, and flooding, have substantial impacts on cereal production in European countries, it is noteworthy that the first two events, droughts and heatwaves, cause a more severe reduction in cereal (wheat, maize, and rye) production, leading to an average yield decrease of 9% and 7.3%, respectively (Brás et al., 2021). Similarly, drought impacts cereal production in Morocco's rainfed areas, leading to fluctuating and low yields ranging from 150 to 3000 kg/ha (Karrou & Oweis, 2014). The winter and spring cereals production in the Central Bohemian region was affected severely, and crop failures were even witnessed during the years 1964, 1976, 2000, and 2003 (Potop et al., 2009) due to severe droughts.

Similarly, in recent decades, the Upper Middle Gangetic Plain (UMGP) region in India has experienced a notable drought condition as a result of deficit rainfall followed by increased evaporation, leading to an expansion of the drought-affected areas (Nath et al., 2017) and thus affected approximately 20–30% of the nation's annual cereal production. As a shred of evidence, Boraiah et al. (2022) reported that almost 4.0 Mha of maize is prone to drought as 80% of maize is grown in the *Kharif* season as a rainfed crop and, as a result, witnessed a yield gap of ~6 Mg ha^{-1} between the potential productivity with assured rainfall conditions (8.0 Mg ha^{-1}) and national average yield (2.1 Mg ha^{-1}). Likewise, during the drought year (1992), southern Africa was able to produce only 12.5 mt compared to 23.5 mt in an average year (1993). As a result, the worldwide annual maize crop yield losses due to drought are about 15% of well-watered yield potential, which is worth $36 billion. Besides climatic factors, edaphic and other crop management factors may also intensify the drought. Quinton et al. (2022) reported that tillage exacerbates the vulnerability of cereal crops to drought and affects the yields of wheat and maize. They also suggested the adoption of no-tillage practices and the avoidance of down-slope cultivation. In general, the water requirement of cereals during the seedling phase remains relatively low due to limited leaf area and reduced evapotranspiration rates. However, the underdeveloped root systems of young plants hinder the extraction of deeper profile water.

Additionally, under water deficit, closing stomata in young crops is an initial response to conserve water, leading to halted vegetative growth. Further, adverse hot and dry conditions can induce wilting, diminishing the plant's photosynthetic capacity and potentially reducing yield. Hence, identifying the yield components contributing to higher yield even under drought conditions is crucial; for instance, drought stress during a seedling stage in spring cereals may result in fewer tillers, while stress at anthesis leads to fewer chaffy kernels (Table 10.1). While the yield components of corn are less affected compared to spring cereals, corn is less susceptible to yield loss from deficit water during the vegetative stage than wheat (Ransom, 2012). However, longer dry soil conditions can affect the development of adventitious roots in both spring cereals and maize and negatively impact water and nutrient absorption, potentially resulting in floppy grains. The effect of drought at the terminal growth stage is also detrimental in wheat as it reduces assimilate production, fertility, and grain weight (Tiwari et al., 2017). Water stress conditions also impact hormonal levels in wheat, with an increase observed in certain hormones

Table 10.1 Yield components of cereals are affected by abiotic stresses stress at various growth stages

Crops	Growth stage	Yield component	References
Wheat and other spring cereals like rye, barley, and oats	Emergence to tillering	Tillers per plant	Slafer and Rawson (1994)
	Late tillering to early flowering	Spike survival per plant	
	Stem elongation to Boot stage	Number of spikelets per spike	
	Mid-stem elongation to grain filling	Number of kernels per spikelet	
	Grain filling to maturity	Kernel size	
Maize	V1–V5	Plant height and overall reproductive traits	Aslam et al. (2015)
	V5–7	Number of kernel rows	Ransom (2012)
	V9–VT	Number of potential kernels per row	
	VT–R1	Number of potential kernels that are fertilized	
	R2–R3	Kernel number to be filled. Stress at R2 and R3 can cause fertilized kernels to be aborted	
	R4–R5	Kernel size	
Rice	Flowering stage	Reduction in grain size, spikelet per panicle	Moonmoon and Islam (2017)
		Incomplete panicle exertion, 30% spikelet sterility, and a 20–46% reduction in seed set	Bahuguna et al. (2018)
	Grain filling	Increased chaffy grains	Yang et al. (2019)
	Onset of anthesis	Reduced the grain yield and harvest index	Ahmad et al. (2022)

like abscisic acid (ABA), salicylic acid, jasmonic acid, and ethylene. In contrast, the endogenous levels of some hormones decrease (Singh et al., 2022).

Drought exerts a significant impact on the yield-attributing traits of cereals, influencing their growth, physiology, and grain development. The reproductive phase is susceptible to drought stress, leading to substantial effects on yield and its contributing factors, as outlined in Table 10.1. Moisture deficiency during seeding stages alters growth patterns, nutrient distribution, and carbohydrate metabolism. In contrast, water deficit in the vegetative growth stage hampers photosynthetic rates and disrupts the crucial source–sink relationship, affecting reproductive efficiency and grain filling (Aslam et al., 2015).

The consequences of water deficit during specific growth stages are noteworthy. For instance, a 5-day drought during pre- and post-pollination stages in maize results in abnormal embryo formation and reduced grain numbers, especially in apical ear regions (Setter et al., 2001). Similarly, drought during wheat's early growth stage adversely affects seedling establishment and tiller production. Terminal drought stress in wheat can be particularly detrimental, lowering fertility, assimilating production, and grain weight (Gull et al., 2019; Tiwari et al., 2015). The crown root initiation stage is considered a critical period for irrigation in wheat, as witnessed 39.35% yield loss in a 56-day drought at this stage (Akter et al., 2023). Other pivotal growth stages, including tillering, booting, anthesis, and grain filling, are also susceptible to water shortage, adversely affecting wheat growth, development, and grain yield. Similarly, late-terminal drought stress during the grain-filling stage proves detrimental to winter barley crops in semiarid Mediterranean regions (Samarah & Alqudah, 2011). Seed development stage drought stress reduces leaf photosynthetic rates, shortens grain filling duration, induces early plant maturity, and severely reduces yield components, grain yield, and quality (Rani et al., 2018). Barley, experiencing drought stress at different growth stages, demonstrates varied yield reductions ranging from 12.9% to 33.6% (Haddadin et al., 2013).

In summary, comprehending the consequences of drought stress at different growth stages is imperative for the effective management of cereal production under water stress conditions. Drought-stressed cereals exhibit a shorter grain-filling period and reduced yield, manifested through fewer tillers, spikes, and grains per plant, along with lower individual grain weight. Besides affecting the yields (Table 10.2) and its attributing traits (Table 10.1), drought also affects the chief components of cereals, namely, starch and carbohydrates. Additionally, protein content and quality, including biochemical composition, are important determinants of cereal grain quality, and they are sensitive to environmental stresses and crop management practices. Hence, drought and other abiotic stresses significantly alter the nutritional and other quality traits of cereal crops through modifications in morphological, physiological, and biochemical aspects (Table 10.3).

10.2.2 Waterlogging/Submergence (Excess Moisture)

The prevailing factor responsible for the emergence of diverse abiotic stresses, such as drought, waterlogging, salinity, and heat, can be traced back to the unpredictability of rainfall. Directly or indirectly, drought, salinity, and heat stress are consequences of deficit rainfall, whereas excessive rainfall leads to issues like waterlogging, submergence, lodging, and heightened vulnerability to pests and diseases. The impact of waterlogging, mainly persuaded by excessive rainfall, on cereal crop yield, is multifaceted and linked to the distinct shoot and root architecture inherent in cereals, making them particularly susceptible to such challenges.

Table 10.2 Effect of drought on cereals yield

Crop	Growth stage[a]	Details of abiotic stress	Decrease in yield (%)	References
Maize	Vegetative	Induced progressive drought	18.6–26.2	Mi et al. (2018)
	Pre-reproductive	Drought for 21 days	15–25	Nesmith and Ritchie (1992)
	Reproductive	–	80–90	Dass et al. (2011)
		Water deficit at flowering	50	Denmead and Shaw (1960)
		Induced progressive drought	41.6–46.6	Mi et al. (2018)
		40% water deficit	39.3	Daryanto et al. (2016)
	Grain filling	Deficit irrigation	20–50	Dass et al. (2011)
			21	Denmead and Shaw (1960)
	At different stages of crop	In sensitive genotypes	34.28–6.15	Sah et al. (2020)
		Intolerant genotypes	38.48–55.95	
		Mild drought	13–63	Ge et al. (2012), Earl and Davis (2003)
		Severe drought	26–85	
Wheat	Crown root initiation	Drought for 56 days	39.35	Akter et al. (2023)
	Pre-reproductive	Irrigation at 55–75% of field capacity	2.03–64.39	Mu et al. (2021)
	Grain filling	Deficit irrigation	10	Tabassam et al. (2014)
	At different stages of crop	40% water deficit	20	Daryanto et al. (2016)
Barley	Seedling to tillering	30% deficit irrigation (of field capacity)	12.9	Haddadin et al. (2013)
	Late boot stage to ripening		26.1	
	Early and late-stage		33.6	
	Pre-heading	30–50 days	56	Eser et al. (2024)
	Grain filling	Glasshouse experiment with 25% field capacity	85–87	Samarah et al. (2009)
		Field experiment at 50% field capacity	73–85	
Rice	Reproductive stage	Field Experiments	48–94	Lafitte et al. (2004)
	Grain-filling stage	Field Experiments	60	

[a]The stage of the crop may vary depending on the crops

Table 10.3 Abiotic stresses affecting the grain quality in different cereals

Crops	Abiotic stress	Effects	References
Wheat	Drought	Reduction in the glutenin	Balla et al. (2011)
		Reduced arabinoxylan in a drought-tolerant variety	Rakszegi et al. (2014)
	Drought and heat	Reduced β-glucan but increased protein and arabinoxylan content	
	Heat	Increase in sucrose, reducing sugars, sugar phosphates, and protein while reducing starch	Gooding et al. (2003), Jenner (1991)
		It affects grain weight and flour quality	Wardlaw et al. (2002)
		Alters in protein composition	Wrigley et al. (1994)
	Elevated atmospheric CO_2	Reducing the overall protein	Blumenthal et al. (1996)
	High N + water availability	Increase protein content	Hawkesford (2014)
		Reduction in β-glucan	Rakszegi et al. (2014)
	Low S	Reduction in sulfur-rich glutenins	Zhao et al. (1999)
Barley	N+ drought	Altered β-Glucan content	Güler (2003)
	Low S	Reduction in sulfur-rich storage proteins-hordeins	Shewry et al. (1983)
	Salinity	Accumulation of free amino acids—asparagine, and proline	Garthwaite et al. (2005)
Maize	Drought	Reduction in total protein content	Ali et al. (2010)
		Reduction in grain oil content (40%) with more oleic and less linoleic acid	
Rice	Drought	Increase in total protein content	Crusciol et al. (2008), Fofana et al. (2010)
	Temperature stress	Reduction in protein content	Ziska and Manalo (1996)
	Elevated temperatures	Increase in α-tocotrienol and/or α-tocopherol	Britz et al. (2007)
Pearl millet	Drought	Increase in free amino acids (free asparagine and proline)	Kusaka et al. (2005)

Waterlogging, characterized by excessive soil moisture, leads to delayed farm operations, poor crop establishment, root damage affecting water and nutrient uptake, impaired photosynthesis, and increased nitrogen loss through leaching and denitrification. Submergence, where roots and parts of the shoot are underwater, further exacerbates the challenges by limiting gas diffusion and reducing light availability. Waterlogging, an abiotic stress, poses a significant threat to crop yield in various global regions. A comprehensive meta-analysis (Tian et al., 2021a, b) encompassing 2419 comparisons to evaluate the impact of waterlogging on crop yield globally indicated an average crop yield reduction of 32.9% due to waterlogging, affecting yield attributing traits, namely, grain weight, biomass, plant height, net photosynthetic rate, and leaf area index. The yield reduction depends on crop type, growth stage, and waterlogging duration, which significantly influence the overall effect, with wheat, a major stable crop, experiencing a 25.53% reduction in yield, which underscores the importance of considering the impact and management of waterlogging stress in cereals.

Excessive soil moisture in the early growing season affects germination, root growth, crop stand (plant population), and crop establishment. The impact of poor crop establishment on yield and yield-attributing traits varies among cereal crops. For instance, maize and wheat are more sensitive to excess moisture compared to sorghum, with stunted and slow growth in the former. However, recent studies indicate that sorghum is sensitive to waterlogging at the five-leaf stage (Zhang et al., 2023a, b). Similarly, barley is also more sensitive to waterlogging than wheat and other cereals, particularly between leaf emergence and the booting stage, and yield loss of more than 80% under natural waterlogging is reported (Table 10.4)—the mechanisms of yield loss different from wheat. For instance, a greater reduction in the number of spikes per plant is common in barley, while a decrease was observed in wheat with respect to the number of fertile florets per spike.

In general, the vulnerability of cereal crops is most pronounced during early vegetative and flowering stages, where submergence of roots leads to reduced intracellular oxygen availability, inhibiting aerobic respiration and energy metabolism. Though cereal crops adapt to develop the formation of adventitious and aerenchyma roots, as observed in maize, wheat, and barley, they respond to the anoxic conditions through fermentation, triggering a cascade of events that impact ATP production, cytosolic pH, and glycolysis. However, cereals, including maize and rice, maintain substrate-level ATP production through anaerobic fermentation, with increased expression of genes related to alcohol dehydrogenase and pyruvate dehydrogenase. Prolonged waterlogging results in the accumulation of toxic chemicals, such as Mn_2^+, Fe_2^+, H_2S, and reactive oxygen species (ROS), adversely affecting photosynthetic rate, dry matter accumulation, plant growth, and, ultimately, grain yield. The grain yield reduction in all the cereals is associated with reductions in the number of tillers and spikes per plant, the number of grains per spike, and the thousand kernel weight.

Along with lodging, waterlogging causes colossal losses both in grain and fodder. For instance, the yield loss in cereals has been reported from Western Australia (WA), which varies between 15% and 85% (Setter, 2000). Waterlogging affects

Table 10.4 Effect of waterlogging on cereals yield

Crops	Growth stage[a]	Details of Abiotic stress	Decrease in yield (%)	References
Maize	V3 stage	1–2 days of continuous waterlogging	7–35	Ren et al. (2014)
	Flowering	Waterlogging for 0–3; 3–6; 6–9, and 9–15 days	22, 40, 53, and 76	Tian et al. (2021a, b)
Wheat	Sowing to grain-filling	Waterlogging for 30 days	50–70	Pais et al. (2022)
	Vegetative stage	Waterlogging for 12 days	11–27	Singh and Setter (2017)
		Waterlogging for 21 days	10–15	Araki et al. (2012)
	Tillering	Waterlogging for 21 days	37–60	Cotrozzi et al. (2021)
		Waterlogging durations for 14 and 35 days	45 and 64	
	Pre-reproductive	From the appearance of the 7th leaf to the anthesis	34–92	De San Celedonio et al. (2014)
		20 days of waterlogging from pre-anthesis to anthesis	35% in wheat	Becheran et al. (2022)
	Flowering	5, 10, 15, 20, 25, and 50-day waterlogging	3.9–88.6	Olgun et al. (2008)
	–	In clay soils	24	Cannell et al. (1984)
	–	Sandy loam soils	21	
	–	Under natural conditions	19–82	Setter et al. (1999)
	Early stages	Waterlogging for 15 days	70	de San Celedonio et al. (2018)
	At different stages of crop	21 days waterlogging at jointing and anthesis	22–35	Araki et al. (2012)
			51–84	
Barley	Vegetative stage	3 and 4 leaf stages for 20 days	10	Masoni et al. (2016)
		7th leaf appearance on to anthesis	40–79	De San Celedonio et al. (2014)
	Tillering	Waterlogging during 25–45 days after emergence	25% (seed weight)	De San Celedonio et al. (2014)
	Pre-reproductive	20 days pre-anthesis to anthesis	44	Becheran et al. (2022)
	Pre-flowering	Waterlogging during 65–85 days after emergence	75% (seed mass)	De San Celedonio et al. (2014)
		Waterlogging was initiated one day before heading for 15 days	70	Liu et al. (2020)
	–	Waterlogging in clay soils	30	Cannell et al. (1984)
	Reproductive stage	Waterlogging for 15 days	70	Liu et al. (2020)
	Early stages	Waterlogging for 15 days	60	De San Celedonio et al. (2018)

(continued)

Table 10.4 (continued)

Crops	Growth stage[a]	Details of Abiotic stress	Decrease in yield (%)	References
Millets				
Proso millet	Pre and post-reproductive stage	Waterlogging treatment was initiated 17 days after sowing to harvesting (TT)	16	Matsuura et al. (2016)
Yellow foxtail			18	
Foxtail millet			4	
Little millet			210	
Rice	Vegetative growth stage	43 cm depth of tap water for about 2 weeks	49.6	Chen et al. (2021)
	Vegetative growth stage	Reducing numbers of tillers, plant leaf area, and dry weight above ground, and in the worst case, by causing high mortality	–	Huang et al. (2020)

10–15 million ha of wheat cultivated area, causing annual yield losses of 20–50% (Pais et al., 2022). The high rainfall zone (HRZ) of Victoria in Australia witnessed wheat yield losses of up to 78%. Waterlogging also affected pasture production across southwestern Australia, and it was reported that the loss was valued at about $ 100 m during wet years (Wallace et al., 1992). The menace of waterlogging is particularly pronounced in South and Southeast Asia, where approximately 80% of maize is cultivated as a rainfed crop. In these regions, waterlogging is becoming a major challenge in maize production, leading to significant and recurrent yield losses. Submergence is a crucial limiting abiotic factor of rice production in South- and Southeast Asia, where heavy monsoon rainfall induces complete submergence (Mittal et al., 2022). However, yield loss can vary (0–100%), depending on the stage of the crop experiencing the submergence (Mottaleb et al., 2015). Waterlogging from the beginning of the flowering stage for various durations reduced the wheat crop yield by 3.9–88.6% (Olgun et al., 2008). Waterlogging for 15 days at a spike growth period decreased the wheat grain yield by 70% (de San Celedonio et al., 2018). Waterlogging for 20 days at 3–4 leaf stages reduced barley yield by 10% (Masoni et al., 2016). Meanwhile, waterlogging for 15 days at the spike growth stage reduced the yield by 70% in barley (de San Celedonio et al., 2018).

Indirectly, waterlogging contributed to poor yield by shifting the sowing window. For instance, in Ohio, excessive rainfall during the usual planting windows from April to May often forces farmers to plant their maize in June. Late planting leads to shorter growing seasons and harsher growing conditions (i.e., hotter and drier) during reproductive stages and eventually witnesses a 12–22% yield reduction (Lindsey et al., 2015).

Further, waterlogging, at the crop maturity stage can affect the timely harvest of crops by diminishing the maneuverability of agricultural machinery. The consequences of delayed harvest are manifold, exposing plants to an extended period of vulnerability to adverse weather conditions. This prolonged exposure heightens the risks of pest infestations, diseases, preharvest sprouting, and lodging, ultimately leading to yield losses. Notably, the impact of harvest delays on wheat yield in Western Australia was observed to decline by up to 0.5% per day of delay (Bolland, 1984). Similarly, for maize cultivation in Ohio, harvest delays post-mid-November contribute to decreased yields, primarily attributed to heightened instances of stalk rot and lodging, especially at higher plant populations (Thomison et al., 2011).

The occurrence of waterlogging and other stresses together simultaneously or consequently is common and leads to huge losses. For instance, waterlogging and sunlight shortage often exist simultaneously in major wheat growing areas under persistent rainy and cloudy weather conditions and affect the yield synergistically (Jiang et al., 2022). Similarly, waterlogging with a low temperature of water in the soil reduced the combined tiller and secondary roots in wheat (Ghobadi et al., 2017). The combined effect of waterlogging and salinity for 2 weeks significantly reduced the plant growth, maximum quantum efficiency, PSII activity, chlorophyll, and relative water content of barley varieties (Zeng et al., 2013). Further, continuous waterlogging causes mineral toxicity and nutrient deficiency. Waterlogging causes Mn^{2+} toxicity in cereals.

The morphological and biochemical traits enable cereal crops to tolerate waterlogging conditions and extract oxygen directly from the air, contributing to waterlogging tolerance. Genetic variability in maize allows breeders to identify waterlogging-tolerant genotypes, with teosinte and its subspecies identified as potential sources for developing tolerant varieties/hybrids. Sagittario and Basribey identified waterlogging-tolerant varieties in Turkey (Yavas et al., 2012). Since waterlogging is always associated with other biophysical processes/es, understanding the crop responses related to these stressors is essential for reducing yield losses in cereal crops. The responses may vary with crops and biomolecules. For instance, waterlogging at the flowering stage in wheat increased the proline accumulation linearly up to 20 days with decreased protein content. In contrast, chlorophyll a and b decreased linearly with the increase in duration of waterlogging up to 50 days (Olgun et al., 2008). However, tolerant genotypes showed 60% higher chlorophyll content in barley (Huang et al., 2015). However, with increasing water table depth from 30 to 90 cm, plant growth parameters and water use efficiency were increased in wheat. However, crop water use was decreased as root development was affected by the higher water table depth (Odili et al., 2023). Meanwhile, barley varieties that are tolerant to waterlogged reduced soil showed better root growth in hypoxia (Abiko et al., 2023).

The tolerance to waterlogging is regulated by various genes involved in transcriptional and translational regulation, signal transduction, and ion transport. Anoxic conditions trigger the release of Ca2+ from mitochondria, activating anoxic genes that impart tolerance. Transcription factors (TFs), such as bZIP, AP2/ERF, bHLH, NAC, and MYB play crucial roles, with specific genes like ZmEREB180 from the ERF-VII TF family identified for enhancing tolerance. Noncoding RNAs and phytohormones, particularly ethylene, also contribute to adaptive changes in response to waterlogging stress. Studying the characterization and validation of genes responsible for hypoxia tolerance in model plants like *Arabidopsis* and rice provides insights into the waterlogging tolerance mechanisms in cereals. Similarly, genotypes with aerenchyma formation QTL (Macquarie+) showed nonsignificant yield reduction compared with nonwaterlogged controls, preventing 23% yield loss under early phenological waterlogging stress (Liu et al., 2020).

10.2.3 Heat Stress (High Temperature)

Temperature is a critical factor influencing the growth, development, and yield of cereal crops. The optimum temperature for each stage of the crop is crucial for better growth and reproduction. However, extreme temperature conditions pose significant challenges to global food security under changing climate scenarios. Cereals are vulnerable to elevated temperatures during their different growth stages. High temperatures can hinder seedling emergence during the germination stage and reduce crop stand establishment. During the vegetative growth stage, heat stress can disrupt photosynthesis and lead to leaf wilting and decreased leaf area. Elevated temperatures disrupt the photosynthetic machinery in cereal crops, leading to a reduction in photosynthetic efficiency and hindering the conversion of sunlight into energy for growth and grain development (Jones et al., 2009). Simultaneously, high temperatures during the reproductive stage result in impaired fertilization, reduced pollen germination, and poor pollen viability, collectively contributing to a decline in grain formation (Hatfield & Prueger, 2015). Heat stress further accelerates the grain filling period, causing premature termination of grain development and resulting in smaller and chappy grains (Porter & Gawith, 1999). Elevated temperatures also elevate respiration rates in cereals, increasing energy consumption by the plants and diverting energy away from growth and grain development (Atkin et al., 2005).

As a consequence of all these adverse effects on plant growth and development (Table 10.5), yield losses were observed in different cereals (Table 10.6). Furthermore, heat stress can also affect the quality of cereal grains, starch composition, protein content, and nutritional profiles, impacting both the economic and nutritional value of the harvested grains (Table 10.3). High temperatures can cause damage to the proteins and enzymes in cereal grains, leading to reduced nutritional quality and decreased storage stability. Moreover, heat stress can increase the susceptibility of cereals to pests and diseases. For example, high temperatures can create favorable conditions for the growth and spread of fungal pathogens. Overall, the impact of heat stress on cereals is multifaceted, affecting various growth stages and ultimately leading to reduced crop production and compromised grain quality. This exploration delves into key findings from various studies to elucidate the primary effects of heat stress on germination, vegetative and reproduction growth, and yield and its attributing traits of different cereal crops (Table 10.5). From compiled data, it is evident that elevated temperatures have a substantial impact, causing yield reduction across cereals due to decreased grain number and reduced grain weight when heat stress occurs during flowering and grain filling, respectively. Despite the vast amount of information on the effect of heat on yield and its attributing traits, it is indispensable to pinpoint the common mechanisms describing the response of crops to heat stress through modeling to enhance our understanding of processes leading to heat stress impacts on yields.

High temperatures negatively affect maize growth and physiological processes and accelerate plant growth, shortening both vegetative and reproductive periods, particularly affecting anthesis, silking, and grain-filling phases. In African regions, high temperatures hasten plant development, reducing essential growth periods and causing midseason crop failure during maize anthesis. Early flowering induced by high temperatures at the vegetative growth phase increases the risk of heat and drought stress. The primary impact is on the impairment of photosynthetic and reproductive machinery, with maize experiencing decreased net photosynthesis at leaf temperatures above 38 °C, mainly when temperature increases are abrupt. Maize pollen viability is found to decline at temperatures above 35 °C, leading to pollination failure and kernel abortion when exposed to 33–40 °C. Elevated temperatures (above 30 °C) impede cell division and amyloplast replication in maize kernels, resulting in reduced grain size and subsequent yield loss (Table 10.6).

Additionally, the growth, development, dry-matter production, and digestibility of forage maize are adversely affected by high temperatures. Barely is also affected by heat and reported a 15% yield reduction due to the impact of heat on grain

Table 10.5 Effect of heat stress on plant growth and development in major cereal crops

Crop	Stage	Stress details	Effect on plant growth and development	References
Maize	Pre-anthesis	40/30 °C (Day/night)	Hastens tasselling, shortened pollen shedding period, extended the ASI and decreased the number of viable pollens	Zenda et al. (2022)
	Anthesis	36/26 °C (Day/night)		
	Post-anthesis		Advanced leaf senescence and reduces the grain filling period	
	Grain filling	above 35 °C	Reduced kernel number	Jones et al. (1984)
	Silking	30–36 °C	Reduced kernel number	Edreira and Otegui (2013)
	Pre-anthesis to silking	33–40 °C	More kernel abortion in temperate hybrids	
Wheat	Germination	Around 45 °C	Impairing seed germination and emergence	Essemine et al. (2010)
		28–30 °C	Reducing seed germination and maturity periods	Yamamoto et al. (2008)
		45 °C, 2 h	Reduced length and dry weight of shoot and root; decreased chlorophyll and membrane stability index	Gupta et al. (2013)
	Seedling stage	42 °C, 24 h	Impair roots and first leaves development; increased ROS and LP	Savicka and Škute (2010)
	Pre-anthesis	Heat stress	Reduced grain weight and number	Ugarte et al. (2007)
	Mid-anthesis	27 and 31 °C	Affect pollen fertility and grain number	Rezaei et al. (2015)
		6 °C above average temperature	10–22% yield reduction	Khatun and Ahmed (2015)
	Post-anthesis	35 °C for 5 days	27% yield reduction	Vignjevic et al. (2015)
		37/28 °C (day/night)	Reduced grain filling period and maturity; drastically reduced fresh weight, dry weight, protein, and starch in grain	Hurkman et al. (2009)
	At the grain-filling stage	34/26 °C (day/night), 16 days	Increased canopy temperature and decreased leaf chlorophyll and maximum quantum yield of photosystem-II	Pradhan and Prasad (2015)
		Above 35 °C	23–37% decline grain weight	Stone and Nicolas (1994)
	Early grain development	38 °C	Grain weight declines by 13% higher in susceptible genotypes	Hays et al. (2007)
		Daily temperature differences (34 °C)	Accelerated the senescence of flag leaves	Zhao et al. (2007)

Rice	Pre-flowering	High night temperatures 27–32 °C	Two days early panicle emergence	Mohammed and Tarpley (2009)
	Flowering		20% reduction in pollen germination and 72% decrease in panicle fertility	
	Flowering	High day temperature (up to 36 °C)	Flowering earlier in the day (by approximately 45 min) for both indica and japonica varieties	Jagadish et al. (2007)
		Elevated day temperatures to 37.5 °C	Reduced pollen germination	Matsui et al. (2001)
		Elevated day temperatures to 40 °C	Reduction in seed set	
		Night temperatures (22–28 °C)	Reduced spikelet number, reduced seed set, and lower grain weight were all reported when	Shi et al. (2013)
Barley		Shift in ambient temperature from 20 °C/16 °C (day/night) to 28/24 °C	Delayed inflorescence and spikelet development and decreased floret and grain number	Ejaz and von Korff (2017)
	Grain filling	Terminal heat stress due to late sowing	42% yield reduction	Oraki et al. (2016)

Table 10.6 Effect of heat/high temperature on cereals yield

Crops	Growth stage[a]	Details of Abiotic stress	Decrease in yield (%)	References
Maize	Silking	Ambient canopy temperature above 6 °C for 3 days	13	Waqas et al. (2021)
	Tasseling stage	30–38 °C for 15 days	14–17	
	Grain filling	35 °C	31	
		28–32 °C	10	
		Every degree above 30 °C in rainfed, drought, and combined stress	1, 1.7, and 40	Lobell et al. (2011)
	Pre-and post-flowering	Temperatures 33–36 °C	10–45	Neiff et al. (2016)
		Heat stress for 2 weeks	20–50	Nielsen (2007)
	Reproductive stage	Variation from normal temperature	80–90	Hatfield and Prueger (2015)
		Heat + drought	Up to100	Heiniger (2001)
Wheat	Reproductive stage	Elevated temperature 6 ± 1 °C higher than mean air temperature of 23 °C	10–22	Khatun and Ahmed (2015)
	Heading to maturity	35 °C for 5 days	27	Vignjevic et al. (2015)
		Hot-dry-windy event	4% per 10 h	Zhao et al. (2022)
	Flowering	Heat waves	40	Templ and Calanca (2020)
	At different stages of crop	5 °C above normal temperature	35	Rivelli et al. (2024)
		40 °C	53.8	Schittenhelm et al. (2020)
		5 °C greater than the optimum temperature	46	Prasad (2011)
Barley	Grain filling	Terminal heat stress due to late sowing	42	Oraki et al. (2016)
Sorghum	Booting and flowering	Early flowering genotypes	87	Craufurd and Peacock (1993)
	Early +late stage	Increasing the duration of severe stress	50–60	
Rice	Pre-flowering stage	For 15 days	86	Soda et al. (2018)

formation and development. Although sorghum is known to be highly heat-tolerant, it is prone to yield losses during the flowering and/or post-flowering phases. Remarkably, the periods between 10 and 5 days before anthesis (coincides with micro- or mega-sporogenesis stage) and between 5 days before and 5 days after anthesis were most sensitive to heat, causing maximum decreases in floret fertility, leading to a poor seed set. It is predicted that temperature above 33 °C may cause yield reduction in sorghum, and 17% and 44% yield reductions were reported due to moderate and higher warming scenarios of 2 °C and 4 °C, respectively. The average reduction across warming scenarios from 1 °C to 5 °C is 10% per degree Celsius (Tack et al., 2017). The mechanisms of heat stress impact on cereals vary from crop to crop, and severity magnitude depends on the crop, stage of the crop, severity of stress, etc. For instance, in wheat, heat stress impacts both grain number and filling, while in maize, heat stress impacts primarily on grain number. The magnitude of crop yield loss depends on the severity and magnitude of variation in temperature in the region. For instance, a 1 °C rise in global mean ambient temperature is predicted to result in average global yield reductions of 6.0%, 3.2%, and 7.4% in wheat, rice, and maize, respectively (Zhao et al., 2017). However, these heat stress-induced crop yield losses vary across regions around the globe. An estimated wheat yield losses of 5.5% and 6% per 1 °C increase for the United States and France, respectively.

Similarly, wheat-growing regions of India and Russia are more susceptible to an increase in temperature, with estimated yield reductions of 9.1% and 7.8%, respectively. In comparison, China is estimated to lose around 2.6% yield for each 1 °C increase in global mean temperature. The elevation of average growing-season temperatures of 2 °C is estimated to cause a 50% and 20% reduction in wheat and maize yield in Australia and sub-Saharan Africa, respectively (Hill & Li, 2022). Similarly, South Africa witnessed a substantial 12.5% reduction in wheat yield with just 24 hours of exposure to temperatures exceeding 30 °C.

In recent decades, terminal heat stress has become a significant problem in wheat and barley cultivation. If the mean temperature during the grain filling stage rises above 31 °C, terminal heat stress starts to occur. According to a study conducted by Singh et al. (2011b), a 1 °C increase in the mean temperature during March and April shortens the wheat crop's growing season by 7 days and decreases yield by 400 kg per hectare. The simulated result shows that terminal heat stress will reduce wheat yield by 18.1, 16.1, and 11.1% in the present, 2020, and 2050, respectively (Dubey et al., 2020). Terminal heat stress due to late sowing reduced barley grain yield by 42% (Oraki et al., 2016).

Similarly, the effect of high temperatures on rice yield formation appears to be related to changes in flowering dynamics, reduced seed set, and lowered grain weight, with more sensitivity to both elevated day and night time temperatures. Further, based on computer modeling, it was found that global warming induces premature aging in wheat (Lobell et al., 2012). Rezaei et al. (2015) elucidated the magnitude and relative impact on physiological processes and yield thereof under high temperatures in three important cereals, namely, wheat, rice, and maize (Table 10.5). Similarly, Farhad et al. (2023) extensively discussed the impact of heat stress on wheat at different stages. In contrast, Zenda et al. (2022) elaborated on the effects of high temperatures in cereals during the reproductive stage.

Table 10.7 Effect of cold/low temperature on cereals yield

Crops	Growth stage[a]	Details of abiotic stress	Decrease in yield (%)	References
Maize	Germination stage	8–10 °C temperature	8–10	Waqas et al. (2021)
	Seedling stage	8–10 °C temperature	20–30	Yan et al. (2017)
	Flowering stage	Less than 12 °C Temperature	40–50	Hayashi (2016)
	Grain filling stage	8–10 °C Temperature	60–75	Waqas et al. (2021)
Wheat	Jointing stage	Temperature of min, max, and mean (6 °C/16 °C/11 °C)	4.6–56.4% and 3.1–44.6%	Ji et al. (2017)
	Booting stage		13.9–85.2% and 3.2–85.9%	
	Anther differentiation	−2 °C for 48 h and 72 h,	100	Zhang et al. (2022b)
		At 2 °C for 24 h	50–90	
	At different stages of crop	For every 1 day increase in freezing duration	3.3% at 14 °C and 8.4% at 17 °C	Zheng et al. (2018)
Rice	Flowering stages	10–12 days prior to heading	19–29%	Ghadirnezhad and Fallah (2014)

10.2.4 Low Temperature (Cold Chilling)

The rising threat of global warming has heightened the volatility of the climate system, manifesting in a surge of frequent extreme events such as both low and high temperatures. Projections from the Intergovernmental Panel on Climate Change (2023) indicate an anticipated global temperature increase exceeding 1.5 °C by the end of the twenty-first century. Additionally, there has been a noticeable increase in the occurrence of prolonged periods of prevailing low temperatures in recent decades, which has affected crop production, particularly winter crops, including cereals. Like heat stress, low-temperature stress (cold/chilling) also causes substantial yield reductions in cereals across the world. For instance, the Kansas region of the United States reported a wheat yield reduction of 31% in 1981, while Australia witnessed 10% yield reductions during 2000–2002 due to low temperatures. Similarly, in China, more than 10% of the wheat planting area (total area of more than 2.92 million ha) was damaged in five growing seasons between 2000 and 2008.

Low temperatures exert a comprehensive impact on cereal crops, influencing vital aspects of their growth and productivity. One of the foremost consequences is a marked decline in photosynthesis, as the efficiency of enzymatic processes diminishes with decreasing temperatures. This compromises the plant's ability to convert sunlight into energy, hampering overall growth and vitality. Concurrently, low temperatures disrupt floral development, leading to spikelet and kernel sterility, thereby affecting reproductive success and causing yield losses. Furthermore, cereal crops exposed to low temperatures experience reductions in grain yield and individual grain weight, impacting both the quantity and quality of the harvest. The induction of water stress is another significant effect, with low temperatures reducing soil moisture and increasing atmospheric water demand, thereby compromising water uptake and various physiological processes. Indirectly, low temperatures accelerate phenology, shortening the growing period and limiting the days available for crop water use within a season.

Additionally, these temperature fluctuations can have ripple effects, influencing the frequency of heat waves and potentially altering the dynamics of weeds, pests, and diseases. Overall, the impact of low temperatures on cereals is a complex interplay of direct and indirect consequences that pose challenges to their resilience and productivity under varying climatic conditions. Understanding and addressing these challenges are crucial for developing resilient cereal varieties and sustainable agricultural practices.

The magnitude and effect of low temperature on plant growth and development depends on the species and its genetics, stage, and other factors. For instance, low temperature at the stem elongation stage had visible effects on spike characteristics, grain development, accumulation of starch, and starch granule development in wheat grains (Yu et al., 2020). Low-temperature stress also affects source–sink organs (i.e., leaves, roots, and spikes) in wheat (Xu et al., 2022a). Thus, several studies reported the effect of low temperature on grain filling and the quality of grain in different cereals. Zhang et al. (2021), using logistic functions, described the dynamics of protein and starch accumulation in wheat grain under low temperatures

Table 10.8 Effect of salinity on cereals yield

Crops	Growth stage[a]	Details of abiotic stress	Decrease in yield (%)	References
Maize	Seedling stage	Salinity of 2 dSm^{-1}	20	Khodarahmpour et al. (2012)
	Reproductive sage	Salinity of 2 dSm^{-1}	10	Huang et al. (2019)
		Salinity of 4 dSm^{-1}	25.2	Azizian and Sepaskhah (2014)
		Salinity of 3.8–7.4 dSm^{-1}	34	Cucci et al. (2019)
Wheat	Vegetative stage	EC > 3 dSm^{-1}	10	Gupta et al. (2020)
	At different stages of crop	15 dSm^{-1} throughout the duration	19.03–49.87	Ghogdi et al. (2012)
Barley		Irrigation water salinity 5–15 dSm^{-1}	27	Hammami et al. (2020)
		Salinity at 15 dSm^{-1}	31.2–64	Pakar et al. (2016)
Rice	At different stages of crop	Salinity ranged from 2 to 8 dSm^{-1}	21–47	Rad et al. (2011)
	Seedling stage	Salinity level 75 mM NaCl	80	Shereen et al. (2005)
	Reproductive stages	34.2 mM NaCl	20	Zhang et al. (2022a)

in two contrasting genotypes. The contents of total amino acids, essential amino acids, and nonessential amino acids increased in the cold-sensitive wheat cultivar (Yangmai16). In contrast, the cold-tolerant cultivar (Xumai30) experienced a decrease in amino acid content, particularly during the jointing and double stages. Therefore, it is very crucial to undertake protective measures during low temperatures to enhance wheat protein quality, particularly considering the observed variations in amino acid composition between cold-sensitive and cold-tolerant wheat cultivars (Hu et al., 2022a, b).

Based on models, several studies predicted yield loss due to low temperatures in cereals. For instance, Xiao et al. (2022) indicated that total aboveground biomass and grain yield were reduced by 63–77% under low-temperature. As a shred of evidence, Liu et al. (2020) reported that severe low temperatures killed the parts of tillers. Frost causes a spike abortion and damages the developing grains (Zheng et al., 2015).

Wheat is an important staple food crop in the world, mainly grown during winter seasons, and thus often experiences frost damage resulting from low temperatures in some parts of the world. The survival of wheat under such conditions was studied with different models. Recent advances in understanding the responses of cereals to low temperatures led to the development of a more robust winter survival simulation model with a more comprehensive geographic application that now also considers cultivar acclimation threshold induction temperature, respiration stress, photoperiod, and other developmental factors. Fowler et al. (2014) developed a Winter Cereal Survival Model and described acclimation, vernalization, dehardening, and damage due to low temperature (LT) stress in cereals. The model has been field-validated and provides the opportunity for the simulation of a wide range of cereal crops. Several quantitative trait loci (QTL) related to freezing tolerance (FT) have been identified on wheat chromosomes, with chromosome 5A harboring several crucial loci, including those associated with C-Repeat Binding Factor (CBF) genes.

Additionally, genes like Vrn-A1, responsible for vernalization-induced flowering, play a pivotal role in FT. Elevated soluble sugar content, regulated by specific genes, has been correlated with improved FT. Various other genes, such as TaNAC2, Wabi5, and TabZIP96, also contribute to enhanced FT by influencing different pathways. In essence, wheat FT is a polygenic trait jointly controlled by multiple genes involved in diverse regulatory mechanisms, offering potential avenues for breeding resilient wheat varieties against freezing damage. Ten stable loci for freezing tolerance (FT) in wheat were detected by genome-wide association analysis. The putative candidate gene TaRPM1-7BL underlying the primary locus QFT.ahau-7B.2 was identified and validated (Pan et al., 2024). Further, understanding the molecular mechanism of brassinolide in winter wheat under low-temperature stress provides insight into the cold tolerance mechanism, which aids in breeding cold-tolerant varieties.

Maize, suited initially for tropical climates, has adapted to temperate regions over time by developing tolerance to low temperatures. Prolonged exposure to cold stress, especially between 8 and 10 °C, disrupts various growth stages in maize, affecting processes from seed germination to reproductive phases. Notably, cold

Table 10.9 Various approaches to the management of drought in cereals

Crops	Approaches	Details	References
Rice	Agronomic and cultural management	Alternate wetting and drying conserve water by about 38% in rice	Lampayan et al. (2015), Akbar et al. (2023)
		Drip irrigation (>249% water productivity)	Akbar et al. (2023)
		Brown manuring with *Sesbania* + bispyribac sodium @25 gha^{-1}	Aravinth et al. (2023)
		Adoption of aerobic rice cultivation benefits in 32–88% higher water productivity	Bouman et al. (2005), Champnes et al. (2023)
		System of Rice Intensification (SRI) practice (14% less water)	Thakur et al. (2014)
		Saturated Soil Culture increased water productivity by 30–115%	Lampayan and Bouman (2005)
		Direct-Seeded Rice (DSR) increased water productivity by 35.83%	Aravinth et al. (2023)
		Seed priming with *Trichoderma* resulted in 52%, 70%, and 66% higher water productivity	Das et al. (2020)
	Application of nutrients/ bio-regulators/ amendments	Balanced and efficient nutrient management	Vijayakumar et al. (2021)
		Nano fertilizers, slow-release fertilizers, and organic nitrification inhibitors	Gobinath et al. (2021)
		Foliar application of rice phyllosphere bacteria increased panicle weight by 22.6%	Devarajan et al. (2021)
		100 μM Melatonin foliar application increased plant growth by improving relative water content and other physiological parameters	Silalert and Pattanagul (2021)
		Exogenous application of salicylic acid @ 100 mg l^{-1} improved the cellular membrane integrity and photosynthesis	Farooq et al. (2009)
		100 mM Glycine betaine enhanced overall growth, physio-biochemical adaptation, and yield attributes	Tisarum et al. (2019)
Wheat	Agronomic and cultural management	Hydropriming or osmopriming with $CaCl_2$ improved the productivity of bread wheat	Farooq et al. (2015)
		Advancing sowing date (Nov. 1st), additional dose of N-fertilizer (150 kg ha^{-1}), and demand-driven irrigations	Dubey et al. (2020)
		Raised-bed sowing (42.6% water saving)	Lamichhane and Soltani (2020)
		Night-time sowing due to the absence of evaporation and soil desiccation	
	Application of nutrients/ bio-regulators/ amendments	30 times diluted *Moringa oleifera* leaf extract sprayed at heading stage	Yasmeen et al. (2012)
		0.5 mM Ascorbic acid in the rooting medium followed by 1 mM foliar application was most effective	Malik et al. (2015)
		32 ppm ortho-silicic acid foliar spray increased 10% yield	Ratnakumar et al. (2016)

Barely	Agronomic	Early sowing	Samarah and Al-Issa (2006)
	Genetic approach	Four QTLs, including WGSQ and WGFL1Q1, were identified	Gous et al. (2016)
Maize	Agronomic and cultural management	Ridge planting increased yield by 15%	Dong et al. (2017)
		Partial and complete film cover mulching increased yield by 9 and 16%	Dong et al. (2017)
		Planting on an appropriate date (April 16–April 28) increased yield by 2.3–24.7%	Lu et al. (2017)
		Conservation tillage with subsoiling and straw increased yield by 9.59%	Deng et al. (2023)
		Shallow and narrow furrow method of planting and irrigation at 50% depletion saved water by 34%	Halli et al. (2021)
		Intercropping and cover cropping with soybean, peanut, etc.	Han et al. (2022)
		Seed priming with $CaCl_2$ for 24 h at a 1:5 (w/v) ratio	Khan et al. (2015)
		Night-time sowing due to the absence of evaporation and soil desiccation	Khan et al. (2017)
	Application of nutrients/ bio-regulators/ amendments	Application of potassium fertilizers	Ebrahimi et al. (2011)
		Sulfur-based fertilizer application	Usmani et al. (2020)
		Biochar with silicon application increased seedlings' growth by 27%	Sattar et al. (2020)
		Foliar application of moringa leaf extract (30 times diluted with water) @ 25 mL/ plant and benzyl amino purine (50 mg L^{-1})	Ali et al. (2011)
		Soil application of 50 g arbuscular mycorrhiza inoculum increased maize seedlings' growth	Gong et al. (2023)
		Silicon (2 mM) and chitosan (200 mg l^{-1}) foliar spray increased relative water content by 30%	Ahmed et al. (2023)
	Genetic and molecular approaches	Screening, evaluation, and characterization of maize genetic resources	Chen et al. (2012), Yue et al. (2018)
		Genetics and inheritance of genes/traits associated with drought tolerance	Haq et al. (2015), Khan et al. (2016)
		Identifying and characterizing the genes/QTLs associated with drought tolerance	Almeida et al. (2014), Zaidi et al. (2016)
		Physio-biochemical and molecular mechanism	Hasibuzzaman et al. (2021)
		Transcriptomics, proteomics, genomic approaches	Min et al. (2016), Wang et al. (2019)
		Development and adoption of drought-tolerant varieties/hybrids	Ribaut and Ragot (2007), McFadden et al. (2019), Lunduka et al. (2019)
Sorghum	Agronomic	Sowing of early maturing and drought-tolerant varieties	Yahaya and Shimelis (2022)
	Genetic	Exploitation of stay green trait	
		Development of high biomass varieties	

Table 10.10 Approaches to better understand and manage waterlogging in cereals

Crops	Approaches	Details	References
Rice	Agronomic and cultural management	Ultrasonic seed treatment at 20 kHz, 220 W increased the seedling emergence rate by 35.07–64.39%	Huang et al. (2020)
	Application of nutrients/bio-regulators/amendments	Potassium application @ 150 kg ha^{-1} as basal increased the plant survival up to >80%	Gautam et al. (2016)
		Biochar @ 10.5 t ha^{-1} improved grain yield by 8.5–10.7%	Liu et al. (2016)
		Split application of fertilizers	Gribaldi et al. (2017)
		Additional post-flood nutrient doses of N and K_2O	Dwivedi et al. (2018)
Wheat	Agronomic and cultural management	Lesser yield reductions were recorded in tolerant genotypes	Herzog et al. (2016)
		Controlled drainage	Deichmann et al. (2019)
	Application of nutrients/bio-regulators/amendments	Exogenous ACC @ 10 mM application recorded lesser yield reduction under 2 days waterlogging at the anthesis stage	Wang et al. (2016)
		Melatonin spray 100 μmol l^{-1} lesser yield reduction under 7 days waterlogging at 50% flowering	Ma et al. (2022)
		1-NAA spray (barley) improved shoot and root growth and reduced leaf senescence	Pang et al. (2007)

Maize	Agronomic and cultural management	Raised bed planting at 50 cm height increased baby corn yield by 11.3% and fodder yield by 10.7%	Panwar and Shivay (2015)
		Ridge tillage increased yield by 39–50%	Ren et al. (2016)
		Drainage increased N uptake by 14%	Motarjemi et al. (2021)
		Increased N fertilizer application	Tian et al. (2021a)
		Subsurface drainage increased grain yield by 50%	Balusamy et al. (2019)
		Split application of fertilizers at different stages	Gribaldi et al. (2017)
		0.1% H_2O_2 seed priming for 12 h before sowing increased yield by 48%	Wang et al. (2022)
	Application of nutrients/bio-regulators/amendments	Spraying 6-BA @ 100 mg l^{-1} after waterlogging increased yield by 31.5%	Hu et al. (2022a, b)
		Sodium nitroprusside @ 500 μmol L^{-1} alleviated the harmful effect of waterlogging.	Jaiswal and Srivastava (2015)
	Genetic, molecular, and genomic approaches	Genetic variability and genetics	Yu et al. (2019), Zaidi et al. (2010)
		Genes/QTLs associated with waterlogging tolerance	Mano and Nakazono (2021), Zaidi et al. (2015), Guo et al. (2021)
		Biochemical studies	Srivastava et al. (2007)
		Identifying genes	Yu et al. (2018)
		Development of tolerant varieties/hybrids	Edmeades et al. (2017)
Sorghum	Agronomic approaches	Development of tolerant varieties/hybrids	
	Genetic approaches	Genetic variability for tolerance	Zhang et al. (2023b)
		Deployment of aquaporin genes (AQP)	Kadam et al. (2017)

Table 10.11 List of genes/QTs in cereals associated with heat stress tolerance

Crop	Gene/QTLs Name	Associated traits/mechanisms
Rice	OsHSP18.6[a]	Enhance antioxidant enzymes
	OsANN1[a]	SOD and CAT activities
	OsNTL3[a]	Regulates transmembrane/ER protein folding
	SNAC3[a]	Oxidative stresses; NAC3 significantly reduced H_2O_2 and MDA contents and increased ROS homeostasis in transgenics
	MnSOD1[a]	Better chaperone activity and grain quality under heat stress
	Rab7[a]	RWC, proline content, and antioxidant enzyme activities, but decreased MDA and ROS scavenging
	fad7[a]	Photosynthetic efficiency and chlorophyll content
	SPW[a]	Cleistogamy
	SGR[b]	Binding of light-harvesting complex during photosystem-II
	NYC[b]	Responsible for the activity of chlorophyll reductase to convert chl-b into chl-a
	PPH[b]	Responsible for the activity pheophytinase for dephytylation to phaeophytin
	qPSLht4.1[c]	Heat tolerance at anthesis
	qTL3.4 and RLPC3.1[c]	HS tolerance at the booting stage
	qhs2, qSF4; SSPF4, qPF4; qHTB10-2 and qHTB11[c]	spikelet fertility under HS
	qSF4 and qPSLht7[c]	stress tolerance index (STI)-spikelet fertility
Maize	ZmHsf05[a]	Influencing germination and shoot growth and associated with basal and acquired thermotolerances
	ZmCDPK7[a]	Enhance photosynthetic rates and antioxidant enzyme activity.
	ZmMAPK1[a]	Increasing proline content, decreasing MDA content, and increasing ROS scavenging
	ZmWRKY106[a]	ABA-signaling and reducing ROS by enhancing antioxidant enzyme activity
	DREB2A[a]	Expression of thermotolerance genes
	ZmPEPC[a]	Enhance photochemical and antioxidant enzyme activities and regulate proline content
	EF-Tu[b]	reduces the thermal aggregation of leaf protein's photosynthetic membrane and increases CO_2 fixation
	TaFER-5B[b]	Reduces oxidative stress by scavenging ROS and improves leaf iron content
	QTL on chr 3[c]	HS tolerance in the tassel and leaf
	S10_1,905,273; S10_1,905,274; and S10_7,132,845[c]	SNPs associated with Grain yield under heat stress

Wheat	TaHsfA6f[a]	ABA accumulation
	TabZIP60[a]	Regulation of ABA
	TaFER-5B[a]	ROS scavenging
	ELF3[a]	Temperature sensitivity at the late reproductive stage
	Ppd-1[a]	Early inflorescence development
	TaGASR1[b]	Reduces ROS and hormonal signal transduction pathway
	TaHSP23.9[b]	Upregulation under heat stress facilitates seed development during the grain-filling phase
	TaFBA1[b]	Upregulation improves photosynthesis and the antioxidant enzyme activity
	TaHsfA2-1[b]	Overexpression of heat shock proteins and chlorophyll content
	stay-green QTL (44 loci)[c]	Stay-green
	QHst.cph-3B.1, QHst.cph-3B.2, and QHst.cph-1D[c]	Photosynthetic efficiency
Barely	ODDSOC2[a]	Reproductive phase duration
	Hsf6A[b]	Regulation of heat shock protein genes
	HvSUT1[b]	Increase sucrose transport into grains
	HSPFLQ1, HSPFLQ2, HGSQ[c]	Stay-green trait
Sorghum	[b]MYB78 and ATAF1	Chaperones, including unique heat shock proteins (HSPs)
	SNPs (SNP2_76939617, SNP4_34147673, S NP2_76939617)	Associated with leaf firing traits
	SNPs (SNP4_34147673, SNP2_76939617, SNP4_34147673)	Associated with leaf blotching traits
	[b]SbHSF5 and SbHSF13	Induced and overexpressed in tolerant varieties

Source: Modified from Hill & Li (2022) and Zenda et al. (2022)
[a]genes
[b]*TF* transcription factor
[c]QTLs

Table 10.12 Approaches to better understand and manage high temperature in cereals

Crops	Approaches	Details	References
		Early planting reduced the risk of day and night time heat stress	Acharjee et al. (2019)
		Delaying the sowing date to achieve mean daily temperature at 24.9–26.4 °C during the first 20 days after heading	Yang et al. (2021)
	Application of nutrients/bio-regulators/ amendments	100 mol L^1 Jasmonates spray recorded a higher opened-spikelet rate of 86.9%	Yang et al. (2021)
		Brassinosteroid @ 5 ppm increased yield by 18.8%	Raghunath et al. (2021)
		10 ppm Ascorbic acid +1.3% Citric acid increased yield by 21.56%	Das et al. (2020)
		0.2% Boron foliar spray increased yield by 8.1%.	Shahid et al. (2018)
		Foliar cytokinin (10^{-5} M) recorded a higher relative tolerance index of 97.69%	Pantoja-Benavides et al. (2022)
		Foliar spraying of calcium acetate (5 mM) alleviated the combined heat and ozone stress by 60.9%	Lakaew et al. (2021)
Wheat	Agronomic and cultural management	The no-tillage technique recorded higher grain yield by 12.93% and straw yield by 15.84%	Singh et al. (2011a)
		Zero tillage (ZT), permanent bed planting (PBP), and furrow irrigated raised bed systems (FIRB)	Yadav et al. (2022)
		Advancing the sowing of wheat by 15–20 days in North-Western India	Yadav et al. (2022)
		Maintaining proper time and methods of sowing	Ortiz et al. (2008)
		Straw mulch reduced soil evaporation by 21–40%	Chen et al. (2007)
		Plastic mulch increased grain yield by 13%	He et al. (2017)
		PGPR strains and endophytes	Hubbard et al. (2014), Yadav et al. (2022)
	Application of nutrients/bio-regulators/ amendments	Foliar spray of KNO_3, KCl, 1 Methylcyclopropene (1-MCP), and GA_3	Kajla et al. (2015)
		Soil application of 450 μM Mg increased grain yield by 15%	Mengutay et al. (2013)
		Soil application of zinc sulphate @ 30 kg.ha^{-1} increased grain yield by 127%	Moshatati and Mousavi (2019)
		Exogenous polyamines (spermine and spermidine) increased grain yield by 5–19%	Jing et al. (2020)

Maize	Agronomic and cultural management	Seed priming with Salicylic acid (50 mg L^{-1}) increased grain yield by 34.9%	Rehman et al. (2015)
		3.3% *moringa* leaf extract (osmopriming) increased grain yield by 38.7%	
		2.2% $CaCl_2$ (osmopriming) increased grain yield by 17.9%	
		Hydropriming (water) increased grain yield by 4.7%	
		Seed priming Kinetin @ 100 mg L^{-1} increased grain yield by 53.2%	Bakhtavar et al. (2015)
	Application of nutrients/bio-regulators/ amendments	A foliar spray of Agrenine and Glycine at 0.006–9.0 mM increased the shoot and root weight by 55–59%	Matysiak et al. (2020)
		Foliar application of Ascorbic Acid (20 mg L^{-1}) increased grain yield by 23.96%	Ahmad et al. (2017)
		Seed priming with H_2O_2 (20 mg L^{-1}) increased grain yield by 21.21%	
	Genetic and genomic approaches	Pre-breeding	Chen et al. (2012), Nelimor et al. (2019)
		Genetics of drought tolerance	Naveed et al. (2016)
		In vitro gamete selection for heat-stress tolerance	Petolino et al. (1990)
		Identifying genes, QTLs, and genomic regions	Frey et al. (2016), Inghelandt et al. (2019), Longmei et al. (2021)
		Morpho-physiological studies	Noor et al. (2019)
		Transcriptomic analysis, Omics	Wang et al. (2020), Li et al. (2020)

Table 10.13 Approaches to better understand and manage low temperature in cereals

Crops	Approaches	Details	References
Rice	Agronomic and cultural management	Optimizing sowing dates increased grain yield by 5.6–8.6%	Liu et al. (2023)
		Right transplanting time first September transplanting increased the grain yield by 4.44, 8.88, and 15.55% as compared to 10, 20, and 30 September, respectively, in Bangladesh	Nahar et al. (2009)
	Application of nutrients/bio-regulators/amendments	Spermidine seedling treatment (1 mM, 2 mM) increased grain yield by 28–57%	Lakshmi et al. (2023)
		GA_3 (10 mg L^{-1}) and Brassinosteroid (0.3 mg L^{-1}) seed priming recorded a maximum germination rate of 98%	Zhang et al. (2023a)
		Zinc (0.08 μM, 0.15 μM, and 0.31 μM $ZnSO_4{\cdot}7H_2O$) improved tiller number and tiller growth rate by 11.28% and 17.78%	Liu et al. (2022)
Wheat	Agronomic and cultural management	Psychrotolerant *Pseudomonads* inoculation increased relative water content, reduced membrane injury, and electrolyte leakage	Mishra et al. (2011)
		Earlier sowing on 22nd October recorded higher yield in Iran	Habibi et al. (2012)
		Tolerant cultivars (Unnat PBW 343) and optimum date of sowing (15th November) increased grain yield by 15.29%	Upadhyay et al. (2023)
	Application of nutrients/bio-regulators/amendments	Optimized phosphorus application increased grain yield by 8.6–20.5%	Xu et al. (2022b)
		Melatonin increases the photosynthetic carbon assimilation and membrane stability.	Sun et al. (2018)
		Salicylic acid spray @ 100 μM increased grain yield by 7.9%	Wang et al. (2021)

Maize	Agronomic and cultural management	Seed priming with 0.5 mM Salicylic acid +50 mM H_2O_2 recorded higher germination percentage of 80% under chilling stress	Li et al. (2017)
		Priming seeds with micronutrients (4 mM Zn + 2.5 mM Mn solution as $ZnSO_4{\cdot}H_2O$ and as $MnSO_4{\cdot}7H_2O$) increased fresh grain yield per plant by 15.15%	Imran et al. (2013)
		Magnetic seed stimulation (150 mT for 3 minutes)	Afzal et al. (2015)
		Sowing on the southern side of the ridge where it gets maximum sunlight during the daytime. Planting tall barrier crops on the northern side of the field protects the crop from cold winds	Dass et al. (2011)
	Application of nutrients/bio-regulators/amendments	Foliar spray increased grain yield by 3% Moringa leaf extract—12% 0.02% Thiourea—10% 0.1% Salicylic acid—7% 3% sorghum water extract—11%	Waqas et al. (2017)
		5-aminolevulinic acid enhanced the superoxide dismutase, catalase, and ribulose-1,5-bisphosphate (RuBP) carboxylase activities	Wang et al. (2018)
	Genetic and genomic approaches	Evaluation of inbred maize using RGB imaging for cold	Enders et al. (2019)
		Genetics/inheritance of genes/traits associated with cold tolerance	Revilla et al. (2000), Costa silva neta et al. (2020)
		QTLs, alleles associated with cold tolerance	Hu et al. (2017), Li et al. (2018)
		Transcriptomics and genomic approaches	Frascaroli and Revilla (2018), Sowiński et al. (2020)

Table 10.14 Approaches to better understand and manage salinity in cereals

Crops	Approaches	Details	References
Rice	Agronomic and cultural management	Riboflavin (10 μM)–treated seedlings exhibited higher biomass accumulation	Jiadkong et al. (2023)
		10%, 20%, and 30% (w/w) biochar increased the active tillers by 69%, 68%, and 90%	Sudratt and Faiyue (2023)
		Humic acid (40 mg l^{-1}) seed priming reduced the salinity-induced plant damage, improved metabolism, and maintained active growth	Shukry et al. (2023)
		Application of green manure and gypsum	Siam et al. (2015)
	Application of nutrients/bio-regulators/amendments	Foliar spray of rice straw-based silicon nanoparticles (20 mgL^{-1})	Ijaz et al. (2023)
		Antioxidant microbial consortia spray increased grain yield by 15.77%	Janaki et al. (2021)
		Organic manures	Chowdhury et al. (2019)
		Poultry manure @ 4 t ha^{-1} increased grain yield by 14.98% FYM @ 5 t ha^{-1} increased grain yield by 10%	
	Genetics and Genomics	Introgression of Saltol QTL	Gregorio (1997)
		Utilization of Wild rice species in breeding	Tian et al. (2011)
		Deployment of transgenes like OsPP1	Liao et al. (2016)
		Genome editing of OsRR22, OsNAC3, and OsmiR535 genes	Yue et al. (2020)
Wheat	Agronomic and cultural management	Salicylic acid seed priming increased tillers by 6–29%	Fardus et al. (2018)
		Gypsum application (8.5–15.5 t ha^{-1}) in ridges	Khan et al. (2010)
		FYM application enhanced the dry matter yield of 78% over the control treatment	Ahmed et al. (2010)
		Hydropriming and halopriming (50 mmol, KCl, and NaCl) improved the final emergence percentage (FEP) emergence index (EI), and reduced the emergence time	Saddiq et al. (2019)
		Plant growth -promoting rhizobacteria and arbuscular mycorrhizal fungi	Seleiman et al. (2022)
	Application of nutrients/bio-regulators/amendments	Foliar applied thiourea (10 mM) increased the EC tolerance by 6–11%	Anjum et al. (2011)
		0.8 mM Ascorbic acid enhanced growth characteristics under salinity stress	Siddiqui et al. (2018)
		Gypsum amendment (6 t ha^{-1}) increased grain yield by 8% under salinity stress	Morsy et al. (2022)

Maize	Agronomic and cultural management	The sowing of seeds on the side of ridges helps to escape the salinity stress, as salt moves in an upward direction due to evapotranspiration	Dass et al. (2011)
		Irrigation with brackish water (at the vegetative stage) and regular water (at the reproductive stage) creates subsurface drainage	Huang et al. (2019)
		Use of organic manures and soil amendments	Khatun et al. (2019)
		27.9% additional yield was produced due to an additional dose of nitrogen	Azizian and Sepaskhah (2014)
	Application of nutrients/bio-regulators/amendments	Seed priming with NaCl (50 mmol L^{-1}) increased grain yield by 36%	Bakht et al. (2011)
		0.8% EDTA-chelated biochar	Huang et al. (2024)
		Seedling pretreatment with 1 mM Salicylic acid and a spray of Salicylic acid	El-Katony et al. (2019)
		Nitric acid (3 mg L^{-1}) and thiourea (500 mg L^{-1})	Kaya et al. (2017)
		1 μM Melatonin spray increased the photosynthetic rate, stomata conductance, and transpiration rate by 23.7%, 40.2%, and 50.6%.	Ren et al. (2020)
		Use of Chitosan and yeast extract	Latef et al. (2019)
		Foliar application of auxin, cytokinin, gibberellins, kinetin (KIN), and indoleacetic acid (IAA)	Davani et al. (2017)
		Seed priming 1.5 mM sodium metasilicate enhanced the leaf's relative water content, photosynthetic pigments, and antioxidant enzyme activities.	Latef and Tran (2016)
	Genetic and genomic approaches	Genetic variability	Batool et al. (2020)
		Genetics of salinity tolerance	Li et al. (2021)
		Development and adoption of salinity-tolerant varieties/hybrids	Carpýcý et al. (2009)
		Identifying genes/QTLs associated with salinity tolerance	Zhang et al. (2019), Sandhu et al. (2020), Ma et al. (2021)
		Physio-biochemical mechanism	Shahzad et al. (2019)

stress during the V6–V9 stages delays flowering and leads to abnormal tassel growth, impacting pollination and grain formation (Table 10.7).

10.2.5 Salinity

Salinity is one of the important abiotic stresses that occur when the levels of Na^+ and Cl^- in soil surpass a standard threshold of 4 dSm^{-1} electrical conductivity. Still, even 2 dSm^{-1} may cause some yield penalty in cereals. In general, salinity is naturally caused by a high rate of evaporation in arid areas, the inundation of salt water in coastal regions, and inappropriate irrigation, mining, and industrialization through human interference. Salinity affects global food security as it influences the crop-based agricultural production system. Presently, about 20% of the world's cultivated areas face salinity stress, a number projected to rise to 50% or more by 2050, posing substantial concerns for agricultural production. Salinity affects crop yields as it interferes with cellular functions, growth, and other physiological processes. The primary physiological impacts of salinity stress include osmotic stress, nutrient absorption imbalance, and ionic toxicity. These factors lead to decreased growth, attributed to the differential allocation of metabolites and hormones, causing various physiological and morphological disturbances.

Cereals such as maize, rice, and wheat exhibit heightened sensitivity to salinity stress, resulting in diminished growth and productivity under saline conditions. Maize, while moderately sensitive to salt stress, faces increased vulnerability during the vegetative stage, especially in arid and semiarid regions with prevalent drought conditions. Although maize is less susceptible to salinity during germination and the initial vegetative stage, it can still experience up to a 20% yield loss due to reduced seedling vigor, whereas causes witnessed huge yield loss of more than 30% if the crop is exposed to salinity during reproductive stage (Table 10.8). In the case of wheat, elevated salt levels gradually reduce germination percentage and impede seedling growth, with higher accumulation of malondialdehyde (MDA) causing oxidative damage in parenchyma cells (Al-Quraan et al., 2013). Salinity causes forced maturity with decreased yield (Ahmad et al., 2013). Irrigation with water having an electrical conductivity (EC) greater than 3 dS m^{-1} during the vegetative stage results in a 10% reduction in wheat yield (Gupta et al., 2020), while continuous irrigation with 15 dS m^{-1} water throughout the crop duration can cause grain yield reductions ranging from 19.03% to 49.87% (Ghogdi et al., 2012). Despite barley's moderate salinity tolerance, irrigation with water up to an EC of 15 dS m^{-1} may lead to a substantial crop yield reduction of up to 64% (Pakar et al., 2016). Additionally, in rice, even low to moderate levels of salinity (2–4 dSm^{-1}) can result in yields being more than halved compared to normal conditions (Zheng et al., 2015), highlighting the significant impact of salinity stress on diverse cereal crops.

Cereals employ several adaptive strategies, including stomatal regulation, osmotic adjustment, maintenance of tissue water potential, ionic exclusion, and antioxidant defense mechanisms to cope with salinity stress. Salt-tolerant maize genotypes exhibit less leaf damage, reduced reduction in leaf water content, and maintained membrane stability. These responses are regulated by the production of ROS in response to salinity stress, neutralized by antioxidative metabolites and enzymes like catalase, ascorbate peroxidase, and superoxide dismutase. For instance, to mitigate sodium accumulation, maize employs strategies like vacuolar compartmentalization, limiting shoot translocation, and maintaining the sodium/potassium ratio for cell turgor maintenance and stomatal regulation. Hormones, particularly abscisic acid (ABA), play a crucial role in sensing salinity stress, modifying plant metabolism, and reducing water loss while promoting growth. Under salt stress, maize adjusts the synthesis of growth-promoting and retarding hormones. Apoplastic acidification, crucial for cell growth and expansion, becomes susceptible to salt stress, with varieties showing less ionic efflux and pumping being more prone to damage. Understanding cereals' responses to salt stress involves considering the intricate interplay of physiological processes, adaptive mechanisms, and hormonal regulation, providing insights for developing salt-tolerant varieties.

10.3 Management of Abiotic Stresses in Cereal Crops

The adverse impacts of abiotic stresses on cereal crop yield and quality are discussed in the previous section. In this section, an attempt has been made to understand the different strategies to minimize the stress effects on cereal yield to ensure global food security. This involves employing strategies for stress avoidance and tolerance and considering genetic resources and agronomic solutions for enhanced production under abiotic stress conditions. Besides, these short-term crop management practices and long-term breeding approaches are aimed at creating resilient and high-yielding cereal varieties capable of withstanding diverse and multiple stresses. In the pursuit of sustainable food production, a holistic approach is essential, encompassing plant physiology, genetics, and agronomy to address the multifaceted challenges posed by abiotic stresses in cereal crops.

Appropriate agronomic management practices aid in adapting and thriving in diverse abiotic stresses like deficit and excess moisture, extreme temperatures (heat and cold), salinity, and other various stresses. These practices encompass selecting suitable varieties and soils, determining the optimal time and method of sowing, managing water and nutrients efficiently, implementing need-based contingent practices, and appropriate timing of harvesting to minimize the losses (Tables 10.9, 10.10, 10.12, 10.13 and 10.14). Implementing these practices has the potential to narrow the yield gap between drought-affected and optimal conditions by 20–25% in maize (Edmeades, 2013).

The cultivation of climate-resilient varieties, coupled with region-specific production technologies, is crucial for sustaining cereal productivity in abiotic stress-prone areas. Recent advancements in genomics, biotechnology, and high-throughput phenotyping offer promising avenues for developing climate-smart cultivars capable of performing well under both stress and normal conditions. Sah et al. (2020) emphasize that such cultivars exhibit reduced yield penalties compared to nonstress-tolerant varieties. Identification of adaptive traits associated with abiotic stress tolerance is essential for breeding resilient cultivars, with efforts focusing on genetically diverse resources like landraces and wild relatives (Nelimor et al., 2019). For instance, utilizing maize hybrids with superior root architecture and biomass has shown promise in sustaining moderate drought stress, yielding higher than non-drought-tolerant hybrids (Adee et al., 2016). While modern hybrids may exhibit increased sensitivity to drought (Lobell et al., 2014), they can adapt to climatic changes through trait optimization (Chen et al., 2013). Furthermore, understanding the genetics and inheritance of stress tolerance traits is crucial, with recent molecular tools aiding in identifying genes, QTLs, and molecular markers associated with stress tolerance. Despite advancements, challenges remain in developing stable transgenic lines exhibiting abiotic stress tolerance. However, conventional and molecular breeding efforts, including marker-assisted selection, have led to the release of climate-resilient maize varieties worldwide (Cooper et al., 2014; Prasanna et al., 2021). Hence, considering the importance of both agronomic and genetic interventions, this section discussed various approaches developed for the management of diverse abiotic stresses in cereals.

10.3.1 Drought Management in Cereals

Agronomically, drought can be managed through planting or sowing at appropriate times and sowing windows to escape from deficit moisture conditions. Similarly, adjustments in the date of sowing, such as early sowing, can be adopted to help crops escape drought in terminal stages. Foliar application of plant hormones such as auxin, gibberellins, cytokinins, abscisic acid, salicylic acid, and brassinosteroids modulate drought responses in rice. Modern techniques in rice cultivation, such as drip irrigation, saturated soil culture, a system of rice intensification, alternate wetting and drying, and brown manuring, could be adopted to save water. Aerobic rice cultivation reduced total water usage by 51% and increased water productivity by 32–88%. Balanced application of nutrients at the right time in the right form is essential for the stress tolerance of rice. Foliar application hormones such as melatonin, salicylic acid, and glycine betaine enhance physio-biochemical adaptation, overall growth, and yield attributes under drought-stress conditions.

Seed treatment techniques such as hydropriming or osmopriming harden the seed for drought tolerance with improved grain quality in wheat. Cultural practices such as cultivation in raised beds conserve soil moisture with a water saving of 42.6%, which could save irrigation water. Adoption of deficit irrigation strategies can prevent water loss without a significant reduction in yield. Potassium application during drought conditions increases the resistance to stress, thereby contributing to a nonsignificant decrease in yield. Wheat is susceptible to water deficit stress during the crown root initiation stage. Hence, irrigation should be provided during this stage. The exogenous application of growth-regulating hormones such as methyl jasmonates, abscisic acid, and salicylic acid and osmotic balancing solutions like glycine betaine improved the water use efficiency by effectively increasing the photosynthetic efficacy and antioxidative capability under deficit irrigation conditions. Foliar application of silicon formulations prevents drought and increases crop yield. The application of zinc, copper, and boron micronutrients alleviates the water deficit stress in wheat by controlling the physiological traits.

Maize faces drought during summer and terminal drought in the rabi season as it is cultivated in residual soil moisture. Results reveal that ridge sowing permits the efficient utilization of irrigation water and nutrients, creating better root growth under drought conditions. Shallow and narrow furrow methods of planting and irrigation at 50% depletion of available soil moisture resulted in a reduction in the number of irrigations and 34% water saving. Likewise, ridge planting increased maize yield by 15%. The mulching practices with plastic films, straws, and cover crops conserved the soil moisture and increased yield. The inter cropping and cover cropping with wide canopy crops such as groundnut and cowpea conserved the soil moisture with additional yield. The macronutrients such as potassium and sulfur and beneficial nutrients such as silicon make the defense mechanism against drought stress. The foliar spray of chitosan with silicon increased the membrane stability index and relative water content, which made the maize plants better able to sustain themselves under drought conditions.

The various physiological, genetic, molecular, and genomic studies collectively contributed to the development of drought-tolerant varieties. For instance, Chen et al. (2012) and Yue et al. (2018) provide insights into maize genetic diversity, aiding in parental selection, whereas Haq et al. (2015) and Khan et al. (2016) elucidate inheritance patterns of drought tolerance traits, guiding breeding efforts. Hasibuzzaman et al. (2021) reveal physiological and molecular mechanisms, informing targeted breeding strategies. Min et al. (2016) and Wang et al. (2019) uncover molecular pathways and candidate genes for genetic manipulation. Almeida et al. (2014) and Zaidi et al. (2016) identify genes and QTLs linked to drought tolerance, facilitating marker-assisted selection to speed up the breeding program. Finally, Ribaut and Ragot (2007), McFadden et al. (2019), and Lunduka et al. (2019) focus on developing and adopting drought-tolerant varieties, integrating findings into breeding programs to enhance maize resilience to drought stress.

10.3.2 Waterlogging Management in Cereals

Under waterlogged conditions, split application of fertilizer with an additional dose of nutrients after stress helps in the speed recovery of plants. Application of 150% of the recommended dose of K as basal increased the survival rate of rice to waterlogging by 46.44%, 82.60%, 42.22%, and 78.22% in IR64, IR64-Sub1, Swarna, and Swarna-Sub1 varieties, respectively. Likewise, the cultivation of waterlogging tolerant genotypes reduced yield reduction. The percent reduction of yield was less in the wheat genotypes KRL 3–4 (12%), NW 1076 (9.8%), KRL 146 (9.0%), KRL 200 (3.2%), and KRL 144 (1.7%) under waterlogging stress, and this may be attributed to the different tolerance mechanisms. The exogenous application of ACC @ 10 mM and melatonin @ 100 μmol l^{-1} reduced the yield reduction under 2 days of waterlogging at the anthesis stage and 7 days of waterlogging at 50% flowering, respectively.

The raised bed method of planting at 50 cm height increased baby corn yield by 11.3% and fodder yield by 10.7%. Ridge tillage effectively alleviated the decline of dry matter accumulation induced by waterlogging, which was conducive to supplying enough source materials to grain filling for waterlogged summer maize, which increased the grain yield by 39–50%. The provision of subsurface drainage at the lateral spacing of 15 m intervals increased maize yield by 50% in waterlogged saline-alkaline soil. Also, drainage increased N uptake by 14% and seed priming (0.1% H_2O_2) for 12 h before sowing increased yield by 48%. Spraying of synthetic growth regulating hormone, namely, 6-BA @ 100 mg l^{-1} after waterlogging increased yield by 31.5% by stimulating cell division, hormone and sugar metabolism in plants, establishing the sink strength and improving the carbon metabolism.

Various genetic, molecular, and genomic approaches have propelled efforts to develop waterlogging-tolerant varieties. Biochemical studies conducted by Srivastava et al. (2007) have further elucidated the physiological mechanisms underlying waterlogging tolerance in maize, whereas studies by Yu et al. (2019), Zaidi et al. (2010), and Zhang et al. (2023a, b) have explored genetic variability and genetics, providing insights into the inheritance of waterlogging tolerance traits in maize and sorghum, respectively. Additionally, research by Yu et al. (2018), Mano and Nakazono (2021), Zaidi et al. (2015), and Guo et al. (2021) has focused on identifying genes and quantitative trait loci (QTLs) associated with waterlogging tolerance, offering potential targets for breeding programs.

10.3.3 Heat Stress Management in Cereals

Crop management practices play a crucial role in addressing the impact of biotic and abiotic factors, including heat stress (HS), on cereals. Strategies such as advancing sowing dates and demand-driven irrigations can mitigate heat (H) plus drought-induced yield losses. Early planting helps avoid terminal HS, aligning the grain-filling period with cooler temperatures in crops like wheat and rice. Proper plant nutrient management and exogenous supply of micronutrients, such as zinc, enhance crop tolerance to moderate heat increases.

Conservation agriculture preserves soil moisture, improving water and nutrient use efficiencies under heat or H+D conditions. Heat priming during early stages enhances thermo tolerance, imprinting short-term and transgenerational HS memory in plants. Management strategies involve exogenous supplementation of phytohormones, osmoprotectants, and bioregulators. Fine-tuning brassinosteroids (BRs) enhances growth promoters, amplifying HS responses for improved acclimation. Melatonin suppresses HS-induced damage and enhances stress-responsive gene expression.

Osmoprotectants like spermidine, glycine betaine, and proline reduce oxidative damage, increasing photosynthesis and improving plant growth under HS. Boosting growth-promoting bacteria via seed treatment enhances seedling survival and plant

HS tolerance. These agronomic interventions augment genotypic capabilities, supporting cereal crop cultivars in coping with HS conditions. The combined approach of adjusting genotype, growth environment, and management practices offers a comprehensive strategy for enhancing crop resilience and productivity under stress conditions.

Delaying the sowing of rice to achieve a mean daily temperature of 24.9–26.4 °C during the first 20 days after the heading is an effective strategy to mitigate heat stress in rice (Yang et al., 2021). The selection of suitable irrigation methods is important. Foliar application of heat stress mitigating micronutrients and hormones such as jasmonates, brassinosteroids, salicylic acid, ascorbic acid, citric acid, and calcium acetate alleviates heat stress and improves grain yield.

The application of organic manures increases the soil's moisture-holding capacity and reduces volatilization loss of nitrogenous fertilizers under high-temperature stress. Mulching with organic residues such as straw and plastic mulching reduces soil evaporation and keeps the soil microclimate in cooler conditions. Further, the adoption of conservation agriculture techniques such as zero tillage, no-tillage, permanent bed planting, and furrow irrigated raised bed systems prevents heat stress by higher soil moisture conservation. Advancing the sowing of wheat by 15–20 days in North-Western India escapes the wheat from terminal heat stress. Plant growth-promoting rhizobacteria (PGPR) increases the wheat crop tolerance toward heat stress. Likewise, foliar nutrition with stress-relieving nutrients such as KNO_3, KCl, magnesium, and zinc is the on-time mitigation strategy for heat stress management. Seed priming with salicylic acid, *moringa* leaf extract, $CaCl_2$, and kinetin increased the grain yield from 17.9% to 53.2% under heat stress and could maintain the crop yield. Foliar application of arginine, glycine, and ascorbic acid during heat stress improves the membrane stability index, relative water contents, chlorophyll contents, and grain yield by acting as the osmoprotectants.

Efforts to develop heat-tolerant varieties in cereals leverage genetic and genomic approaches. Pre-breeding studies by Chen et al. (2012) and Nelimor et al. (2019) focus on harnessing genetic diversity for heat tolerance traits. Understanding the genetics of drought tolerance, as explored by Naveed et al. (2016), provides insights into related stress responses. In vitro gamete selection for heat stress tolerance, pioneered by Petolino et al. (1990), offers a novel approach for breeding heat-tolerant varieties. Molecular approaches, including genome-wide association studies (GWAS) in rice (Ye et al., 2015; Lafarge et al., 2017; Bheemanahalli et al., 2021), maize (Gao et al., 2019; Longmei et al., 2021), and wheat (Pradhan et al., 2019), transcriptomics, and functional genomics in different cereals play a crucial role in dissecting quantitative trait loci (QTLs) or genes (Table 10.11), heat shock proteins (HSPs), and regulatory pathways associated with thermotolerance in cereals. A number of heat shock proteins (HSPs) in cereal plants act as molecular chaperones that aid damaged proteins to facilitate their repair or removal or prevent the irreversible aggregation of denaturing proteins. For instance, HSP70, HSP90, and HSP100 genes exhibit upregulation to confer HS tolerance in barley at the R stage. Similarly, in rice (OsHSP101) (TaHSP23.9) and maize (ZmHsf06), HSP genes function as a chaperone to positively regulate plant responses heat stress (Zenda et al., 2022).

Hence, the information on QTLs/genes (Table 10.11) may help to exploit and utilize them in molecular breeding as well as to unravel the mechanism of heat tolerance through expression and transgenic studies. Together, these genetic and genomic approaches contribute to the development of heat-tolerant cereal varieties, which are crucial for ensuring food security in the face of climate change-induced temperature fluctuations (Table 10.12).

10.3.4 Low-Temperature Management

Early sowing on 22nd October prevented the chilling stress and recorded a higher yield of wheat. Tolerant cultivars with the optimum date of sowing (15th November) increased grain yield by 15.29% in India (Upadhyay et al., 2023). Seed treatment and soil inoculation of psychrotolerant *Pseudomonads* increased the relative water content, reduced membrane injury electrolyte leakage, and prevented chilling stress in wheat. Optimized application of nutrients increases the stress tolerance in wheat. Foliar application of melatonin and salicylic acid increased the photosynthetic carbon assimilation, membrane stability, and grain yield of wheat under chilling stress.

Similarly, maize tolerance to chilling stress varies with inherent adaptive traits and antioxidative defense mechanisms, etc. Cultivars with vigorous root architecture, root biomass, and efficient antioxidant defense exhibit better growth and greater stress tolerance under chilling stress. Seed priming with 0.5 mM Salicylic acid +50 mM H_2O_2 recorded a higher germination percentage of 80% under chilling stress (Li et al., 2017). Seed priming with micronutrients (4 mM Zn + 2.5 mM Mn solution as $ZnSO_4 \cdot H_2O$ and as $MnSO_4 \cdot 7H_2O$) increased fresh grain yield per plant by 15.15%. Sowing the seeds on the southern side of the ridge prevents the low-temperature stress by increasing the duration of maximum sunlight in the daytime. A foliar spray of 5-aminolevulinic acid enhanced the superoxide dismutase, catalase, and ribulose-1,5-bisphosphate (RuBP) carboxylase activities (Table 10.13).

Adaptation strategies are essential to mitigate yield losses of cereals in regions with extremely low temperatures, especially during the reproductive stage. Incorporating cold-tolerant varieties into crop production represents a sustainable and cost-efficient approach to addressing low-temperature stress in cereals. However, the development of such varieties requires a diverse gene pool and a deeper understanding of genetic and molecular mechanisms. Maize cultivars show inherent variations in response to low temperatures, with traits such as root morphology and antioxidative defense mechanisms playing a crucial role. Understanding these genetic variations is essential for breeding tolerant maize hybrids or varieties against different levels of cold stress. Enders et al. (2019) evaluated inbred maize using RGB imaging for cold tolerance assessment. Revilla et al. (2000) and Neta et al. (2020) investigated the genetics and inheritance of genes and traits associated with cold tolerance. Hu et al. (2017) and Li et al. (2018) identified QTLs and alleles linked to cold tolerance, providing valuable genetic markers for breeding programs. Furthermore, Frascaroli and Revilla (2018) and Sowiński et al. (2020) utilized transcriptomics and genomic approaches to unravel molecular mechanisms underlying cold tolerance in maize.

Winter cereals exhibit varying low-temperature tolerances, and rye stands out as highly responsive to low-temperature induction, showing rapid acclimatization and up-regulation of related genes. Despite attempts to transfer rye's superior tolerance to wheat through interspecific gene transfers, such efforts have been unsuccessful. The exceptional low-temperature tolerance of rye does not consistently manifest in tetraploid and hexaploid wheat backgrounds. Further, Triticeae genomes in wheat and interspecific crosses operate as competitively balanced systems, influenced by chromosome dosage ratios favoring the parent with the greater chromosome number. Molecular studies indicate that cold tolerance-associated gene expression is regulated across genomes, with the higher ploidy parent controlling expression levels at the transcriptional level in many interspecific crosses (Fowler et al., 1999). These findings suggest that the key to enhancing LT tolerance in cold-sensitive species lies in the primary regulation of native LT genes rather than through the introduction of exotic structural genes.

Collectively, these studies contribute to the development of cold-tolerant maize varieties by enhancing our understanding of genetic factors and molecular pathways associated with cold stress response.

10.3.5 Salinity Management of in Cereals

Soil amendments replenish the soil salinity, and the application of green manure @ 10 t ha−1 and gypsum @ 1 g kg−1 improved the grain yield to 4.00 t ha−1, with a yield reduction of just 10.91%. Humic acid (40 mg l^{-1}) seed priming reduced the salinity-induced plant damage, improved metabolism, and maintained active growth. Organic amendments such as farm yard manure, vermicompost, poultry manures, and biochar application reduce salinity stress. The application of good-quality water improves the properties of saline soil. Submergence of the field for 2–4 weeks before transplanting leaches down the salts in addition to optimum fertilizer application. Neutralization of sodic water by the addition of gypsum @ 75% and adopting the sprinkler method of irrigation optimized the yield in wheat. Hydropriming and halopriming improve the emergence percentage emergence index and promote early emergence by causing physiological changes in seeds under salinity stress. Plant growth-promoting rhizobacteria and arbuscular mycorrhizal fungi favor the rhizospheric soil by the production of root exudates and enhanced root metabolism. Foliar application of thiourea (10 mM) increased the EC tolerance by 6–11%. Soaking maize seeds in water before sowing increases the germination percentage in maize. Elements like K provide some cellular signaling alone or in association with phytohormones under stress situations. Melatonin, chitosan, yeast extract, auxin, cytokinin, gibberellins, kinetin, indoleacetic acid and salicylic acid spray increased the photosynthetic activities and grain yield under salinity stress in maize.

Efforts to address salinity challenges primarily concentrate on enhancing salt tolerance in cereal crops through molecular breeding methods. While notable progress has been achieved in improving salt tolerance in wheat and rice (Boraiah et al., 2023), similar advancements for other cereals are limited. Salinity tolerance is genetically complex, encompassing various growth and physiological processes such as root architecture, nutrient uptake, and hormonal regulation (Wani et al., 2020). Maize and its wild relatives exhibit genetic variation for salinity tolerance, offering valuable resources for genetic exploration (Table 10.14). Understanding the genetic basis of salinity tolerance has facilitated the identification of underlying genes and molecular mechanisms, including component traits and QTLs (Ma et al., 2021; Krishnamurthy et al., 2022; Singh & Devi, 2023). This knowledge enables the quantification of trait contributions to salinity tolerance, thereby aiding the development of superior, salt-tolerant varieties.

10.4 Strategies for Development Climate Smart-Abiotic Stress Tolerant Varieties: Way Forward

The impact of abiotic stresses on crops, including cereals, was significantly visible and well documented, along with management strategies. The advanced computing and inclusion of artificial intelligence (AI) in weather-crop modeling help in better predictions of the likely occurrence of weather abnormalities and abiotic stresses and their impact on crops thereof. Despite the availability of prediction models and management approaches, yield losses due to abiotic stresses cannot be stopped completely. However, losses can be minimized by developing climate-smart crop management practices and climate-smart and abiotic stress-tolerant varieties. Despite crop experts having proposed various strategies to achieve the targets, it is essential to reorient and integrate these strategies to develop appropriate approaches for better management of the abiotic stresses in cereals. Hence, understanding the response of crops and their tolerance mechanisms against the diverse abiotic stresses is the prime strategy followed by unrevealing the traits/genes/QTLs for introgression into elite breeding lines.

The importance of abiotic stress tolerance mechanisms in cereal crops, particularly regarding traits associated with drought tolerance, i.e., water use efficiency and stomatal architecture, has been a subject of critical discussion among crop experts (Ingram, 2005). Despite the susceptibility of significant cereal crops to various abiotic stresses, they exhibit notable drought tolerance, primarily attributed to their unique stomatal architecture and rapid response mechanisms, which enhance photosynthesis efficiency and overall water use efficiency. Schafer et al. (2018) highlighted the evolution of nitrate sensitivity in stomatal closure as a crucial factor contributing to the resilience of cereals under environmental stresses. They demonstrated that a specific amino acid motif in the guard cells, found in the anion channel AtSLAC1 of grasses, enables the control of stomatal aperture by nitrate, playing a pivotal role in the evolution of monocots from dicots. Similarly, the vital mechanisms for salinity tolerance include excluding Na^+ from the transpiration route, sequestering Na^+ and Cl^- in vacuoles within root and leaf cells, and cellular and metabolic processes that facilitate vigorous growth despite the osmotic stress caused by external salt. For instance, wheat varieties Sakha 8, S-24, W4909, and W4910 possess positive characteristics closely associated with effective discrimination of Na^+ and K^+ in leaves and a favorable Chlorophyll Content Index (CCI) in leaves (Saddiq et al., 2021). Hence, emphasizing the advanced screening through phenomics facilities for such traits and other surrogated traits is crucial for the successful identification and incorporation of new tolerance genes into crop plants.

Breeders have successfully developed numerous varieties tolerant to salinity and waterlogging by incorporating tolerance genes/QTLs sourced from landraces (Boraiah et al., 2023). Notably, the genotype Kharchia 65 has been extensively utilized in India for breeding wheat varieties with salt tolerance and has served as a donor parent for various wheat improvement programs worldwide. Likewise, the development of several submergence-tolerant rice varieties has been achieved by leveraging submergence-tolerance QTLs derived from a landrace called FR13A. Besides these traits, considering new crop genetic traits (CGT), it is inevitable that abiotic stress-tolerant varieties will develop with maximum genetic gain during the breeding program. For instance, along with terminal heat tolerance, traits like early vigor to increase early crop water use and late flowering to reverse an earlier anthesis in warmer conditions should be considered in wheat abiotic resistance breeding.

Hence, a better understanding of mechanisms and uncovering the genomic regions and biomolecules associated with the response to abiotic stresses aids in developing climate-smart with abiotic stress tolerant varieties. The recent advances in genome editing and other molecular tools, including transgenic approaches, supplement and hasten conventional and molecular breeding. Further, speed breeding approaches may accelerate abiotic stress resistance breeding.

10.5 Conclusion

Addressing the impact of climate change-driven abiotic stresses on cereals, which are a significant component of the global food supply, requires a deep and holistic understanding of the mechanisms enabling grass flora to adapt, escape, and tolerate different stressors. Despite the recent evolution of cereals among diverse flora, their environmental stress response mechanisms are notably advanced and prominent compared to animals. Extensive studies have elucidated the molecular and cellular mechanisms underlying abiotic stress tolerance in cereal crops such as maize, rice, and wheat, including responses to water deficit, heat, submergence, and salinity. Despite the progress made in developing abiotic stress-tolerant varieties, there is an urgent need to connect the dots and unravel the complex mechanisms associated with tolerance to multiple abiotic stresses in order to develop multiple stress-tolerant and climate-smart varieties. While genomics has made significant strides, further exploration of other omics levels, such as transcriptomic and proteomic profiling, is crucial for achieving a comprehensive understanding at the molecular level.

Additionally, image-based phenomics approaches play a vital role in understanding plant responses to environmental stresses by measuring surrogate traits or mechanisms associated with abiotic stress tolerance. An integrated omics approach, encompassing genomics, phenomics, transcriptomics, and proteomics, has been instrumental in unraveling the traits and genes associated with multiple abiotic stress tolerance, thereby aiding in the development of climate-smart cereal crop varieties through the introgression of such components via molecular and conventional breeding methods. This comprehensive approach is essential for ensuring food security in the face of evolving environmental challenges that trigger abiotic stresses.

References

Acharjee, T.K., van Halsema, G., Ludwig, F., et al., 2019. Shifting planting date of Boro rice as a climate change adaptation strategy to reduce water use[J]. *Agricultural Systems, 168*, 131–143.

Abiko, T., Todoroki, T., Pham Thi Thanh, H., Nakamura, T., Haraguchi, Y., Tanaka, T., Saisho, D., & Kai, H. (2023). Barley varieties tolerant to waterlogged reduced soil show the better root growth in hypoxia. *Plant Production Science, 26*(4), 429–439.

Adee, E., Roozeboom, K., Balboa, G. R., Schlegel, A., & Ciampitti, I. A. (2016). Drought-tolerant corn hybrids yield more in drought-stressed environments with no penalty in non-stressed environments. *Frontiers in Plant Science, 7*, 210680. https://doi.org/10.3389/fpls.2016.01534

Afzal, I., Noor, M. A., Bakhtavar, M. A., Ahmad, A., & Haq, Z. (2015). Improvement of spring maize performance through physical and physiological seed enhancements. *Seed Science and Technology, 43*(2), 238–249.

Ahmad, M., Shahzad, A., Iqbal, M., Asif, M., & Hirani, A. H. (2013). Morphological and molecular genetic variation in wheat for salinity tolerance at germination and early seedling stage. *Australian Journal of Crop Science, 7*(1), 66–74.

Ahmad, I., Basra, S. M. A., Akram, M., Wasaya, A., Ansar, M., Hussain, S., Iqbal, A., & Hussain, S. A. (2017). Improvement of antioxidant activities and yield of spring maize through seed priming and foliar application of plant growth regulators under heat stress conditions. *Semina: Ciências Agrárias, 38*(1), 47–56.

Ahmad, H., Zafar, S. A., Naeem, M. K., Shokat, S., Inam, S., Rehman, M. A. U., Naveed, S. A., Xu, J., Li, Z., Ali, G. M., & Khan, M. R. (2022). Impact of pre-anthesis drought stress on physiology, yield-related traits, and drought-responsive genes in green super rice. *Frontiers in Genetics, 13*, 832542.

Ahmed, B. O., Inoue, M., & Moritani, S. (2010). Effect of saline water irrigation and manure application on the available water content, soil salinity, and growth of wheat. *Agricultural Water Management, 97*(1), 165–170.

Ahmed, S., Iqbal, M., Ahmad, Z., Iqbal, M. A., Artyszak, A., Sabagh, A. E., Alharby, H. F., & Hossain, A. (2023). Foliar application of silicon-based nanoparticles improve the adaptability of maize (Zea mays L.) in cadmium contaminated soils. *Environmental Science and Pollution Research, 30*(14), 41002–41013.

Akbar, G., Islam, Z., Hameed, S., & Ahmad, B. (2023). Yield and water productivity of rice as influenced by responsive drip irrigation, alternate wetting and drying versus conventional flooding under silty loam soil texture. *Water Resources and Irrigation Management-WRIM, 12*(1–3), 1–13.

Akter, N., Brishty, T. A., Karim, M. A., Ahmed, M. J. U., & Islam, M. R. (2023). Leaf water status and biochemical adjustments as a mechanism of drought tolerance in two contrasting wheat (Triticum aestivum L.) varieties. *Acta Physiologiae Plantarum, 45*(3), 50.

Ali, Q., Ashraf, M., & Anwar, F. (2010). Seed composition and seed oil antioxidant activity of maize under water stress. *Journal of the American Oil Chemists' Society, 87*, 1179–1187.

Ali, Z., Basra, S. M. A., Munir, H. A. S. S. A. N., Mahmood, A. R. S. H. A. D., & Yousaf, S. H. A. H. I. D. A. (2011). Mitigation of drought stress in maize by natural and synthetic growth promoters. *Journal of Agriculture and Social Sciences, 7*(2), 56–62.

Almeida, G. D., Nair, S., Borém, A., Cairns, J., Trachsel, S., Ribaut, J. M., Bänziger, M., Prasanna, B. M., Crossa, J., & Babu, R. (2014). Molecular mapping across three populations reveals a QTL hotspot region on chromosome 3 for secondary traits associated with drought tolerance in tropical maize. *Molecular Breeding, 34*, 701–715.

Al-Quraan, N. A., Sartawe, F. A. B., & Qaryouti, M. M. (2013). Characterization of γ-aminobutyric acid metabolism and oxidative damage in wheat (Triticum aestivum L.) seedlings under salt and osmotic stress. *Journal of Plant Physiology, 170*(11), 1003–1009.

Anjum, F., Wahid, A., Farooq, M., & Javed, F. (2011). Potential of foliar applied thiourea in improving salt and high temperature tolerance of bread wheat (Triticum aestivum). *International Journal of Agriculture and Biology, 13*(2), 251–256.

Anonymous. (2018). Wheat initiative. In *Annual report: Coordinating global wheat research* (p. 60). Federal Research Centre for Cultivated Plants. Available online: https://www.wheatinitiative.org

Araki, H., Hossain, M. A., & Takahashi, T. (2012). Waterlogging and hypoxia have permanent effects on wheat root growth and respiration. *Journal of Agronomy and Crop Science, 198*(4), 264–275.

Aravinth, K. V., Avudaithai, S., Ramesh, T., Balasubramaniam, P., & Sundar, M. (2023). Impact of alternate wetting and drying irrigation and brown manuring on water use, weed control and yield of drum seeded rice. *Journal of Applied and Natural Science, 15*(1), 75–84.

Aslam, M., Maqbool, M. A., & Cengiz, R. (2015). Effects of drought on maize. In *Drought stress in maize (Zea mays L)* (pp. 5–17). Springer.

Atkin, O. K., et al. (2005). Effects of temperature on the respiration of roots and the partitioning of recent carbon to respiration in Lolium perenne. *New Phytologist, 168*(1), 179–188.

Azizian, A., & Sepaskhah, A. R. (2014). Maize response to different water, salinity and nitrogen levels: Agronomic behavior. *International Journal of Plant Production, 8*(1), 107–130.

Bahuguna, R. N., Tamilselvan, A., Muthurajan, R., Solis, C. A., & Jagadish, S. V. K. (2018). Mild preflowering drought priming improves stress defences, assimilation and sink strength in rice under severe terminal drought. *Functional Plant Biology, 45*(8), 827–839. https://doi.org/10.1071/FP17248

Bakht, J., Shafi, M., Jamal, Y., & Sher, H. (2011). Response of maize (Zea mays L.) to seed priming with NaCl and salinity stress. *Spanish Journal of Agricultural Research, 9*(1), 252–261.

Bakhtavar, M. A., Afzal, I., Basra, S. M. A., Ahmad, A. U. H., & Noor, M. A. (2015). Physiological strategies to improve the performance of spring maize (Zea mays L.) planted under early and optimum sowing conditions. *PLoS One, 10*(4), e0124441.

Balla, K., Rakszegi, M., Li ZhonGy, L. Z., Bekes, F., Bencze, S., & Veisz, O. (2011). Quality of winter wheat in relation to heat and drought shock after anthesis. *Czech Journal of Food Sciences, 29*(2), 117–128.

Balusamy, A., Udayasoorian, C., & Jayabalakrishnan, R. (2019). Effect of subsurface drainage system on maize growth, yield and soil quality. *International Journal of Current Microbiology and Applied Sciences, 8*, 1206–1215.

Batool, M., Saqib, M., Qayyum, M. A., Jan, M., Hameed, S. A., Khan, A. K., Khaliq, A., Naz, T., Hussain, A., & Nawaz, S. (2020). 106. Screening of different maize (Zea mays L.) genotypes against salinity under hydroponics. *Pure and Applied Biology (PAB), 9*(1), 1118–1129.

Becheran, D. E., Miralles, D. J., Abeledo, L. G., Alvarez Prado, S., & de San Celedonio, R. P. (2022). Source-sink limitations for grain weight in wheat and barley under waterlogging conditions during pre-anthesis. *Journal of Agronomy and Crop Science, 208*(1), 76–88. https://doi.org/10.1111/jac.12571

Bheemanahalli, R., Knight, M., Quinones, C., Doherty, C. J., & Jagadish, S. K. (2021). Genome-wide association study and gene network analyses reveal potential candidate genes for high night temperature tolerance in rice. *Scientific Reports, 11*(1), 6747.

Blumenthal, C., Rawson, H. M., McKenzie, E., Gras, P. W., Barlow, E. W. R., & Wrigley, C. W. (1996). Changes in wheat grain quality due to doubling the level of atmospheric CO2. *Cereal Chemistry, 73*, 762–766.

Bolland, M. D. A. (1984). Grain losses due to delayed harvesting of barley and wheat. *Australian Journal of Experimental Agriculture, 24*(126), 391–395.

Boraiah, K. M., Basavaraj, P. S., Halli, H. M., Mukri, G., Yathish, K. R., Prakash, R. N., & Kumar, S. (2022). Maize: Impacts and management of abiotic stresses. In *Abiotic stresses in agriculture: Impacts and management* (pp. 116–150). ICAR-NIASM.

Boraiah, K. M., Basavaraj, P. S., Kakade, V. D., Harisha, C. B., Khapte, P., Halagundegowda, G. R., Krishnamurthy, D., Kulshreshtha, N., Vijayakumar, H. P., Naik, B., & Pathak, H. (2023). Abiotic stress-tolerant crop varieties in India: Status and a way forward. In *Recent trends in plant breeding and genetic improvement*. IntechOpen. Available at: https://doi.org/10.5772/intechopen.1001916

Bouman, B. A. M., Peng, S., Castaneda, A. R., & Visperas, R. M. (2005). Yield and water use of irrigated tropical aerobic rice systems. *Agricultural Water Management, 74*(2), 87–105.

Brás, T. A., Seixas, J., Carvalhais, N., & Jägermeyr, J. (2021). Severity of drought and heatwave crop losses tripled over the last five decades in Europe. *Environmental Research Letters, 16*(6), 065012.

Britz, S. J., Prasad, P. V. V., Moreau, R. A., Allen, L. H., Kremer, D. F., & Boote, K. J. (2007). Influence of growth temperature on the amounts of tocopherols, tocotrienols, and γ-oryzanol in brown rice. *Journal of Agricultural and Food Chemistry, 55*(18), 7559–7565.

Cannell, R. A., Belford, R. K., Gales, K., Thomson, R. J., & Webster, C. P. (1984). Effects of waterlogging and drought on winter wheat and winter barley grown on a clay and a sandy loam soil: I. Crop growth and yield. *Plant and Soil, 80*, 53–66.

Carpýcý, E. B., Celýk, N., & Bayram, G. (2009). Effects of salt stress on germination of some maize (Zea mays L.) cultivars. *African Journal of Biotechnology, 8*(19).

Champness, M., Ballester, C., & Hornbuckle, J. (2023). Effect of soil moisture deficit on aerobic rice in temperate Australia. *Agronomy, 13*(1), 168.

Chen, S. Y., Zhang, X. Y., Pei, D., Sun, H. Y., & Chen, S. L. (2007). Effects of straw mulching on soil temperature, evaporation and yield of winter wheat: Field experiments on the North China Plain. *Annals of Applied Biology, 150*(3), 261–268.

Chen, J., Xu, W., Velten, J., Xin, Z., & Stout, J. (2012). Characterization of maize inbred lines for drought and heat tolerance. *Journal of Soil and Water Conservation, 67*(5), 354–364.

Chen, C., Huang, J., Zhu, L., Shah, F., Nie, L., Cui, K., & Peng, S. (2013). Varietal difference in the response of rice chalkiness to temperature during ripening phase across different sowing dates. *Field Crops Research, 151*, 85–91.

Chen, Y., Song, J., Yan, C., & Hong, X. (2021). Effects of submergence stress at the vegetative growth stage on hybrid rice growth and grain yield in China. *Chilean Journal of Agricultural Research, 81*(2), 191–201.

Chowdhury, S., Bhusan, D., Hashem, M. A., & Hoque, M. A. (2019). Organic amendments for mitigating soil salinity in rice. *Research in Agriculture Livestock and Fisheries, 6*(1), 11–17.

Cooper, M., Gho, C., Leafgren, R., Tang, T., & Messina, C. (2014). Breeding drought-tolerant maize hybrids for the US corn-belt: Discovery to product. *Journal of Experimental Botany, 65*(21), 6191–6204.

Costa Silva Neta, I, Vilela de Resende Von Pinho, É., de Abreu, V. M., Rezende Vilela, D., Santos, M. C., Oliveira dos Santos, H., Diniz Cabral Ferreira, R. A., Garcia Von Pinho, R., & Coelho de Castro Vasconcellos, R. (2020). Gene expression and genetic control to cold tolerance during maize seed germination. *BMC Plant Biology, 20*, 1–14.

Cotrozzi, L., Lorenzini, G., Nali, C., Pisuttu, C., Pampana, S., & Pellegrini, E. (2021). Transient waterlogging events impair shoot and root physiology and reduce grain yield of durum wheat cultivars. *Plants, 10*(11), 2357.

Craufurd, P. Q., & Peacock, J. M. (1993). Effect of heat and drought stress on sorghum (Sorghum bicolor). II. Grain yield. *Experimental Agriculture, 29*(1), 77–86.

Crusciol, C. A. C., Arf, O., Soratto, R. P., & Mateus, G. P. (2008). Grain quality of upland rice cultivars in response to cropping systems in the Brazilian tropical savanna. *Scientia Agricola, 65*, 468–473.

Cucci, G., Lacolla, G., Boari, F., Mastro, M. A., & Cantore, V. (2019). Effect of water salinity and irrigation regime on maize (Zea mays L.) cultivated on clay loam soil and irrigated by furrow in Southern Italy. *Agricultural Water Management, 222*, 118–124.

Daryanto, S., Wang, L., & Jacinthe, P. A. (2016). Global synthesis of drought effects on maize and wheat production. *PLoS One, 11*(5), e0156362.

Das, S., Mohanty, S., Dash, D., & Muduli, K. C. (2020). Enhancement of growth and seed yield of rice (Oryza sativa L.) through foliar spray of osmoprotectants under high temperature stress. *Indian Journal of Traditional Knowledge*. https://doi.org/10.56042/ijtk.v19i1.30846

Dass, S., Singh, I., Chikkappa, G. K., Parihar, C. M., Kaul, J., Singode, A., Manivannan, A., & Singh, D. K. (2011). *Abiotic stresses in maize: Some issues and solutions* (p. 16). Directorate of Maize Research, Indian Council of Agricultural Research.

Davani, D., Nabipour, M., & Roshanfekr Dezfouli, H. (2017). Effects of different concentrations of cytokinin and auxin hormones on yield and yield components of grain maize (Zea mays L.) in salinity conditions. *Plant Productions, 40*(1), 69–80.

de San Celedonio, R. P., Abeledo, L. G., & Miralles, D. J. (2014). Identifying the critical period for waterlogging on yield and its components in wheat and barley. *Plant and Soil, 378*, 265–277.

de San Celedonio, R. P., Abeledo, L. G., & Miralles, D. J. (2018). Physiological traits associated with reductions in grain number in wheat and barley under waterlogging. *Plant and Soil, 429*, 469–481. https://doi.org/10.1007/s11104-018-3708-4

Deichmann, M. M., Andersen, M. N., Thomsen, I. K., & Børgesen, C. D. (2019). Impacts of controlled drainage during winter on the physiology and yield of winter wheat in Denmark. *Agricultural Water Management, 216*, 118–126. ISSN 0378-3774. https://doi.org/10.1016/j.agwat.2019.01.013

Deng, Z., Huang, M., Zhang, W., Wang, G., Huang, X., Liang, G., & Li, N. (2023). Effects of five years conservation tillage for hedging against drought, stabilizing maize yield, and improving soil environment in the drylands of northern China. *PLoS One, 18*(3), e0282359.

Denmead, O. T., & Shaw, R. H. (1960). The effects of soil moisture stress at different stages of growth on the development and yield of corn 1. *Agronomy Journal, 52*(5), 272–274.

Devarajan, A. K., Muthukrishanan, G., Truu, J., Truu, M., Ostonen, I., Kizhaeral, S. S., Panneerselvam, P., & Kuttalingam Gopalasubramanian, S. (2021). The foliar application of rice phyllosphere bacteria induces drought-stress tolerance in Oryza sativa (L.). *Plants, 10*(2), 387.

Dong, W., Zhang, L., Duan, Y., Sun, L., Zhao, P., van der Werf, W., Evers, J. B., Wang, Q., Wang, R., & Sun, Z. (2017). Ridge and furrow systems with film cover increase maize yields and mitigate climate risks of cold and drought stress in continental climates. *Field Crops Research, 207*, 71–78.

Dubey, R., Pathak, H., Chakrabarti, B., Singh, S., Gupta, D. K., & Harit, R. C. (2020). Impact of terminal heat stress on wheat yield in India and options for adaptation. *Agricultural Systems, 181*, 102826.

Durowoju, O. S., Ologunorisa, T. E., & Akinbobola, A. (2023). Impact of drought on cereal crop yields in the Savanna Region of Nigeria. *African Geographical Review, 42*(3), 273–293.

Dwivedi, S. K., Kumar, S., Bhakta, N., Srivastava, A. K., Mishra, J. S., Kumar, V., Kumara, B. H., Bhatt, B. P., & Singh, S. (2018). Physiological mechanism and nutrient management strategies for flood tolerance in rice grown in lowland flood prone ecosystem. *Journal of Crop Science and Biotechnology, 21*, 321–331.

Earl, H. J., & Davis, R. F. (2003). Effect of drought stress on leaf and whole canopy radiation use efficiency and yield of maize. *Agronomy Journal, 95*(3), 688–696.

Ebrahimi, S. T., Yarnia, M., Benam, M. K., & Tabrizi, E. F. M. (2011). Effect of potassium fertilizer on corn yield (Jeta cv.) under drought stress condition. *American-Eurasian Journal of Agricultural and Environmental Science, 10*(2), 257–263.

Edmeades, G. O. (2013). *Progress in achieving and delivering drought tolerance in maize-an update* (p. 130). ISAAA.

Edmeades, G. O., Trevisan, W., Prasanna, B. M., & Campos, H. (2017). Tropical maize (Zea mays L.). In *Genetic improvement of tropical crops* (pp. 57–109). Springer.

Edreira, J. R., & Otegui, M. E. (2013). Heat stress in temperate and tropical maize hybrids: a novel approach for assessing sources of kernel loss in field conditions. *Field Crops Research, 142*, 58–67.

Ejaz, M., & von Korff, M. (2017). The genetic control of reproductive development under high ambient temperature. *Plant Physiology, 173*(1), 294–306.

El-Katony, T. M., El-Bastawisy, Z. M., & El-Ghareeb, S. S. (2019). Timing of salicylic acid application affects the response of maize (Zea mays L.) hybrids to salinity stress. *Heliyon, 5*(4).

Enders, T. A., St. Dennis, S., Oakland, J., Callen, S. T., Gehan, M. A., Miller, N. D., Spalding, E. P., Springer, N. M., & Hirsch, C. D. (2019). Classifying cold-stress responses of inbred maize seedlings using RGB imaging. *Plant Direct, 3*(1), e00104.

Eser, C., Soylu, S., & Ozkan, H. (2024). Drought responses of traditional and modern wheats in different phenological stages. *Field Crops Research, 305*, 109201. ISSN 0378-4290. https://doi.org/10.1016/j.fcr.2023.109201

Essemine, J., Ammar, S., & Bouzid, S. (2010). Impact of heat stress on germination and growth in higher plants: Physiological, biochemical and molecular repercussions and mechanisms of defence. *Journal of Biological Sciences, 10*(6), 565–572.

FAO. (2023). *Crop prospects and food situation—Triannual Global Report No. 3*, November 2023. https://doi.org/10.4060/cc8566en

Fardus, J., Matin, M. A., Hasanuzzaman, M., Hossain, M. A., & Hasanuzzaman, M. (2018). Salicylic acid-induced improvement in germination and growth parameters of wheat under salinity stress. *The Journal of Animal and Plant Sciences, 28*(1), 197–207.

Farhad, M., Kumar, U., Tomar, V., Bhati, P. K., Krishnan, J. N., Barek, V., Brestic, M., & Hossain, A. (2023). Heat stress in wheat: A global challenge to feed billions in the current era of the changing climate. *Frontiers in Sustainable Food Systems, 7*, 1203721.

Farooq, M., Basra, S. M. A., Wahid, A., Ahmad, N., & Saleem, B. A. (2009). Improving the drought tolerance in rice (Oryza sativa L.) by exogenous application of salicylic acid. *Journal of Agronomy and Crop Science, 195*(4), 237–246.

Farooq, S., Shahid, M., Khan, M. B., Hussain, M., & Farooq, M. (2015). Improving the productivity of bread wheat by good management practices under terminal drought. *Journal of Agronomy and Crop Science, 201*(3), 173–188.

Fofana, M., Chérif, M., Kone, B., Futakuchi, K., & Audebert, A. (2010). Effect of water deficit at grain repining stage on rice grain quality. *Journal of Agricultural Biotechnology and Sustainable Development, 2*, 100–107.

Fowler, D. B., Limin, A. E., & Ritchie, J. T. (1999). Low-temperature tolerance in cereals: Model and genetic interpretation. *Crop Science, 39*(3), 626–633.

Fowler, D. B., Byrns, B. M., & Greer, K. J. (2014). Overwinter low-temperature responses of cereals: Analyses and simulation. *Crop Science, 54*(6), 2395–2405.

Frascaroli, E., & Revilla, P. (2018). Genomics of cold tolerance in maize. In *The maize genome* (pp. 287–303). Springer.

Frey, F. P., Presterl, T., Lecoq, P., Orlik, A., & Stich, B. (2016). First steps to understand heat tolerance of temperate maize at adult stage: Identification of QTL across multiple environments with connected segregating populations. *Theoretical and Applied Genetics, 129*, 945–961.

FSIN and Global Network Against Food Crises. (2023). GRFC 2023. Rome.

Gao, J., Wang, S., Zhou, Z., Wang, S., Dong, C., Mu, C., Song, Y., Ma, P., Li, C., Wang, Z., & He, K. (2019). Linkage mapping and genome-wide association reveal candidate genes conferring thermotolerance of seed-set in maize. *Journal of Experimental Botany, 70*(18), 4849–4864.

Garthwaite, A. J., von Bothmer, R., & Colmer, T. D. (2005). Salt tolerance in wild Hordeum species is associated with restricted entry of Na^+ and Cl^- into the shoots. *Journal of Experimental Botany, 56*(419), 2365–2378.

Gautam, P., Lal, B., Tripathi, R., Shahid, M., Baig, M. J., Maharana, S., Puree, C., & Nayak, A. K. (2016). Beneficial effects of potassium application in improving submergence tolerance of rice (Oryza sativa L.). *Environmental and Experimental Botany, 128*, 18–30.

Ge, T., Sui, F., Bai, L., Tong, C., & Sun, N. (2012). Effects of water stress on growth, biomass partitioning, and water-use efficiency in summer maize (Zea mays L.) throughout the growth cycle. *Acta Physiologiae Plantarum, 34*, 1043–1053.

Ghadirnezhad, R., & Fallah, A. (2014). Temperature effect on yield and yield components of different rice cultivars in flowering stage. *International Journal of Agronomy, 2014*, 1–4.

Ghobadi, M. E., Ghobadi, M., & Zebarjadi, A. (2017). Effect of waterlogging at different growth stages on some morphological traits of wheat varieties. *International Journal of Biometeorology, 61*(4), 635–645. https://doi.org/10.1007/s00484-016-1240-x

Ghogdi, E. A., Izadi-Darbandi, A., & Borzouei, A. (2012). Effects of salinity on some physiological traits in wheat (Triticum aestivum L.) cultivars. *Indian Journal of Science and Technology, 5*, 1901–1906.

Gobinath, R., Manasa, V., Surekha, K., Vijayakumar, S., & Bandeppa, B. (2021). New age nutrient carriers for rice based cropping systems. *Indian Farming, 71*(4), 12–15.

Gong, M., Bai, N., Wang, P., Su, J., Chang, Q., & Zhang, Q. (2023). Co-Inoculation with arbuscular mycorrhizal fungi and dark septate endophytes under drought stress: Synergistic or competitive effects on maize growth, photosynthesis, root hydraulic properties and aquaporins? *Plants, 12*(14), 2596.

Gooding, M. J., Ellis, R. H., Shewry, P. R., & Schofield, J. D. (2003). Effects of restricted water availability and increased temperature on the grain filling, drying and quality of winter wheat. *Journal of Cereal Science, 37*(3), 295–309.

Gous, P. W., Hickey, L., Christopher, J. T., Franckowiak, J., & Fox, G. P. (2016). Discovery of QTL for stay-green and heat-stress in barley (Hordeum vulgare) grown under simulated abiotic stress conditions. *Euphytica, 207*(2), 305–317.

Gregorio, G. B. (1997). *Tagging salinity tolerant genes in rice using Amplified Fragment Length Polymorphism (AFLP)* [PhD Dissertation, University of the Philippines Los Baños College], 118 p.

Gribaldi, G., Nurlaili, N., Dewi, N., Danial, E., Sakalena, F., & Suwignyo, R. A. (2017). Modified application of nitrogen fertilizer for increasing rice variety tolerance toward submergence stress. *International Journal of Agronomy, 2017*, 9734036.

Güler, M. (2003). Barley grain β-glucan content as affected by nitrogen and irrigation. *Field Crops Research, 84*(3), 335–340.

Gull, A., Lone, A. A., & Wani, N. U. I. (2019). Biotic and abiotic stresses in plants. In *Abiotic and biotic stress in plants* (pp. 1–19). IntechOpen.

Guo, Z., Zhou, S., Wang, S., Li, W. X., Du, H., & Xu, Y. (2021). Identification of major QTL for waterlogging tolerance in maize using genome-wide association study and bulked sample analysis. *Journal of Applied Genetics, 62*, 405–418.

Gupta, N. K., Agarwal, S., Agarwal, V. P., Nathawat, N. S., Gupta, S., & Singh, G. (2013). Effect of short-term heat stress on growth, physiology and antioxidative defence system in wheat seedlings. *Acta Physiologiae Plantarum, 35*, 1837–1842.

Gupta, A., Singh, S. K., Singh, M. K., Singh, V. K., Modi, A., Singh, P. K., & Kumar, A. (2020). Plant growth–promoting rhizobacteria and their functional role in salinity stress management. In *Abatement of environmental pollutants* (pp. 151–160). Elsevier.

Habibi, F., Normohamady, G., Abad, H. H. S., Eivazi, A., & Heravan, E. M. (2012). Effect of sowing date on cold tolerance and some agronomic traits in bread wheat genotypes at west Azerbaijan conditions. *World Applied Sciences Journal, 16*(2), 232–239.

Haddadin, M. F., Abdel-Ghani, A., & Al-Majali, N. (2013). Response of barely varieties to drought stress imposed at different developmental stages. *Jordan Journal of Agricultural Sciences, 9*(4).

Halli, H. M., Angadi, S., Kumar, A., Govindasamy, P., Madar, R., Baskar, V. D. C., Elansary, H. O., Tamam, N., Abdelbacki, A. M., & Abdelmohsen, S. A. (2021). Assessment of planting method and deficit irrigation impacts on physio-morphology, grain yield and water use efficiency of maize (Zea Mays L.) on vertisols of semi-arid tropics. *Plants, 10*(6), 1094.

Hammami, Z., Qureshi, A. S., Sahli, A., Gauffreteau, A., Chamekh, Z., Ben Azaiez, F. E., Ayadi, S., & Trifa, Y. (2020). Modeling the effects of irrigation water salinity on growth, yield and water productivity of barley in three contrasted environments. *Agronomy, 10*(10), 1459.
Han, F., Guo, S., Naseer, M. A., Guo, R., Cai, T., Zhang, P., Jia, Z., Yang, D., Chen, X., & Ren, X. (2022). Potential use of maize–peanut intercropping to adapt to drought and nitrogen–shortage impacts. *Plant and Soil*, 1–20.
Haq, A. U., Tahir, M. H. N., Ahsan, M., Ahmad, R., & Akram, H. M. (2015). Screening and inheritance pattern studies of maize seedlings under normal and water stress conditions. *Pakistan Journal of Life and Social Sciences, 13*, 97–103.
Hasibuzzaman, A. S. M., Akter, F., Bagum, S. A., Hossain, N., Akter, T., & Uddin, M. S. (2021). Morpho-physiological mechanisms of maize for drought tolerance. In *Plant stress physiology* (p. 229). IntechOpen.
Hatfield, J. L., & Prueger, J. H. (2015). Temperature extremes: Effect on plant growth and development. *Weather and Climate Extremes, 10*, 4–10.
Hawkesford, M. J. (2014). Reducing the reliance on nitrogen fertilizer for wheat production. *Journal of Cereal Science, 59*(3), 276–283.
Hayashi, T. (2016). Varietal difference in the effects of low temperature on tassel development in hybrid maize. *Plant Production Science, 19*(2), 230–237.
Hays, D. B., Do, J. H., Mason, R. E., Morgan, G., & Finlayson, S. A. (2007). Heat stress induced ethylene production in developing wheat grains induces kernel abortion and increased maturation in a susceptible cultivar. *Plant Science, 172*(6), 1113–1123.
He, G., Wang, Z., Ma, X., He, H., Cao, H., Wang, S., Dai, J., Luo, L., Huang, M., & Malhi, S. S. (2017). Wheat yield affected by soil temperature and water under mulching in dryland. *Agronomy Journal, 109*(6), 2998–3006.
Heiniger, R. W. (2001). *The impact of early drought on corn yield.* North Carolina State University. http://www.ces.ncsu.edu/plymouth/cropsci/docs/early_drought_impact_on_corn.html
Herzog, M., Striker, G. G., Colmer, T. D., & Pedersen, O. (2016). Mechanisms of waterlogging tolerance in wheat–a review of root and shoot physiology. *Plant, Cell & Environment, 39*(5), 1068–1086.
Hill, C. B., & Li, C. (2022). Genetic improvement of heat stress tolerance in cereal crops. *Agronomy, 12*(5), 1205.
Hu, G., Li, Z., Lu, Y., Li, C., Gong, S., Yan, S., Li, G., Wang, M., Ren, H., Guan, H., & Zhang, Z. (2017). Genome-wide association study identified multiple genetic loci on chilling resistance during germination in maize. *Scientific Reports, 7*(1), 10840.
Hu, J., Ren, B., Chen, Y., Liu, P., Zhao, B., & Zhang, J. (2022a). Exogenous 6-benzyladenine improved the ear differentiation of waterlogged summer maize by regulating the metabolism of hormone and sugar. *Frontiers in Plant Science, 13*, 848989.
Hu, X., Ma, J., Qian, W., Cao, Y., Zhang, Y., Liu, B., Tang, L., Cao, W., Zhu, Y., & Liu, L. (2022b). Effects of low temperature on the amino acid composition of wheat grains. *Agronomy, 12*(5), 1171.
Huang, X., Shabala, S., Shabala, L., Rengel, Z., Wu, X., Zhang, G., & Zhou, M. (2015). Linking waterlogging tolerance with Mn2+ toxicity: A case study for barley. *Plant Biology, 17*(1), 26–33.
Huang, M., Zhang, Z., Zhu, C., Zhai, Y., & Lu, P. (2019). Effect of biochar on sweet corn and soil salinity under conjunctive irrigation with brackish water in coastal saline soil. *Scientia Horticulturae, 250*, 405–413.
Huang, C. T., Klos, K. E., & Huang, Y. F. (2020). Genome-wide association study reveals the genetic architecture of seed vigor in oats. *G3: Genes, Genomes, Genetics, 10*(12), 4489–4503.
Huang, P., Huang, S., Ma, Y., Danish, S., Hareem, M., Syed, A., Elgorban, A. M., Eswaramoorthy, R., & Wong, L. S. (2024). Alleviation of salinity stress by EDTA chelated-biochar and arbuscular mycorrhizal fungi on maize via modulation of antioxidants activity and biochemical attributes. *BMC Plant Biology, 24*(1), 63.
Hubbard, M., Germida, J. J., & Vujanovic, V. (2014). Fungal endophytes enhance wheat heat and drought tolerance in terms of grain yield and second-generation seed viability. *Journal of Applied Microbiology, 116*(1), 109–122.

Hurkman, W. J., Vensel, W. H., Tanaka, C. K., Whitehand, L., & Altenbach, S. B. (2009). Effect of high temperature on albumin and globulin accumulation in the endosperm proteome of the developing wheat grain. *Journal of Cereal Science, 49*(1), 12–23.

Ijaz, U., Ahmed, T., Rizwan, M., Noman, M., Shah, A. A., Azeem, F., Alharby, H. F., Bamagoos, A. A., Alharbi, B. M., & Ali, S. (2023). Rice straw based silicon nanoparticles improve morphological and nutrient profile of rice plants under salinity stress by triggering physiological and genetic repair mechanisms. *Plant Physiology and Biochemistry, 201*, 107788.

Imran, M., Mahmood, A., Römheld, V., & Neumann, G. (2013). Nutrient seed priming improves seedling development of maize exposed to low root zone temperatures during early growth. *European Journal of Agronomy, 49*, 141–148.

Inghelandt, D. V., Frey, F. P., Ries, D., & Stich, B. (2019). QTL mapping and genome-wide prediction of heat tolerance in multiple connected populations of temperate maize. *Scientific Reports, 9*(1), 14418.

Ingram, K. T. (2005). Drought-Related characteristics of important cereal crops. In *Monitoring and predicting agricultural drought: A global study*. Oxford Academic. https://doi.org/10.1093/oso/9780195162349.003.0008

Intergovernmental Panel on Climate Change (IPCC). (2023). Climate Change 2023: Synthesis Report. In Core Writing Team, H. Lee, J. Romero (Eds.), *A Report of the Intergovernmental Panel on Climate Change in: Contribution of Working Groups I, II and III to the Sixth Assessment Report of the Intergovernmental Panel on Climate Change*. IPCC. https://doi.org/10.59327/IPCC/AR6-9789291691647.001

Jagadish, S. K., Craufurd, P. Q., & Wheeler, T. (2007). High temperature stress and spikelet fertility in rice (Oryza sativa L.). *Journal of Experimental Botany, 58*(7), 1627–1635.

Jaiswal, A., & Srivastava, J. P. (2015). Effect of nitric oxide on some morphological and physiological parameters in maize exposed to waterlogging stress. *African Journal of Agricultural Research, 10*(35), 3462–3471.

Janaki, P., Alagesan, A., Ejilane, J., Nithila, S., Balasubramaniam, P., & Santhy, P. (2021). Effect of soil and crop management practices on sodicity stress alleviation and rice productivity under water scarce condition. *Journal of Applied and Natural Science, 13*(4), 1238–1248.

Jenner, C. F. (1991). Effects of exposure of wheat ears to high temperature on dry matter accumulation and carbohydrate metabolism in the grain of two cultivars. I. Immediate responses. *Functional Plant Biology, 18*(2), 165–177.

Ji, H., Xiao, L., Xia, Y., Song, H., Liu, B., Tang, L., Cao, W., Zhu, Y., & Liu, L. (2017). Effects of jointing and booting low temperature stresses on grain yield and yield components in wheat. *Agricultural and Forest Meteorology, 243*, 33–42.

Jiadkong, K., Nampei, M., Wangsawang, S., & Ueda, A. (2023). Riboflavin seed priming activates OsNHXs expression to alleviate salinity stress in rice seedlings. *Journal of Plant Growth Regulation, 42*(5), 3032–3042.

Jiang, M., Xuan, S., & Muneer, M. A. et al. (2022). Response of dry matter partition and yield components to waterlogging and sunlight shortage in different growth stages of wheat. *Natural Hazards, 110*, 1133–1152. https://doi.org/10.1007/s11069-021-04984-3

Jing, J., Guo, S., Li, Y., & Li, W. (2020). The alleviating effect of exogenous polyamines on heat stress susceptibility of different heat resistant wheat (Triticum aestivum L.) varieties. *Scientific Reports, 10*(1), 7467. https://doi.org/10.1038/s41598-020-64468-5

Jones, R. J., Quattar, S., & Crookston, R. K. (1984). Thermal environment during endosperm cell division and grain filling in maize: Effects on Kernel growth and development in vitro 1. *Crop Science, 24*(1), 133–137.

Jones, H. G., Serraj, R., Loveys, B. R., Xiong, L., Wheaton, A., & Price, A. H. (2009). Thermal infrared imaging of crop canopies for the remote diagnosis and quantification of plant responses to water stress in the field. *Functional Plant Biology, 36*(11):978–989. https://doi.org/10.1071/FP09123

Kadam, S., Abril, A., Dhanapal, A. P., Koester, R. P., Vermerris, W., Jose, S., & Fritschi, F. B. (2017). Characterization and regulation of aquaporin genes of sorghum [Sorghum bicolor (L.) Moench] in response to waterlogging stress. *Frontiers in Plant Science, 8*, 228710.

Kajla, M., Yadav, V. K., Chhokar, R. S., & Sharma, R. K. (2015). Management practices to mitigate the impact of high temperature on wheat. *Journal of Wheat Research, 7*(1), 1–12.

Karrou, M., & Oweis, T. (2014). Assessment of the severity and impact of drought spells on rainfed cereals in Morocco. *African Journal of Agricultural Research, 9*(49), 3519–3530.

Kaya, C., Ashraf, M., Osman, S., Tahir, P., & Tuna, A. T. (2017). The Combined effects of nitric oxide and thiourea on plant growth and mineral nutrition of salt-stressed plants of two maize cultivars with differential salt tolerance. *Journal of Plant Nutrition*. https://doi.org/10.1080/01904167.2016.1270314

Khan, M. J., Jan, M. T., Khan, A. U., Arif, M., & Shafi, M. (2010). Management of saline sodic soils through cultural practices and gypsum. *Pakistan Journal of Botany, 42*(6), 4143–4155.

Khan, M. B., Hussain, M., Raza, A., Farooq, S., & Jabran, K. (2015). Seed priming with CaCl2 and ridge planting for improved drought resistance in maize. *Turkish Journal of Agriculture and Forestry, 39*(2), 193–203.

Khan, N. H., Ahsan, M., Naveed, M., Sadaqat, H. A., & Javed, I. (2016). Genetics of drought tolerance at seedling and maturity stages in Zea mays L. *Spanish Journal of Agricultural Research, 14*(3), e0705.

Khan, A., Jamal, Y., Gul, F., Ullah, H., Ali, J., Naeem, I., & Khan, M. A. (2017). Evaluation of vegetative performance of maize with plowing management at Buner. *Pure and Applied Biology (PAB), 6*(1), 125–139.

Khatun, S., & Ahmed, J. U. (2015). Response of elevated temperature on carbohydrate accumulation and grain yield in different wheat cultivars. *Bangladesh Journal of Agricultural Research, 40*(2), 205–215.

Khatun, M., Shuvo, M. A. R., Salam, M. T. B., & Rahman, S. H. (2019). Effect of organic amendments on soil salinity and the growth of maize (Zea mays L.). *Plant Science Today, 6*(2), 106–111.

Khodarahmpour, Z., Ifar, M., & Motamedi, M. (2012). Effects of NaCl salinity on maize (Zea mays L.) at germination and early seedling stage. *African Journal of Biotechnology, 11*(2), 298–304.

Krishnamurthy, S. L., Lokeshkumar, B. M., Rathor, S., Warraich, A. S., Yadav, S., Gautam, R. K., Singh, R. K., & Sharma, P. C. (2022). Development of salt-tolerant rice varieties to enhancing productivity in salt-affected environments. *Environmental Sciences Proceedings, 16*(1), 30.

Kusaka, M., Ohta, M., & Fujimura, T. (2005). Contribution of inorganic components to osmotic adjustment and leaf folding for drought tolerance in pearl millet. *Physiologia Plantarum, 125*(4), 474–489. https://doi.org/10.1111/j.1399-3054.2005.00578.x

Lafarge, T., Bueno, C., Frouin, J., Jacquin, L., Courtois, B., & Ahmadi, N. (2017). Genome-wide association analysis for heat tolerance at flowering detected a large set of genes involved in adaptation to thermal and other stresses. *PLoS One, 12*(2), e0171254.

Lafitte, H., Ismail, A., & Bennett, J. (2004). Abiotic stress tolerance in rice for Asia: Progress and the future. In *Proceedings of the 4th international crop science congress*.

Lakaew, K., Akeprathumchai, S., & Thiravetyan, P. (2021). Foliar spraying of calcium acetate alleviates yield loss in rice (Oryza sativa L.) by induced anti-oxidative defence system under ozone and heat stresses. *Annals of Applied Biology, 178*(2), 414–426.

Lakshmi, G., Beena, R., Soni, K. B., Viji, M. M., & Jha, U. C. (2023). Exogenously applied plant growth regulator protects rice from heat-induced damage by modulating plant defense mechanism. *Journal of Crop Science and Biotechnology, 26*(1), 63–75. https://doi.org/10.1007/s12892-022-00162-4

Lamichhane, J. R., & Soltani, E. (2020). Sowing and seedbed management methods to improve establishment and yield of maize, rice and wheat across drought-prone regions: A review. *Journal of Agriculture and Food Research, 2*, 100089.

Lampayan, R. M., & Bouman, B. A. M. (2005). Management strategies for saving water and increase its productivity in lowland rice-based ecosystems. In *Proceedings of the First Asia-Europe workshop on sustainable resource management and policy options for rice ecosystems* (pp. 1–32).

Lampayan, R. M., Rejesus, R. M., Singleton, G. R., & Bouman, B. A. (2015). Adoption and economics of alternate wetting and drying water management for irrigated lowland rice. *Field Crops Research, 170*, 95–108.

Latef, A. A., & Tran, L. S. P. (2016). Impacts of priming with silicon on the growth and tolerance of maize plants to alkaline stress. *Frontiers in Plant Science, 7*, 182842.

Latef, A. A. H., Mostofa, M. G., Rahman, M. M., Abdel-Farid, I. B., & Tran, L. S. P. (2019). Extracts from yeast and carrot roots enhance maize performance under seawater-induced salt stress by altering physio-biochemical characteristics of stressed plants. *Journal of Plant Growth Regulation, 38*(3), 966–979.

Li, Z., Xu, J., Gao, Y., Wang, C., Guo, G., Luo, Y., Huang, Y., Hu, W., Sheteiwy, M. S., Guan, Y., & Hu, J. (2017). The synergistic priming effect of exogenous salicylic acid and H2O2 on chilling tolerance enhancement during maize (Zea mays L.) seed germination. *Frontiers in Plant Science, 8*, 1153.

Li, X., Wang, G., Fu, J., Li, L., Jia, G., Ren, L., Lubberstedt, T., Wang, G., Wang, J., & Gu, R. (2018). QTL mapping in three connected populations reveals a set of consensus genomic regions for low temperature germination ability in Zea mays L. *Frontiers in Plant Science, 9*, 65.

Li, Z., Tang, J., Srivastava, R., Bassham, D. C., & Howell, S. H. (2020). The transcription factor bZIP60 links the unfolded protein response to the heat stress response in maize. *Plant Cell, 32*(11), 3559–3575.

Li, P., Yang, X., Wang, H., Pan, T., Wang, Y., Xu, Y., Xu, C., & Yang, Z. (2021). Genetic control of root plasticity in response to salt stress in maize. *Theoretical and Applied Genetics, 134*, 1475–1492.

Liao, Y. D., Lin, K. H., Chen, C. C., & Chiang, C. M. (2016). Oryza sativa protein phosphatase 1a (OsPP1a) involved in salt stress tolerance in transgenic rice. *Molecular Breeding, 36*, 1–19.

Lindsey, A. J., Thomison, P. R., Mullen, R., & Geyer, A. B. (2015). Corn response to planting date as affected by plant population and hybrid in continuous corn cropping systems. *Crop, Forage & Turfgrass Management, 1*(1), 1–7. https://doi.org/10.2134/cftm2014.0073

Liu, Y., Lu, H., Yang, S., & Wang, Y. (2016). Impacts of biochar addition on rice yield and soil properties in a cold waterlogged paddy for two crop seasons. *Field Crops Research, 191*, 161–167.

Liu, K., Harrison, M. T., Ibrahim, A., Manik, S. N., Johnson, P., Tian, X., Meinke, H., & Zhou, M. (2020). Genetic factors increasing barley grain yields under soil waterlogging. *Food and Energy Security, 9*(4), e238.

Liu, Z., Meng, J., Sun, Z., Su, J., Luo, X., Song, J., Li, P., Sun, Y., Yu, C., & Peng, X. (2022). Zinc application after low temperature stress promoted rice tillers recovery: Aspects of nutrient absorption and plant hormone regulation. *Plant Science, 314*, 111104.

Liu, K., Song, C., Ye, P., Liu, H., Xia, Z., Waseem, H., Deng, Y., Liu, Z., Wang, L., Wang, B., & Yang, X. (2023). Optimizing rice sowing dates for high yield and climate adaptation in Central China. *Agronomy, 13*(5), 1339.

Lobell, D. B., Bänziger, M., Magorokosho, C., & Vivek, B. (2011). Nonlinear heat effects on African maize as evidenced by historical yield trials. *Nature Climate Change, 1*(1), 42–45.

Lobell, D. B., Sibley, A., & Ivan Ortiz-Monasterio, J. (2012). Extreme heat effects on wheat senescence in India. *Nature Climate Change, 2*(3), 186–189. https://doi.org/10.1038/nclimate1356

Lobell, D. B., Roberts, M. J., Schlenker, W., Braun, N., Little, B. B., Rejesus, R. M., & Hammer, G. L. (2014). Greater sensitivity to drought accompanies maize yield increase in the US Midwest. *Science, 344*(6183), 516–519.

Longmei, N., Gill, G. K., Zaidi, P. H., Kumar, R., Nair, S. K., Hindu, V., Vinayan, M. T., & Vikal, Y. (2021). Genome wide association mapping for heat tolerance in sub-tropical maize. *BMC Genomics, 22*, 1–14.

Lu, H. D., Xue, J. Q., & Guo, D. W. (2017). Efficacy of planting date adjustment as a cultivation strategy to cope with drought stress and increase rainfed maize yield and water-use efficiency. *Agricultural Water Management, 179*, 227–235.

Lunduka, R. W., Mateva, K. I., Magorokosho, C., & Manjeru, P. (2019). Impact of adoption of drought-tolerant maize varieties on total maize production in south Eastern Zimbabwe. *Climate and Development, 11*(1), 35–46.

Ma, L., Zhang, M., Chen, J., Qing, C., He, S., Zou, C., Yuan, G., Yang, C., Peng, H., Pan, G., & Lübberstedt, T. (2021). GWAS and WGCNA uncover hub genes controlling salt tolerance in maize (Zea mays L.) seedlings. *Theoretical and Applied Genetics, 134*, 3305–3318.

Ma, S., Gai, P., Geng, B., Wang, Y., Ullah, N., Zhang, W., Zhang, H., Fan, Y., & Huang, Z. (2022). Exogenous melatonin improves waterlogging tolerance in wheat through promoting antioxidant enzymatic activity and carbon assimilation. *Agronomy, 12*(11), 2876. https://doi.org/10.3390/agronomy12112876

Malik, S., Ashraf, M., Arshad, M., & Malik, T. A. (2015). Effect of ascorbic acid application on physiology of wheat under drought stress. *Pakistan Journal of Agricultural Sciences, 52*(1), 209–217.

Mano, Y., & Nakazono, M. (2021). Genetic regulation of root traits for soil flooding tolerance in genus Zea. *Breeding Science, 71*(1), 30–39.

Masoni, A., Pampana, S., & Arduini, I. (2016). Barley response to waterlogging duration at tillering. *Crop Science, 56*(5), 2722–2730.

Matsui, T., Omasa, K., & Horie, T. (2001). The difference in sterility due to high temperatures during the flowering period among japonica-rice varieties. *Plant Production Science, 4*(2), 90–93.

Matsuura, A., An, P., Murata, K., & Inanaga, S. (2016). Effect of pre-and post-heading waterlogging on growth and grain yield of four millets. *Plant Production Science, 19*(3), 348–359.

Matysiak, K., Kierzek, R., Siatkowski, I., Kowalska, J., Krawczyk, R., & Miziniak, W. (2020). Effect of exogenous application of amino acids l-arginine and glycine on maize under temperature stress. *Agronomy, 10*(6), 769.

McFadden, J., Smith, D., Wechsler, S., & Wallander, S. (2019). *Development, adoption, and management of drought-tolerant corn in the United States*. EIB-204, U.S. Department of Agriculture, Economic Research Service.

Mengutay, M., Ceylan, Y., Kutman, U. B., & Cakmak, I. (2013). Adequate magnesium nutrition mitigates adverse effects of heat stress on maize and wheat. *Plant and Soil, 368*, 57–72.

Mi, N., Cai, F., Zhang, Y., Ji, R., Zhang, S., & Wang, Y. (2018). Differential responses of maize yield to drought at vegetative and reproductive stages. *Plant, Soil & Environment, 64*(6).

Min, H., Chen, C., Wei, S., Shang, X., Sun, M., Xia, R., Liu, X., Hao, D., Chen, H., & Xie, Q. (2016). Identification of drought tolerant mechanisms in maize seedlings based on transcriptome analysis of recombination inbred lines. *Frontiers in Plant Science, 7*, 1080.

Mishra, P. K., Bisht, S. C., Ruwari, P., Selvakumar, G., Joshi, G. K., Bisht, J. K., Bhatt, J. C., & Gupta, H. S. (2011). Alleviation of cold stress in inoculated wheat (Triticum aestivum L.) seedlings with psychrotolerant Pseudomonads from NW Himalayas. *Archives of Microbiology, 193*, 497–513.

Mittal, L., Tayyeba, S., & Sinha, A. K. (2022). Finding a breather for Oryza sativa: Understanding hormone signalling pathways involved in rice plants to submergence stress. *Plant, Cell & Environment, 45*(2), 279–295.

Mohammed, A. R., & Tarpley, L. (2009). High nighttime temperatures affect rice productivity through altered pollen germination and spikelet fertility. *Agricultural and Forest Meteorology, 149*(6–7), 999–1008.

Moonmoon, S., & Islam, M. T. (2017). Effect of drought stress at different growth stages on yield and yield components of six rice (Oryza sativa L.) genotypes. *Fundamental and Applied Agriculture, 2*(3), 285–289.

Morris, P. C., Reuveni, M., & Ortiz, R. (2024). Impact and mitigation of abiotic stress in cereals. *Frontiers in Plant Science, 15*, 1374631.

Morsy, S., Elbasyoni, I. S., Baenziger, S., & Abdallah, A. M. (2022). Gypsum amendment influences performance and mineral absorption in wheat cultivars grown in normal and saline-sodic soils. *Journal of Agronomy and Crop Science, 208*(5), 675–692.

Moshatati, A., & Mousavi, S. H. (2019). Effect of zinc sulfate application on grain yield of bread wheat (Triticum aestivum L.) cv. Chamran under terminal heat stress conditions in Ahvaz. *Iranian Journal of Crop Sciences, 21*(3), 254–267.

Motarjemi, S. K., Rosenbom, A. E., Iversen, B. V., & Plauborg, F. (2021). Important factors when simulating the water and nitrogen balance in a tile-drained agricultural field under long-term monitoring. *Science of the Total Environment, 787*, 147610. ISSN 0048-9697. https://doi.org/10.1016/j.scitotenv.2021.147610

Mottaleb, K. A., Gumma, M. K., Mishra, A. K., & Mohanty, S. (2015). Quantifying production losses due to drought and submergence of rainfed rice at the household level using remotely sensed MODIS data. *Agricultural Systems, 137*, 227–235.

Mu, Q., Cai, H., Sun, S., Wen, S., Xu, J., Dong, M., & Saddique, Q. (2021). The physiological response of winter wheat under short-term drought conditions and the sensitivity of different indices to soil water changes. *Agricultural Water Management, 243*, 106475.

Nahar, K., Hasanuzzaman, M., & Majumder, R. R. (2009). Effect of low temperature stress in transplanted aman rice varieties mediated by different transplanting dates. *Academic Journal of Plant Sciences, 2*(3), 132–138.

Nath, R., Nath, D., Li, Q., Chen, W., & Cui, X. (2017). Impact of drought on agriculture in the Indo-Gangetic Plain, India. *Advances in Atmospheric Sciences, 34*, 335–346.

Naveed, M., Ahsan, M., Akram, H. M., Aslam, M., & Ahmed, N. (2016). Genetic effects conferring heat tolerance in a cross of tolerant× susceptible maize (Zea mays L.) genotypes. *Frontiers in Plant Science, 7*, 192159.

Neiff, N., Trachsel, S., Valentinuz, O. R., Balbi, C. N., & Andrade, F. H. (2016). High temperatures around flowering in maize: Effects on photosynthesis and grain yield in three genotypes. *Crop Science, 56*(5), 2702–2712.

Nelimor, C., Badu-Apraku, B., Tetteh, A. Y., & N'guetta, A. S. (2019). Assessment of genetic diversity for drought, heat and combined drought and heat stress tolerance in early maturing maize landraces. *Plants, 8*(11), 518.

NeSmith, D. S., & Ritchie, J. T. (1992). Effects of soil water-deficits during tassel emergence on development and yield component of maize (Zea mays). *Field Crops Research, 28*(3), 251–256.

Nielsen, R. L. (2007). *Assessing effects of drought on corn grain yield*. Purdue University.

Noor, J. J., Vinayan, M. T., Umar, S., Devi, P., Iqbal, M., Seetharam, K., & Zaidi, P. H. (2019). Morpho-physiological traits associated with heat stress tolerance in tropical maize ('Zea mays' L.) at reproductive stage. *Australian Journal of Crop Science, 13*(4), 536–545.

Odili, F., Bhushan, S., & Hatterman-Valenti, H. et al. (2023). Water table depth effect on growth and yield parameters of hard red spring wheat (Triticum aestivum L.): a lysimeter study. *Applied Water Science, 13*, 65 (2023). https://doi.org/10.1007/s13201-023-01868-8

OECD/FAO. (2023). *OECD-FAO agricultural outlook 2023-2032*. OECD Publishing. https://doi.org/10.1787/08801ab7-en

Ojebisi, S. M., & Kolade-Oje, T. Y. (2021). Responses of cereal farmers to drought in Guinea savannah ecological zone of Nigeria. In J. M. Luetz & D. Ayal (Eds.), *Handbook of climate change management*. Springer. https://doi.org/10.1007/978-3-030-57281-5_263

Olgun, M., Metin Kumlay, A., Cemal Adiguzel, M., & Caglar, A. (2008). The effect of waterlogging in wheat (T. aestivum L.). *Acta Agriculturae Scandinavica Section B–Soil and Plant Science, 58*(3), 193–198.

Oraki, A., Siahpoosh, M. R., Rahnama, A., & Lakzadeh, I. (2016). The effects of terminal heat stress on yield, yield components and some morpho-phenological traits of barley genotypes (Hordeum vulgare L.) in Ahvaz weather conditions. *Iranian Journal of Field Crop Science, 47*(1), 29–40.

Ortiz, R., Sayre, K. D., Govaerts, B., Gupta, R., Subbarao, G. V., Ban, T., Hodson, D., Dixon, J. M., Ortiz-Monasterio, J. I., & Reynolds, M. (2008). Climate change: Can wheat beat the heat? *Agriculture, Ecosystems & Environment, 126*(1–2), 46–58.

Pais, I. P., Moreira, R., Semedo, J. N., Ramalho, J. C., Lidon, F. C., Coutinho, J., Maçãs, B., & Scotti-Campos, P. (2022). Wheat crop under waterlogging: Potential soil and plant effects. *Plants, 12*(1), 149.

Pakar, N., Pirasteh-Anosheh, H., Emam, Y., & Pessarakli, M. (2016). Barley growth, yield, antioxidant enzymes, and ion accumulation affected by PGRs under salinity stress conditions. *Journal of Plant Nutrition, 39*(10), 1372–1379.

Pan, X., Nie, X., Gao, W., Yan, S., Feng, H., Cao, J., Lu, J., Shao, H., Ma, C., Chang, C., & Zhang, H. (2024). Identification of genetic loci and candidate genes underlying freezing tolerance in wheat seedlings. *Theoretical and Applied Genetics, 137*(3), 1–15.

Pang, J., Ross, J., Zhou, M., Mendham, N., & Shabala, S. (2007). Amelioration of detrimental effects of waterlogging by foliar nutrient sprays in barley. *Functional Plant Biology, 34*(3), 221–227.

Pantoja-Benavides, A. D., Garces-Varon, G., & Restrepo-Díaz, H. (2022). Foliar cytokinins or brassinosteroids applications influence the rice plant acclimatization to combined heat stress. *Frontiers in Plant Science, 13*, 983276.

Panwar, A. S., & Shivay, Y. S. (2015). Production potential of baby corn (Zea mays) on raised bed in waterlogged lowland rice fallow in North East India. *The Indian Journal of Agricultural Sciences, 85*(11), 1424–1427.

Pathak, H., Rane, J., Kurade, P. N., Nangare, D. D., & Kochewad, S. A. (2022). *Abiotic stresses in agriculture: Impacts and management*. ICAR-National Institute of Abiotic Stress Management. xii + 583 pages. ISBN: 978-81-949091-0-1.

Petolino, J. F., Cowen, N. M., Thompson, S. A., & Mitchell, J. C. (1990). Gamete selection for heat-stress tolerance in maize. *Journal of Plant Physiology, 136*(2), 219–224.

Porter, J. R., & Gawith, M. (1999). Temperatures and the growth and development of wheat: A review. *European Journal of Agronomy, 10*(1), 23–36.

Potop, V., Türkott, L., & Kožnarová, V. (2009). Drought impact on variability crop yields in Central Bohemia. *Cereal Research Communications, 37*(2), 295–304.

Pradhan, G. P., & Prasad, P. V. (2015). Evaluation of wheat chromosome translocation lines for high temperature stress tolerance at grain filling stage. *PLoS One, 10*(2), e0116620.

Pradhan, S., Babar, M. A., Robbins, K., Bai, G., Mason, R. E., Khan, J., Shahi, D., Avci, M., Guo, J., Maksud Hossain, M., & Bhatta, M. (2019). Understanding the genetic basis of spike fertility to improve grain number, harvest index, and grain yield in wheat under high temperature stress environments. *Frontiers in Plant Science, 10*, 1481.

Prasad, R. (2011). Aerobic rice systems. *Advances in Agronomy, 111*, 207–247.

Prasanna, B. M., Cairns, J. E., Zaidi, P. H., Beyene, Y., Makumbi, D., Gowda, M., Magorokosho, C., Zaman-Allah, M., Olsen, M., Das, A., & Worku, M. (2021). Beat the stress: Breeding for climate resilience in maize for the tropical rainfed environments. *Theoretical and Applied Genetics, 134*(6), 1729–1752.

Quinton, J. N., Öttl, L. K., & Fiener, P. (2022). Tillage exacerbates the vulnerability of cereal crops to drought. *Nature Food, 3*(6), 472–479.

Rad, H. E., Farshid, A. M., Amiri, R. E., & Khaledian, M. R. (2011). The effects of salinity at different growth stage on rice yield. *Ecology, Environment and Conservation, 17*(3), 455–462.

Raghunath, M., Beena, R., Mohan, V., Viji, V., Manju, R., & Stephen, R. (2021). High temperature stress mitigation in rice (*Oryza sativa* L.): Foliar application of plant growth regulators and nutrients. *Journal of Crop and Weed, 17*(1), 34–47.

Rakszegi, M., Lovegrove, A., Balla, K., Láng, L., Bedő, Z., Veisz, O., & Shewry, P. R. (2014). Effect of heat and drought stress on the structure and composition of arabinoxylan and β-glucan in wheat grain. *Carbohydrate Polymers, 102*, 557–565.

Rani, S., Chaudhary, A., & Rani, K. (2018). Management strategies for abiotic stresses in barley. *Wheat and Barley Research, 10*(3), 151–165.

Ransom, J. (2012). *Early season drought stress impacts wheat and corn development*. North Dakota State University Crop & Pest Report. Available: https://www.ag.ndsu.edu/cpr/plant-science/early-seasondrought-stress-impacts-wheat-and-corn-development-5-17-12

Ratnakumar, P., Deokate, P. P., Rane, J., Jain, N., Kumar, V., Berghe, D. V., & Minhas, P. S. (2016). Effect of ortho-silicic acid exogenous application on wheat (*Triticum aestivum* L.) under drought. *Journal of Functional and Environmental Botany, 6*, 34–42.

Rehman, H., Iqbal, H., Basra, S. M., Afzal, I., Farooq, M., Wakeel, A., & Wang, N. (2015). Seed priming improves early seedling vigor, growth and productivity of spring maize. *Journal of Integrative Agriculture, 14*(9), 1745–1754.

Ren, B., Zhang, J., Li, X., Fan, X., Dong, S., Liu, P., & Zhao, B. (2014). Effects of waterlogging on the yield and growth of summer maize under field conditions. *Canadian Journal of Plant Science, 94*(1), 23–31. https://doi.org/10.4141/cjps2013-175

Ren, B., Dong, S., Liu, P., Zhao, B., & Zhang, J. (2016). Ridge tillage improves plant growth and grain yield of waterlogged summer maize. *Agricultural Water Management, 177*, 392–399.

Ren, J., Ye, J., Yin, L., Li, G., Deng, X., & Wang, S. (2020). Exogenous melatonin improves salt tolerance by mitigating osmotic, ion, and oxidative stresses in maize seedlings. *Agronomy, 10*(5), 663.

Revilla, P., Malvar, R. A., Cartea, M. E., Butrón, A., & Ordás, A. (2000). Inheritance of cold tolerance at emergence and during early season growth in maize. *Crop Science, 40*(6), 1579–1585.

Rezaei, E. E., Webber, H., Gaiser, T., Naab, J., & Ewert, F. (2015). Heat stress in cereals: Mechanisms and modelling. *European Journal of Agronomy, 64*, 98–113.

Ribaut, J. M., & Ragot, M. (2007). Marker-assisted selection to improve drought adaptation in maize: The backcross approach, perspectives, limitations, and alternatives. *Journal of Experimental Botany, 58*(2), 351–360.

Rivelli, G. M., Calderini, D. F., Abeledo, L. G., Miralles, D. J., & Rondanini, D. P. (2024). Yield and quality traits of wheat and rapeseed in response to source-sink ratio and heat stress in post-flowering. *European Journal of Agronomy, 152*, 127028.

Saddiq, M. S., Iqbal, S., Afzal, I., Ibrahim, A. M., Bakhtavar, M. A., Hafeez, M. B., Jahanzaib, & Maqbool, M. M. (2019). Mitigation of salinity stress in wheat (Triticum aestivum L.) seedlings through physiological seed enhancements. *Journal of Plant Nutrition, 42*(10), 1192–1204.

Saddiq, M. S., Iqbal, S., Hafeez, M. B., Ibrahim, A. M., Raza, A., Fatima, E. M., Baloch, H., Jahanzaib, Woodrow, P., & Ciarmiello, L. F. (2021). Effect of salinity stress on physiological changes in winter and spring wheat. *Agronomy, 11*(6), 1193.

Sah, R. P., Chakraborty, M., Prasad, K., Pandit, M., Tudu, V. K., Chakravarty, M. K., Narayan, S. C., Rana, M., & Moharana, D. (2020). Impact of water deficit stress in maize: Phenology and yield components. *Scientific Reports, 10*(1), 2944.

Samarah, N. H., & Al-Issa, T. A. (2006). Effect of planting date on seed yield and quality of barley grown under semi-arid Mediterranean conditions. *Journal of Food Agriculture and Environment, 4*(2), 222.

Samarah, N., & Alqudah, A. (2011). Effects of late-terminal drought stress on seed germination and vigor of barley (Hordeum vulgare L.). *Archives of Agronomy and Soil Science, 57*(1), 27–32.

Samarah, N. H., Alqudah, A. M., Amayreh, J. A., & McAndrews, G. M. (2009). The effect of late-terminal drought stress on yield components of four barley cultivars. *Journal of Agronomy and Crop Science, 195*(6), 427–441.

Sandhu, D., Pudussery, M. V., Kumar, R., Pallete, A., Markley, P., Bridges, W. C., & Sekhon, R. S. (2020). Characterization of natural genetic variation identifies multiple genes involved in salt tolerance in maize. *Functional & Integrative Genomics, 20*, 261–275.

Sattar, A., Sher, A., Ijaz, M., Ul-Allah, S., Butt, M., Irfan, M., Rizwan, M. S., Ali, H., & Cheema, M. A. (2020). Interactive effect of biochar and silicon on improving morpho-physiological and biochemical attributes of maize by reducing drought hazards. *Journal of Soil Science and Plant Nutrition, 20*, 1819–1826.

Savicka, M., & Škute, N. (2010). Effects of high temperature on malondialdehyde content, superoxide production and growth changes in wheat seedlings (Triticum aestivum L.). *Ekologija, 56*(1), 26–33.

Schäfer, N., Maierhofer, T., Herrmann, J., Jørgensen, M. E., Lind, C., von Meyer, K., Lautner, S., Fromm, J., Felder, M., Hetherington, A. M., & Ache, P. (2018). A tandem amino acid residue motif in guard cell SLAC1 anion channel of grasses allows for the control of stomatal aperture by nitrate. *Current Biology, 28*(9), 1370–1379.

Schittenhelm, S., Langkamp-Wedde, T., Kraft, M., Kottmann, L., & Matschiner, K. (2020). Effect of two-week heat stress during grain filling on stem reserves, senescence, and grain yield of European winter wheat cultivars. *Journal of Agronomy and Crop Science, 206*(6), 722–733.

Seleiman, M. F., Aslam, M. T., Alhammad, B. A., Hassan, M. U., Maqbool, R., Chattha, M. U., Khan, I., Gitari, H. I., Uslu, O. S., Roy, R., & Battaglia, M. L. (2022). Salinity stress in wheat: Effects, mechanisms and management strategies. *Phyton, 91*(4), 667–694. https://doi.org/10.32604/phyton.2022.017365

Setter, T. L. (2000). *Farming systems for waterlogging prone sandplain soils of the south coast: Grains Research and Development Corporation: GRDC project number DAW292: A research project of Agriculture Western Australia/research supervisor T.L. Setter*. Agriculture Western Australia.

Setter, T. L., Burguess, P., Waters, I., & Kuo, J. (1999). Genetic diversity of barley and wheat for waterlogging tolerance in Western Australia. In *Proceedings of the Australian barley technical symposium*.

Setter, T. L., Flannigan, B. A., & Melkonian, J. (2001). Loss of kernel set due to water deficit and shade in maize: Carbohydrate supplies, abscisic acid, and cytokinins. *Crop Science, 41*(5), 1530–1540.

Shahid, M., Nayak, A. K., Tripathi, R., Katara, J. L., Bihari, P., Lal, B., & Gautam, P. (2018). Boron application improves yield of rice cultivars under high temperature stress during vegetative and reproductive stages. *International Journal of Biometeorology, 62*, 1375–1387.

Shahzad, H., Ullah, S., Iqbal, M., Bilal, H. M., Shah, G. M., Ahmad, S., Zakir, A., Ditta, A., Farooqi, M. A., & Ahmad, I. (2019). Salinity types and level-based effects on the growth, physiology and nutrient contents of maize (Zea mays). *Italian Journal of Agronomy, 14*(4), 199–207.

Shereen, A., Mumtaz, S., Raza, S., Khan, M. A., & Solangi, S. (2005). Salinity effects on seedling growth and yield components of different inbred rice lines. *Pakistan Journal of Botany, 37*(1), 131–139.

Shewry, P. R., Franklin, J., Parmar, S., Smith, S. J., & Miflin, B. J. (1983). The effects of sulphur starvation on the amino acid and protein compositions of barley grain. *Journal of Cereal Science, 1*(1), 21–31.

Shi, W., Muthurajan, R., Rahman, H., Selvam, J., Peng, S., Zou, Y., & Jagadish, K. S. (2013). Source–sink dynamics and proteomic reprogramming under elevated night temperature and their impact on rice yield and grain quality. *New Phytologist, 197*(3), 825–837.

Shukry, W. M., Abu-Ria, M. E., Abo-Hamed, S. A., Anis, G. B., & Ibraheem, F. (2023). The efficiency of humic acid for improving salinity tolerance in salt sensitive rice (Oryza sativa): Growth responses and physiological mechanisms. *Gesunde Pflanzen, 75*(6), 2639–2653.

Siam, M. Z., Hossain, S. S., Hassan, A. K., & Kader, M. A. (2015). Amelioration of salinity stress on transplant Aman rice through green manure and gypsum. *Bangladesh Agronomy Journal, 17*(1), 1–10.

Siddiqui, M. H., Alamri, S. A., Al-Khaishany, Y. Y., Al-Qutami, M. A., & Ali, H. M. (2018). Ascorbic acid application improves salinity stress tolerance in wheat. *Chiang Mai Journal of Science, 45*(3), 1296–1306.

Silalert, P., & Pattanagul, W. (2021). Foliar application of melatonin alleviates the effects of drought stress in rice (Oryza sativa L.) seedlings. *Notulae Botanicae Horti Agrobotanici Cluj-Napoca, 49*(3), 12417–12417.

Singh, R. B., & Devi, R. (2023). Achieving salinity-tolerance in cereal crops: Major insights into Genomics-Assisted Breeding (GAB). In *Making plant life easier and productive under salinity - Updates and prospects* [Working Title]. IntechOpen. Available at: https://doi.org/10.5772/intechopen.112570

Singh, S. P., & Setter, T. L. (2017). Effect of waterlogging on element concentrations, growth and yield of wheat varieties under farmer's sodic field conditions. *Proceedings of the National Academy of Sciences, India Section B: Biological Sciences, 87*, 513–520. https://doi.org/10.1007/s40011-015-0607-9

Singh, A., Singh, D., Kang, J. S., & Aggarwal, N. (2011a). Management practices to mitigate the impact of high temperature on wheat: A review. *IIOAB Journal, 2*(7), 11–22.

Singh, et al. (2011b). Effect of high temperature on yield attributing traits in bread wheat. *Bangladesh Journal of Agricultural Research, 36*(3), 415–426.

Singh, A. K., Gurumurthy, S., Rane, J., Mamrutha, H. M., & Singh, G. P. (2022). Wheat: Impact and management of abiotic stresses. In H. Pathak et al. (Eds.), *Abiotic stresses in agriculture: Impacts and management* (pp. 97–115). ICAR-National Institute of Abiotic Stress Management. eISBN No. 978-81-949091 0-1.

Slafer, G. A., & Rawson, H. M. (1994). Sensitivity of wheat phasic development to major environmental factors: A re-examination of some assumptions made by physiologists and modellers. *Functional Plant Biology, 21*(4), 393–426.

Soda, N., Gupta, B. K., Anwar, K., Sharan, A., Govindjee, Singla-Pareek, S. L., & Pareek, A. (2018). Rice intermediate filament, OsIF, stabilizes photosynthetic machinery and yield under salinity and heat stress. *Scientific Reports, 8*(1), 4072.

Sowiński, P., Fronk, J., Jończyk, M., Grzybowski, M., Kowalec, P., & Sobkowiak, A. (2020). Maize response to low temperatures at the gene expression level: A critical survey of transcriptomic studies. *Frontiers in Plant Science, 11*, 576941.

Srivastava, J. P., Gangey, S. K., & Shahi, J. P. (2007). Waterlogging resistance in maize in relation to growth, mineral compositions and some biochemical parameters. *Indian Journal of Plant Physiology, 12*(1), 28.

Stone, P. J., & Nicolas, M. E. (1994). Wheat cultivars vary widely in their responses of grain yield and quality to short periods of post-anthesis heat stress. *Functional Plant Biology, 21*(6), 887–900.

Sudratt, N., & Faiyue, B. (2023). Biochar mitigates combined effects of soil salinity and saltwater intrusion on rice (Oryza sativa L.) by regulating ion uptake. *Agronomy, 13*(3), 815.

Sun, L., Li, X., Wang, Z., Sun, Z., Zhu, X., Liu, S., Song, F., Liu, F., & Wang, Y. (2018). Cold priming induced tolerance to subsequent low temperature stress is enhanced by melatonin application during recovery in wheat. *Molecules, 23*(5), 1091.

Tabassam, M., Hussain, M., Sami, A., Shabbir, I., Bhutta, A. N., Mubusher, M., & Ahmad, S. (2014). Impact of drought on the growth and yield of wheat. *Science in Agriculture, 7*(1), 11–18.

Tack, J., Lingenfelser, J., & Jagadish, S. K. (2017). Disaggregating sorghum yield reductions under warming scenarios exposes narrow genetic diversity in US breeding programs. *Proceedings of the National Academy of Sciences, 114*(35), 9296–9301.

Templ, B., & Calanca, P. (2020). Critical increase in the occurrence of heat stress during reproductive growth in Russian wheat beyond 1.5 C global warming. *Weather and Climate Extremes, 30*, 100281.

Thakur, A. K., Mohanty, R. K., Patil, D. U., & Kumar, A. (2014). Impact of water management on yield and water productivity with system of rice intensification (SRI) and conventional transplanting system in rice. *Paddy and Water Environment, 12*(4), 413–424.

Thomison, P. R., Mullen, R. W., Lipps, P. E., Doerge, T., & Geyer, A. B. (2011). Corn response to harvest date as affected by plant population and hybrid. *Agronomy Journal, 103*(6), 1765–1772.

Tian, L., Tan, L., Liu, F., Cai, H., & Sun, C. (2011). Identification of quantitative trait loci associated with salt tolerance at seedling stage from Oryza rufipogon. *Journal of Genetics and Genomics, 38*(12), 593–601.

Tian, G., Qi, D., Zhu, J., & Xu, Y. (2021a). Effects of nitrogen fertilizer rates and waterlogging on leaf physiological characteristics and grain yield of maize. *Archives of Agronomy and Soil Science, 67*(7), 863–875. https://doi.org/10.1080/03650340.2020.1791830

Tian, L. X., Zhang, Y. C., Chen, P. L., Zhang, F. F., Li, J., Yan, F., Dong, Y., & Feng, B. L. (2021b). How does the waterlogging regime affect crop yield? A global meta-analysis. *Frontiers in Plant Science, 12*, 634898.

Tisarum, R., Theerawitaya, C., Samphumphung, T., Takabe, T., & Cha-Um, S. (2019). Exogenous foliar application of glycine betaine to alleviate water deficit tolerance in two Indica rice genotypes under greenhouse conditions. *Agronomy, 9*(3), 138.

Tiwari, R., Sheoran, S., & Rane, J. (2015). Wheat improvement for drought and heat tolerance. In R. S. Shukla, P. C. Mishra, R. Chatrath, R. K. Gupta, S. S. Tomar, & I. Sharma (Eds.), *Recent trends on production strategies of wheat in India* (pp. 39–58). Directorate of Wheat Research.

Tiwari, V., Mamrutha, H. M., Sareen, S., Sheoran, S., Tiwari, R., Sharma, P., Singh, C., Singh, G., & Rane, J. (2017). Managing abiotic stresses in wheat. In *Abiotic stress management for resilient agriculture* (pp. 313–337). Springer.

Trnka, M., Feng, S., Semenov, M. A., Olesen, J. E., Kersebaum, K. C., Rötter, R. P., Semerádová, D., Klem, K., Huang, W., Ruiz-Ramos, M., & Hlavinka, P. (2019). Mitigation efforts will not fully alleviate the increase in water scarcity occurrence probability in wheat-producing areas. *Science Advances, 5*(9), eaau2406.

Ugarte, C., Calderini, D. F., & Slafer, G. A. (2007). Grain weight and grain number responsiveness to pre-anthesis temperature in wheat, barley and triticale. *Field Crops Research, 100*(2–3), 240–248. ISSN 0378-4290. https://doi.org/10.1016/j.fcr.2006.07.010

Upadhyay, R., Chandra, G., Gambhir, L., & Kumar, P. (2023). Performance of wheat (Triticum aestivum L.) varieties under cold stress condition. *Indian Journal of Ecology, 50*(2), 319–324.

Usmani, M. M., Nawaz, F., Majeed, S., Shehzad, M. A., Ahmad, K. S., Akhtar, G., Aqib, M., & Shabbir, R. N. (2020). Sulfate-mediated drought tolerance in maize involves regulation at physiological and biochemical levels. *Scientific Reports, 10*(1), 1147.

Vignjevic, M., Wang, X., Olesen, J. E., & Wollenweber, B. (2015). Traits in spring wheat cultivars associated with yield loss caused by a heat stress episode after anthesis. *Journal of Agronomy and Crop Science, 201*(1), 32–48.

Vijayakumar, S., Kumar, D., Jinger, D., Bussa, B., & Panda, B. B. (2021). 4R nutrient stewardship based potassium. *Indian Farming, 71*(01), 22–25.

Wallace, J. F., Wheaton, G. A., McFarlane, D. J. (1992). *Effects of waterlogging on crop and pasture production in the Upper Great Southern, Western Australia*. Department of Primary Industries and Regional Development, Western Australia, Perth. Technical Bulletin 86.

Wang, X., Huang, M., Zhou, Q., Cai, J., Dai, T., Cao, W., & Jiang, D. (2016). Physiological and proteomic mechanisms of waterlogging priming improves tolerance to waterlogging stress in wheat (Triticum aestivum L.). *Environmental and Experimental Botany, 132*, 175–182.

Wang, Y., Li, J., Gu, W., Zhang, Q., Tian, L., Guo, S., & Wei, S. (2018). Exogenous application of 5-aminolevulinic acid improves low-temperature stress tolerance of maize seedlings. *Crop and Pasture Science, 69*(6), 587–593.

Wang, Y., Tao, H., Tian, B., Sheng, D., Xu, C., Zhou, H., Huang, S., & Wang, P. (2019). Flowering dynamics, pollen, and pistil contribution to grain yield in response to high temperature during maize flowering. *Environmental and Experimental Botany, 158*, 80–88.

Wang, H. Q., Liu, P., Zhang, J. W., Zhao, B., & Ren, B. Z. (2020). Endogenous hormones inhibit differentiation of young ears in maize (Zea mays L.) under heat stress. *Frontiers in Plant Science, 11*, 533046.

Wang, W., Wang, X., Huang, M., Cai, J., Zhou, Q., Dai, T., & Jiang, D. (2021). Alleviation of field low-temperature stress in winter wheat by exogenous application of salicylic acid. *Journal of Plant Growth Regulation, 40*, 811–823.

Wang, S., Hu, J., Ren, B., Liu, P., Zhao, B., & Zhang, J. (2022). Effects of hydrogen peroxide priming on yield, photosynthetic capacity and chlorophyll fluorescence of waterlogged summer maize. *Frontiers in Plant Science, 13*, 1042920.

Wani, S. H., Kumar, V., Khare, T., Guddimalli, R., Parveda, M., Solymosi, K., Suprasanna, P., & Kavi Kishor, P. B. (2020). Engineering salinity tolerance in plants: progress and prospects. *Planta, 251*, 1–29.

Waqas, M. A., Khan, I., Akhter, M. J., Noor, M. A., & Ashraf, U. (2017). Exogenous application of plant growth regulators (PGRs) induces chilling tolerance in short-duration hybrid maize. *Environmental Science and Pollution Research, 24*, 11459–11471.

Waqas, M. A., Wang, X., Zafar, S. A., Noor, M. A., Hussain, H. A., Azher Nawaz, M., & Farooq, M. (2021). Thermal stresses in maize: Effects and management strategies. *Plants, 10*(2), 293.

Wardlaw, I. F., Blumenthal, C., Larroque, O., & Wrigley, C. W. (2002). Contrasting effects of chronic heat stress and heat shock on kernel weight and flour quality in wheat. *Functional Plant Biology, 29*(1), 25–34.

Wrigley, C. W., Blumenthal, C., Gras, P. W., & Barlow, E. W. R. (1994). Temperature variation during grain filling and changes in wheat-grain quality. *Functional Plant Biology, 21*(6), 875–885.

Xiao, L., Asseng, S., Wang, X., Xia, J., Zhang, P., Liu, L., Tang, L., Cao, W., Zhu, Y., & Liu, B. (2022). Simulating the effects of low-temperature stress on wheat biomass growth and yield. *Agricultural and Forest Meteorology, 326*, 109191.

Xu, H., Hassan, M. A., Sun, D., Wu, Z., Jiang, G., Liu, B., Ni, Q., Yang, W., Fang, H., Li, J., & Chen, X. (2022a). Effects of low temperature stress on source–sink organs in wheat and phosphorus mitigation strategies. *Frontiers in Plant Science, 13*, 807844.

Xu, H., Wu, Z., Xu, B., Sun, D., Hassan, M. A., Cai, H., Wu, Y., Yu, M., Chen, A., Li, J., & Chen, X. (2022b). Optimized phosphorus application alleviated adverse effects of short-term low-temperature stress in winter wheat by enhancing photosynthesis and improved accumulation and partitioning of dry matter. *Agronomy, 12*(7), 1700.

Yadav, M. R., Choudhary, M., Singh, J., Lal, M. K., Jha, P. K., Udawat, P., Gupta, N. K., Rajput, V. D., Garg, N. K., Maheshwari, C., & Hasan, M. (2022). Impacts, tolerance, adaptation, and mitigation of heat stress on wheat under changing climates. *International Journal of Molecular Sciences, 23*(5), 2838.

Yahaya, M. A., & Shimelis, H. (2022). Drought stress in sorghum: Mitigation strategies, breeding methods and technologies—A review. *Journal of Agronomy and Crop Science, 208*(2), 127–142.

Yamamoto, Y., Aminaka, R., Yoshioka, M., Khatoon, M., Komayama, K., Takenaka, D., Yamashita, A., Nijo, N., Inagawa, K., Morita, N., & Sasaki, T. (2008). Quality control of photosystem II: Impact of light and heat stresses. *Photosynthesis Research, 98*, 589–608.

Yan, J., Wu, Y., Li, W., Qin, X., Wang, Y., & Yue, B. (2017). Genetic mapping with testcrossing associations and F2: 3 populations reveals the importance of heterosis in chilling tolerance at maize seedling stage. *Scientific Reports, 7*(1), 3232. https://doi.org/10.1038/s41598-017-03585-0

Yang, X., Wang, B., Chen, L., Li, P., & Cao, C. (2019). The different influences of drought stress at the flowering stage on rice physiological traits, grain yield, and quality. *Scientific Reports, 9*(1), 3742.

Yang, J., Miao, W., & Chen, J. (2021). Roles of jasmonates and brassinosteroids in rice responses to high temperature stress–a review. *The Crop Journal, 9*(5), 977–985.

Yasmeen, A., Basra, S. M. A., Ahmad, R., & Wahid, A. (2012). Performance of late sown wheat in response to foliar application of Moringa oleifera Lam. leaf extract. *Chilean Journal of Agricultural Research, 72*(1), 92.

Yavas, I., Unay, A., & Aydin, M. (2012). The waterlogging tolerance of wheat varieties in western of Turkey. *The Scientific World Journal, 2012*, 529128.

Ye, C., Tenorio, F. A., Argayoso, M. A., Laza, M. A., Koh, H. J., Redoña, E. D., Jagadish, K. S., & Gregorio, G. B. (2015). Identifying and confirming quantitative trait loci associated with heat tolerance at flowering stage in different rice populations. *BMC Genetics, 16*, 1–10.

Yu, F., Liang, K., Zhang, Z., Du, D., Zhang, X., Zhao, H., Ul haq, B., & Qiu, F. (2018). Dissecting the genetic architecture of waterlogging stress-related traits uncovers a key waterlogging tolerance gene in maize. *Theoretical and Applied Genetics, 131*, 2299–2310.

Yu, F., Liang, K., Han, X., Du, D., Pan, Z., & Qiu, F. (2019). Major natural genetic variation contributes to waterlogging tolerance in maize seedlings. *Molecular Breeding, 39*, 1–13.

Yu, X., Hao, D., Yang, J., Ran, L., Zang, Y., & Xiong, F. (2020). Effects of low temperature at stem elongation stage on the development, morphology, and physicochemical properties of wheat starch. *PeerJ, 8*, e9672.

Yue, H., Chen, S., Bu, J., Wei, J., Peng, H., Li, Y., Li, C., & Xie, J. (2018). Response of main maize varieties to water stress and comprehensive evaluation in Hebei Province. In *IOP Conference Series: Earth and Environmental Science* (Vol. 108, No. 4, p. 042002). IOP Publishing.

Yue, E., Cao, H., & Liu, B. (2020). OsmiR535, a potential genetic editing target for drought and salinity stress tolerance in Oryza sativa. *Plants, 9*(10), 1337.

Zaidi, P. H., Maniselvan, P., Srivastava, A., Yadav, P., & Singh, R. P. (2010). Genetic analysis of water-logging tolerance in tropical maize (Zea mays L.). *Maydica, 55*(1), 17–26.

Zaidi, P. H., Rashid, Z., Vinayan, M. T., Almeida, G. D., Phagna, R. K., & Babu, R. (2015). QTL mapping of agronomic waterlogging tolerance using recombinant inbred lines derived from tropical maize (Zea mays L) germplasm. *PLoS One, 10*(4), e0124350.

Zaidi, P. H., Seetharam, K., Krishna, G., Krishnamurthy, L., Gajanan, S., Babu, R., Zerka, M., Vinayan, M. T., & Vivek, B. S. (2016). Genomic regions associated with root traits under drought stress in tropical maize (Zea mays L.). *PLoS One, 11*(10), e0164340.

Zenda, T., Wang, N., Dong, A., Zhou, Y., & Duan, H. (2022). Reproductive-stage heat stress in cereals: Impact, plant responses and strategies for tolerance improvement. *International Journal of Molecular Sciences, 23*(13), 6929.

Zeng, F., Shabala, L., Zhou, M., Zhang, G., & Shabala, S. (2013). Barley responses to combined waterlogging and salinity stress: Separating effects of oxygen deprivation and elemental toxicity. *Frontiers in Plant Science, 4*, 47814. https://doi.org/10.3389/fpls.2013.00313

Zhang, M., Liang, X., Wang, L., Cao, Y., Song, W., Shi, J., Lai, J., & Jiang, C. (2019). A HAK family Na+ transporter confers natural variation of salt tolerance in maize. *Nature Plants, 5*(12), 1297–1308.

Zhang, C., Gu, K., Gu, D., Zhang, S., & Wu, J. (2021). Quantifying the effect of low-temperature events on the grain quality formation of wheat. *Journal of Cereal Science, 100*, 103257. ISSN 0733-5210. https://doi.org/10.1016/j.jcs.2021.103257

Zhang, R., Wang, Y., Hussain, S., Yang, S., Li, R., Liu, S., Chen, Y., Wei, H., Dai, Q., & Hou, H. (2022a). Study on the effect of salt stress on yield and grain quality among different rice varieties. *Frontiers in Plant Science, 13*, 918460.

Zhang, Y., Liu, L., Chen, X., & Li, J. (2022b). Effects of low-temperature stress during the anther differentiation period on winter wheat photosynthetic performance and spike-setting characteristics. *Plants, 11*(3), 389.

Zhang, H., Zhang, X., Gao, G., Ali, I., Wu, X., Tang, M., Chen, L., Jiang, L., & Liang, T. (2023a). Effects of various seed priming on morphological, physiological, and biochemical traits of rice under chilling stress. *Frontiers in Plant Science, 14*, 1146285.

Zhang, R., Yue, Z., Chen, X., Huang, R., Zhou, Y., & Cao, X. (2023b). Effects of waterlogging at different growth stages on the photosynthetic characteristics and grain yield of sorghum (Sorghum bicolor L.). *Scientific Reports, 13*(1), 7212.

Zhao, F. J., Hawkesford, M. J., & McGrath, S. P. (1999). Sulphur assimilation and effects on yield and quality of wheat. *Journal of Cereal Science, 30*(1), 1–17.

Zhao, H., Dai, T., Jing, Q., Jiang, D., & Cao, W. (2007). Leaf senescence and grain filling affected by post-anthesis high temperatures in two different wheat cultivars. *Plant Growth Regulation, 51*, 149–158.

Zhao, C., Liu, B., Piao, S., Wang, X., Lobell, D. B., Huang, Y., Huang, M., Yao, Y., Bassu, S., Ciais, P., & Durand, J. L. (2017). Temperature increase reduces global yields of major crops in four independent estimates. *Proceedings of the National Academy of Sciences, 114*(35), 9326–9331.

Zhao, H., Zhang, L., Kirkham, M. B., Welch, S. M., Nielsen-Gammon, J. W., Bai, G., Luo, J., Andresen, D. A., Rice, C. W., Wan, N., & Lollato, R. P. (2022). US winter wheat yield loss attributed to compound hot-dry-windy events. *Nature Communications, 13*(1), 7233.

Zheng, B., Chapman, S. C., Christopher, J. T., Frederiks, T. M., & Chenu, K. (2015). Frost trends and their estimated impact on yield in the Australian wheatbelt. *Journal of Experimental Botany, 66*(12), 3611–3623.

Zheng, D., Yang, X., Mínguez, M. I., Mu, C., He, Q., & Wu, X. (2018). Effect of freezing temperature and duration on winter survival and grain yield of winter wheat. *Agricultural and Forest Meteorology, 260*, 1–8.

Ziska, L. H., & Manalo, P. A. (1996). Increasing night temperature can reduce seed set and potential yield of tropical rice. *Functional Plant Biology, 23*(6), 791–794.

Zulkiffal, M., Ahsan, A., Ahmed, J., Musa, M., Kanwal, A., Saleem, M., Anwar, J., ur Rehman, A., Ajmal, S., Gulnaz, S., & Javaid, M. M. (2021). Heat and drought stresses in wheat (Triticum aestivum L.): Substantial yield losses, practical achievements, improvement approaches, and adaptive. In *Plant stress physiology* (Vol. 3). Intech.

Varietal Interventions and Their Significance in Rice-Based Cropping Systems of Drought-Prone Ecologies in Odisha

11

Swati Nayak, Anirban Nath, S. K. Mosharaf Hossain, Varsha Singh, and Subhasmita Mohapatra

Abstract

The eastern state of Odisha comprises of mostly smallholder farmers who remain extremely vulnerable to climatic stress including drought. Out of the last 41 episodes of natural calamities that hit the state in the last 50 years, 19 of these were severe droughts. This is aggravated by the fact that the state is dependent mostly on rainfed systems and only 38% of the cultivated area has coverage or access to some degree of irrigation. Securing sustainable rice-based cropping systems for smallholders in Odisha necessitates the availability of quality seeds for drought-tolerant rice varieties and short-duration non-paddy crops during the dry seasons. Through multi-year rigorous varietal on-farm testings (OFTs), several products have been identified and also scaled. Some of these are now highly popular and demanded varieties like Sahabhagi Dhan, DRR Dhan 42, CR Dhan 801. The yield performance in both stress and non-stress conditions, the grain quality and maturity duration and strategic demand creation efforts and supply have played an integrated role in large-scale adoption of these products. Going beyond stress-tolerant varieties, the recent development of national breeding pipelines around direct-seeded rice systems (DSRs) carries significant potential for drought-coping mechanisms. DSR as a system is directed towards saving of water and labour as well reducing greenhouse gas emissions. Under the breeding pipeline and market segment of DELS-I/R (Direct seeded early maturity long slender-irrigated/rainfed) many varieties have been tested through multi-location on-farm trials in various drought-prone and upland regions like Bargarh, Sambalpur. Newer early maturity varieties such as ADT 56, CO 56, MTU 1001, PR 126 and Tripura Hakuchuk 2 were extensively tested and compared with established checks like MTU 1010 and Sahbhagi Dhan. This requires a comprehensive eval-

S. Nayak (✉) · A. Nath · S. K. M. Hossain · V. Singh · S. Mohapatra
International Rice Research Institute, Varanasi, Uttar Pradesh, India
e-mail: s.nayak@irri.org

R. Dubey et al. (eds.), *Drought and Heat Stress in Agriculture*,
https://doi.org/10.1007/978-981-96-5735-3_11

uation and identification of a greater number of suitable products (varieties) for drought ecologies (not necessarily having drought tolerant genes, but tailored enough for drought escape through early maturity or low water responsiveness) and scaling those for impact. To enhance farm returns in post-kharif periods, the effectiveness of short-duration oilseeds (sunflower, sesame) and pulses (green gram, black gram, lentils, etc.) remains potential. The deployment of early maturing rice varieties ensures a second crop in otherwise dry-season fallow areas, contributing to sustainable agricultural practices. Finally, in order to ensure a steady supply of seeds for the rice and non-paddy crops, robust varietal and seed system interventions for drylands, carry potential for creating resilient model for combating the drought and its effects among the smallholders.

Keywords

Drought · Drought-tolerant · Variety · Seed · Rice · On-farm testing · Early maturity

11.1 Introduction

In India, drought primarily occurs due to spatio-temporal variation in the Southwest Monsoon along with the regional Physiographic conditions (Sam et al., 2020). Furthermore, according to the Indian Meteorological Department (IMD), a rainfall deficit of 25% over the normal range from June to September (Southwest Monsoon season) can be considered a meteorological drought (Bera et al., 2021). Owing to the extensive fluctuations in Southwest monsoon along with the poor water holding capacity in areas predominated by sandy or red lateritic soil of Eastern and Central India, the meteorological followed by hydrological and agricultural droughts are recurrent in these regions. In Odisha, agriculture is the main source of livelihood for about 4.8 million farmers the majority of whom (91%) are classified as smallholders and marginal. Rice being the major cereal consumed by the population is cultivated across approximately 35.85 lakh ha of land or 65.83% of the total net sown area during the kharif (wet seasons) (Odisha Agriculture Statistics, 2018–2019). Twenty-five per cent (25%) of the rice-based cropping system is prone to frequent droughts of different degrees and causes regular crop damage. The average rainfall in the state is about 1451 mm, higher than the national average of 1128 mm. However, it is the inter-month variation in rainfall, especially during monsoon months (June–Sept) makes the farmers more vulnerable to episodic drought. The late onset of monsoon in June prevents many farmers from timely sowing rice is a recurrent phenomenon. Over 41 years (from 1960 to 2012), a total of 19 years has witnessed severe droughts across multiple districts of the state during the kharif season (wet season), which is the primary cropping season for rice in the state (OSDMA, Govt. of Odisha, India, 2023). According to government reports, the state experiences severe droughts once every five years which results in disruption of agriculture in the state. The situation is further compounded and by the limited coverage of areas of irrigation facilities. Currently, only around 38% of the net sown area has assured irrigation coverage. The drought-prone areas of the state include 47 administrative blocks spanning across the districts of Bolangir, Bargarh, Nuapada, Kalahandi, and Phulbani. IRRI's former STRASA (Stress Tolerant Rice for Africa and South Asia) program has achieved significant success in the past decade by developing multiple rice varieties resilient to varying levels of meteorological droughts (Dar et al., 2020). However, due to poor varietal awareness among farmers along with limited access to quality seeds of drought-tolerant varieties, resilience in terms of the rice-based cropping system in Odisha during the past decades has been suboptimal. The situation is further exacerbated owing to post-kharif rice fallows in the state, the majority of which are spread across the rainfed drought-prone areas mentioned above (Chandna & Mondal, 2020). Considering the above-mentioned challenges, the current research sheds light on various approaches focussing on varietal testing, dissemination and seed system resilience coupled with rice fallow management across the various intervention zones in Odisha.

11.2 Overview of Drought and Its Impact on the Rice-Based Cropping System in Odisha

Odisha receives an annual rainfall of 1451.2 mm, predominantly from June to September, crucial for the kharif crop. Subsequent post-monsoon rains in October and November support short-duration crop cultivation following the rice harvest. Historical rainfall patterns show significant fluctuations (Fig. 11.1), leading to mild to severe droughts in some years, along with instances of heavy rainfall causing extreme flooding in the state. However, it is worthwhile to note that, agricultural drought in Odisha is mostly triggered by erratic rainfall during critical months of the monsoon months.

In most instances, mild to moderate droughts occur due to the early withdrawal of the monsoon, leading to yield losses caused by moisture stress during critical

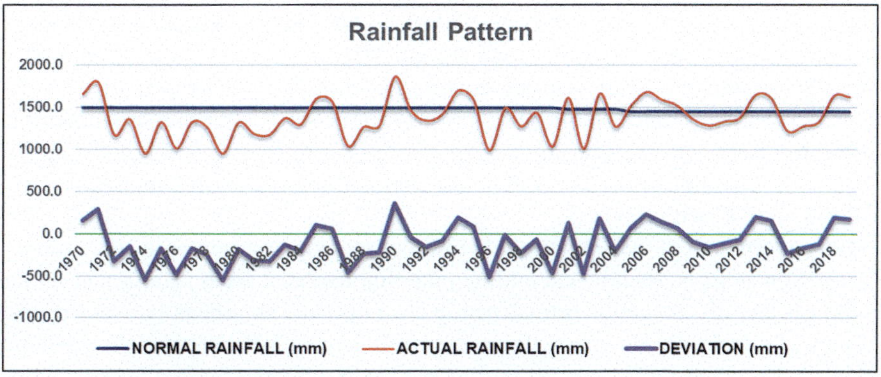

Fig. 11.1 Rainfall pattern in Odisha. (Source: DAFE, Govt. of Odisha)

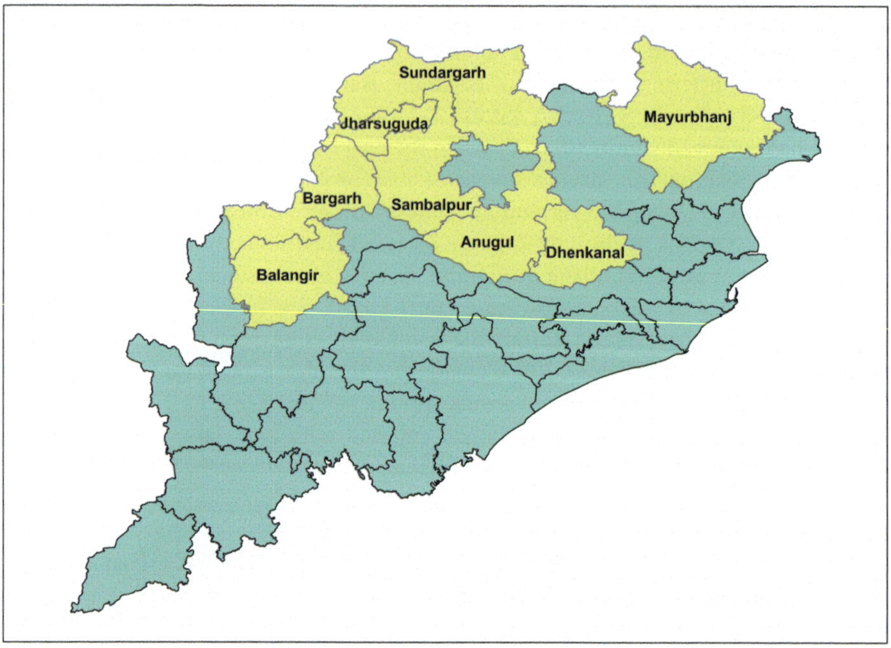

Fig. 11.2 District-level frequency of agricultural drought from 2000 to 2019 in India and Odisha. (Source: DM Division, DAC&FW and MNCFC, New Delhi, September 2020)

crop stages such as active tillering, panicle initiation, booting, heading and flowering (Fig. 11.2). Such instances necessitate the availability of early maturing varieties (less than 90 days) especially in the drought-prone rainfed areas. Varieties like Khandagiri, Konark, Surendra, Sathia, Heera, Mandakini and Vandana with very early flowering (50–60 days) were popular among the rainfed drought-prone ecosystems. These varieties primarily relied on drought escape as a mechanism for avoiding moisture stress. However, the major drawback of these varieties involved a lower yield potential of 3 tha^{-1} under optimum situations. Furthermore, in rainfed environments, since farmers initiate the kharif sowing either in a seedbed or through broadcasting, typically with the second rainfall of the monsoon season, a temporary pause in the monsoon after seedling emergence often leads to crop failure for the mentioned varieties due to their limited ability to recover from drought conditions. During the Rabi season, after harvesting rice, many of these areas remain fallow because of insufficient water availability and limited access to improved early maturing post-kharif crops.

The government of Odisha, through different schemes, has increased the number of irrigation sources, thereby expanding potential irrigated areas. Major and minor (both flow and lift) irrigation infrastructure has improved significantly over the past decades (Table 11.1). However, they are insufficient to fully counter the episodic droughts that occur during the monsoon seasons. Resultantly, resource-poor farmers, especially smallholders remain vulnerable to water scarcity.

Table 11.1 Source-wise irrigation potential created (IPC) from 2016–2017 to 2021–2022

	2016–2017		2017–2018		2018–2019		2019–2020		2020–2021		2021–2022	
Irrigation sources	Kharif	Rabi	Kharif	Rabi	Kharif	Rabi	Kharif	Rabi	Kharif	Rabi	Kharif	Rabi
Major and medium	14.04	6.44	14.26	6.52	14.46	6.62	14.65	6.73	14.86	6.86	15.1	6.99
Minor (flow)	6.63	0.86	7.04	0.9	7.18	0.91	7.3	0.92	7.36	0.93	7.36	0.93
Minor (lift)	9.98	4.84	11.05	5.34	11.67	5.65	12.23	5.93	12.97	6.29	14.05	6.74
Mega lift	0.7	0	0.54	0	0.95	0	1.25	0	1.55	0	2.08	0
Other Sources	7.52	6.54	7.59	5.93	7.64	5.92	7.64	5.93	7.64	5.93	7.68	5.95
Total	38.87	18.68	40 48	18.69	41.9	19.1	43.07	19.51	44.38	20	46.28	20.61

Source: Economic Survey, Odisha 2022–2023

11.3 Dissemination and Impact of First-Generation Drought Tolerant Varieties

IRRI has been consistently dedicated to innovation and creating drought-tolerant rice varieties, focusing on disseminating these advancements to farmers across Africa and the Asia-Pacific regions. The STRASA Project, spanning from 2005 to 2019 and funded by the Bill and Melinda Gates Foundation (BMGF), stands out as a highly successful research initiative. This project played a pivotal role in the successful formulation and release of over 30 high-yielding rice varieties resilient to drought conditions in both Asia and Africa. Notably, significant efforts were undertaken during this project to integrate key drought-tolerant Quantitative Trait Loci (QTLs) such as *qDTY1.1, qDTY2.1, qDTY2.2, qDTY3.1, qDTY4.1, qDTY12.1, qDTY6.3*, among others, into widely adaptable rice varieties like IR64, Swarna and TDK-1 (Dixit et al., 2017; Khanna et al., 2022; Yadaw et al., 2013). This strategic incorporation led to the development of numerous high-yielding rice varieties with enhanced drought tolerance.

The first generation of improved drought-tolerant varieties tested in Odisha involved CR 2624, IR 74371-70-1-1 (Sahbhagi Dhan), IR 74371-3-1-1 and IR 55419-04 (Mall et al., 2015); along with DRR42 and DRR44. Among the first introduced varieties, Sahbhagi Dhan, DRR42 and DRR44 gained popularity owing to their strong drought recovery potential under moderate to severe water stress conditions compared to local checks like Khandagiri (Figs. 11.3, 11.4 and 11.5).

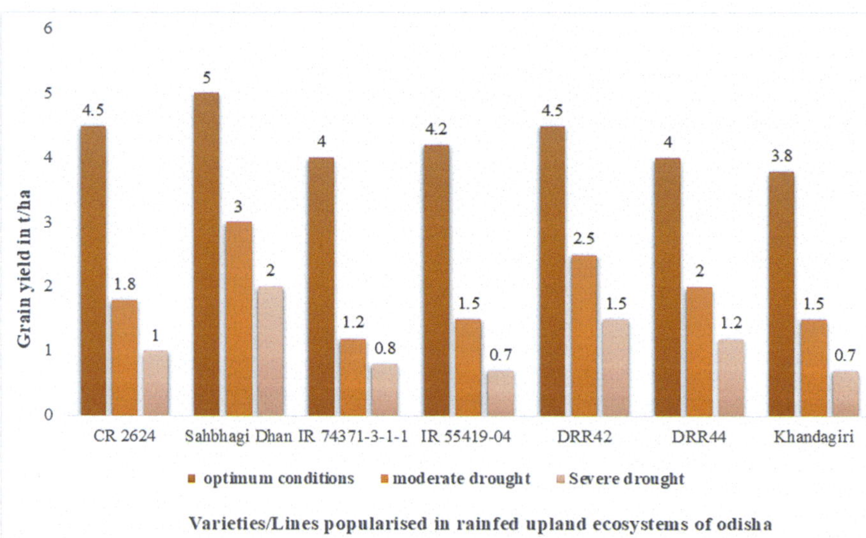

Fig. 11.3 Comparative yield of first-generation drought tolerant varieties in Odisha compared to local check "Khandagiri" during optimum, moderate drought, and severe drought conditions. (Source: IRRI)

Fig. 11.4 Comparative yield of Sahbhagi Dhan and MTU 1010 under moderate drought conditions. (Source: IRRI)

Fig. 11.5 Comparative yield of Sahbhagi Dhan and hybrid rice under moderate drought conditions. (Source: IRRI)

In 2019, as part of its on-farm trials focusing on stress-tolerant varieties, IRRI selected Nuagaon and Bisra blocks in Sundergarh district and Maneswar in Sambalpur district, regions frequently affected by drought. Collaborating with local NGOs DISHA and ADARSHA, IRRI initiated a project involving 100 farmers. The aim was to introduce two promising stress-tolerant varieties, namely DRR 44 and Sahbhagi Dhan. Sahbhagi Dhan, being a 105-day variety, was advised for cultivation in upland areas, whereas DRR 44, with a duration of 120 days, was advocated for mid-land locations, where life-saving irrigation could be ensured in case of drought. The capacity of Sahabhagi Dhan and DRR 44 to withstand water stress conditions was apparent in their yields when compared to other conventional varieties in cultivation (Pradhan et al. 2020). Despite the scarcity and uneven distribution of rainfall, both these varieties demonstrated resilience to stress, leading to a successful harvest for the farmers. The farmers experienced an average yield advantage of nearly 1 tha^{-1} even under moderate drought conditions. Additionally, empirical evidence indicated that cultivation of early maturing Sahbhagi Dhan encouraged farmers to plant second crops in the otherwise fallow areas of the rainfed upland areas of Odisha.

Although Sahbhagi Dhan shows significant potential for drought resilience in Odisha, relying on a limited number of varieties can result in a genetic bottleneck. Furthermore, since the variety was first introduced in India in 2010, it is crucial to consider alternatives in terms of varietal turnover to ensure sustained yield performance over the long term. Thus, testing and dissemination of new drought-tolerant varieties followed by infusion of the variety into the rice seed chain is of immense importance.

11.4 Drought-Prone Ecosystem and Varietal Choices

The effect of water scarcity is more pronounced for rice in the upland ecosystem. IRRI between 2016 and 2019 demonstrated and promoted suitable drought-tolerant varieties in the upland areas of Mayurbhanj, Bargarh, Sambalpur, Sundergarh, Kalahandi and Koraput. The demonstration conducted in farmers' fields showed a reasonable yield advantage over other traditionally grown varieties, especially in areas that confronted water shortage.

IRRI, through this project introduced and evaluated Sahabhagi D (105 days) DRR Dhan (120 days), and DRR 44 (120 days) against traditionally grown varieties—MTU 1001, Naveen, Lalat, MTU 1010. Figures 11.6, 11.7 and 11.8 show comparative yield advantages and percentage of yield gain of Sahabhagi Dhan, DRR Dhan 42 and DRR Dhan 44 over the years.

11.4.1 Green Super Rice and Other High-Yielding Products

Green super rice (GSR) is another segment to manage the impact of drought. This is a special group of rice varieties that are highly responsive to inputs and require

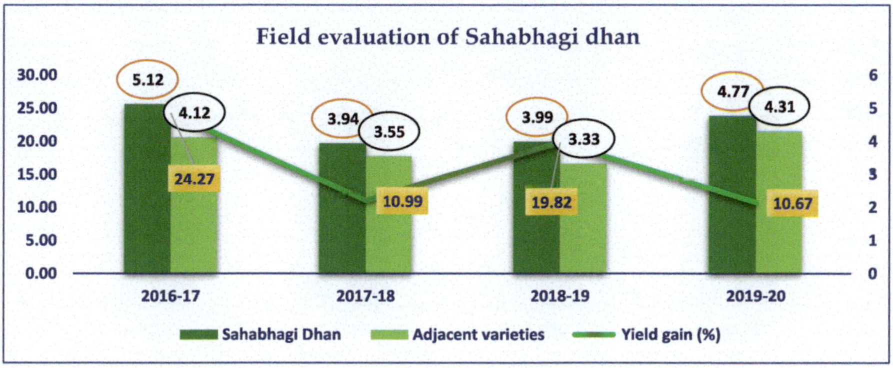

Fig. 11.6 Field evaluation of Sahbhagi Dhan (2017–2020) (Source: IRRI)

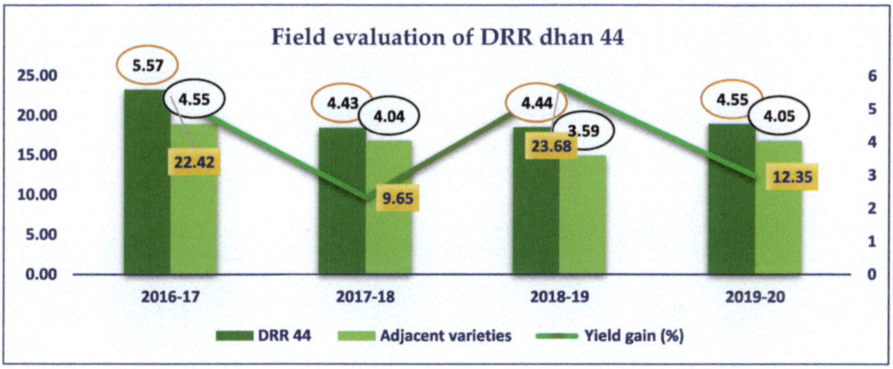

Fig. 11.7 Field evaluation of DRR 44 (2017–2020). (Source: IRRI)

less amount of water. Thus, in water-scarce areas, such varieties are found suitable as they produce significant yield advantages over other popularly grown varieties. IRRI, in Odisha, field-trialled three new GSR varieties—BINA Dhan 17, BRRI Dhan 69 and BRRI Dhan 75 during 2020–2022 and evidenced their superiority over the farmers' checks. Figure 11.9 provides a visual representation of the advantages of green super rice against the checks in respect of yield.

11.4.2 Development of Rice Varieties Suited for DSR Cultivation

Going beyond stress-tolerant varieties, the recent development of national breeding pipelines around direct-seeded rice systems (DSRs) carries significant potential for drought coping mechanisms. DSR as a system is directed towards saving water and labour as well as reducing greenhouse gas emissions (Joshi et al., 2013; Laing et al., 2018; Pathak et al., 2013). The higher water requirement in transplanted rice can be compensated by incorporating a combination of dry DSR cultivation of

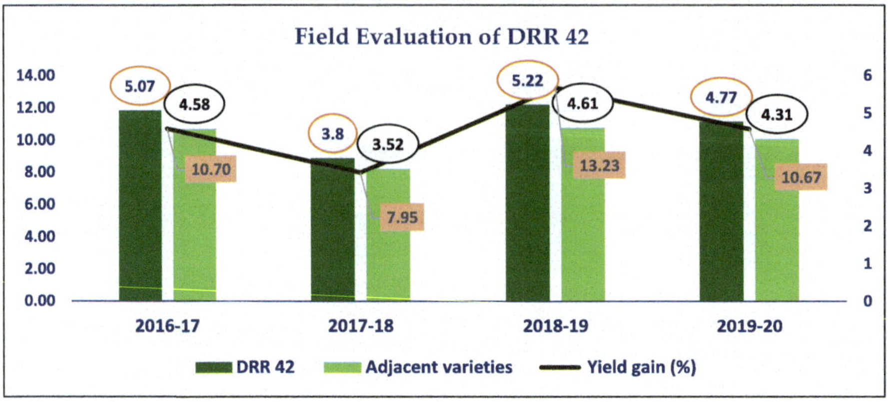

Fig. 11.8 Field evaluation of DRR 42 (2017–2020). (Source: IRRI)

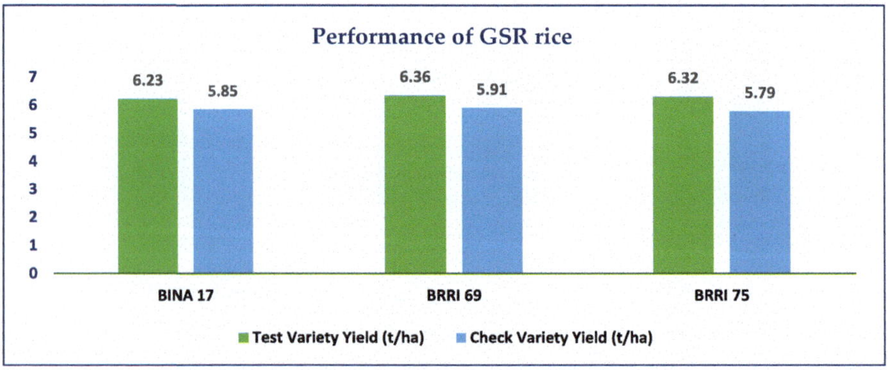

Fig. 11.9 Field evaluation of GSR varieties (2020–2022). (Source: IRRI)

drought-tolerant varieties in the drought-prone upland ecosystems of Odisha. However, to achieve a superior phenotype of the crop under dry DSR conditions, the identification and development of superior rice varieties suited to dry DSR conditions is of utmost importance. Recent advances in Genomic Breeding have paved the way for identifying QTLs which are important for superior performances of rice under DSR conditions (Fig. 11.10). Leveraging modern breeding tools, IRRI has identified rice varieties that are endowed with the traits necessary for the DSR system of cultivation designated as DELS-I/R (Direct seeded, early duration, long slender soft grain rice under irrigated or rainfed) market segments suited for the upland ecology (Sagare et al., 2020).

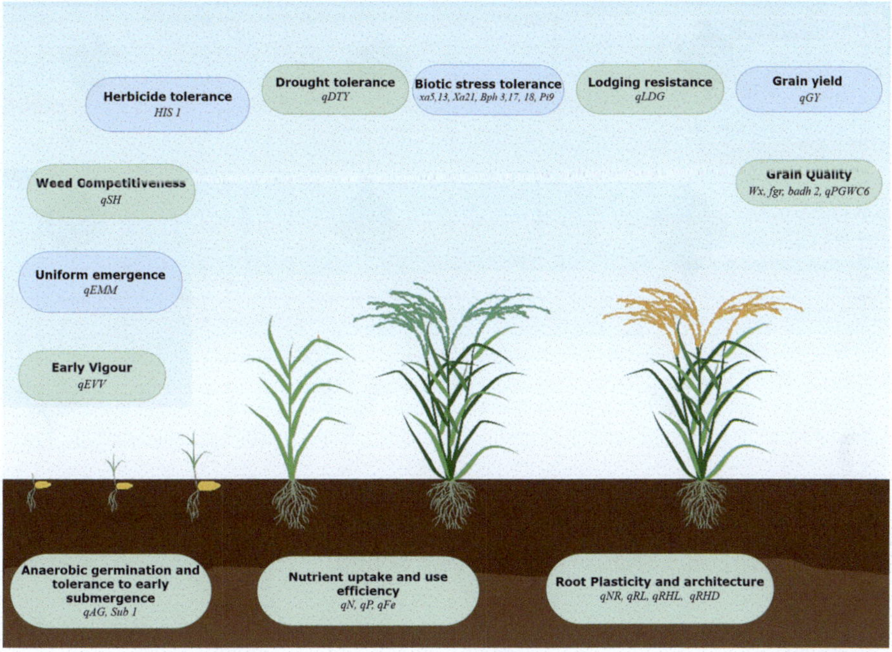

Fig. 11.10 Important traits in rice for superior performance under Dry DSR conditions along with underlying QTLs listed on the right

11.5 On-Farm Trials for DELS-R/I Segment in Odisha

To identify suitable varieties for diverse rainfed upland ecologies in Odisha during the kharif season of 2022, On-Farm Trials (OFTs) were undertaken. Nine varieties, including two popular checks, were shortlisted based on their performance in multi-locational testing under the IRRI–NARES (Indian National Agricultural Research & Education System) program and AICRPR (All India Coordinated Research Project on Rice) trials. The selected varieties—Sabour Harshit, Rajendra Saraswati, CG Barani 2, Swarna Shakti Dhan, Tripura Hakuchuk 2, Co51, PR 126 and the checks DRR42 and Sahbhagi Dhan were subjected to strip trials in farmers' fields across 15 locations in Bolangir, Bargarh, Nuapada, Kalahandi and Phulbani districts, covering various rainfed ecologies and sown using tractor driven seed drills under dry DSR conditions. Phenotypic data related to yield and yield-contributing traits, along with farmers' perceptions, were documented at the trial sites.

To assess the impact of genotype × location interaction, a pooled analysis of variance (ANOVA) was conducted. The results indicated a significant impact of the interaction component, suggesting that regional differences in environmental conditions can influence the performance of the varieties, even when tested under seemingly uniform rainfed upland ecologies in Odisha. Additionally, significant

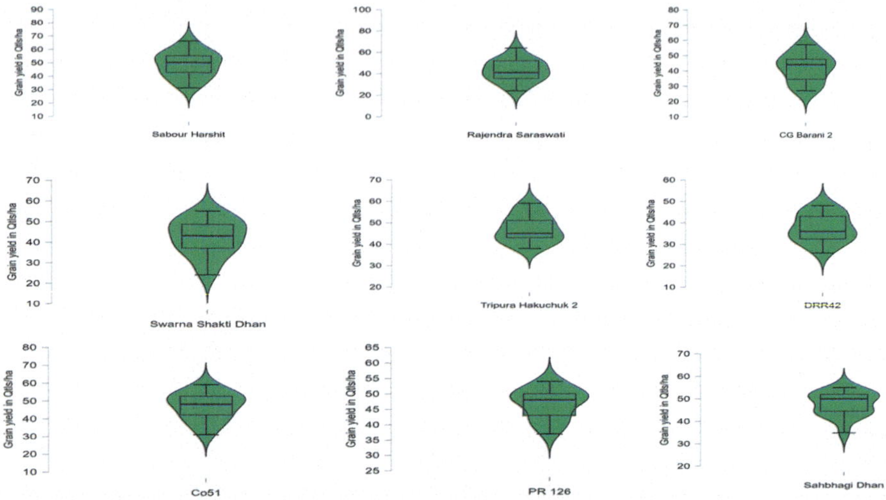

Fig. 11.11 Violin plots showing grain yield in q/ha for nine varieties tested across 15 OFTs (n = 15) in upland rainfed ecologies of Odisha. (Source: IRRI)

variations in terms of yield distribution were recorded among the varieties (Fig. 11.11) suggesting the influence of localised environmental conditions on grain yield. The experiment also revealed that all the varieties exhibited significantly higher average yield (at 5% level) over the check DRR 42 (Table 11.2). However, a significant yield advantage over Sahbhagi Dhan could not be recorded for the varieties under evaluation, indicating the adaptive advantage the variety possesses when cultivated across the rainfed ecologies of the region.

Among the varieties tested, Sabour Harshit, Tripura Hakuchuk 2, Co51 and PR 126 exhibited significantly better yield performance compared to the other varieties. However, the yield performance of the four varieties was statistically at par with each other. The experiments further indicated that under optimum conditions Sabour Harshit has considerably higher yield potential. Whereas, in terms of yield stability, Tripura Hakuchuk 2, Sahbhagi Dhan and PR 126 exhibited stronger resilience towards environmental instabilities, which can be further corroborated through the smaller standard deviation in yield recorded for the varieties when compared to Sabour Harshit (Table 11.2).

As discussed earlier, dependence on a single variety for a specific ecology leads to vulnerability in case of severe stress. Thus, a PCA was performed to ensure that significant variability exists among the varieties being evaluated across the 15 locations. The analysis indicated two principal components showing eigenvalue greater than 1. The PCI which explained 53.06% of the total variance in the population, comprised the varieties- Swarna Shakti Dhan, Tripura Hakuchuk 2, PR 126, DRR42, Sahbhagi Dhan and Co51 (Fig. 11.12). Whereas the three varieties, namely, Sabour Harshit, CG Barani 2 and Rajendra Saraswati belonging to PC2 explained 15.28% of the total variance observed. Thus, significant variation can be attributed to the

Table 11.2 Mean, range and standard deviation for six varieties tested in OFTs during kharif 2022 in Odisha

Sl. No.	Varieties	Mean (q/ha)	Standard deviation	Range
1	Sabour Harshit	48.53[a]	9.841	31.00–66.00
2	Rajendra Saraswati	42.93	10.573	24.00–64.00
3	CG Barani 2	42.00	9.335	27.00–57.00
4	Swarna Shakti Dhan	42.20	8.946	24.00–55.00
5	Tripura Hakuchuk 2	47.33[a]	6.444	38.00–59.00
6	Sahbhagi Dhan (Check 1)	47.66[a]	5.486	35.00–55.00
7	DRR42 (Check 2)	37.60	6.577	26.00–48.00
8	Co51	46.80[a]	7.921	31.00–59.00
9	PR 126	46.00[a]	4.980	37.00–54.00
	LSD at 5%	3.58		

[a]Varieties showing significantly higher yield over check two and other varieties under evaluation (at 5% level of significance) (Source: IRRI)

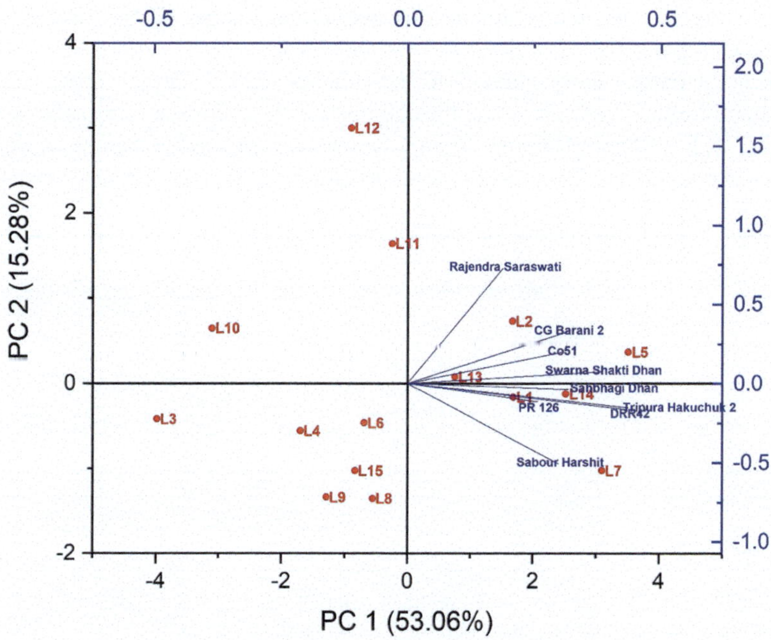

Fig. 11.12 Biplot showing the distribution of varieties and locations across two principal components. (Source: IRRI)

varieties belonging to PC1, which can be leveraged in terms of localised adaptation across the different upland ecosystems.

11.6 Cluster Demonstrations for the DELS-I/R Segment in the Upland Ecosystems in Odisha

The significance of cluster demonstrations in farmers' fields cannot be overstated, as they play a crucial role in raising awareness among farmers regarding the effectiveness of specific agricultural technologies. Nayak et al. (2022) previously documented the importance of frontline demonstrations, highlighting their substantial impact on generating awareness about submergence-tolerant varieties in various lowland ecologies of Odisha. The research identified three superior varieties—Sabour Harshit, Tripura Hakuchuk 2 and Co 51 which were actively demonstrated across intervention areas, particularly under Dry DSR conditions.

The cluster demonstrations involved the cultivation of these varieties using tractor-operated seed drills across one hectare of land at multiple locations within the intervention areas. Farmer participatory programs were conducted in collaboration with local Krishi Vigyan Kendras (KVKs), State Agriculture Universities (SAUs), farmers' collectives such as Farmers' Producer Organizations (FPOs) and Farmers' Producer Companies (FPCs), along with the Odisha State Seed Corporations (OSSC).

During the participatory evaluations in the fields in 2022, farmers were surveyed about their likelihood of cultivating the demonstrated varieties in successive seasons (Fig. 11.13). For Sabour Harshit, 70% of the farmers engaged in the cluster demonstrations expressed a willingness to grow the variety in successive kharif seasons. In

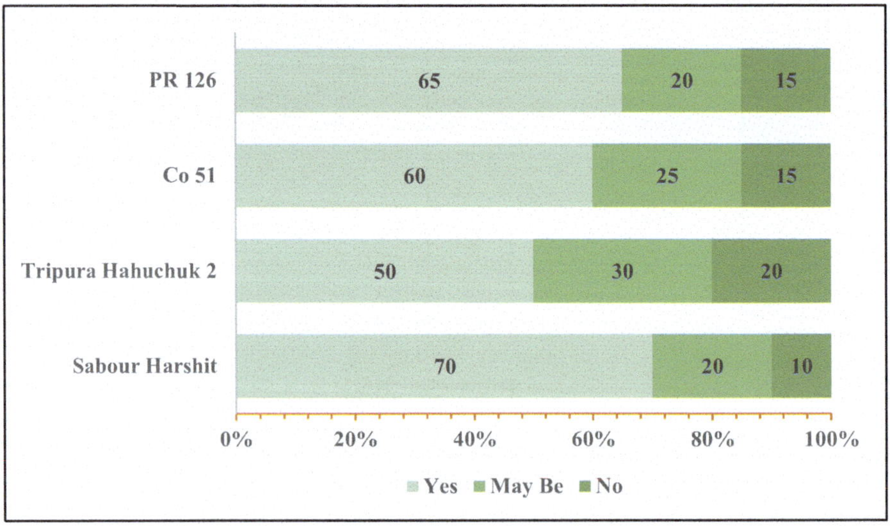

Fig. 11.13 Farmers' response on the likelihood of adoption of newly introduced varieties of DELS-I/R segment in frontline cluster demonstrations areas (2022). (Source: IRRI)

the case of Tripura Hakuchuk 2, half of the participating farmers indicated their willingness to cultivate the variety in subsequent seasons.

This observation aligned with expectations, considering the higher yield potential of Sabour Harshit under optimal conditions. It"s worth noting that, during the experimentation period, severe drought conditions were not recorded at any of the trial sites, with only moderate water stress observed in a few locations. As a result, the preference for Sabour Harshit was anticipated to be higher compared to Tripura Hakuchuk 2, which is known for its exceptional drought tolerance traits under severe water stress conditions. Nevertheless, it is crucial to consider the trade-off between achieving maximum yield gain and ensuring resilience against drought stress when assessing vulnerability to abiotic stress at the trial locations (Sinebo, 2005). In other words, while maximising yield is important, it"s equally essential to select and promote varieties that demonstrate resilience and adaptability to withstand adverse conditions such as drought. The balance between these factors becomes particularly significant in agricultural contexts where the susceptibility to abiotic stresses, like drought, is a prevalent concern. Selecting crop varieties that not only offer high-yield potential but also exhibit robustness in the face of environmental challenges is vital for ensuring sustainable and resilient agricultural practices.

11.7 Adoption of Drought-Fit Varieties in Eastern India

IRRI studied the adoption of three types of varieties, recommended in drought-prone rice ecology. The shorter duration rice varieties (<125 days) that escape drought by early harvest, drought-tolerant varieties that are resilient to water scarcity, and green super rice varieties are known for their lower water requirement. It also evaluated DSR-suitable varieties, which by establishment method, use less water for crop growth and maturity. A study was employed to understand their adoption pattern after community-led trials and demonstrations. The proportion of farmers who consecutively cultivated the varieties for three years was assessed to be adaptors. The experiment underscored significant adoption among the participants who participated in trials and demonstrations (Fig. 11.14). The reasonable yield even in water stress conditions, drought resilience, DSR suitability, agronomic and quality traits such as crop height, high head rice recovery, grain size and high-test weight were found to be adoption triggers for these varieties.

11.8 Cluster Demonstrations for Non-paddy Crops in Rice Fallows in Odisha

Expanding on the utilisation of residual soil moisture and implementing life-saving irrigation along with other water-conserving practices, a focus on both crop diversification and intensification through community engagement has been targeted in the otherwise rice fallows in Odisha.

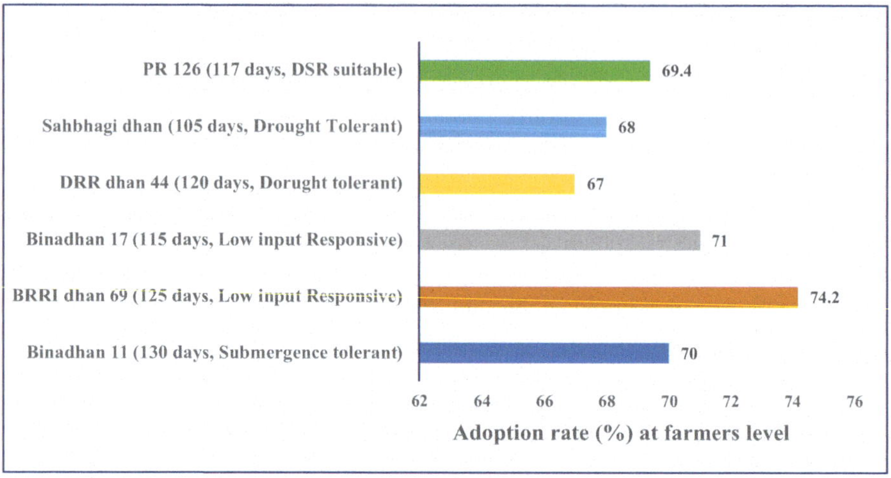

Fig. 11.14 Adoption rate of drought-fit varieties (2020–2022) in demonstration sites. (Source: IRRI)

In an earlier study, IRRI evaluated short-duration non-paddy crops namely green gram and black gram and oilseeds namely toria (sarson or lahi) and sunflower in potentially utilisable rice fallows across Bhadrak, Cuttack and Mayurbhanj districts (Peramaiyan et al., 2023). The study indicated better performances from the pulses (green gram and black gram) compared to the oilseeds under the rainfed conditions of the rice fallows.

Likewise, the potential of employing sesame and pigeon peas (Arhar) as post-kharif crops in rice fallows has been previously documented by Ramesh et al. (2019) and Mula (2012), respectively. Nevertheless, a considerable obstacle in the successful implementation of post-kharif crops customised for specific ecosystems stems from the lack of awareness and limited access to suitable varieties of these crops. To address this issue and raise awareness among farmers, cluster demonstrations were conducted in otherwise fallow lands within the same intervention areas where rice varieties of the DELS-I/R segment were deployed. Given that short-duration varieties with drought resilience are particularly well-suited for fallow ecosystems, the key varieties selected for these demonstrations are of significant importance (Table 11.3).

11.9 Integrated Technology Diffusion: Scaling of Seed Systems and DSR Cultivation in Odisha

Presently, the Odisha State Seed Corporation supplies around 30,000 metric tons of certified paddy seeds annually, distributed through a network of over 3000 dealers (IRRI 2019). Additionally, the Department of Agriculture contributes another 1500 metric tons through various development schemes. Unfortunately, this combined

Table 11.3 Non-paddy crop varieties elected for cluster demonstration in the rice fallow areas of the rainfed upland ecosystems in Odisha

Crop	Variety names and duration	Remarks
Green Gram	IPM 2-5-7 (VIRAT) (52–56 days); SIKHA (IPM 410-3) (65–70 days); IPM 02-14 (62–70 days), IPM 99-125 (Meha) (65–72 days), IPM 02-3 (70–72 days), and MH 421(60–65 days)	The varieties used are resistant towards MYMV and shows an average yield of 10–11 q/ha under optimum situations.
Black Gram	VBN 8 (65–75 days), PU-31 (75–80 days), and PU-35 (75–80 days)	Varieties are resistant against MYMV and show an average yield of 9–11 q/ha.
Rapeseed-toria	Uttara (93–97 days)	Can be recommended for areas where one or two life-saving irrigation can be facilitated.
Sesame	Uma (75–80 days), Shubhra (78–84 days), Nirmala (80–85 days)	Pale white to white seeds for maximum marketability along with tolerance to *Macrophomina* sp. and *Alternaria* sp
Sunflower	MSFH-17 (85–88 days)	Can be recommended for areas where one or two lifesaving irrigation can be facilitated
Pigeon Pea (Arhar)	ICPL 20326 (96–100 days) ICPL 20338 (90–100 days)	Early maturing varieties suitable for rice fallows

Source: IRRI

supply is insufficient to meet the seed requirements for the 3.5 million hectares dedicated to rice cultivation in the wet season.

With an ideal Seed Replacement Rate (SRR) of 33% for rice (Hossain & Nayak, 2020), the demand stands at least 60,000 metric tons of seeds each year. However, challenges related to infrastructure, inadequate production capacity and inefficient distribution networks make it impractical to achieve this target solely through state channels. Therefore, there is a need to encourage and empower farmers to produce high-quality seeds, which can significantly enhance farmers" seed security and augment the SRR (Kumar et al., 2016). In this regard, strengthening the community-based informal seed system can play an important role in the dissemination of STRVs in the state. The importance of the informal/farmers' seed system has been evinced in the earlier works of Almekinders et al. (2002) and Hossain and Nayak (2023).

In line with the purview, IRRI over the years has focussed on training and capacity building of various farmers' collectives (SHG, FPO and FPCs) in terms of quality seed production (QSP), storage, and marketing (Habib & Nayak, 2023). Consistent efforts have resulted in bridging the gap between the adoption of stress-tolerant rice varieties (STRVs) and non-STRVs in the state as reflected by the annual breeder seed indent of rice from Odisha (Fig. 11.15). The growing trend of STRVs adoption coupled with the adoption of the DSR method of cultivation, can play a vital role in mitigating abiotic stress like drought across the various rainfed ecosystems in the state.

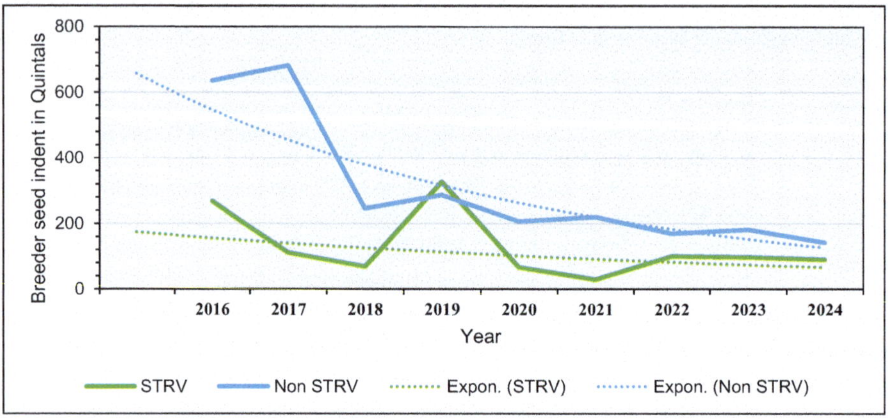

Fig. 11.15 Annual BS (Breeder Seed) indent for STRVs and non-STRVs in Odisha. (Source: https://seednet.gov.in/)

11.9.1 Quality Seed Production with Farmers' Groups

IRRI in 2023, supported 11 farmers collectives and capacitated them for quality seed production, processing, and marketing. Two new varieties—Swarna Samriddhi and BRRI Dhan 69 that require less irrigation water, were chosen and early generation seeds (Breeder Seed and Foundation Seed) were provided. The collectives, including all women FPCs, were also supported with the seed sale registration process and certification and marketing. A total of 100 tons of seeds were produced through a community-led seed network.

Extension programs that furnish information and resources could be advantageous for smallholder women farmers who often play a crucial role in agriculture (Shrestha et al., 2019). The availability of drought-suitable varieties can thus be expanded by leveraging these strong community seed channels. Table 11.4 presents the details of seed production by farmers' collectives.

11.10 Conclusion

In the face of frequent climatic stress, particularly severe droughts, smallholder farmers in Odisha confront significant challenges in sustaining rice-based cropping systems in the rainfed upland ecosystems of the state. With a reliance on rainfed systems and limited irrigation coverage, the vulnerability of these farmers is evident. The state"s history of natural calamities, with a substantial number being severe droughts, underscores the urgent need for sustainable solutions. The introduction and widespread adoption of drought-tolerant rice varieties and short-duration non-paddy crops have emerged as crucial strategies. Following the initial success of drought-tolerant varieties like Sahabhagi Dhan, DRR Dhan 42 and DRR Dhan 44; new varieties with drought tolerance and phenotypic suitability for dry DSR cultivation

Table 11.4 Seed production by farmers' Collectives

Sl. No	Location	FPO name	Area (Ha.)	Variety
1	Balangir(Agalpur, Puintala)	Sairaj Farmers Producer Company Limited	2	Swarna Samridhi
2		Abhaya Farmers Producer Company Limited	2	
3		Bancharen Farmer Producer Company Limited	2	
4		Puintala Farmer Producer Company Limited	2	
5		Pruthunna Farmers Producer Company Limited	2	
6	Mayurbhanj(Sukruli, Jashpur)	Shaktimayee Producer Company Ltd.	2	Swarna Samridhi/ BRRI Dhan 69
7		Matrushakti Producer Company Limited	2	
8		Jashipur Farmers Producer Company Limited	2	
9		Swayam Sampurna Producer Company ltd.	2	
10		Maa Santoshi FPC	3	
11		Shri Jagannath FPC	3	

Source: IRRI

practices have been identified through robust on-farm trials in Odisha. Varieties, namely, Sabour Harshit, Tripura Hakuchuk 2, Co51 and PR 126 have shown great promise in terms of yield stability under water stress conditions along with adaptability towards dry DSR mode of cultivation. Furthermore, the exploration of short-duration oilseeds and pulses for the post-kharif period holds promise for enhancing farm returns. The deployment of early-maturing rice varieties not only addresses drought challenges but also contributes to sustainable land use by cultivating a second crop in fallow areas during the dry season. For awareness generation, large-scale cluster demonstrations in farmers' fields were ensured, displaying the impact of the integrated approach of DSR varieties, dry DSR cultivation, and early maturing non-paddy crops during post-kharif period in the rice fallows. Ultimately, training in terms of quality seed production among the farming communities, coupled with varietal infusion can ensure access to new technologies and long-term resilience against drought stress across the rainfed ecosystems in Odisha.

References

Almekinders, C. J., & Louwaars, N. P. (2002). The importance of the farmers' seed systems in a functional national seed sector. *Journal of New Seeds, 4*(1–2), 15–33.

Bera, B., Shit, P. K., Sengupta, N., Saha, S., & Bhattacharjee, S. (2021). Trends and variability of drought in the extended part of Chhota Nagpur plateau (Singbhum Protocontinent), India applying SPI and SPEI indices. *Environmental Challenges, 5*, 100310.

Chandna, P. K., & Mondal, S. (2020). Analyzing multi-year rice-fallow dynamics in Odisha using multi-temporal Landsat-8 OLI and Sentinel-1 Data. *GIScience & Remote Sensing, 57*(4), 431–449.

Dar, M. H., Waza, S. A., Shukla, S., Zaidi, N. W., Nayak, S., Hossain, M., Kumar, A., Ismail, A. M., & Singh, U. S. (2020). Drought tolerant rice for ensuring food security in Eastern India. *Sustainability, 12*(6), 2214.

Dixit, S., Yadaw, R. B., Mishra, K. K., & Kumar, A. (2017). Marker-assisted breeding to develop the drought-tolerant version of Sabitri, a popular variety from Nepal. *Euphytica, 213*, 1–16.

Habib, A. M., Nayak, S., & Mohapatra, S. (2023). *Fostering gender-friendly quality seed production trainings in Bangladesh*. GREAT. https://greatagriculture.org/news/fostering-gender-friendly-quality-seed-production-trainings-in-bangladesh/

Hossain, S. M., & Nayak, S. (2020). *Paradox of rice seed system in India* (Vol. 3, p. 29).

Hossain, M., Nayak, S., Gupta, D., Raja Sekhar, G., Shukla, S., Nath, A., & Singh, V. (2023). Farmer-led popularization and dissemination of new rice varieties in Telangana help close the seed gap. *Rice Today*. 18 Sep. https://ricetoday.irri.org/farmer-led-popularization-and-dissemination-of-new-rice-varieties-in-telangana-help-close-the-seed-gap/

Joshi, E., Kumar, D., Lal, B., Nepalia, V., Gautam, P., & Vyas, A. K. (2013). Management of direct seeded rice for enhanced resource-use efficiency. *Plant Knowledge Journal, 2*(3), 119–134.

Khanna, A., Anumalla, M., Catolos, M., Bartholomé, J., Fritsche-Neto, R., Platten, J. D., et al. (2022). Genetic trends estimation in IRRIs rice drought breeding program and identification of high yielding drought-tolerant lines. *Rice, 15*(1), 14.

Kumar, S. P. J., Prasad, S. R., Kumar, M., Chandusingh, Sinha, A. K., & Pathak, A. (2016). Seed quality markers: A review. *Research & Reviews: Journal of Botany Science, S3*, 13–17.

Laing, A. M., Roth, C. H., Chialue, L., Gaydon, D. S., Grünbühel, C. M., Inthavong, T., Phengvichith, V., Schiller, J., Thiravong, K., & Williams, L. J. (2018). Mechanised dry seeding is an adaptation strategy for managing climate risks and reducing labour costs in rainfed rice production in lowland Lao PDR. *Field Crops Research, 225*, 32–46.

Mall, A. K., Swain, P., Baig, M. J., & Singh, O. N. (2015). Identifying drought tolerant rice genotypes using participatory research approach for resource poor farmer's of Odisha. *ORYZA—An International Journal on Rice, 52*(3), 181–185.

Mula, M. G. (2012). *Introduction and Expansion of Improved Pigeonpea (Arhar) Production Technology in Rainfed Upland Ecosystems of Odisha*. Technological Empowerment and Sustainable Livelihood.

Nayak, S., Das, K., Bhowmick, M. K., Hossain, S. M., Singh, S., Singh, U. M., & Singh, V. K. (2022). Flood stress prevalence and assessment of farmers' preparedness with resilient rice varieties: A review in the context of Eastern India. In S. Choudhury & D. Moulick (Eds.), *Response of field crops to abiotic stress: current status and future prospects* (p. 277). CRC Press.

Pathak, H., Sankhyan, S., Dubey, D. S., Bhatia, A., & Jain, N. (2013). Dry direct-seeding of rice for mitigating greenhouse gas emission: Field experimentation and simulation. *Paddy and Water Environment, 11*, 593–601.

Peramaiyan, P., Srivastava, A. K., Kumar, V., Seelan, L. P., Banik, N. C., Khandai, S., et al. (2023). Crop establishment and diversification strategies for intensification of rice-based cropping systems in rice-fallow areas in Odisha. *Field Crops Research, 302*, 109078.

Pradhan, S., Rahman, F. H., Sethy, S., Pradhan, G., & Sen, J. (2020). Evaluation of short duration drought tolerant rice varieties in drought prone areas of Subarnapur District of Odisha. *International Journal of Plant & Soil Science, 32*(8), 21–26.

Ramesh, K., Ratnakumar, P., Harisudan, A., & Bhaskar, S. (2019). Sesame (Sesamum indicum) in the rice fallow environments—A critical appraisal. *Journal of Oilseeds Research, 36*(4).

Sagare, D. B., Abbai, R., Jain, A., Jayadevappa, P. K., Dixit, S., Singh, A. K., et al. (2020). More and more of less and less: Is genomics-based breeding of dry direct-seeded rice (DDSR) varieties the need of hour? *Plant Biotechnology Journal, 18*(11), 2173–2186.

Sam, A. S., Padmaja, S. S., Kächele, H., Kumar, R., & Müller, K. (2020). Climate change, drought and rural communities: Understanding people's perceptions and adaptations in rural eastern India. *International Journal of Disaster Risk Reduction, 44*, 101436.

Shrestha, R. B., & Thapa, Y. B. (2019). Policy and program priorities for agricultural research and development in South Asia. *Agricultural Policy and Program Framework: Priority Areas for Research, 1*, 1–29.

Sinebo, W. (2005). Trade off between yield increase and yield stability in three decades of barley breeding in a tropical highland environment. *Field Crops Research, 92*(1), 35–52.

Yadaw, R. B., Dixit, S., Raman, A., Mishra, K. K., Vikram, P., Swamy, B. M., et al. (2013). A QTL for high grain yield under lowland drought in the background of popular rice variety Sabitri from Nepal. *Field Crops Research, 144*, 281–287.

Adapting Soil and Nutrient Management to a Changing Climate: Emerging Innovations and Policy Priorities

12

Abhijit Rai and Sayantan Sarkar

Abstract

Climate change presents significant challenges to global agriculture, impacting food security, altering crop yields, and affecting the resilience of agricultural systems. With the global population projected to reach nearly 9 billion by 2050, the urgency to sustainably increase agricultural productivity amidst climatic uncertainties has never been greater. This chapter investigates the intersection of climate change and agriculture, guided by the latest Intergovernmental Panel on Climate Change (IPCC) assessments, which underscore the necessity for immediate action to mitigate the adverse effects of climate change. It explores innovative, sustainable agricultural management practices that hold the potential to feed the growing population while mitigating environmental degradation. Highlighting advancements in soil and nutrient management, the chapter emphasizes the critical role of these innovations in adapting to and mitigating the impacts of climate change on agriculture. Through an examination of soil health indicators and resource conservation strategies, it presents forward-thinking approaches aimed at enhancing soil health and optimizing nutrient use. This discussion underscores the importance of aligning agricultural practices with contemporary climate policies, showcasing how innovative management practices are essential for ensuring food security and environmental sustainability in the face of a changing climate.

A. Rai
Department of Plant Sciences, College of Agriculture and Bioresources, University of Saskatchewan, Saskatoon, SK, Canada
e-mail: abr922@mail.usask.ca

S. Sarkar (✉)
Texas A&M AgriLife Research and Extension Center, Corpus Christi, TX, USA

R. Dubey et al. (eds.), *Drought and Heat Stress in Agriculture*,
https://doi.org/10.1007/978-981-96-5735-3_12

Keywords

Food security · Climate policies · Environmental sustainability · Optimizing nutrient use

12.1 Introduction

The Intergovernmental Panel on Climate Change (IPCC) highlights that climate change presents significant challenges to global ecosystems, human health, and economies, with urgent action required to mitigate its impacts. According to the IPCC's assessments, key concerns for the future include increasing global temperatures, rising sea levels, and more frequent extreme weather events such as hurricanes, droughts, and heavy rainfall (IPCC, 2021). These changes are expected to exacerbate food and water scarcity, threaten coastal communities and lead to loss of biodiversity. This goal necessitates a rapid transition toward renewable energy, enhanced energy efficiency, and sustainable land use practices.

Climate change, marked by shifts in temperature, precipitation patterns, and an increased frequency of extreme weather events, has ushered in a new era of uncertainty for agriculture. The scientific consensus, supported by the Intergovernmental Panel on Climate Change (IPCC), unequivocally affirms that climate change is already affecting agriculture, manifesting in altered crop yields and an unsettling specter of food security (Shukla et al., 2022). It is a global concern that imperils the stability and resilience of our agricultural systems.

Simultaneously, the global population continues to surge, adding a layer of urgency to the equation. By 2050, the world is projected to be home to nearly 9 billion people (United Nations, Department of Economic and Social Affairs, Population Division, 2019), each requiring sustenance. The challenge lies in meeting the dietary needs of this burgeoning population, which necessitates not only increased agricultural productivity but also sustainability in the face of climate-related disruptions. Innovative sustainable agriculture management practices are at the forefront of addressing the pressing challenge of feeding an ever-expanding global population while simultaneously mitigating the escalating threats of environmental degradation and climate change.

These solutions are poised not only to enhance crop productivity but also to mitigate and adapt to the adverse impacts of climate change on agriculture. This chapter embarks on a comprehensive exploration of the emerging innovations that are reshaping soil and nutrient management practices in the context of a changing climate. It is within these pages that we delve into key soil health indicators to assess climate change impact, and forward-thinking strategies designed to address critical aspects such as soil health, nutrient optimization, and resource conservation. We shine a spotlight on the latest advancements in soil and nutrient management, emphasizing their significance in aligning agricultural practices with contemporary climate policies.

12.2 Climate Change and Agriculture

Changes in temperature, precipitation patterns, and extreme weather events due to climate change have profound impacts on agricultural systems worldwide, affecting crop yields, soil health, and pest patterns. Temperature fluctuations significantly impact crop yields. An increase in average temperatures can accelerate crop maturation, reducing the growth period and thus potentially decreasing yields (Zhao et al., 2017; Rasul et al., 2011). For example, wheat and rice, staple crops for billions, have shown reduced yields with increased temperature stress (Wang et al., 2018). Additionally, extreme heat can lead to heat stress, adversely affecting crop physiology. Heat stress during flowering periods, for instance, can drastically reduce yields. This is particularly evident in crops like maize, where high temperatures during the flowering stage can lead to reduced kernel formation (Edreira et al., 2011; Dong et al., 2021). However, the impact varies with geography and crop type. In some cooler regions, a moderate increase in temperature might initially benefit certain crops by extending the growing season (Hatfield et al., 2011, Mueller et al., 2012). On the other hand, precipitation changes manifest as either droughts or floods, both detrimental to crop yields. Drought conditions lead to water stress, impairing plant growth and reducing yields. A study in the United States revealed how the 2012 drought significantly decreased corn and soybean yields (Westcott & Jewison, 2013). Floods, conversely, can oversaturate soils, damage plant roots, and interrupt the nutrient uptake necessary for healthy crop growth (Filipović, 2020).

Climate change can also impact soil structure, moisture levels, and nutrient content, influencing the overall fertility of the soil (Bronick & Lal, 2005; Singh et al. 2002). Increased temperatures can enhance soil evaporation rates, leading to drier soils and, subsequently, desertification in extreme cases. Drier soils are more susceptible to erosion, and this loss of topsoil depletes the land of its nutrient-rich layer, critical for plant growth. These changes can lead to the accumulation of salts on the soil surface, affecting plant growth, soil microbial communities, and overall soil health. Conversely, increased precipitation can lead to soil saturation, which disrupts the balance of oxygen and nutrients in the soil, affecting plant and microbial life. Additionally, intense rainfall can cause soil erosion, removing the fertile top layer of soil and potentially leading to long-term reductions in soil productivity (Martínez-Mena et al., 2020; Lal, 2015). Climate change also alters the carbon cycle in soils. Warmer temperatures accelerate organic matter decomposition, releasing more carbon dioxide into the atmosphere and reducing the soil's capacity to store carbon (Davidson and Janssens, 2006). This not only contributes further to climate change but also decreases soil fertility. Furthermore, the warming of tropical soils due to climate change leads to a decline in microbial diversity and an increase in soil respiration rates. This adaptation of the microbial community to warmer temperatures can exacerbate the release of carbon dioxide, thereby contributing to further climate change (Nottingham et al., 2022).

Climate change significantly alters pest and disease patterns in agriculture, posing new challenges for crop management. Rising temperatures can expand the geographical range of many pests and pathogens, exposing crops to pests they have not previously encountered. Warmer climates can also increase the number of pest generations per season, leading to more frequent and severe infestations (Skendžić et al., 2021). For example, studies have shown an expansion of the range of crop pests, like the Colorado potato beetle and the European corn borer, attributed to increasing temperatures (Bebber et al., 2015; Porter et al., 1991). Furthermore, changes in humidity and precipitation patterns can create environments more conducive to fungal and bacterial diseases, impacting crop health (Singh et al. 2023). Temperature and moisture changes also affect the efficacy of pesticides. Higher temperatures can increase the degradation rate of certain pesticides, reducing their effectiveness. On the other hand, increased rainfall can wash pesticides off plants, necessitating more frequent applications, which further increases costs and potential environmental impacts (Bloomfield et al., 2006; Delcour et al., 2015).

The effects of climate change on agriculture encompass a broad spectrum of concerns. While prior research has predominantly concentrated on the implications for aboveground ecosystems, analyzing potential changes in plant biodiversity and biochemistry, recent evidence underscores the critical impact of climate change on subterranean systems. Studies by Bardgett et al. (2008) highlight those climatic stressors, including increased temperatures, significantly influence the rhizosphere, the structure of soil microbial communities, and key soil processes such as respiration, nitrogen mineralization, and overall carbon dynamics within ecosystems. Consequently, it is essential to explore and implement novel soil and nutrient management techniques to mitigate these impacts. Such measures are crucial for enhancing the resilience of agricultural practices against the negative consequences of climatic shifts. By adopting innovative management strategies, agriculture can better cope with and recover from climate-induced challenges, including heightened erosion rates, soil salinization, changes in rainfall patterns, and the occurrence of extreme weather conditions. Prioritizing the health of the soil and optimizing nutrient management not only supports sustained agricultural productivity but also contributes to increased carbon storage and the preservation of vital ecosystem services.

12.3 Key Soil Health Indicators and Climate Change

Soil health plays a pivotal role in bolstering the resilience of agricultural systems when confronted with the challenges posed by our swiftly changing climate. To effectively navigate the intricate web of climate change impacts on soil health, we must leverage a comprehensive set of key indicators rooted in the physical, chemical, and biological aspects of soils. These indicators serve as indispensable tools for evaluating the ability of soil ecosystems to withstand and adapt to the evolving pressures exerted by climate change. However, there are limited studies that have tried to directly link soil health indicators to climate change effects, but the broader scientific literature acknowledges the impact of climate change on soil health through changes in temperature, precipitation patterns, and increased frequency of extreme weather events (Allen et al., 2011; Raghavendra et al., 2020; Lehmann et al., 2020). These key soil health indicators are intricately linked to climate change and not only offer vital insights into gauging the crop productivity and repercussions of climate change but also serve as tools for assessing, embracing, and implementing innovative strategies aimed at bolstering resilience in agricultural systems in a sustainable and environmentally responsible manner (Table 12.1).

12.3.1 Physical Health Indicators of Soil

Soil Structure Soil structure refers to the arrangement of soil particles into aggregates. Aggregate stability measures the resistance of these aggregates to external forces, and porosity quantifies the void spaces within soil. Soil structure influences organic carbon accumulation, water storage, root growth, and microbial activity, all crucial components of soil health (Dalal and Moloney 2000). Changes in climate, such as increased rainfall intensity, can affect aggregate stability, potentially impacting soil erosion rates and water storage capacity. Elevated temperatures and variable rainfall patterns may alter soil porosity and affect root development and soil biological activities (Piglai & De Nobili, 1993).

Soil Infiltration Soil infiltration measures the rate at which water enters and moves through the soil. Soil available water capacity is a key property governing water availability for plants and soil processes. Infiltration is vital for soil health assessments, while available water capacity affects plant growth and various soil functions (Arias et al. 2005). Climate change, including variable and intense rainfall or drought events, can alter soil infiltration rates and water availability for plants, necessitating adaptive management strategies (O'Farrell et al., 2010).

Bulk Density Bulk density assesses soil compactness and aeration, with lower values indicating better soil structure. Bulk density influences soil aeration and infiltration, both key components of soil health (Dalal & Moloney, 2000; Pattison et al. 2008). Elevated temperatures and extreme rainfall events may increase bulk density due to organic carbon loss, making soil more prone to compaction, particularly in response to land management activities (Birkás et al., 2009).

Rooting Depth Rooting depth represents how deeply plant roots extend into the soil. Changes in rooting depth can affect plant available water, subsoil properties, and soil constraints, all influencing soil health (Arias et al. 2005; Birkás et al., 2009). Under prolonged drought, changes in rooting depth can have a significant impact on plant productivity, particularly in the presence of subsoil constraints (Dang et al., 2008).

Table 12.1 Soil health indicators and their impact on climate change

Indicator	Description	Impact of climate change
Soil physical health		
Soil structure	Arrangement of soil particles into aggregates; influences organic carbon accumulation, water storage, root growth, and microbial activity	Changes in climate can affect aggregate stability, soil erosion rates, water storage capacity, and root development
Soil infiltration	Rate at which water enters and moves through the soil; crucial for soil health assessments	Climate change can alter infiltration rates and water availability for plants, necessitating adaptive management strategies
Bulk density	Soil compactness and aeration; lower values indicate better soil structure	Elevated temperatures and extreme rainfall events may increase bulk density, making soil more prone to compaction
Rooting depth	Depth of plant roots into the soil; influences plant available water, subsoil properties, and soil constraints	Changes in rooting depth can impact plant productivity, particularly under prolonged drought and subsoil constraints
Soil surface cover	Presence of materials like crop residues or biological soil crusts on the soil surface	Surface cover provides soil protection, erosion reduction, and organic matter input; mitigates impacts of severe rainfall, drought, or erosion events
Soil chemical health		
Soil pH	Measures soil acidity or alkalinity; influences nutrient availability, biological activity, and crop performance	Climate-induced changes in organic matter, nutrient cycling, and plant productivity can indirectly affect soil pH
Electrical conductivity (EC)	Measures salt concentration in the soil; indicator of soil quality	Temperature and precipitation changes can affect EC levels, with decreased EC in response to increased temperature and decreased precipitation
Cation exchange capacity (CEC)	Determine soil's ability to retain essential nutrient cations; critical for nutrient retention and soil quality	Elevated temperatures can lead to loss of CEC, potentially impacting nutrient leaching in response to intense rainfall events
Plant available nutrients	Extractable nutrients provide insights into plant nutrient availability and environmental risks	Climate change can affect nutrient cycling, particularly nitrogen, which is closely linked to soil organic carbon cycling; impacts on other nutrients require further investigation

(continued)

Table 12.1 (continued)

Indicator	Description	Impact of climate change
Soil biological health		
Soil organic matter (SOM)	Living and non-living components within soils; influences nutrient cycling, organic carbon storage, microbial habitat, and aggregate stability	Climate change may impact SOM turnover rates and accessibility to micro-organisms
Soil carbon (SOC)	Subset of SOM; essential indicator linked to ecosystem performance	Predicted to decline with increasing global temperatures, impacting soil functions and processes
Light Fraction & Macro-Organic Matter	Mineral-free particulate plant and animal residues within SOM; responsive to management practices	Elevated temperatures can deplete labile soil organic C, affecting light fraction and macro-organic matter
Potentially Mineralizable C & N	Measure amount of mineralizable organic matter; insights into organic matter quality and nutrient dynamics	Can assess soil health under climate change, affecting nutrient dynamics within growing seasons and over time
Soil respiration	Measures release of CO_2 or consumption of O_2; linked to microbial activity and nutrient cycling	Critical link between climate change and global C cycle; responds to temperature and rainfall changes
Soil microbial biomass	Living portion of SOM; sensitive indicator of soil processes	Responds to short-term environmental changes, with potential for long-term assessments
Enzyme activity	Reflects nutrient cycling and microbial status; integrates information on microbial and physicochemical soil conditions	Climate change factors like elevated CO_2 and atmospheric N deposition may impact microbial enzyme activities, affecting organic C turnover and nutrient cycling

Soil Surface Cover Soil surface cover refers to the presence of materials like crop residues or biological soil crusts on the soil surface. Surface cover provides soil protection, erosion reduction, and organic matter input, affecting soil health (Box & Bruce, 1996). Soil surface cover can dissipate raindrop impact energy, stabilize soil, and retain water and nutrients, making it a key factor in mitigating the impacts of severe rainfall, drought, or erosion events (Lal 1995; Salvador Sanchis et al., 2008).

12.3.2 Chemical Health Indicators of Soil

Soil pH Soil pH measures the acidity or alkalinity of the soil. Soil pH influences various soil functions, including nutrient availability, biological activity, and crop performance (Dalal & Moloney, 2000). While direct pH changes from climate drivers like elevated temperatures are expected to be slow, climate-induced changes in organic matter, nutrient cycling, and plant productivity can indirectly affect soil pH (Reth et al., 2005).

Electrical Conductivity (EC) EC measures salt concentration in the soil. EC is an indicator of soil quality, informing about salinity, nutrient cycling, and biological activity (Arnold et al. 2005). Climate change factors like temperature and precipitation can affect EC levels, with studies showing decreased EC in response to increased temperature and decreased precipitation in specific environments (Pariente 2001).

Sorption Capacity and Cation Exchange Capacity (CEC) Sorption capacity and CEC determine a soil's ability to retain essential nutrient cations like Ca, Mg, and K and immobilize potentially toxic cations like Al and Mn. These properties are critical for nutrient retention and soil quality, making them valuable soil health indicators (Dalal & Moloney, 2000). Elevated temperatures can lead to the loss of CEC in soils, especially those rich in organic matter, potentially impacting nutrient leaching in response to intense rainfall events (Davidson & Janssens, 2006).

Plant Available Nutrients Extractable nutrients like N, phosphorus, and potassium provide insights into plant nutrient availability and potential environmental risks. These nutrients are crucial for plant growth and environmental quality assessment, serving as indicators of soil health (Dalal & Moloney, 2000; Doran et al., 1996). Climate change can affect nutrient cycling, especially nitrogen (N), which is closely linked to soil organic carbon cycling. The precise impacts on other plant-available nutrients like phosphorus and sulfur require further investigation (Weil & Magdoff, 2004).

12.3.3 Biological Health Indicators of Soil

Soil Organic Matter (SOM) SOM includes living and non-living components within soils with diverse properties and functions. SOM is crucial for soil quality as it influences various soil functions, such as nutrient cycling, organic carbon storage, microbial habitat, and aggregate stability (Haynes, 2008; Weil & Magdoff, 2004). Changes in SOM can affect fertility, biodiversity, soil structure, water retention, and resilience. Climate change may impact SOM turnover rates and accessibility to micro-organisms (Davidson & Janssens, 2006; Kuzyakov & Gavrichkova, 2010).

Soil Carbon (SOC) SOC is a subset of SOM that refers to soil carbon content. SOC is an essential indicator linked to ecosystem performance, though it may not encompass all ecosystem traits (Janzen, 2006). SOC is predicted to decline with increasing global temperatures, impacting soil functions and processes (Lal et al., 2007).

Light Fraction and Macro-Organic Matter These components consist of mineral-free particulate plant and animal residues within SOM. They are responsive to management practices and may serve as early indicators of adaptive responses to climate change (Gregorich et al., 1994). Elevated temperatures can deplete labile soil organic C, a major source of light fraction and macro-organic matter (Davidson & Janssens, 2006).

Potentially Mineralizable C and N These indicators measure the amount of mineralizable organic matter in the soil. They provide insights into organic matter quality and nutrient dynamics (Haynes, 2008). They can assess soil health under climate change, affecting nutrient dynamics within growing seasons and over time (Gregorich et al., 1994).

Soil Respiration Soil respiration measures the release of CO_2 or consumption of O_2, positively correlated with SOM content. It serves as a biological indicator linked to microbial activity and nutrient cycling (Dalal & Moloney, 2000). Soil respiration is a critical link between climate change and the global C cycle, with responses to temperature and rainfall changes (Wixon & Balser, 2009).

Soil Microbial Biomass Microbial biomass is the living portion of SOM and a sensitive indicator of soil processes. It influences nutrient and energy dynamics, mediating transfers between SOC fractions (Haynes, 2008). Microbial biomass responds to short-term environmental changes, with the potential for long-term assessments (Rinnan et al. 2008).

Enzyme Activity Soil enzyme activities reflect nutrient cycling and microbial status. Enzyme activities integrate information on microbial and physicochemical soil conditions (Aon et al. 2001). Climate change factors like elevated CO_2 and atmospheric N deposition may impact microbial enzyme activities, affecting organic C turnover and nutrient cycling (Dorodnikov et al., 2009).

These indicators play essential roles in assessing soil health and are sensitive to changes in climate conditions, including temperature, rainfall patterns, and extreme weather events, although the direct impact on some properties may be gradual or indirect. Understanding these relationships is vital for developing effective climate adaptation strategies in agriculture. Improving one indicator can often lead to enhancement in other metrics of soil health and help develop a more resilient agricultural system.

12.4 Health Improvement Practices for Soil

12.4.1 Conservation Tillage

Conservation tillage, a set of soil management practices designed to minimize soil disturbance and maintain organic soil cover, plays a crucial role in sustainable agriculture and climate change mitigation. This approach includes various techniques such as zero tillage (ZT), reduced tillage, strip tillage, and no-till farming, each characterized by the degree to which the soil is disturbed and the amount of crop residue left on the field (Kumari et al. 2023). Conservation tillage practices provide several benefits over conventional tillage, but one of the most direct benefits of conservation tillage in mitigating climate change is fewer passes over the field, leading to reductions in fuel consumption and labor costs (Desouky et al. 2022). This economic benefit can make sustainable practices more attractive to farmers. In addition, conservation tillage improves soil structure, increases organic matter content, enhances microbial activity, and enhances water infiltration and retention in the soil, improving drought resilience and water use efficiency.

12.4.1.1 Climate Change Mitigation and Adaptations Opportunities

Conservation tillage practices, characterized by minimal soil disturbance and the retention of crop residues on the field, are increasingly recognized for their substantial contributions to climate change mitigation and adaptation. Through a range of studies across diverse cropping systems and geographical locations, the effectiveness of conservation tillage in reducing soil erosion, improving water retention, and enhancing water use efficiency has been well documented. These practices play a critical role in creating more resilient agricultural systems capable of withstanding the challenges posed by climate change. Conservation tillage significantly mitigates soil erosion, a critical aspect of soil health that is directly impacted by climate change. Busari et al. (2015) observed marked reductions in soil erosion rates in no-till fields in Nigeria, highlighting the protective effect of crop residue cover. Similarly, Triplett and Dick (2008) review highlighted that reduced tillage and no-till practices effectively decreased wind and water erosion in cropping systems across South America, Australia, the United States, Canada, and Europe by maintaining soil cover and structure. Baumhardt et al. (2015) further demonstrated the efficacy of conservation tillage in the Mid-Atlantic region of the United States, showing significant reductions in runoff and soil erosion alongside improvements in water infiltration.

By minimizing soil disturbance, conservation tillage contributes to climate change mitigation by enhancing soil organic carbon sequestration. Research by Gómez-Muñoz et al. (2021) in Spain and Hussain et al. (2021) in Pakistan highlighted the potential of conservation tillage to enhance carbon sequestration. Gómez-Muñoz et al., observed an increase in soil organic carbon stocks in olive orchards practicing reduced tillage, while Hussain et al., reported that conservation tillage in wheat fields led to higher soil organic matter accumulation, acting as a carbon sink and mitigating greenhouse gas emissions. Dubey et al. (2021) focused on the carbon sequestration potential of conservation tillage in temperate climates. Their longitudinal study revealed that conservation tillage practices, particularly no-till and reduced tillage, significantly increased soil organic carbon stocks over time. The accumulation of organic matter not only sequesters carbon, thereby mitigating greenhouse gas emissions but also contributes to the overall fertility and health of the soil. Studies comparing conservation tillage practices with conventional tillage have found significant benefits. For instance, a field study by Yadav et al. (2020) in India compared conservation tillage practices with conventional tillage and found significant benefits in terms of soil structure and crop yield. Sadiq et al. (2021) examined the long-term effects of conservation tillage on crop yield and soil structure in a tropical setting. Their findings indicate that conservation tillage leads to an improvement in soil physical properties, including bulk density, porosity, and water infiltration rates. These changes result in a more favorable growing environment for crops, evidenced by increased yields and reduced vulnerability to erosion and compaction. The study revealed that conservation tillage increased the percentage of water-stable macro-aggregates, enhancing soil health and promoting sustainable crop production. () conducted research in Mediterranean environments, showing that conservation tillage, including reduced tillage and crop residue retention, improved soil physical, chemical, and biological properties. These improvements contributed to increased soil health and agroecosystem resilience, making cropping systems more adaptable to climate variability and change.

The impact of conservation tillage on water retention and use efficiency is particularly relevant under the current challenges of climate variability and water scarcity. The presence of organic cover increases soil porosity and aggregate stability, enhancing the soil's water retention capacity and reducing evaporation rates. Bescansa et al. (2006) conducted a study in Spain, illustrating how conservation tillage increased soil water retention through improved soil structure and porosity. Li et al. (2022) further supported these findings with their research in the Loess Plateau of China, showing that the implementation of conservation tillage practices significantly enhanced soil moisture content by reducing evaporation and increasing water infiltration rates. Dinkar et al., 2024 found that rainfed cropping systems under conservation tillage retained more moisture, reducing evaporation rates and enhancing water infiltration. In dryland farming systems, Smith et al. (2019) showed that the residue cover minimized water loss, making more water available for crops, a crucial adaptation strategy in water-scarce environments. Zhang et al. (2008) observed improved water use efficiency in irrigated systems under no-till practices, highlighting the benefits of reduced soil disturbance in preserving soil structure, enhancing water infiltration, and reducing the need for irrigation. Avila-Salem et al. (2023) explored the impact of conservation tillage on soil water retention in arid environments. By comparing fields managed under conventional tillage with those under conservation tillage, they found that the latter maintained higher soil moisture levels throughout the growing season. This result is attributed to the improved soil structure and increased organic matter content, which enhances the soil's capacity to hold water, a critical factor in drought-prone areas.

In addition, conservation tillage has been shown to improve soil biodiversity. Sharma et al. (2023) through comprehensive analysis of the effects of conservation tillage on soil biodiversity and nutrient cycling, situated in a semi-arid region, demonstrated a significant increase in microbial diversity and enzymatic activity within soils managed under conservation tillage practices. This enhancement in soil life contributes to more robust nutrient cycling and improved soil structure, leading to healthier crops and reduced need for chemical fertilizers.

12.4.1.2 Considerations and Challenges

The transition to conservation tillage practices, while beneficial for climate change mitigation and adaptation, encompasses several agronomic, environmental, and socio-economic challenges that must be addressed to realize its full potential. The shift away from traditional tillage can lead to increased weed pressure due to reduced soil disturbance. The reliance on herbicides for weed control might increase, raising concerns about environmental impacts and resistance development (Vencill et al., 2012). Developing integrated weed management strategies that are compatible with conservation tillage is essential for sustainable crop production (Bajwa, 2014). Retaining crop residues on the field may also create habitats conducive to certain pests and diseases, potentially affecting crop health and yields (Fanadzo et al. 2010). Monitoring and managing pest and disease risks are crucial to maintaining the benefits of conservation tillage (Holland, 2004).

In addition, the rate of carbon sequestration under conservation tillage can vary, influenced by factors such as soil type, climate, and cropping system (Hussain et al., 2021). Understanding these dynamics is vital for accurately assessing the climate mitigation potential across different agricultural landscapes. Also, increased use of nitrogen fertilizers in conservation tillage systems can elevate nitrous oxide emissions, a potent greenhouse gas, offsetting some of the climate benefits (Huang et al. 2004). Optimizing nitrogen management to minimize emissions is a significant consideration. A comprehensive research approach is essential to address the challenges associated with conservation tillage. Collaborative efforts can lead to the development of effective management strategies, tailor practices to local conditions, and promote the adoption of conservation tillage (Costantini et al., 2020).

12.4.2 Cover Cropping

Cover cropping stands as a foundational practice in sustainable agriculture, heralded for its significant contributions to mitigating and adapting to climate change. The integration of diverse plants, including legumes, grasses, and other green manures within crop rotation cycles, is strategically designed to improve soil health, enhance biodiversity, and boost crop productivity. Bourgeois et al. (2022) have documented how these practices not only improve soil conditions but also increase above- and below-ground biodiversity, leading to more resilient ecosystems. Moreover, Teasdale et al. (2007) have highlighted the positive impact of cover cropping on overall crop productivity, showcasing the multifaceted benefits of these strategies. Moreover, cover crops serve multiple ecological roles, including improving soil structure (Clark, 2007), fixing atmospheric nitrogen (Peoples et al., 2009), suppressing weeds (Nichols et al. 2020), and providing habitats for beneficial insects (Bottenberg et al. 1999). Additionally, Fernandez et al. (2016) have emphasized how cover cropping enhances bacterial functional diversity, further benefiting nutrient cycling and plant growth. Through these actions, cover cropping contributes significantly to both the mitigation of climate change effects and the adaptation of agricultural systems to evolving environmental conditions (Tonitto et al., 2006).

12.4.2.1 Climate Change Mitigation and Adaptations Opportunities

One of the primary ways cover cropping contributes to climate change mitigation is through carbon sequestration. The addition of organic matter from cover crops to the soil captures carbon dioxide from the atmosphere and stores it in the soil as carbon, thereby reducing greenhouse gas concentrations. Poeplau and Don (2015) conducted a comprehensive meta-analysis revealing that the practice of cover cropping leads to significant increases in soil organic carbon stocks. This study underscores the potential of cover crops to contribute to climate change mitigation by enhancing the soil's carbon storage capacity. West and Post (2002) provided early evidence of the benefits of agricultural practices, including cover cropping, on carbon sequestration. Their analysis suggested that changes in management practices have the potential to significantly increase soil carbon levels, thereby contributing to the reduction of atmospheric CO_2 levels. Lal (2004) emphasizes the role of soil carbon sequestration in mitigating climate change, highlighting that practices like cover cropping can enhance soil organic carbon (SOC) and reduce the overall carbon footprint of agricultural activities. By enhancing the capacity of soils to sequester carbon, cover cropping not only contributes to the mitigation of climate change but also improves soil health and agricultural sustainability.

Cover cropping also plays a critical role in nutrient management by reducing nitrogen losses in agricultural fields. By leveraging the natural processes of both leguminous and nonleguminous cover crops, these practices offer a dual benefit: reducing the dependency on synthetic fertilizers—known for their high greenhouse gas emissions—and enhancing soil health, thereby contributing to a more resilient agricultural ecosystem. For example, Tonitto et al. (2006) examined the role of cover crops in nitrogen cycling and their potential to reduce nitrate leaching. Their findings suggest that cover crops can significantly mitigate nitrogen losses, thereby reducing the environmental impact of agricultural nitrogen use. Schipanski et al. (2014) explored the multifunctional benefits of cover crops, including their role in enhancing nitrogen retention in agricultural systems. Their research emphasizes the importance of cover crops in creating more sustainable farming practices that align with climate change mitigation efforts. Abdalla et al. (2019) conducted a meta-analysis on the effects of cover cropping on soil nitrogen dynamics. The study highlighted that cover crops, especially legumes, are effective in capturing residual nitrogen, thus preventing leaching and enhancing soil nitrogen availability for subsequent crops. However, Lacey et al. (2020) highlighted that while cover crops effectively reduce nitrogen losses, there is a competition for soil nitrogen between cover crops and subsequent cash crops, underscoring the importance of strategic management to maximize environmental benefits.

The benefits of cover cropping extend beyond mere soil enhancement; they fundamentally alter soil properties to improve water infiltration and retention, reduce erosion, and prevent soil compaction. These changes have profound implications for agricultural sustainability and resilience. Blanco-Canqui et al. (2015) conducted a study focusing on the impact of cover crops on soil physical properties. They found that cover crops significantly increase soil porosity and aggregate stability, which are vital for enhancing water infiltration and storage in the soil profile, thereby mitigating the effects of erratic rainfall patterns. Poeplau and Don (2015), in their meta-analysis, also highlighted the role of cover crops in increasing soil organic carbon stocks. Higher organic matter not only contributes to carbon sequestration but also the improvement of soil structure and water-holding capacity, offering better resistance against drought conditions. () explored the effects of cover cropping on soil moisture availability during dry periods. Their research indicates that cover crops can maintain higher levels of soil moisture, which is critical for crop survival during drought spells. Feng et al. (2021) examined the long-term effects of cover cropping on soil health and resilience. They observed that continuous use of cover crops leads to significant improvements in soil health indicators, such as increased biodiversity and nutrient cycling efficiency, which in turn enhances the system's resilience to climate change. In addition, Alliaume et al. (2014) found that cover cropping effectively reduces runoff and soil erosion under different agricultural management systems, highlighting its role in conserving soil and water resources.

12.4.2.2 Considerations and Challenges

While the benefits of cover cropping for climate change mitigation and adaptation are clear, certain challenges need to be addressed. For instance, the successful establishment of cover crops requires careful selection based on climate, soil type, and main crop rotation (Kaye & Quemada, 2017). Additionally, managing cover crops, including termination and integration into existing cropping systems, poses challenges that can impact the efficacy of these practices. Additionally, trade-offs in weed suppression and nutrient management emphasize the importance of strategic planning and integration of cover crops into broader farm management practices (Bourgeois et al., 2022). Moreover, cover crops can compete with primary crops for water, nutrients, and light if not managed properly, potentially reducing yields (Davis et al. 2012; Deines et al. 2017), while cover crops can improve soil moisture retention, they may also use water that would otherwise be available for the subsequent cash crop, particularly in water-scarce regions (Strauss et al. 1996). Understanding the water dynamics of cover cropping systems is essential for their sustainable implementation (Moore, 2023). Additionally, leguminous cover crops fix atmospheric nitrogen, which can benefit subsequent crops. However, the timing of nitrogen release may not always align with the nitrogen demand of the following cash crops, requiring careful management to avoid nitrogen losses (Nyabami et al., 2023). Therefore, strategic planning and management are necessary to optimize the benefits of cover cropping without negatively impacting main crop productivity.

12.4.3 Crop Rotation

Crop rotation is a crucial agricultural method involving the planting of different types of crops in a recurring sequence on the same plot of land over different seasons or years. This strategy is integral for breaking pest cycles (Tariq et al., 2019), improving soil health (Agomoh et al. 2021), and reducing reliance on chemical inputs (Shah et al., 2021), among other benefits. A diverse cropping system is less likely to suffer total failure under adverse weather conditions, as different crops respond differently to stress factors like drought, heat, and excess moisture. The varied root systems and organic matter inputs associated with different crops can increase soil organic carbon, a key component of soil health (Yang et al., 2020). Also, including nitrogen-fixing legumes in the rotation cycle can naturally replenish soil nitrogen levels (Kakraliya et al., 2018), diminishing greenhouse gas emissions associated with fertilizer production and application (Jensen and Hauggaard-Nielsen, 2003).

12.4.3.1 Climate Change Mitigation and Adaptations Opportunities

Crop rotations provide benefits, in terms of improving crop productivity, and soil health, and breaking pest and disease cycles. Crop rotation offers tangible benefits in the context of climate change mitigation and adaptation. The practice contributes to carbon sequestration through enhanced soil organic matter. Liu et al. (2022) explored the effects of diversifying crop rotations on agroecosystem services and resilience, finding enhanced crop productivity and sustainability. These benefits are pivotal as agriculture faces the dual challenge of contributing to and combating climate change. The carbon sequestration potential of rotated crops, especially when integrated with conservation agriculture practices, as shown by Chandra et al. (2020), leads to lower carbon footprints and higher system outputs, demonstrating a clear path toward reducing agriculture's environmental impact. Moreover, the diversification of crops within rotational systems increases agricultural resilience to extreme weather events, reducing the risk of total crop failure under adverse conditions. This resillence is particularly crucial as climate variability intensifies, threatening food security and livelihoods.

A core benefit of crop rotation is the marked improvement in soil health. Singh (2023), Yang et al. (2020), and Chahal et al. (2021) demonstrate that incorporating diverse crops, including cover crops and legumes, into rotation schemes enhances soil structure, increases organic matter content, and boosts nutrient cycling. Singh (2023) found that incorporating cover crops in rotation significantly improved soil structure and nutrient content, which are critical for subsequent crops' productivity. This aligns with Yang et al. (2020), who underscored the importance of diverse cropping systems in maintaining soil health across different agroecosystems. The introduction of nitrogen-fixing legumes within rotations, as highlighted by Chahal et al. (2021), naturally replenishes soil nitrogen levels, reducing the necessity for synthetic fertilizers and thereby cutting down on greenhouse gas emissions. These improvements are attributed to the varied root structures and organic matter inputs from different crops, which enhance soil aeration, moisture retention, and microbial diversity.

Crop rotation has been shown to significantly impact yield stability and productivity. Studies by Manns and Martin (2018) and Zhang et al. (2008) highlight the positive correlation between crop diversity and yield stability. Manns and Martin (2018) conducted a comprehensive study revealing that plant diversity, including diverse crop rotations, enhances crop yield stability by improving soil organic carbon levels. This finding is echoed in Zhang et al. (2008), where soybean–corn rotation yielded higher crop outputs and economic returns compared to continuous cropping systems in Northeast China's Corn Belt. These examples illustrate how rotational practices contribute to improved soil conditions, which in turn bolster crop productivity. Furthermore, the residual effects of crop rotations, as investigated by Khurana et al. (2007), suggest that previous crops in a rotation can significantly influence the productivity of subsequent crops, highlighting the long-term benefits of rotational planning. Rotating crops, especially with the inclusion of pulse crops or leguminous cover crops, leads to improved soil health, which in turn supports higher yields. This yield improvement is not just a result of enhanced nutrient availability but also due to the reduction in pest and disease pressures, as diverse rotations interrupt the lifecycle of many crop-specific pests and pathogens.

The strategic use of crop rotation breaks pest and disease cycles, a benefit underscored by the works of Chan and Heenan (1996) and Saprykin et al. (2021). Chan and Heenan (1996) discussed how continuous cropping deteriorates soil health and exacerbates pest pressures, whereas rotational practices mitigate these issues. Similarly, the study by Saprykin et al. (2021) on incorporating honey plants into rotations found that such diversification not only improved soil fertility but also reduced soil degradation and pest incidence by alternating crop species, especially those with different susceptibilities to pests and diseases, farmers can naturally reduce the populations of these agricultural nuisances without the heavy reliance on chemical controls. This biological pest management approach not only lowers production costs but also mitigates the environmental and health risks associated with pesticide use.

The economic and social benefits of crop rotation are evident in the increased yields and profits. He et al. (2021) reported that a potato–rice rotation in China led to higher yields, profits, and improved soil traits compared to continuous rice cropping. Yang et al. (2020) highlighted significant benefits to groundwater sustainability and economic viability, underscoring the environmental and financial advantages of crop diversification. Diversified rotations can lead to more stable income streams for farmers by providing multiple harvest opportunities and reducing the need for expensive inputs. Additionally, the practice supports rural communities by promoting sustainable land use and long-term agricultural productivity.

12.4.3.2 Considerations and Challenges

Crop rotation, celebrated for its multifaceted advantages in fostering sustainable agriculture—ranging from bolstering soil health and ensuring yield stability to mitigating pest pressures—presents a nuanced terrain of implementation challenges. According to Kaye and Quemada's seminal study in 2017, aligning crop selections with prevailing environmental and soil dynamics holds the key to optimizing rotation benefits, notably augmenting soil health and curbing pest burdens. They underscore the imperative of a nuanced comprehension of local climatic nuances, soil compositions, and crop requisites to finely calibrate resource utilization and curtail adverse ecological footprints.

Davis et al. (2012) analysis underscores the intricate orchestration demanded by crop rotation management, particularly emphasizing the requisite adjustments in farm machinery, labor allocation, and temporal alignment of agricultural practices. The review underscores that diversifying crops within rotation frameworks necessitates a comprehensive skill set, encompassing familiarity with tailored care protocols, harvesting methodologies, and postharvest handling techniques across diverse crop varieties.

Strauss et al.'s (1996) discourse delves into the intricate interplay between crop rotations, soil moisture dynamics, and water usage efficiency. While acknowledging the potential for rotations to enhance soil moisture retention, the authors caution against potential water consumption spikes by certain crops, particularly in arid regions, posing a risk to water availability for subsequent cycles. Moreover, Deines et al. (2017) investigation explores the conundrum of synchronizing nitrogen release from leguminous cover crops with the nitrogen requisites of ensuing cash crops. This synchronization emerges as pivotal in averting nitrogen losses and ensuring timely nutrient availability to meet crop demands effectively.

In a parallel vein, Baldos et al.'s 2021 inquiry delves into the economic ramifications underpinning the adoption of crop rotation and diversification strategies. Their findings unveil a dichotomy wherein the enduring benefits of crop rotation, such as amplified soil health and yield consistency, stand juxtaposed against pronounced initial investments and labor outlays. These encompass procurement expenses for novel seeds, customization or acquisition of machinery tailored to diverse crops, and potential requisites for augmented labor or expertise to navigate the intricacies of a more diverse cropping regime.

Despite these challenges, the manifest advantages of crop rotation for fostering agricultural sustainability remain unequivocal. However, surmounting these hurdles mandates strategic foresight, investments in knowledge dissemination, infrastructural enhancements, and perpetual innovation in agricultural methodologies. Effectively addressing these imperatives holds the key to unlocking the full potential of crop rotation in fortifying agricultural sustainability and resilience against future challenges.

12.4.4 Precision Agriculture

Precision agriculture holds significant potential in both mitigating and adapting to climate change, offering multifaceted opportunities to address the challenges posed by shifting environmental conditions. By optimizing resource use and minimizing inputs, such as water, fertilizers, and pesticides, precision agriculture contributes to reducing greenhouse gas emissions associated with conventional farming practices (Sarkar 2020, 2021). Additionally, precise monitoring and management of crop health and soil conditions enable farmers to adapt to changing climate patterns, such as altered precipitation levels and temperature fluctuations (Chapu et al., 2022; Baath et al., 2023). Furthermore, the integration of advanced technologies, including remote sensing and data analytics, allows for early detection of climate-related risks, such as droughts, floods, and pest outbreaks, facilitating proactive decision-making and risk mitigation strategies (Balota and Sarkar, 2020; Bennett et al., 2021; Rai et al., 2022).

12.4.4.1 Climate Change Mitigation and Adaptations Opportunities

In the quest to tackle climate change, the agricultural sector stands on the cusp of a technological revolution, where specific precision agriculture (PA) technologies are being deployed to enhance sustainability and resilience. These technologies, ranging from soil moisture sensors to variable rate irrigation systems and from GPS-guided tractors to remote sensing drones, offer targeted solutions to the multifaceted challenges posed by a changing climate.

Soil moisture sensors and variable rate irrigation systems, as highlighted in studies by Aasen et al. (2018) and Say et al. (2017), exemplify the strides made in water conservation. These technologies enable a reduction in water use by 20–30% and up to 25%, respectively, by precisely applying water only where and when it is needed. This not only conserves water but also significantly reduces the energy footprint associated with irrigation, directly contributing to climate change mitigation by lowering greenhouse gas emissions related to energy use.

The strategic application of fertilizers through variable rate technology (VRT) is another beacon of efficiency. Research by Sarkar et al. (2025) demonstrates how VRT can tailor fertilizer application rates to the specific needs of different field zones, reducing fertilizer usage by 10–15%. This precision not only cuts down on the excess fertilizer that often leads to harmful runoff and greenhouse gas emissions but also ensures that crops receive an optimal nutrient supply, enhancing yield without the environmental toll.

The role of GPS and sensor technologies in pesticide management is critical for environmental sustainability (Sarkar and Jha, 2020). Christensen et al. (2009) and Pedersen et al. (2004) showcase how these technologies facilitate a reduction in herbicide use by up to 25% by enabling farmers to target pesticide applications precisely where weed infestations occur. This targeted approach minimizes the ecological impact of farming, reducing pesticide runoff into waterways and lowering the overall chemical load on the environment.

Precision seeding technologies, which optimize planting depth and spacing according to detailed soil and environmental data, are at the forefront of seed use optimization. Studies like Lowenberg-DeBoer and Erickson (2020) reveal that such technologies can lead to more efficient seed use by 5–10%, enhancing crop emergence and uniformity. This not only conserves valuable seed resources but also promotes higher yield stability, an essential factor for farming in the face of climate uncertainty.

Remote sensing, including the use of UAVs (Unmanned Aerial Vehicles) for monitoring crop health and environmental conditions, plays a pivotal role in enabling data-driven decision-making (Sarkar et al., 2020, 2021). Torres-Sánchez et al. (2018) demonstrate how UAV-based remote sensing can identify variability in tree health and guide the targeted application of treatments, resulting in a 15% increase in yield and improvements in fruit quality. Such technologies allow for the early detection of climate-related risks like droughts and pest outbreaks, empowering farmers with the information needed to implement proactive strategies (Sarkar et al., 2021). Thermal imaging and remote sensing technologies, as discussed by Rains et al. (2011), offer early warning systems for water stress and disease, reducing disease incidence by up to 20%. By providing timely data, these technologies enable targeted interventions, minimizing potential yield losses and enhancing the adaptability of crops to climate variability (Sarkar et al., 2022a, b).

PA also facilitates data-driven decision-making and operational efficiency, as noted by Griffin et al. (2007), who reported an average input savings of 20% and a 15% increase in farm operational efficiency, respectively. These advancements enable farmers to respond more agilely to climate-induced risks, such as unpredictable weather patterns and emerging pest pressures, enhancing the resilience of agricultural systems (Adhikari et al., 2023).

12.4.4.2 Considerations and Challenges

While precision agriculture (PA) offers transformative potential for enhancing sustainability and resilience in farming systems, its adoption is not without challenges. These considerations span technological, economic, and social domains, each requiring careful attention to fully realize the benefits of PA. One of the primary technological challenges in PA is the high initial cost and complexity of implementing advanced systems. Schimmelpfennig (2016) discusses the economic barriers to adopting PA technologies, noting that the upfront costs for equipment and software can be prohibitive for many farmers, particularly those in developing countries or operating small-scale farms. Furthermore, the complexity of PA systems requires a significant knowledge base and technical proficiency, which can be a substantial barrier to entry. The economic considerations extend beyond the initial investment to include the ongoing costs associated with data management and analysis. Griffin et al. (2007) highlight the need for robust decision support systems to process and analyze the vast amounts of data generated by PA technologies. The development and maintenance of such systems represent additional costs that can impact the overall economic viability of PA for some farmers.

While PA aims to enhance environmental sustainability, there are considerations regarding the potential unintended consequences of technology use. For instance, Robertson et al. (2000) caution against the over-reliance on technological solutions that may lead to increased energy consumption and greenhouse gas emissions, counteracting some of the environmental benefits of PA. The authors argue for a balanced approach that considers the environmental footprint of manufacturing, operating, and disposing of PA technologies. In addition, social challenges can present barriers, including the digital divide and the need for training and education. Azevedo et al. (2019) discuss the importance of addressing the digital divide in agriculture, emphasizing that access to and familiarity with digital technologies vary widely among farmers. This gap can prevent equitable access to PA technologies, particularly for older farmers or those in less developed regions. Overcoming this challenge requires targeted educational programs and support services to build digital literacy and technical skills within the farming community.

12.4.5 Integrated Nutrient Management

Integrated Nutrient Management (INM) stands as a critical strategy in the agricultural sector's efforts to both mitigate and adapt to climate change. This approach, which harmonizes the use of organic, inorganic, and biological nutrients within an ecologically and economically sustainable framework, aims to not only boost crop productivity but also to enhance environmental sustainability. Janssen (1998) introduced this concept, emphasizing the need for synchronizing nutrient supply with crop demand to minimize losses through leaching, runoff, volatilization, and emissions, thereby significantly reducing the greenhouse gas emissions linked with conventional fertilizer usage and enhancing the resilience of agricultural ecosystems to climate variabilities.

12.4.5.1 Climate Change Mitigation and Adaptation Through INM

INM's role in mitigating climate change is multifaceted. By enhancing nutrient-use efficiency and curbing reliance on chemical fertilizers, INM directly cuts down the emissions of potent greenhouse gases such as nitrous oxide. The addition of organic matter to soils not only boosts carbon sequestration but also ameliorates soil structure and fertility, contributing to a decrease in carbon dioxide emissions. Gruhn et al. (2000) stress the importance of fine-tuning nutrient cycles to prevent environmental nitrogen losses while achieving optimal efficiency. Innovative techniques within the INM framework, such as the deep placement of fertilizers and the adoption of urea coatings (Zhang et al., 2012), aim to curtail nutrient losses and bolster plant uptake, thereby fostering the environmental sustainability of agricultural practices.

Lal (2015) underscores the vital role of soil health in agricultural resilience, indicating that enhancing soil's physical, chemical, biological, and hydrological properties significantly contributes to farm productivity and sustainability. By incorporating organic amendments like compost and manure with chemical fertilizers, INM not only enriches soil organic matter but also improves soil structure and water retention, key for boosting soil resilience against erosion and degradation. Furthermore, the diversification of nutrient sources, including organic amendments, green manures, and cover crops, stabilizes and enriches the soil ecosystem, as highlighted by Drinkwater et al. (1998), thus making it more resilient to climate variability.

Cassman et al. (2002) emphasize the critical need to align nutrient application with crop nutrient demands to optimize nutrient use efficiency and minimize environmental losses. Practices such as split applications of nitrogen fertilizer and the utilization of slow-release formulations ensure that nutrients are available to crops at crucial growth stages, enhancing their resilience to drought and pest pressures.

12.4.5.2 Challenges in Implementing INM

Implementing INM to combat climate change entails overcoming numerous challenges, spanning knowledge dissemination, economic constraints, and policy support. Kiptot et al. (2007) highlight the knowledge and training gap among farmers and agricultural extension services as a significant barrier. The complexity of INM necessitates specialized training to understand soil nutrient dynamics and the interaction between different nutrient sources.

Economic viability poses another hurdle, especially where organic inputs are costlier than chemical fertilizers. Behera and Panda (2009) discuss the economic challenges of adopting organic farming practices integral to INM, pointing out that initial investments and the transition period can be financially daunting for farmers. Moreover, the lack of supportive policy and regulatory frameworks significantly hinders INM adoption. Pretty and Bharucha (2014) call for policy shifts to support integrated approaches like INM, advocating for policies that provide financial incentives for sustainable practices and invest in research and extension services to spread knowledge on INM.

12.4.5.3 Biofertilizers and Soil Inoculants

Biofertilizers and soil inoculants represent an innovative and sustainable approach within the agricultural sector, aiming to enhance soil fertility and plant health while mitigating the environmental impacts associated with synthetic fertilizers (Meena et al., 2015; Singh & Jha, 2017). These biologically based solutions introduce beneficial microorganisms into the soil, which facilitate nutrient solubilization, fixation, and mineralization, thereby improving nutrient availability to crops (Sashidhar & Podile, 2010). This method aligns with the broader goals of sustainable agriculture by reducing chemical inputs, enhancing soil biodiversity, and contributing to climate change mitigation and adaptation.

12.4.5.4 Climate Change Mitigation and Adaptation Opportunities

The integration of biofertilizers and soil inoculants into agricultural systems is gaining traction as a potent strategy for mitigating and adapting to climate change. Through the enhancement of crop productivity, resource use efficiency, and soil health, these biological amendments stand at the forefront of sustainable agricultural practices.

The study by Adesemoye et al. (2009) serves as a prime example of how biofertilizers can elevate crop yields, demonstrating a notable 14% increase in soybean pod number and an 18% rise in total biomass. This yield enhancement, attributed to the synergistic effects of co-inoculating with *Bradyrhizobium japonicum* and *Bacillus* spp., underscores biofertilizers' capability to bolster nutrient availability and plant health, directly contributing to climate change mitigation by potentially reducing the need for synthetic fertilizers and the associated greenhouse gas emissions.

Vessey (2003) elucidates the significant role of plant growth-promoting rhizobacteria (PGPR) in enhancing water and nutrient uptake efficiency. This review highlights how PGPR can improve water use efficiency, especially under drought conditions, by augmenting root water uptake capabilities. Furthermore, the review notes up to a 20% reduction in nitrogen fertilizer requirements due to PGPR inoculation, showcasing biofertilizers' potential to optimize resource use efficiency and adapt to climate variability. Yanni et al. (2001) investigated the effects of inoculating rice with diazotrophic rhizobacteria, documenting a 25% increase in root biomass. This improvement in root development not only facilitates better water and nutrient uptake but also stabilizes soil structure, crucial for enhancing agricultural resilience to extreme weather events like floods and droughts, thereby aiding in climate change adaptation. Canbolat et al. (2006) explored the impact of biofertilizers on soil health in wheat cultivation, finding a 30% increase in available phosphorus and a 40% rise in soil enzyme activities. These enhancements in soil microbial activity and nutrient content play a pivotal role in sustainable nutrient cycling and soil fertility, essential for maintaining productive soils in the face of changing climate conditions. A study by Bhatia et al. (2013) highlighted the role of *Azospirillum brasilense* in reducing methane emissions from rice fields by 20%, attributed to enhanced plant root oxygen release. This alteration in soil redox conditions underscores the potential of biofertilizers and soil inoculants to contribute to climate change mitigation by reducing greenhouse gas emissions from agricultural soils.

In addition, recent research demonstrates that biofertilizers and soil inoculants can significantly reduce pest and disease incidence in various crops by enhancing plant health and soil quality. For example, a study by Singh et al. (2020) found that inoculation with rhizobacteria-based biofertilizers led to a notable decrease in the occurrence of root-knot nematodes in tomato crops, attributed to the biofertilizers' ability to improve plant resistance mechanisms. Similarly, research by Gomez et al. (2021) highlighted the effectiveness of Trichoderma spp. as a soil inoculant in suppressing Fusarium wilt in banana plants, with results showing a reduction in disease severity and an increase in plant growth parameters. These findings are supported by Zhou and Wu (2018), who reported that the application of mycorrhizal fungi in corn fields enhanced nutrient uptake and reduced susceptibility to pest attacks, leading to higher yields and improved crop health.

12.4.5.5 Considerations and Challenges

The adoption of biofertilizers and soil inoculants, while offering substantial benefits for climate change mitigation and adaptation, also faces several challenges and considerations. These challenges can hinder the widespread implementation of these sustainable agricultural practices, impacting their effectiveness in contributing to climate resilience and sustainability goals.

A fundamental challenge in the broader adoption of biofertilizers and soil inoculants is the knowledge and awareness gap among farmers and agricultural stakeholders. Despite the proven benefits, a lack of understanding about the application, effectiveness, and potential return on investment of biofertilizers can limit their use. Farmers may be hesitant to switch from conventional practices due to uncertainties about the outcomes or the perception of biofertilizers as less effective than chemical alternatives. Enhanced extension services and education programs are crucial to bridge this gap, as emphasized by Kiptot et al. (2007), who advocate for participatory approaches like farmer field schools to improve farmers' knowledge about integrated soil fertility management.

The variability in the quality and consistency of biofertilizer products is a significant challenge, affecting their effectiveness and farmers' confidence. Issues with microbial content, viability, and formulation can yield inconsistent field results. Quality control and standardization are essential for the successful adoption of these products. Herridge et al. (2008) discuss the importance of ensuring the efficacy of biofertilizers, noting that rigorous testing and validation are crucial for gaining farmers' trust and encouraging wider use. Regulatory frameworks and certification standards could play a pivotal role in maintaining product quality and reliability. Moreover, the initial cost and perceived risk of transitioning to biofertilizer-based systems pose significant economic challenges for many farmers, particularly smallholders. While biofertilizers can be cost-effective in the long term due to reduced chemical inputs and improved soil health, the upfront investment and the need for new knowledge and practices can be barriers. Financial incentives, subsidies for biofertilizer products, and support for risk-sharing mechanisms could facilitate wider adoption, as discussed by Behera and Panda (2009), who highlight the economic aspects of adopting the practices.

While biofertilizers and soil inoculants present a viable strategy for sustainable agriculture in the context of climate change, addressing the challenges of knowledge dissemination, product quality, economic viability, and supportive policies are crucial. Overcoming these barriers requires concerted efforts from researchers, policymakers, industry stakeholders, and the farming community to fully realize the potential of biofertilizers and soil inoculants in contributing to climate change mitigation and adaptation.

12.5 Policy Priorities

I. *Enhancing Research and Development (R&D):* Policymakers should prioritize R&D in sustainable agricultural technologies and practices. The IPCC highlights the need for innovation in climate-resilient crop varieties and sustainable farming practices that reduce greenhouse gas emissions and adapt to changing climatic conditions. R&D investments can facilitate the development of new biofertilizers, soil inoculants, and precision agriculture technologies that optimize nutrient use efficiency and reduce environmental impacts, as demonstrated by Hochman et al. (2017), who underscore the potential of such innovations in improving water use efficiency and reducing emissions.

II. *Implementing Education and Extension Services:* Education and capacity building are critical for the adoption of sustainable agricultural practices. FAO emphasizes the role of extension services in bridging the knowledge gap among farmers, particularly in developing countries. Policies should support comprehensive education programs to train farmers in the application of biofertilizers, soil health management, and conservation practices, aligning agricultural productivity with climate resilience, as evidenced by the effectiveness of farmer field schools highlighted by Kiptot et al. (2007).

III. *Promoting Sustainable Soil and Nutrient Management:* The FAO advocates for integrated soil fertility management (ISFM) as a key strategy for enhancing soil health and agricultural productivity. Policies should encourage the adoption of INM and ISFM practices, including the use of organic amendments, biofertilizers, and precision nutrient management techniques. Subsidies for organic inputs, tax incentives for sustainable practices, and certification schemes for sustainably produced foods can incentivize farmers to adopt these practices, supported by the findings of Drinkwater et al. (1998) and Gruhn et al. (2000) on the benefits of integrated nutrient management.

IV. *Fostering Climate-Smart Agriculture (CSA):* Climate-smart agriculture, as endorsed by both the IPCC and FAO, offers a holistic approach to addressing the challenges of climate change in the agricultural sector. Policy frameworks should integrate CSA principles, promoting practices that increase productivity, enhance resilience, and reduce emissions. This includes support for conservation agriculture, cover cropping, crop rotation, and the development of resilient crop varieties, as supported by the work of Lipper et al. (2014) and the FAO (2013) on the benefits of CSA and conservation agriculture.

V. *Strengthening Policy Integration and Coordination:* Effective policy integration across agricultural, environmental, and climate policy domains is essential. The IPCC stresses the importance of coherent policies that address climate change mitigation and adaptation in an integrated manner. National and international policies should be aligned to provide a consistent regulatory framework that supports sustainable agricultural practices and climate goals.

VI. *Enhancing Access to Financing:* Access to finance is a critical enabler for the adoption of sustainable agricultural practices. Policies should establish financial mechanisms, such as green loans, grants, and insurance products, that support farmers in adopting climate-resilient practices. The FAO advocates for financial instruments that lower the risk and increase the affordability of sustainable technologies and inputs for farmers, as illustrated by the FAO (2017) on the importance of climate finance in agriculture.

12.6 Conclusion

The exploration of soil and nutrient management practices within the context of climate change unveils a complex but hopeful vista. As underscored by the Intergovernmental Panel on Climate Change (IPCC), the exigency of addressing climate change's multifaceted impacts on agriculture cannot be overstated. The path forward necessitates a synergistic blend of innovative practices, policy support, and a deepened understanding of soil health dynamics. This chapter has traversed the landscape of emerging solutions—from conservation tillage to cover cropping, crop rotation, and the deployment of precision agriculture technologies—each harboring the potential to mitigate climate-induced adversities while bolstering agricultural resilience and productivity. Also, the evidence presented reinforces the imperative for a paradigm shift towards sustainable agricultural practices that not only ensure food security for a burgeoning global population but also enhance the environmental sustainability of farming systems. Integrated Nutrient Management (INM), biofertilizers, and soil inoculants emerge as pivotal in this transformation, offering avenues to enhance nutrient use efficiency, reduce dependency on synthetic inputs, and improve soil health—a cornerstone for agricultural sustainability.

However, the journey is fraught with challenges, including economic barriers, knowledge gaps, and the need for robust policy frameworks that incentivize sustainable practices. Addressing these challenges necessitates a concerted effort from all stakeholders—governments, the agricultural sector, research institutions, and communities. The transition towards sustainable soil and nutrient management is not merely a choice but a necessity to navigate the precarious balance between meeting global food demands and preserving the ecological integrity of our planet.

As we stand at the crossroads of climate change and agricultural sustainability, the choices made today will indelibly shape the resilience of our food systems. It is incumbent upon us to harness the insights gleaned from scientific research, technological advancements, and policy innovations to foster an agricultural landscape that is both productive and sustainable. In doing so, we pave the way for a future where agriculture not only thrives in the face of climate change but also plays a pivotal role in mitigating its impacts. The global community must collectively endeavor to implement and scale up the sustainable soil and nutrient management practices that will secure the foundation of global food security and environmental stewardship for generations to come.

Acknowledgments The authors would like to extend their heartfelt gratitude to their respective research laboratories for providing invaluable support and resources throughout the journey of working on this book chapter. The experience gained in these labs has been instrumental in enhancing our understanding of the field and has contributed significantly to the quality of this work. We are also thankful to our colleagues and mentors for their guidance and insightful discussions, which have enriched our perspectives. This book chapter would not have been possible without the collective efforts and support of all those who have contributed to our academic and professional growth.

References

Aasen, H., Burkart, A., Bolten, A., & Bareth, G. (2018). Generating 3D hyperspectral information with lightweight UAV snapshot cameras for vegetation monitoring: From camera calibration to quality assurance. *ISPRS Journal of Photogrammetry and Remote Sensing, 140*, 77–89. https://doi.org/10.1016/j.isprsjprs.2018.03.015

Abdalla, M., Osborne, B., Lanigan, G., Forristal, D., Williams, M., Smith, P., & Jones, M. B. (2019). Meta-analysis of the effects of cover cropping on soil nitrogen dynamics. *Journal of Soil and Water Conservation, 74*(1), 1–13. https://doi.org/10.2489/jswc.74.1.1

Adesemoye, A. O., Torbert, H. A., & Kloepper, J. W. (2009). Plant growth-promoting rhizobacteria allow reduced application rates of chemical fertilizers. *Microbial Ecology, 58*(4), 921–929. https://doi.org/10.1007/s00248-009-9531-y

Adhikari, S., Joshi, A., Chandra, A. K., Bharati, A., Sarkar, S., Dinkar, V., Kumar, A., & Singh, A. K. (2023). SMART plant breeding from pre-genomic to post-genomic era for developing climate-resilient cereals. In *Smart plant breeding for field crops in post-genomics era* (pp. 41–97). Springer.

Agomoh, I. V., Drury, C. F., Yang, X., Phillips, L. A., & Reynolds, W. D. (2021). Crop rotation enhances soybean yields and soil health indicators. *Soil Science Society of America Journal, 85*(4), 1185–1195.

Allen, D. E., Singh, B. P., & Dalal, R. C. (2011). Soil health indicators under climate change: A review of current knowledge. *Soil Science Society of America Journal, 75*(6), 1601. https://doi.org/10.2136/sssaj2011.0026

Alliaume, F., Rossing, W. A. H., Tittonell, P., Jorge, G., & Dogliotti, S. (2014). Reduced tillage and cover crops improve water capture and reduce erosion of fine textured soils in raised bed tomato systems. *Agriculture, Ecosystems & Environment, 183*, 127–137. https://doi.org/10.1016/j.agee.2013.10.028

Aon, M. A., Cabello, M. N., Sarena, D. E., Colaneri, A. C., Franco, M. G., Burgos, J. L., & Cortassa, S. (2001). I. Spatio-temporal patterns of soil microbial and enzymatic activities in an agricultural soil. *Applied Soil Ecology, 18*(3), 239–254.

Arias, M. E., González-Pérez, J. A., González-Vila, F. J., & Ball, A. S. (2005). Soil health: A new challenge for microbiologists and chemists. *International microbiology, 8*(1), 13–21.

Arnold, S. L., Doran, J. W., Schepers, J., Wienhold, B., Ginting, D., Amos, B., & Gomes, S. (2005). Portable probes to measure electrical conductivity and soil quality in the field. *Communications in soil science and plant analysis, 36*(15-16), 2271–2287.

Avila-Salem, L., Díaz-Pérez, J. C., Soto-Rocha, J., & Singh, R. P. (2023). Effects of conservation tillage on soil water retention and crop yield in arid environments: A review. *Agronomy for Sustainable Development, 43*(2), 34. https://doi.org/10.1007/s13593-023-00789-x

Azevedo, V., Strikis, N. M., Santos, R. A., de Souza, J. G., Ampuero, A., Cruz, F. W., ... & Edwards, R. L. (2019). Medieval Climate Variability in the eastern Amazon-Cerrado regions and its archeological implications. *Scientific reports, 9*(1), 20306.

Baath, G. S., Sarkar, S., Northup, B. K., Sapkota, B. R., Gowda, P. H., & Flynn, K. C. (2023). Summer pulses as sources of green manure and soil cover in the US Southern Great Plains. *Crop and Environment, 2*(2), 66–74.

Bajwa, A. A. (2014). Weed management in conservation agriculture systems. *Weed Science, 62*(3), 491–506. https://doi.org/10.1614/WS-D-13-00164.1

Balota, M., & Sarkar, S. (2020, July). *Transpiration of Peanut in the field under Rainfed production*. Paper Presented at the American Peanut Research and Education Society Annual Meeting, Virtual.

Bardgett, R. D., Mommer, L., & De Vries, F. T. (2008). Going underground: Root traits as drivers of ecosystem processes. *Trends in Ecology & Evolution, 29*(12), 692–699. https://doi.org/10.1016/j.tree.2014.10.006

Baumhardt, R. L., Stewart, B. A., & Sainju, U. M. (2015). North American soil conservation practices. *Advances in Agronomy, 133*, 69–118. https://doi.org/10.1016/bs.agron.2015.06.003

Bebber, D. P., Ramotowski, M. A. T., & Gurr, S. J. (2015). Crop pests and pathogens move polewards in a warming world. *Nature Climate Change, 3*(11), 985–988. https://doi.org/10.1038/nclimate1990

Behera, U. K., & Panda, R. K. (2009). Economic analysis of organic and inorganic fertilizer use in rainfed rice systems of India. *Experimental Agriculture, 45*(1), 1–22. https://doi.org/10.1017/S001447970800697X

Bennett, R., Chamberlin, K., Morningweg, D., Wang, N., Sarkar, S., Balota, M., Burow, M., Chagoya, J., & Pham, H. (2021, July). *Response to drought stress in a subset of the US peanut mini-core evaluated in three states*. Paper Presented at the American Peanut Research and Education Society Annual Meeting, Virtual.

Bescansa, P., Imaz, M. J., Virto, I., Enrique, A., & Hoogmoed, W. B. (2006). Soil water retention as affected by tillage and residue management in semiarid Spain. *Soil and Tillage Research, 87*(1), 19–27. https://doi.org/10.1016/j.still.2005.02.028

Bhatia, A., Sasmal, S., Jain, N., Pathak, H., Kumar, R., & Singh, A. (2013). Mitigating nitrous oxide and methane emissions from soil with biochar and Azospirillum brasilense. *Soil Science Society of America Journal, 77*(2), 672. https://doi.org/10.2136/sssaj2012.0188

Birkás, M., Jolánkai, M., Gyuricza, C., & Percze, A. (2009). Tillage effects on compaction, earthworms and other soil quality indicators in Hungary. *Soil and Tillage Research, 104*(2), 145–155. https://doi.org/10.1016/j.still.2009.02.005

Blanco-Canqui, H., Shaver, T. M., Lindquist, J. L., Shapiro, C. A., Elmore, R. W., Francis, C. A., & Hergert, G. W. (2015). Cover crops and ecosystem services: Insights from studies in temperate soils. *Agronomy Journal, 107*(6), 2449–2474. https://doi.org/10.2134/agronj15.0086

Bloomfield, J. P., Williams, R. J., Gooddy, D. C., Cape, J. N., & Guha, P. (2006). Impacts of climate change on the fate and behaviour of pesticides in surface and groundwater—A UK perspective. *Science of the Total Environment, 369*(1–3), 163–177. https://doi.org/10.1016/j.scitotenv.2006.05.019

Bottenberg, H., Frantz, G., & Mellinger, H. C. (1999). Refuge and cover crop plantings for beneficial insect habitats. In *Proceedings-Florida State Horticultural Society* (Vol. 112, pp. 339–341).

Bourgeois, B., Munoz, F., Fried, G., Mahaut, L., Armengot, L., Denelle, P., Storkey, J., Gaba, S., & Violle, C. (2022). What makes a weed a weed? A large-scale evaluation of arable weeds through a functional lens. *American Journal of Botany, 109*(1), 140–151. https://doi.org/10.1002/ajb2.1597

Box, J. E., & Bruce, R. C. (1996). Tiller density and yield: Influence of soil water and nitrogen. *Crop Science, 36*(4), 905–909. https://doi.org/10.2135/cropsci1996.0011183X003600040012x

Bronick, C. J., & Lal, R. (2005). Soil structure and management: A review. *Geoderma, 124*(1–2), 3–22. https://doi.org/10.1016/j.geoderma.2004.03.005

Busari, M. A., Kukal, S. S., Kaur, A., Bhatt, R., & Dulazi, A. A. (2015). Conservation tillage impacts on soil, crop and the environment. *International Soil and Water Conservation Research, 3*(2), 119–129. https://doi.org/10.1016/j.iswcr.2015.05.002

Canbolat, M. Y., Bilen, S., Çakmakçı, R., Şahin, F., & Aydın, A. (2006). Effect of plant growth promoting bacteria (PGPB) on yield, growth and nutrient contents in organically growing raspberry. *Scientia Horticulturae, 111*(1), 38–43. https://doi.org/10.1016/j.scienta.2006.09.002

Cassman, K. G., Dobermann, A., & Walters, D. T. (2002). Agroecosystems, nitrogen-use efficiency, and nitrogen management. *Ambio, 31*(2), 132–140. https://doi.org/10.1579/0044-7447-31.2.132

Chahal, G. S., Jordan, D. L., York, A. C., Brandenburg, R. L., Burton, J. D., Danehower, D., & Monaco, T. J. (2021). Impact of crop rotation and tillage on weed seedbanks and yields in a long-term cropping systems experiment. *Weed Science, 69*(1), 123–132. https://doi.org/10.1017/wsc.2020.73

Chan, K. Y., & Heenan, D. P. (1996). The influence of crop rotation on soil structure and soil physical properties under conventional tillage. *Soil and Tillage Research, 37*(2-3), 113–125.

Chandra, A., Jain, R., Solomon, S., & Shrivastava, A. K. (2020). Potential of conservation agriculture for climate change mitigation in India. *Current Science, 118*(7), 1052–1060. https://doi.org/10.18520/cs/v118/i7/1052-1060

Chapu, I., Okello, D. K., Okello, R. C. O., Odong, T. L., Sarkar, S., & Balota, M. (2022). Exploration of alternative approaches to phenotyping of late leaf spot and groundnut rosette virus disease for groundnut breeding. *Frontiers in Plant Science, 13*, 912332.

Christensen, S., Sørensen, J., & Kudsk, P. (2009). Herbicide management with precision farming technology: Implications for environmental quality. *European Journal of Agronomy, 30*(3), 151–160. https://doi.org/10.1016/j.eja.2008.11.005

Clark, A. (Ed.). (2007). *Managing cover crops profitably* (3rd ed.). Sustainable Agriculture Research and Education (SARE). https://www.sare.org/publications/managing-cover-crops-profitably/

Costantini, E. A. C., Branquinho, C., Nunes, A., Schwilch, G., Stavi, I., Valdecantos, A., & Zucca, C. (2020). Soil indicators to assess the effectiveness of restoration strategies in dryland ecosystems. *Ecological Indicators, 105*, 617–629. https://doi.org/10.1016/j.ecolind.2019.05.109

Dalal, R. C., & Moloney, D. (2000). Soil physical fertility and soil structure. In M. E. Sumner (Ed.), *Handbook of soil science* (pp. E89–E131). CRC Press.

Dang, Y. P., Dalal, R. C., Buck, S. R., Harms, B., Kelly, R., Hochman, Z., Schwenke, G. D., Biggs, A. J. W., Ferguson, N. J., Norrish, S., Routley, R., McDonald, M., Hall, C., Singh, D. K., Daniells, I. G., Farquharson, R., Manning, W., Speirs, S. D., Grewal, H. S., et al. (2008). Subsoil constraints to crop production on clay soils in North-Eastern Australia: A review. *Australian Journal of Soil Research, 46*(4), 265–279. https://doi.org/10.1071/SR07184

Davidson, E. A., & Janssens, I. A. (2006). Temperature sensitivity of soil carbon decomposition and feedbacks to climate change. *Nature, 440*(7081), 165–173. https://doi.org/10.1038/nature04514

Davis, A. S., Hill, J. D., Chase, C. A., Johanns, A. M., & Liebman, M. (2012). Increasing cropping system diversity balances productivity, profitability and environmental health. *PLoS One, 7*(10), e47149. https://doi.org/10.1371/journal.pone.0047149

Deines, J. M., Kendall, A. D., & Hyndman, D. W. (2017). Annual irrigation dynamics in the U.S. Northern High Plains derived from Landsat satellite data. *Geophysical Research Letters, 44*(18), 9350–9360. https://doi.org/10.1002/2017GL074071

Delcour, I., Spanoghe, P., & Uyttendaele, M. (2015). Literature review: Impact of climate change on pesticide use. *Food Research International, 68*, 7–15. https://doi.org/10.1016/j.foodres.2014.09.030

Desouky, H. M., Farag, A. A., Belal, A. A., & Abdel-Rahman, G. N. (2022). Precision farming: A new approach towards sustainable agriculture. *Alexandria Science Exchange Journal, 39*(1), 311–322. https://doi.org/10.21608/asejaiqjsae.2018.25658

Dinkar, V., Sarkar, S., Pandey, S., & Antred, S. H. (2024). Plant stress phenotyping: Current status and future prospects. *Advances in Agronomy, 188*, 247.

Dong, H., Kong, X., Wang, W., & Song, Q. (2021). Heat stress impairs wheat (Triticum aestivum L.) seedlings growth and photosynthesis through inhibition of nitrate reductase and ATP synthase protein expression. *Plant Physiology and Biochemistry, 148*, 263–270. https://doi.org/10.1016/j.plaphy.2020.01.006

Doran, J. W., Sarrantonio, M., & Liebig, M. A. (1996). Soil health and sustainability. *Advances in Agronomy, 56*, 1–54. https://doi.org/10.1016/S0065-2113(08)60178-9

Dorodnikov, M., Knorr, K. H., Kuzyakov, Y., & Wilmking, M. (2009). Plant-mediated CH4 transport and contribution of photosynthates to methanogenesis at a boreal mire: A 14C pulse-labeling study. *Biogeosciences, 6*(12), 3225–3233. https://doi.org/10.5194/bg-6-3225-2009

Drinkwater, L. E., Wagoner, P., & Sarrantonio, M. (1998). Legume-based cropping systems have reduced carbon and nitrogen losses. *Nature, 396*(6708), 262–265. https://doi.org/10.1038/24376

Dubey, A., Malla, M. A., Kumar, A., Yadav, S., Hashem, A., & Abd-Allah, E. F. (2021). Impact of climate change on agriculture and food security. *Vulnerability, Adaptation, and Mitigation Strategies, 1*, 1–19. https://doi.org/10.1007/978-3-030-37425-9_1

Edreira, J. I. R., Otegui, M. E., Uribelarrea, M., & Carcova, J. (2011). Heat stress in temperate and tropical maize hybrids: Kernel growth, water relations and assimilate availability for grain filling. *Field Crops Research, 123*(2), 204–213. https://doi.org/10.1016/j.fcr.2011.05.010

Fanadzo, M., Chiduza, C., Mnkeni, P. N. S., & Van Averbeke, W. (2010). Crop production management practices as a cause for low water productivity at Zanyokwe irrigation scheme. *Water SA, 36*(1), 1–10. https://doi.org/10.4314/wsa.v36i1.50903

Feng, G., Zhang, F., Li, X., Wang, J., Sun, J., & Rengel, Z. (2021). Sustainable crop production in response to global climate change: A review of soil fertility management strategies. *Frontiers in Plant Science, 11*, 584897. https://doi.org/10.3389/fpls.2020.584897

Fernandez, A. L., Sheaffer, C. C., Wyse, D. L., Staley, C., Gould, T. J., & Sadowsky, M. J. (2016). Structure of bacterial communities in soil following cover crop and organic fertilizer incorporation. *Applied Soil Ecology, 107*, 30–38. https://doi.org/10.1016/j.apsoil.2016.05.001

Filipović, V., Coquet, Y., Pot, V., Houot, S., & Benoit, P. (2020). Impact of 17 years of organic residue management on soil structure stability. *Agriculture, Ecosystems & Environment, 290*, 106771. https://doi.org/10.1016/j.agee.2019.106771

Gómez-Muñoz, B., Hatch, D. J., Bol, R., García-Ruiz, R., & Gallardo, A. (2021). The role of cover crops in mitigating N2O emissions from Mediterranean agroecosystems: A review. *European Journal of Soil Science, 72*(1), 32–49. https://doi.org/10.1111/ejss.12998

Gregorich, E. G., Carter, M. R., Angers, D. A., Monreal, C. M., & Ellert, B. H. (1994). Towards a minimum data set to assess soil organic matter quality in agricultural soils. *Canadian Journal of Soil Science, 74*(4), 367–385. https://doi.org/10.4141/cjss94-051

Griffin, T. W., Lowenberg-DeBoer, J., Lambert, D. M., Peone, J., Payne, T., & Daberkow, S. G. (2007). Adoption, profitability, and making better use of precision farming data. *Agribusiness, 23*(1), 79–97. https://doi.org/10.1002/agr.20103

Gruhn, P., Goletti, F., & Yudelman, M. (2000). *Integrated nutrient management, soil fertility, and sustainable agriculture: Current issues and future challenges*. Food, Agriculture, and the Environment Discussion Paper 32. International Food Policy Research Institute.

Hatfield, J. L., Boote, K. J., Kimball, B. A., Ziska, L. H., Izaurralde, R. C., Ort, D., Thomson, A. M., & Wolfe, D. (2011). Climate impacts on agriculture: Implications for crop production. *Agronomy Journal, 103*(2), 351–370. https://doi.org/10.2134/agronj2010.0303

Haynes, R. J. (2008). Soil organic matter and soil function—Review of the literature and underlying mechanisms. *Soil Use and Management, 24*(4), 333–343. https://doi.org/10.1111/j.1475-2743.2008.00169.x

He, W., Yang, J. Y., Drury, C. F., Smith, W. N., Grant, B. B., He, P., Zhou, W., & Qian, B. (2021). Assessing the effects of climate change and tillage practices on water use, water use efficiency and yield of spring wheat in the North China plain. *Science of the Total Environment, 750*, 141630. https://doi.org/10.1016/j.scitotenv.2020.141630

Herridge, D. F., Peoples, M. B., & Boddey, R. M. (2008). Global inputs of biological nitrogen fixation in agricultural systems. *Plant and Soil, 311*(1–2), 1–18. https://doi.org/10.1007/s11104-008-9668-3

Hochman, Z., Gobbett, D. L., & Horan, H. (2017). Climate trends account for stalled wheat yields in Australia since 1990. *Global Change Biology, 23*(5), 2071–2081. https://doi.org/10.1111/gcb.13592

Holland, J. M. (2004). The environmental consequences of adopting conservation tillage in Europe: Reviewing the evidence. *Agriculture, Ecosystems & Environment, 103*(1), 1–25. https://doi.org/10.1016/j.agee.2003.12.018

Huang, Y., Zou, J., Zheng, X., Wang, Y., & Xu, X. (2004). Nitrous oxide emissions as influenced by amendment of plant residues with different C:N ratios. *Soil Biology and Biochemistry, 39*(8), 2362–2370. https://doi.org/10.1016/j.soilbio.2007.04.009

Hussain, M., Farooq, S., Hasan, W., Ul-Allah, S., Tanveer, M., Farooq, M., & Nawaz, A. (2021). Drought stress in sunflower: Physiological effects and its management through breeding and agronomic alternatives. *Agricultural Water Management, 243*, 106467. https://doi.org/10.1016/j.agwat.2020.106467

IPCC. (2021). In V. Masson-Delmotte, P. Zhai, A. Pirani, S. L. Connors, C. Péan, S. Berger, N. Caud, Y. Chen, L. Goldfarb, M. I. Gomis, M. Huang, K. Leitzell, E. Lonnoy, J. B. R. Matthews, T. K. Maycock, T. Waterfield, O. Yelekçi, R. Yu, & B. Zhou (Eds.), *Climate Change 2021: The physical science basis. Contribution of Working Group I to the Sixth Assessment Report of the Intergovernmental Panel on Climate Change*. Cambridge University Press. https://www.ipcc.ch/report/ar6/wg1/

Janzen, H. H. (2006). The soil carbon dilemma: Shall we hoard it or use it? *Soil Biology and Biochemistry, 38*(3), 419–424. https://doi.org/10.1016/j.soilbio.2005.10.008

Janssen, B. H. (1998). Efficient use of nutrients: an art of balancing. *Field Crops Research, 56*(1-2), 197–201.

Jensen, E. S., & Hauggaard-Nielsen, H. (2003). How can increased use of biological N2 fixation in agriculture benefit the environment? *Plant and Soil, 252*(1), 177–186. https://doi.org/10.1023/A:1024189029211

Kakraliya, S. K., Singh, U., Bohra, A., Choudhary, K. K., Kumar, S., & Meena, R. S. (2018). Nitrogen and legume—A meta-analysis. *Legume Research, 41*(4), 493–505. https://doi.org/10.18805/LR-377

Kaye, J. P., & Quemada, M. (2017). Using cover crops to mitigate and adapt to climate change. A review. *Agronomy for Sustainable Development, 37*(1), 4. https://doi.org/10.1007/s13593-016-0410-x

Khurana, H. S., Phillips, S. B., Bijay-Singh, Johnston, A. M., & Gupta, R. K. (2007). Performance of site-specific nutrient management for irrigated, transplanted rice in Northwest India. *Agronomy Journal, 100*(2), 493–501. https://doi.org/10.2134/agronj2007.0134

Kiptot, E., Hebinck, P., Franzel, S., & Richards, P. (2007). Adopting improved farm technology: A study of dairy farmers in Kericho and Nyandarua districts, Kenya. *Experimental Agriculture, 43*(1), 97–112. https://doi.org/10.1017/S0014479706004085

Kumari, M., Singh, V. P., Tripathi, R., & Joshi, P. K. (2023). Conservation agriculture in India: Trends, benefits, and future prospects. *Agriculture, Ecosystems & Environment, 287*, 106823. https://doi.org/10.1016/j.agee.2019.106823

Kuzyakov, Y., & Gavrichkova, O. (2010). Time lag between photosynthesis and carbon dioxide efflux from soil: A review of mechanisms and controls. *Global Change Biology, 16*(12), 3386–3406. https://doi.org/10.1111/j.1365-2486.2010.02179.x

Lacey, J., West, J. S., & Bravo, C. (2020). The impact of cover crops on disease dynamics in sustainable agriculture. *Crop Protection, 127*, 104923. https://doi.org/10.1016/j.cropro.2019.104923

Lal, R. (1995). Global soil erosion by water and carbon dynamics. In R. Lal, J. Kimble, E. Levine, & B. A. Stewart (Eds.), *Soils and global change* (pp. 131–142). CRC Press.

Lal, R. (2004). Soil carbon sequestration impacts on global climate change and food security. *Science, 304*(5677), 1623–1627. https://doi.org/10.1126/science.1097396

Lal, R. (2015). Restoring soil quality to mitigate soil degradation. *Sustainability, 7*(5), 5875–5895. https://doi.org/10.3390/su7055875

Lal, R., Follett, R. F., Stewart, B. A., & Kimble, J. M. (2007). Soil carbon sequestration to mitigate climate change and advance food security. *Soil Science, 172*(12), 943–956. https://doi.org/10.1097/ss.0b013e31815cc498

Lehmann, J., Bossio, D. A., Kögel-Knabner, I., & Rillig, M. C. (2020). The concept and future prospects of soil health. *Nature Reviews Earth & Environment, 1*(10), 544–553. https://doi.org/10.1038/s43017-020-0080-8

Li, P., Chen, J., Zhao, G., Holden, J., Liu, B., Chan, F. K. S., ... & Mu, X. (2022). Determining the drivers and rates of soil erosion on the Loess Plateau since 1901. *Science of the Total Environment, 823*, 153674.

Lipper, L., Thornton, P., Campbell, B. M., Baedeker, T., Braimoh, A., Bwalya, M., Caron, P., Cattaneo, A., Garrity, D., Henry, K., Hottle, R., Jackson, L., Jarvis, A., Kossam, F., Mann, W., McCarthy, N., Meybeck, A., Neufeldt, H., Remington, T., Sen, P. T., Sessa, R., Shula, R., Tibu, A., & Torquebiau, E. F. (2014). Climate-smart agriculture for food security. *Nature Climate Change, 4*(12), 1068–1072. https://doi.org/10.1038/nclimate2437

Liu, X., Tan, S., Song, X., Wu, X., Zhao, G., Li, S., & Liang, G. (2022). Response of soil organic carbon content to crop rotation and its controls: A global synthesis. *Agriculture, Ecosystems & Environment, 335*, 108017.

Lowenberg-DeBoer, J., & Erickson, B. (2020). Setting the record straight on precision agriculture adoption. *Agronomy Journal, 112*(4), 2569–2580. https://doi.org/10.1002/agj2.20216

Manns, H. R., & Martin, R. C. (2018). Crop diversity and stability of revenue on farms in Central Alberta, Canada. *Agricultural Systems, 159*, 82–89. https://doi.org/10.1016/j.agsy.2017.10.010

Martínez-Mena, M., Almagro, M., García-Franco, N., de Vente, J., & Boix-Fayos, C. (2020). Long-term effectiveness of check dams and reforestation for soil erosion control in a semi-arid Mediterranean catchment with badlands. *Geomorphology, 350*, 106933. https://doi.org/10.1016/j.geomorph.2019.106933

Meena, V. S., Maurya, B. R., & Verma, J. P. (2015). Does a rhizospheric microorganism enhance K+ availability in agricultural soils? *Microbiological Research, 169*(5–6), 337–347. https://doi.org/10.1016/j.micres.2013.09.003

Mueller, N. D., Gerber, J. S., Johnston, M., Ray, D. K., Ramankutty, N., & Foley, J. A. (2012). Closing yield gaps through nutrient and water management. *Nature, 490*(7419), 254–257. https://doi.org/10.1038/nature11420

Moore, E. B. (2023). Challenges and opportunities for cover crop mediated soil water use efficiency enhancements in temperate rain-fed cropping systems: a review. *Land, 12*(5), 988.

Nichols, V., English, L., Carlson, S., Gailans, S., & Liebman, M. (2020). Effects of long-term cover cropping on weed seedbanks. *Frontiers in Agronomy, 2*, 591091.

Nottingham, A. T., Meir, P., Velasquez, E., & Turner, B. L. (2022). Soil carbon loss by experimental warming in a tropical forest. *Nature, 584*(7820), 234–237. https://doi.org/10.1038/s41586-020-2566-4

Nyabami, T., Mucheru-Muna, M., Waswa, F., Mairura, F., & Mugendi, D. N. (2023). Integrating legumes to improve nitrogen use efficiency and soil health in smallholder farming systems: A review. *Agriculture, Ecosystems & Environment, 290*, 106777. https://doi.org/10.1016/j.agee.2019.106777

O'Farrell, P. J., Anderson, P. M. L., Le Maitre, D. C., & Holmes, P. M. (2010). Insights and opportunities offered by a rapid ecosystem service assessment in promoting a conservation agenda in an urban biodiversity hotspot. *Ecology and Society, 15*(3), 27. https://doi.org/10.5751/ES-03426-150327

Pariente, S. (2001). Soil salinization and land desertification. *Desertification, 10*(4), 55–66. https://doi.org/10.1007/BF02825962

Pattison, A. B., Versteeg, C., Akiew, S., & Kirkegaard, J. (2008). Resistance of Brassica species to root-knot nematode (Meloidogyne javanica) is associated with the induction of two unique enzymes. *Plant Pathology, 55*(4), 488–494.

Pedersen, S. M., Fountas, S., Have, H., & Blackmore, B. S. (2004). Agricultural robots—System analysis and economic feasibility. *Precision Agriculture, 5*(4), 295–308. https://doi.org/10.1023/B:PRAG.0000040806.39604.8b

Peoples, M. B., Brockwell, J., Herridge, D. F., Rochester, I. J., Alves, B. J. R., Urquiaga, S., Boddey, R. M., Dakora, F. D., Bhattarai, S., Maskey, S. L., Sampet, C., Rerkasem, B., Khan, D. F., Hauggaard-Nielsen, H., & Jensen, E. S. (2009). The contributions of nitrogen-fixing crop legumes to the productivity of agricultural systems. *Symbiosis, 48*(1–3), 1–17. https://doi.org/10.1007/BF03179980

Piglai, A., & De Nobili, M. (1993). Tillage effects on soil hydraulic properties and crop yield in maize (Zea mays L.)-sunflower (Helianthus annuus L.) rotation in Central Italy. *Soil and Tillage Research, 27*(4), 335–346. https://doi.org/10.1016/0167-1987(93)90058-2

Poeplau, C., & Don, A. (2015). Carbon sequestration in agricultural soils via cultivation of cover crops—A meta-analysis. *Agriculture, Ecosystems & Environment, 200*, 33–41. https://doi.org/10.1016/j.agee.2014.10.024

Porter, J. H., Parry, M. L., & Carter, T. R. (1991). The potential effects of climatic change on agricultural insect pests. *Agricultural and Forest Meteorology, 57*(1–3), 221–240. https://doi.org/10.1016/0168-1923(91)90088-8

Pretty, J., & Bharucha, Z. P. (2014). Sustainable intensification in agricultural systems. *Annals of Botany, 114*(8), 1571–1596. https://doi.org/10.1093/aob/mcu205

Raghavendra, A. S., Sage, R. F., & Monson, R. K. (Eds.). (2020). C4 photosynthesis and related CO2 concentrating mechanisms. *Springer*. https://doi.org/10.1007/978-90-481-9407-0

Rai, A., Sarkar, S., & Jha, P. K. (2022). Deficit irrigation: An optimization strategy for a sustainable agriculture. In *Soil-water, agriculture, and climate change: Exploring linkages* (pp. 163–181). Springer.

Rains, G. C., Kays, S. J., & Archer, T. L. (2011). Remote sensing and variable rate nitrogen application: Impacts on corn yield and residual soil nitrate. *Agronomy Journal, 103*(2), 396–406. https://doi.org/10.2134/agronj2010.0325

Rasul, G., Chaudhry, Q. Z., & Mahmood, A. (2011). Global warming and its impacts on the agricultural production system of Pakistan. *Contemporary South Asia, 19*(4), 443–453. https://doi.org/10.1080/09584935.2011.627190

Reth, S., Reichstein, M., & Falge, E. (2005). The effect of soil water content, soil temperature, soil pH-value and the root mass on soil CO2 efflux—A modified model. *Plant and Soil, 268*(1), 21–33. https://doi.org/10.1007/s11104-005-0175-5

Rinnan, R., Michelsen, A., & Jonasson, S. (2008). Effects of litter addition and warming on soil carbon, nutrient pools and microbial communities in a subarctic heath ecosystem. *Applied Soil Ecology, 39*(3), 271–281.

Robertson, G. P., Paul, E. A., & Harwood, R. R. (2000). Greenhouse gases in intensive agriculture: Contributions of individual gases to the radiative forcing of the atmosphere. *Science, 289*(5486), 1922–1925. https://doi.org/10.1126/science.289.5486.1922

Sadiq, M. S., Singh, B., & Demissie, H. (2021). Impact of conservation tillage on soil properties, water use and yield of sugar beet in a Mediterranean environment. *Agricultural Water Management, 243*, 106422. https://doi.org/10.1016/j.agwat.2020.106422

Salvador Sanchis, M. P., Maestre-Valero, J. F., Calatrava-Requena, J., & Martínez-Álvarez, V. (2008). Evaluation of agricultural water-saving practices for the Water Framework Directive: A case study of the Campo de Cartagena Irrigation District (Spain). *Agricultural Water Management, 95*(9), 1031–1042. https://doi.org/10.1016/j.agwat.2008.04.005

Sarkar, S. (2020). *Development of high-throughput phenotyping methods and evaluation of morphological and physiological characteristics of peanut in a sub-humid environment.* (Doctoral). Virginia Tech, Blacksburg, VA. Retrieved from https://vtechworks.lib.vt.edu/server/api/core/bitstreams/2070d20e-6d2b-4a0b-9cd2-d32f3fcb6d1d/content. Available from Virginia Tech Libraries VTechWorks database.

Sarkar, S. (2021). *High-throughput estimation of soil nutrient and residue cover: A step towards precision agriculture* (pp. 581–596). Soil Science: Fundamentals to Recent Advances.

Sarkar, S., Cazenave, A. B., Oakes, J., McCall, D., Thomason, W., Abbot, L., & Balota, M. (2020). High-throughput measurement of peanut canopy height using digital surface models. *The Plant Phenome Journal, 3*(1), e20003.

Sarkar, S., Cazenave, A.-B., Oakes, J., McCall, D., Thomason, W., Abbott, L., & Balota, M. (2021). Aerial high-throughput phenotyping of peanut leaf area index and lateral growth. *Scientific Reports, 11*(1), 21661.

Sarkar, S., & Jha, P. K. (2020). Is precision agriculture worth it? Yes, may be. *Journal of Crop Science and Biotechnology, 9*(14), 4–9.

Sarkar, S., Oakes, J., Cazenave, A.-B., Burow, M. D., Bennett, R. S., Chamberlin, K. D., Wang, N., White, M., Payton, P., & Mahan, J. (2022a). Evaluation of the US peanut germplasm mini-core collection in the Virginia-Carolina region using traditional and new high-throughput methods. *Agronomy, 12*(8), 1945.

Sarkar, S., Rai, A., & Jha, P. K. (2022b). Remote sensing and high-throughput techniques to phenotype crops for drought tolerance. In *Soil-water, agriculture, and climate change: Exploring linkages* (pp. 107–129). Springer.

Sarkar, S., Leyton, J. M. O., Noa-Yarasca, E., Adhikari, K., Hajda, C. B., & Smith, D. R. (2025). Integrating Remote Sensing and Soil Features for Enhanced Machine Learning-Based Corn Yield Prediction in the Southern US. *Sensors (Basel, Switzerland), 25*(2), 543.

Saprykin, S. V., Zolotarev, V. N., Ivanov, I. S., & Ostrikova, M. G. (2021). Creation of new varieties of comb-shaped wheat grass (Agropyron pectinatum (Bieb.) Beauv.) as a factor of increasing the efficiency of grass growing in arid conditions. In IOP Conference Series: *Earth and Environmental Science* (Vol. *663*, No. 1, p. 012029). IOP Publishing.

Sarkar, S., Ramsey, A. F., Cazenave, A.-B., & Balota, M. (2021). Peanut leaf wilting estimation from RGB color indices and logistic models. *Frontiers in Plant Science, 12*, 658621.

Sashidhar, B., & Podile, A. R. (2010). Mineral phosphate solubilization by rhizosphere bacteria and scope for manipulation of the direct oxidation pathway involving glucose dehydrogenase. *Journal of Applied Microbiology, 109*(1), 1–12. https://doi.org/10.1111/j.1365-2672.2009.04645.x

Say, S. M., Keskin, M., Sehri, M., & Sekerli, Y. E. (2017). Precision agriculture practices in fruit and vegetable production. *Journal of Agricultural Science and Technology, 19*(6), 1159–1171. http://jast.modares.ac.ir/article_16497.html

Schimmelpfennig, D. (2016). *Farm profits and adoption of precision agriculture*. Economic Research Report No. ERR-217. U.S. Department of Agriculture, Economic Research Service. https://www.ers.usda.gov/webdocs/publications/80326/err-217.pdf

Schipanski, M. E., Barbercheck, M., Douglas, M. R., Finney, D. M., Haider, K., Kaye, J. P., Kemanian, A. R., Mortensen, D. A., Ryan, M. R., Tooker, J., & White, C. (2014). A framework for evaluating ecosystem services provided by cover crops in agroecosystems. *Agricultural Systems, 125*, 12–22. https://doi.org/10.1016/j.agsy.2013.11.004

Shah, A. N., Tanveer, M., Shahzad, B., Yang, G., Fahad, S., Ali, S., Bukhari, M. A., Tung, S. A., Hafeez, A., & Souliyanonh, B. (2021). Soil compaction: A review of past and present techniques of detection and management. *Land Use Policy, 99*, 104869. https://doi.org/10.1016/j.landusepol.2020.104869

Sharma, A. K., Sharma, K., & Gaur, R. K. (2023). Biofertilizers for sustainable agriculture and environment. *Bioresource Technology Reports, 7*, 100270. https://doi.org/10.1016/j.biteb.2019.100270

Shukla, A. K., Behera, S. K., Chaudhari, S. K., & Singh, G. (2022). Fertilizer use in Indian agriculture and its impact on human health and environment. *Indian Journal of Fertilizer, 18*(3), 218–237.

Singh, A. K., & Jha, S. K. (2017). The effects of biofertilizer application on soil health and agricultural sustainability. *Sustainability, 9*(7), 1189. https://doi.org/10.3390/su9071189

Singh, S., Kang, J. S., & Mulvaney, R. L. (2020). Nitrogen management for sustainable wheat production. *Journal of Crop Improvement, 34*(6), 825–853. https://doi.org/10.1080/15427528.2020.1810382

Singh, B., Singh, Y., Ladha, J. K., Bronson, K. F., Balasubramanian, V., Singh, J., & Khind, C. S. (2002). Chlorophyll meter- and leaf color chart-based nitrogen management for rice and wheat in Northwestern India. *Agronomy Journal, 94*(4), 821–829. https://doi.org/10.2134/agronj2002.0821

Singh, B. K., Delgado-Baquerizo, M., Egidi, E., Guirado, E., Leach, J. E., Liu, H., & Trivedi, P. (2023). Climate change impacts on plant pathogens, food security and paths forward. *Nature Reviews Microbiology, 21*(10), 640–656.

Skendžić, S., Zovko, M., Živković, I. P., Lemić, D., Mikac, K. M., Igrc Barčić, J., & Bažok, R. (2021). The impact of climate change on agricultural insect pests. *Insects, 12*(5), 440. https://doi.org/10.3390/insects12050440

Smith, C. J., Hunt, J. R., Wang, E., Macdonald, B. C., Xing, H., Denmead, O. T., ... & Zhao, Z. (2019). Using fertiliser to maintain soil inorganic nitrogen can increase dryland wheat yield with little environmental cost. *Agriculture, ecosystems & environment, 286*, 106644.

Strauss, S. Y., Conner, J. K., & Rush, S. A. (1996). Foliar herbivory affects floral characters and plant attractiveness to pollinators: Implications for male and female plant fitness. *American Naturalist, 177*(3), 348–357.

Tariq, M., Wright, D., Ross, J., & Naylor, R. (2019). Global agriculture and climate change: A review of impacts, adaptation and mitigation strategies. *International Journal of Agricultural Sustainability, 17*(5), 376–392. https://doi.org/10.1080/14735903.2019.1663081

Teasdale, J. R., Coffman, C. B., & Mangum, R. W. (2007). Potential long-term benefits of no-tillage and organic cropping systems for grain production and soil improvement. *Agronomy Journal, 99*(5), 1297–1305. https://doi.org/10.2134/agronj2006.0364

Tonitto, C., David, M. B., & Drinkwater, L. E. (2006). Replacing bare fallows with cover crops in fertilizer-intensive cropping systems: A meta-analysis of crop yield and N dynamics. *Agriculture, Ecosystems & Environment, 112*(1), 58–72. https://doi.org/10.1016/j.agee.2005.07.003

Torres-Sánchez, J., López-Granados, F., De Castro, A. I., & Peña-Barragán, J. M. (2018). Configuration and specifications of an Unmanned Aerial Vehicle (UAV) for early site specific weed management. *PLoS One, 8*(3), e58210. https://doi.org/10.1371/journal.pone.0058210

Triplett, G. B., & Dick, W. A. (2008). No-tillage crop production: A revolution in agriculture! *Agronomy Journal, 100*(Supplement 3), S-153. https://doi.org/10.2134/agronj2007.0005c

United Nations, Department of Economic and Social Affairs, Population Division. (2019). World Population Prospects 2019: Highlights (ST/ESA/SER.A/423).

Vencill, W. K., Nichols, R. L., Webster, T. M., Soteres, J. K., Mallory-Smith, C., Burgos, N. R., Johnson, W. G., & McClelland, M. R. (2012). Herbicide resistance: Toward an understanding of resistance development and the impact of herbicide-resistant crops. *Weed Science, 60*(Sp1), 2–30. https://doi.org/10.1614/WS-D-11-00196.1

Vessey, J. K. (2003). Plant growth promoting rhizobacteria as biofertilizers. *Plant and Soil, 255*(2), 571–586. https://doi.org/10.1023/A:1026037216893

Wang, J., Wang, E., Feng, L., Yin, H., & Yu, W. (2018). Effects of warming and autonomous breeding on the phenological development and grain yield of double rice systems in China. *Agronomy Journal, 110*(2), 680–689. https://doi.org/10.2134/agronj2017.08.0479

Weil, R. R., & Magdoff, F. (Eds.). (2004). *Soil organic matter in sustainable agriculture*. CRC Press. https://doi.org/10.1201/9780203496374

West, T. O., & Post, W. M. (2002). Soil organic carbon sequestration rates by tillage and crop rotation: A global data analysis. *Soil Science Society of America Journal, 66*(6), 1930–1946. https://doi.org/10.2136/sssaj2002.1930

Westcott, P. C., & Jewison, M. (2013). *Weather effects on expected corn and soybean yields*.

Wixon, D. L., & Balser, T. C. (2009). Toward conceptual clarity: PLFA in soil microbial ecology. *Soil Biology and Biochemistry, 41*(7), 1475–1486. https://doi.org/10.1016/j.soilbio.2009.03.017

Yadav, G. S., Lal, R., Meena, R. S., Babu, S., Das, A., Bhowmik, S. N., Datta, M., Layak, J., & Saha, P. (2020). Conservation tillage and mulching effects on soil properties, water conservation, and crop yields in rice–wheat rotation of Eastern Himalayas, India. *Land, 9*(7), 235. https://doi.org/10.3390/land9070235

Yang, X., Chen, F., Lin, X., Liu, Z., & Zhang, H. (2020). Effects of crop rotation and tillage on soil microbial communities and diversity in a rainfed cropping system. *Soil and Tillage Research, 197*, 104503. https://doi.org/10.1016/j.still.2019.104503

Yanni, Y. G., Rizk, R. Y., Abd El-Fattah, F. K., Squartini, A., Corich, V., Giacomini, A., de Bruijn, F., Rademaker, J., Maya-Flores, J., Ostrom, P., Vega-Hernández, M., Hollingsworth, R. I., Martínez-Molina, E., Mateos, P., Velázquez, E., Wopereis, J., Triplett, E., Umali-Garcia, M., Anarna, J. A., et al. (2001). The beneficial plant growth-promoting association of rhizobium leguminosarum bv. Trifolii with rice roots. *Australian Journal of Plant Physiology, 28*(9), 845–870. https://doi.org/10.1071/PP01045

Zhang, L., van der Werf, W., Bastiaans, L., Zhang, S., Li, B., & Spiertz, J. H. J. (2008). Light interception and utilization in relay intercrops of wheat and cotton. *Field Crops Research, 144*, 212–220.

Zhang, H., Xue, Y., Wang, Z., Yang, J., & Zhang, J. (2012). Recent progress in drought and salt tolerance studies in Brassica crops. *Breeding Science, 62*(1), 29–43. https://doi.org/10.1270/jsbbs.62.29

Zhao, C., Liu, B., Piao, S., Wang, X., Lobell, D. B., Huang, Y., Huang, M., Yao, Y., Bassu, S., Ciais, P., Durand, J. L., Elliott, J., Ewert, F., Janssens, I. A., Li, T., Lin, E., Liu, Q., Martre, P., Müller, C., Peng, S., Peñuelas, J., Ruane, A. C., Wallach, D., Wang, T., Wu, D., Liu, Z., Zhu, Y., Zhu, Z., & Asseng, S. (2017). Temperature increase reduces global yields of major crops in four independent estimates. *Proceedings of the National Academy of Sciences, 114*(35), 9326–9331. https://doi.org/10.1073/pnas.1701762114

Zhou, X., & Wu, F. (2018). Dynamics of the diversity of fungal and Fusarium communities during continuous cropping of cucumber in the greenhouse. *FEMS Microbiology Ecology, 84*(3), 469–483. https://doi.org/10.1111/1574-6941.12273

Utilization of Nanotechnology for Mitigation of Drought and Heat Stress in Crop Plants

13

Alok Singh Jayara, Adarsh Dangwal, Rajeew Kumar, and Priyanka Pandey

Abstract

The looming challenge of feeding a global population projected to reach 9 billion necessitates substantial increases in food grain production. Concurrently, climate change, manifesting through increasing occurrences of drought and heat stress, poses a significant threat to agricultural productivity. Drought and heat stress are critical factors causing substantial reductions in the growth and yield of crop plants. Heat stress, defined as temperature rise beyond a threshold causing irreversible damage to crops, along with decreasing rainfall and its variability, have escalated the frequency and severity of drought events. Recent data indicate a fourfold increase in drought frequency from the 1970s to the 2010s, highlighting the urgency of this issue. The impacts of these stresses are profound, with major crops like wheat, maize, rice, and sugarcane experiencing significant yield reductions. These reductions directly threaten global food security, particularly in countries like India, where a large portion of the workforce relies on agriculture. Addressing these challenges necessitates innovative and cost-effective approaches. Traditional agronomic and breeding methods have provided some mitigation but face limitations in cost and expertise. Emerging technologies, particularly nanotechnology, offer promising solutions. Nanoparticles, due to their unique physical and chemical properties, can enhance crop productivity under stress conditions. They improve seed germination, vegetative growth, and yield components by increasing chlorophyll content, net photosynthesis, and biomass accumulation. Additionally, nanoparticles boost antioxidant enzyme activities,

A. S. Jayara · A. Dangwal · R. Kumar (✉)
Department of Agronomy, G B Pant University of Agriculture & Technology, Pantnagar, Uttarakhand, India

P. Pandey
Department of Molecular Biology and Genetic Engineering, G B Pant University of Agriculture & Technology, Pantnagar, Uttarakhand, India

R. Dubey et al. (eds.), *Drought and Heat Stress in Agriculture*,
https://doi.org/10.1007/978-981-96-5735-3_13

enhance root hydraulic conductance, and improve nutrient use efficiency. The application of nanofertilizers further mitigates abiotic stress by reducing resource use and greenhouse gas emissions. Thus, nanotechnology presents a viable pathway to enhance agricultural resilience against drought and heat stress, ensuring sustainable food production in the face of climate change.

Keywords

Nanotechnology · Drought stress · Heat stress · Sustainable food production · Nano-fertilizer

13.1 Introduction

Global population is about to reach 9 billion by year, which poses increasing demand for raising the food grain production to ensure the food security. Increasing climatic changes is major threat to the food production. Drought and heat stress are the most important among the several factors which pose serious limitations. The temperature rise beyond a certain threshold sufficient enough to cause permanent damage to growth and yield of crop plants is known as heat stress (Lipiec et al., 2013). Decreasing rainfall as well as increase in the variability in terms of its spread increases the probability of drought frequency. Four-time increase in the frequency of drought has been noticed from 40 in 1970s to 150 in 2010s (FAO, 2021). Rising evaporative demand coupled with decrease in rainfall increases the drought incidence.

Global mean temperature for 2020 has been 1.2 °C above that of 1850–1900 (preindustrial), making it one of the warmest years (WMO, 2020). An average global warming of 0.85 °C for land and ocean temperatures has been noticed between 1880 and 2012 (Lamaoui et al., 2018). Drought and heat stress severity have been implication on the growth, yield and reproduction of crop plants. Magnitude, rate of increase and duration of exposure of the heat stress determine the deteriorating effect on the crop plants (Tiwari et al., 2014). The major cereal crops are at major risk with these twin stresses. Agronomic species will face yield decline of 2.5–10% with increase in temperature in twenty-first century (Kohila & Gomathi, 2018). Wheat crop yield can reduce by 3–20% with each degree increase in temperature in tropical and subtropical areas (Tyagi & Pandey, 2022). Sudden increase in temperature resulted in the decline of 9% wheat yields in 2002–2003 in India (Tiwari et al., 2014). The heat stress coupled with drought stress can lead to major decline in maize crop yields, with reduction of 1.7% with each degree day spent above and over 30 °C (Cairns et al., 2013; Alam et al., 2017). Heat and drought stress during the flowering and grain filling can cause large yield reduction of rice crops (Lawas et al., 2018). Yield and quality of sugarcane are reduced under high temperature conditions (Sanghera et al., 2019). Maharashtra and Karnataka have faced reduction of 40% in sugarcane yield due to drought conditions (Misra et al., 2022). Pulses being grown under rainfed conditions, are more susceptible to drought and heat stress (Basu et al., 2022). 1–2 °C increase in temperature may jeopardize the yields of major pulses, including cowpea, chickpea, lentil, common bean, and groundnut, *rabi* pulses are at higher risk than *kharif* one (Basu et al., 2022). This reduction in crop yields has direct implications for the food security of nations across the world. The economic impact for the countries like India is more severe, as it impacts the livelihood of 45.5% workforce dependent on agriculture, of which vulnerable group of women shares 62.9% (Press Information Bureau, 2023). Multifarious approaches have been adopted for the mitigation of the heat and drought stress. The agronomic approaches for the mitigation of heat and drought stress include selection of crops and varieties, varying sowing dates, seed priming, foliar application of various chemicals, conservation agriculture, irrigation management, etc., as well as there are breeding approaches, including both conventional and modern methods (Ahmad et al., 2018; Jayara et al., 2023). However, looking at the gravity of the problem, there is needed to bring in disruptive technologies which can address these problems at physiological level. Moreover, the mentioned measures face the challenges of cost and expertise (Iqbal et al., 2019a, b). Nanotechnology is one such approach that is showing promising results in alleviating drought and heat stresses.

Nanotechnology is science dealing with the particle size below 100 nm, which changes the properties of the material. These changed physical and chemical properties of the material enable crop productivity under stress conditions. The synthesized nanoparticles, also known as engineered nanoparticles (ENs) have ameliorating effect on the seed germination, vegetative growth and yield components under abiotic stress. The increase in various growth parameters, chlorophyll content, and net photosynthesis has been reported under stress conditions with the application of ENs in crop plants. These all lead to higher biomass accumulation in plants, resulting in higher yields. Nanoparticles are also responsible for the increase in activity of anti-oxidant enzymes, including catalase, superoxide dismutase, and peroxidases. At the same time, these are reported to increase in the quality parameters in the crop plants, including the protein content, vitamins, and other crop-specific biomolecules under abiotic stress conditions. Specifically, under drought conditions, the application of nanoparticles leads to an increase in root hydraulic conductance, osmolyte concentration, antioxidant activity, and chlorophyll content for higher photosynthesis. It also increases the protein concentration required by plants for ROS detoxification, hormone pathways, oxidation-reduction, and stress signaling under drought stress. Under heat stress, there is increase in lipid per oxidation in plants, which can be reduced by the application of nanoparticles, thus preventing membrane damage and loss of electrolytes. In addition to these direct effects, the application of fertilizers in nano forms can be helpful in mitigating abiotic stress by reducing resource use. Nutrient use efficiency is comparatively higher, which reduces the loss of nutrients to the atmosphere or any other natural resource. Nanofertilizers are itself the lesser energy intensive as well have the lower carbon footprint per unit of product/nutrient compared to their bulk forms. Thus, a shift toward nanofertilizers can also reduce the share of agriculture in greenhouse gas emissions.

13.2 Brief Overview of Nanoparticle

According to the Swiss Nano Science Institute, the "nano" word in nanotechnology originates from the Greek word "nanos" which means "dwarf." Nano is 60,000 times smaller than the size of Human hair or in scaler nanometer means thousandth of a meter (10^{-9} m) (Reddy, 2019). Nanografi explains that nanotechnology is emerging daily over the past 10–15 years, nanotechnology has become an indispensable part of humanity's daily life. This doesn't mean that this technology is being used for the first time today, but there is evidence of its use historically, such as the use of Sulfide nanocrystals by the Greeks and Romans 2000 years ago to dye their hairs (Minkel, 2006). Nanotechnology is supplying benefits to various Industries like Cosmetics, Pharmaceutical, Agriculture, Communication, etc. (Cardoza et al., 2022; Neme et al., 2021; Hamza & Jaafar, 2022). In agriculture sector, nanotechnology is becoming a promising tool. It helps Agriculture in various aspects like in Precision farming, which can solve numerous issues in traditional agriculture and transform this industry (Samreen et al., 2022), in Post-harvest management of Horticultural crops to reduce the post-harvest losses by using nano-based formulations and Nano biosensors (Upadhyay et al., 2021) (Fig. 13.1).

The nanostructures have distinct properties from the bulk due to high surface area to volume ratio and potential quantum effects at the nanoscale (Guisbiers et al., 2012). This specialty has been utilized for various purposes in agriculture during

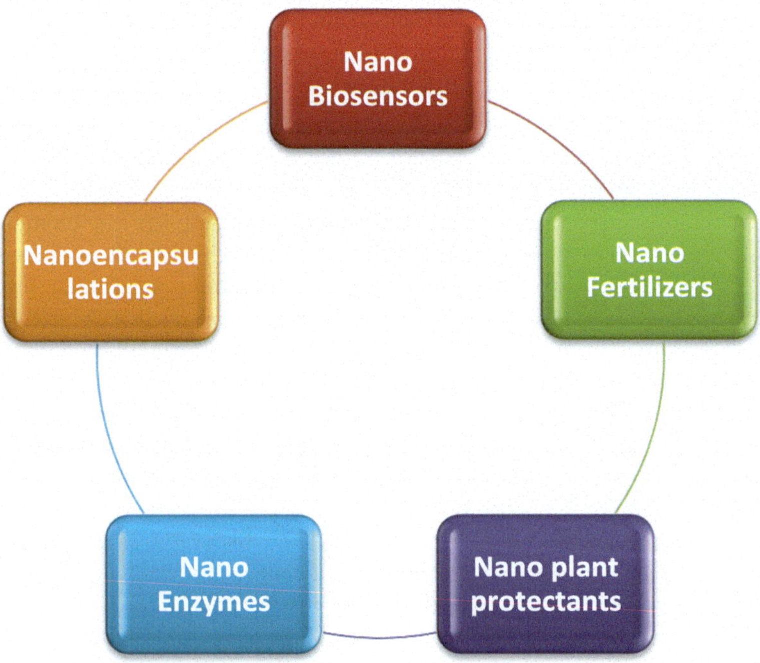

Fig. 13.1 Application of nanotechnology in the agriculture sector

Table 13.1 Positive effects of nanoparticles during cultivation of crops in agriculture

Effect	References
Promotes the seed germination	Zaghdoud and Nagaz (2022)
Improves the drought stress tolerance	Ahsan Ayyaz et al. (2022)
Improves the yield (as a fertilizer)	Benzon et al. (2015)
Helps in plant protection (as a fungicide)	Melanie et al. (2022)
Controls the weed and Disease management	Jalil and Ansari (2020)

cultivation, and it has been observed that there have been positive outcomes. Enhancing crop productivity and growth can be improved by using nanoparticles as nano fertilizers and biostimulants, therefore, it has been identified as a crucial tool for combating the negative effects of Abiotic stress (Gonzalo tortilla et al., 2023). Various applications of nanotechnology are represented in Table 13.1.

13.3 Physiology of Heat and Drought Stress

The major impact of heat stress occurs in disrupting the membrane permeability and bringing changes in the cell membrane stability (Sabagh, 2020). There has been reporting of increased electrical conductivity with an increase in heat stress in wheat crops (Kumar et al., 2012; Sharma et al., 2017). Consequent inward flux of calcium is responsible for the generation of hydrogen peroxide. The negative effect of heat stress on photosynthesis is evident beyond 40 °C (Kaushal et al., 2016). Damage to PSII as well as the inhibition of Rubisco activase causes a reduction in photosynthesis (Naz et al., 2018). The Photosystem II gets dislodged from the thylakoid membrane due to the increased fluidity of the thylakoid membrane (Prasad & Staggenborg, 2008). Further, the electron transport system is also impacted by the heat stress. Respiration rate increased beyond the temperature of 50 °C which causes loss of photosythates and thus can be correlated with reduced crop yields (Prasad & Staggenborg, 2008; Mohammad & Tarpley, 2009). The activation of parp1 and parp 2 enzymes in plant cell nucleus causes the polymerization of ADP-ribose, which uses NAD+ as substrate. This leads to the utilization of ATP leading to increased mitochondrial respiration and negatively impacts energy homeostasis (Jayara et al., 2023). Relative water content and osmotic stress increases sharply in the plants with the heat stress. Heat stress leads to hormonal imbalances in the plants. In general, there is an increased concentration of the abscisic acid in the crop plants and decrease in the other growth hormones (Li et al., 2020; Zheng et al., 2020). There is also an increase in the osmolytes concentration in the crop plants with the heat stress mainly amino acids, sorbitols, glucose, mannitols, polyols, polyamines, etc. (Jayara et al., 2023). However, this act as protectors for crops against heat stress as these increase the osmotic potential, stabilize proteins and helps in ROS detoxification.

A small decrease in the relative water content leads to a reduction in water potential and turgor pressure resulting in the closure of stomata which not only limits the water loss as well as limits the gaseous exchange impacting the process of photosynthesis (Grant, 2012; Ahmad et al., 2018; Bashir et al., 2021; Yang et al., 2021; Hemati et al., 2022). The water loss from the cells is first stage of water stress in plants (Gupta et al., 2020). Loss of chlorophyll occurs due to activation of chlorophyllase and enzyme inactivation, under severe drought stress which contributes in the reduction of photosynthesis (Ahmad et al., 2018; Bashir et al., 2021; Yang et al., 2021). This action of crop plants is regulated by the signal being sent from the drying roots and shoots. Mesophyll conductance is also reduced, which adds to the reduction of the photosynthesis. Stomata movement is regulated by the production of abscisic acid (ABA) (Gupta et al., 2020; Bashir et al., 2021). The reduction of nutrient uptake and translocation causes the reduction in the nutrient concentration in plant tissues (Bashir et al., 2021). Drought encourages production and accumulation of organic and inorganic solutes such as proline, betaine, trehalose, and other soluble carbohydrates, which acts as protectants in form of maintenance of osmotic potential at cellular level and stabilizes cell membrane structure (Ahmad et al., 2018; Gupta et al., 2020). The inorganic solutes include potassium, calcium, silicon, and sodium which maintain osmoregulation, maintains cell wall functioning enhances phosphoenol pyruvate carboxylase activity and enhances photosynthetic gas exchange (Ahmad et al., 2018; Yang et al., 2021). Reactive oxygen species production increases under the drought stress as there is destruction of PSII oxygen-releasing complex and reaction centre which disturbs electron utilization (Ahmad et al., 2018). The activity of antioxidant enzymes such as peroxidases and glutathione in response to the ROS, balance of these two determine the extent of oxidative damage in the plants (Ahmad et al., 2018; Bashir et al., 2021).

13.3.1 Nanotechnology for Heat Stress

Nanoparticles are reported to improve the morphological parameters with nanoparticle treatments compared to the control conditions. Exogenous application of zinc nanoparticles increases the activity of Superoxide dismutase enzyme as well as increase the biosynthesis of the trypthophan, resulting in increased sucrose production, which counters the decreased photosynthesis as result of chlorophyll degradation under heat stress (Kareem et al., 2022). Metal nanoparticles play an important role in increasing seed coat porosity for better germination; cell membrane repair, their better integrity and reduced electrolyte leakage, and scavenges the reactive oxygen species (Thakur et al., 2021). Cerium nanoparticles are reported to unregulated the heat shock proteins in maize crops (Zhao et al., 2012). Increased accumulation of osmolytes, photosynthetic pigments adjustment and further adjustment of stomatal conductance increased the abiotic stress resistance in crop plants through application of nanoparticles (Singh et al., 2021). Pollen viability is increased with application of Selenium nanoparticles as these scavenge the reactive oxygen species, which leads to higher yields under heat stress conditions (Djanaguiraman et al., 2018). Application of nanoparticles leads to enhanced germination as these seed germination, germination rate and seed vigor index in addition to the enhanced water uptake by stimulation of the water channels (Singh et al., 2021). A brief literature review of the effect of application of nanoparticles under heat stress in various crops is represented in Table 13.2.

13.4 Nanotechnology for Drought Stress

Drought is the biggest environmental problem among abiotic stresses, which reduces the production and yield of crops. It is necessary to deal with this to meet the food demand of today's increasing population. It has been seen in many research findings that different types of particles are proving effective in dealing with drought stress. There is a lot of evidence associated with drought stress tolerance, in which metal and metal oxide nanoparticles seem to regulate multiple plant protective mechanisms. Maize plants primed with copper nanoparticles to produce drought-tolerant crops show increased anthocyanin, chlorophyll, and carotenoid content. Moreover, it produced higher seed numbers and grain yield compared to water priming through repression of ROS accumulation in maize plants, it means nano copper priming makes the plants drought tolerant (Dong et al., 2022). Whereas calcium nanoparticles increased the biomass and nutrient acquisition of *Brassica napus* plant by boosting the enzymatic and non-enzymatic antioxidants under drought stress and also boosted the efficiency of gas exchange and photosystem II (Ahsan Ayyaz et al., 2022). Whereas if we talk about metal oxides, then foliar application of zinc oxide nanoparticles may be a promising practice to improve drought stress tolerance of eggplant grown in saline–alkaline soil. These positive effects are mainly due to the improvement of macro and micronutrients, increase of Relative water content, mitigation of cell membrane damage (MSI), and increase of anatomical characteristics of eggplant leaves and stems (Semida et al., 2021). Whereas in wheat plant, SiO_2 nanoparticles reduced the adverse effects of drought stress because SiO_2 nanoparticles can produce various metabolites that reduce transpiration, increase photosynthesis, affect stomatal conductance, and increase leaf chlorophyll content and phytochemical efficiency (Behboudi et al., 2018a, b). Both nanoparticles and ionic micronutrient formulations increase the yield of soybean under drought stress, it effects on crop growth, yield, and grain micronutrient quality during drought, supporting many observations that ion release from nanoparticle control their reactivity (Dimkpa et al., 2017). Brief literature review of the effect of application of nanoparticles under drought stress in various crops is represented in Table 13.3.

13.5 Conclusion

Nanotechnology is being used to mitigate drought and heat stress in crops, which have become a major threat to food production due to climate change. Drought and heat stress can seriously affect crop growth, yield, and reproduction, resulting in reduced yields of major cereal crops, wheat, rice, sugarcane, and pulses. Nanotechnology dealing with particles less than 100 nm in width has shown promise for mitigating these

Table 13.2 Effect of spray of nanoparticles in various crops under heat stress conditions

Crop/Heat stress/stage	Nanoparticle	Dose	Effect	References
Wheat/35–40 °C/trifoliate stage	Silver	50 & 75 mg/l	5–5.4% increase in root length; 6.6–7.5% increase in root number; 22.2–26.1% increase in shoot length; 1.3–2% plant fresh weight; 0.36–0.6% increase in plant dry matter compared to control (without heat stress/without Ag NPs)	Iqbal et al. (2019a)
Mung bean/44.1 °C/flowering stage	Zinc	0, 15, 30, 45 and 60 mg/l	Total phenolic content increase upto 139%; H_2O_2 production reduced by 25% with 60 mg/l dose; Chlorophyll content increased by 25% and grain yield by 30% with 30 mg/l dose compared to control	Kareem et al. (2022)
Wheat/32 °C/seedling stage	Zinc and Titanium	1.5 and 10 ppm	122.5% and 82.8% increase in germination with ZnO and TiO_2, respectively at 1.5 ppm; Lowest H_2O_2 recorded with Zinc nanoparticles treatment; Two-fold increase in phenol and flavnol content with application of nanoparticles	Thakur et al. (2021)
Wheat/45 °C/seedling stage	Silicon	Seed soaking @ 1.66 mM	Normal grana lamellae; improvement in chlorophyll/Cartenoid ratio; improved Fv/Fm values; Around 10% increase in chlorophyll a and b; More integrated structure of Mesophyll cell membrane	Younis et al. (2020)
Corn	Cerium	0,400 and 800 mg/kg	39 and 30 times increase in catalase activity; 3 and 10 times increase in ascorbate peroxidase enzyme activity	Zhao et al. (2012)

Sorghum/38 °C/28 °C/booting stage	Selenium	10 mg/l	18% decrease in the thylakoid membrane leakage; Marginal increase in chlorophyll content, stomatal conductance, and photosynthetic rate; 25% reduction in H_2O_2 production and 30% reduction in MDA content; 22, 24, 11, and 9% increase in the SOD, CAT, POX, and GPX enzyme activity, respectively; 14%, 19% and 26% increase in pollen germination, seed setting, and seed yield, respectively with Se NP spray under heat stress than nonspray.	Djanaguiraman et al. (2018)
Wheat	Silver	25, 50, 75, 100 mg/l	>50 mg/l resulted in an increase of 12.2% in RWC, 10% in Chl a, 16.4% in Chl b; 4% and 5.8% reduction in proline and sugar content, respectively.	Iqbal et al. (2019a)
Wheat/38 °C/Seedling stage	Selenium	10 mg/l	Chlorophyll content increased in all 8 bread wheat genotypes	Omar et al. (2023)

Table 13.3 Effect of spray of nanoparticles in various crops under drought stress conditions

Crop	Nanoparticle	Dose	Effect	References
Rapeseed	Calcium	100 mg/l	Increases plant biomass and Nutrient uptake; Increases gas exchange and photosystem II efficiency; Improves enzymatic and nonenzymatic antioxidants	Ahsan Ayyaz et al. (2022)
Eggplant	Zinc Oxide	0, 50, and 100 ppm	50 and 100 ppm ZnO NP improved growth characteristics and increased fruit yield by 12.2% and 22.6%, respectively; Highest water productivity, i.e., 50.8–66.1% was obtained	Semida et al. (2021)
Wheat	Silicon Oxide	0, 30, 60, and 90 ppm	Increased leaf greenness (SPAD) and Relative water content (RWC) of 12.54 and 84.04%; Wheat yield also increased by 25.35 and 17.81% due to NP application	Behboudi et al. (2018a, b)
Soybean	Zinc Oxide, Copper oxide, Boron trioxide along with their salts	2.8 Zn, 1.3 Cu, and 0.6 B mg/kg	Increases shoot growth by 33% and grain yield by 36%	Dimkpa et al. (2017)
Maize	Copper	450 ml of 69.4 μM	Increases the plant biomass and leaf water content; Increases the anthocyanin, chlorophyll, and caretenoid contents; Increases the total seed counts and grain yield	Dong et al. (2022)
Chickpea	Titanium dioxide	0, 5, 20, 10, and 40 mg/l	Increases the leaf area by 35%; Increases the chlorophyll index; Increases the osmotic potential; Dry weight increased by 10%	Razavizadeh et al. (2023)
Wheat	Titanium dioxide	0.1%	Improved leaf health and growth kinetics	Dawood et al. (2019)
Barley	Chitosan	30, 60, and 90 ppm	Increased relative water content, grain protein, proline content, catalase and superoxide mutase	Behboudi et al. (2018b)

stresses, improving seed germination, vegetative growth, and yield components under abiotic stress conditions. Nanoparticles also increase the activity of antioxidant enzymes and increase root hydraulic conductivity, osmolyte content, and protein content in stressed plants, resulting in higher biomass accumulation and better yield. In addition, nanofertilizers can improve nutrient use and reduce greenhouse gas emissions in agriculture. Nanotechnology has proven effective in improving crop productivity and tolerance to heat and drought stress, providing a disruptive solution that addresses these problems at a physiological level.

References

Ahmad, Z., Waraich, E. A., Akhtar, S., Anjum, S., Ahmad, T., Mahboob, W., & Rizwan, M. (2018). Physiological responses of wheat to drought stress and its mitigation approaches. *Acta Physiologiae Plantarum, 40*, 1–13.

Alam, M. A., Seetharam, K., Zaidi, P. H., Dinesh, A., Vinayan, M. T., & Nath, U. K. (2017). Dissecting heat stress tolerance in tropical maize (Zea mays L.). *Field Crops Research, 204*, 110–119.

Bashir, S. S., Hussain, A., Hussain, S. J., Wani, O. A., Zahid Nabi, S., Dar, N. A., & Mansoor, S. (2021). Plant drought stress tolerance: Understanding its physiological, biochemical and molecular mechanisms. *Biotechnology & Biotechnological Equipment, 35*(1), 1912–1925.

Basu, P. S., Chaturvedi, S. K., Gaur, P. M., Mondal, B., Meena, S. K., Das, K., et al. (2022). Physiological mechanisms of tolerance to drought and heat in major pulses for improving yield under stress environments. In *Advances in plant defense mechanisms*. IntechOpen.

Behboudi, F., Tahmasebi Sarvestani, Z., Kassaee, M. Z., Modares Sanavi, S. A. M., & Sorooshzadeh, A. (2018a). Improving growth and yield of wheat under drought stress via application of SiO2 nanoparticles. *Journal of Agricultural Science and Technology, 20*(7), 1479–1492.

Behboudi, F., Tahmasebi Sarvestani, Z., Kassaee, M. Z., Modares Sanavi, S. A. M., Sorooshzadeh, A., & Ahmadi, S. B. (2018b). Evaluation of chitosan nanoparticles effects on yield and yield components of barley (Hordeum vulgare L.) under late season drought stress. *Journal of Water and Environmental Nanotechnology, 3*(1), 22–39.

Benzon, H. R. L., Rubenecia, M. R. U., Ultra, V. U., Jr., & Lee, S. C. (2015). Nano-fertilizer affects the growth, development, and chemical properties of rice. *International Journal of Agronomy and Agricultural Research, 7*(1), 105–117.

Cairns, J. E., Crossa, J., Zaidi, P. H., Grudloyma, P., Sanchez, C., Araus, J. L., & Atlin, G. N. (2013). Identification of drought, heat, and combined drought and heat tolerant donors in maize. *Crop Science, 53*(4), 1335–1346.

Cardoza, C., Nagtode, V., Pratap, A., & Mali, S. N. (2022). Emerging applications of nanotechnology in cosmeceutical health science: Latest updates. *Health Sciences Review, 4*, 100051.

Dawood, M. F., Abeed, A. H., & Aldaby, E. E. (2019). Titanium dioxide nanoparticles model growth kinetic traits of some wheat cultivars under different water regimes. *Plant Physiology Reports, 24*, 129–140.

Dimkpa, C. O., Bindraban, P. S., Fugice, J., Agyin-Birikorang, S., Singh, U., & Hellums, D. (2017). Composite micronutrient nanoparticles and salts decrease drought stress in soybean. *Agronomy for Sustainable Development, 37*, 1–13.

Djanaguiraman, M., Belliraj, N., Bossmann, S. H., & Prasad, P. V. (2018). High-temperature stress alleviation by selenium nanoparticle treatment in grain sorghum. *ACS Omega, 3*(3), 2479–2491.

Dong, Van Nguyen Huong Mai, Nguyen Nga Thanh, Le Kien Huu, Nguyen Hoa Thi, Nguyen Huong Mai, Le Anh Trung, Nguyen Ngan Thi Thu, Dinh Son Anh, Hoang Chien, Van Ha (2022). Copper nanoparticle application enhances plant growth and grain yield in maize under drought stress conditions. *Journal of Plant Growth Regulation, 41*(1), 364–375. https://doi.org/10.1007/s00344-021-10301-w

FAO. (2021). *The impact of disasters and crises on agriculture and food security* (p. 2021). Food and Agriculture Organization of the United Nations Rome.

Gonzalo, A., Parra-Nunez, P., Bachmann, A. L., Sanchez-Moran, E., & Bomblies, K. (2023). Partial cytological diploidization of neoautotetraploid meiosis by induced cross-over rate reduction. *Proceedings of the National Academy of Sciences, 120*(33), e2305002120.

Grant, O. M. (2012). Understanding and exploiting the impact of drought stress on plant physiology. In *Abiotic stress responses in plants: Metabolism, productivity and sustainability* (pp. 89–104). Springer.

Guisbiers, G., Mejía-Rosales, S., & Deepak, F. L. (2012). Nanomaterial properties: Size and shape dependencies. *Journal of Nanomaterials, 2012*, 20–20.

Gupta, A., Rico-Medina, A., & Caño-Delgado, A. I. (2020). The physiology of plant responses to drought. *Science, 368*(6488), 266–269.

Hamza, E. K., & Jaafar, S. N. (2022). Nanotechnology application for wireless communication system. In *Nanotechnology for electronic applications* (pp. 115–130). Springer Nature Singapore.

Hemati, A., Moghiseh, E., Amirifar, A., Mofidi-Chelan, M., & Asgari Lajayer, B. (2022). Physiological effects of drought stress in plants. In *Plant stress mitigators: Action and application* (pp. 113–124). Springer Nature Singapore.

Iqbal, M., Raja, N. I., Mashwani, Z. U. R., Hussain, M., Ejaz, M., & Yasmeen, F. (2019a). Effect of silver nanoparticles on growth of wheat under heat stress. *Iranian Journal of Science and Technology, Transactions A: Science, 43*, 387–395.

Iqbal, M., Raja, N. I., Mashwani, Z. U. R., Wattoo, F. H., Hussain, M., Ejaz, M., & Saira, H. (2019b). Assessment of AgNPs exposure on physiological and biochemical changes and antioxidative defence system in wheat (Triticum aestivum L) under heat stress. *IET Nanobiotechnology, 13*(2), 230–236.

Jalil, S. U., & Ansari, M. I. (2020). Role of nanomaterials in weed control and plant diseases management. In *Nanomaterials for agriculture and forestry applications* (pp. 421–434). Elsevier.

Jayara, A. S., et al. (2023). Physiological impact of heat stress and their alleviation measures in agriculture: A review. *Journal of Applied and Natural Science, 15*(4), 1741–1749.

Kareem, H. A., Saleem, M. F., Saleem, S., Rather, S. A., Wani, S. H., Siddiqui, M. H., et al. (2022). Zinc oxide nanoparticles interplay with physiological and biochemical attributes in terminal heat stress alleviation in mungbean (Vigna radiata L.). *Frontiers in Plant Science, 13*, 842349.

Kaushal, N., Bhandari, K., Siddique, K. H. M., Nayyar, H. I., & Moral, M. T. (2016). Food crops face rising temperatures: An overview of responses, adaptive mechanisms, and approaches to improve heat tolerance. *Cogent Food & Agriculture, 2*, 1–4.

Kohila, S., & Gomathi, R. (2018). Adaptive physiological and biochemical response of sugarcane genotypes to high-temperature stress. *Indian Journal of Plant Physiology, 23*, 245–260.

Kumar, R. R., Goswami, S., Sharma, S. K., Singh, K., Gadpayle, K. A., Kumar, N., Rai, G. K., Singh, M., & Rai, R. D. (2012). Protection against heat stress in wheat involves change in cell membrane stability, antioxidant enzymes, osmolyte, H2O2 and transcript of heat shock protein. *International Journal of Plant Physiology & Biochemistry, 4*(4), 83–91.

Lamaoui, M., Jemo, M., Datla, R., & Bekkaoui, F. (2018). Heat and drought stresses in crops and approaches for their mitigation. *Frontiers in Chemistry, 6*, 26.

Lawas, L. M. F., Shi, W., Yoshimoto, M., Hasegawa, T., Hincha, D. K., Zuther, E., & Jagadish, S. K. (2018). Combined drought and heat stress impact during flowering and grain filling in contrasting rice cultivars grown under field conditions. *Field Crops Research, 229*, 66–77.

Li, M., Jannasch, A. H., & Jiang, Y. (2020). Growth and hormone alterations in response to heat stress in perennial ryegrass accessions differing in heat tolerance. *Journal of Plant Growth Regulation, 39*, 1022–1029.

Lipiec, J., Doussan, C., Nosalewicz, A., & Kondracka, K. (2013). Effect of drought and heat stresses on plant growth and yield: A review. *International Agrophysics, 27*(4), 463.

Melanie, M., Miranti, M., Kasmara, H., Malini, D. M., Husodo, T., Panatarani, C., et al. (2022). Nanotechnology-based bioactive antifeedant for plant protection. *Nanomaterials, 12*(4), 630.

Minkel, J. (2006). Ancient hair dye harnessed nanotechnology. *Scientific American*. https://www.scientificamerican.com/article/ancient-hair-dye-harnesse/

Misra, V., Solomon, S., Mall, A. K., Abid, M., Abid Ali Khan, M. M., & Ansari, M. I. (2022). Drought stress and sustainable sugarcane production. In *Microbial biotechnology for sustainable agriculture* (Vol. 1, pp. 353–368). Singapore.

Mohammad, A. R., & Tarpley, L. (2009). Impact of high nighttime temperature on respiration, membrane stability, antioxidant capacity, and yield of rice plants. *Crop Science, 49*, 313–322.

Naz, N., Durrani, F., Shah, Z., Khan, N. A., & Ullah, I. (2018). Influence of heat stress on growth and physiological activities of potato (Solanum tuberosum L.). *Phyton, 87*, 225–230.

Neme, K., Nafady, A., Uddin, S., & Tola, Y. B. (2021). Application of nanotechnology in agriculture, postharvest loss reduction and food processing: Food security implication and challenges. *Heliyon, 7*(12), e08539.

Omar, A. A., Heikal, Y. M., Zayed, E. M., Shamseldin, S. A., Salama, Y. E., Amer, K. E., & Mohamed, A. H. (2023). Conferring of drought and heat stress tolerance in wheat (Triticum aestivum L.) genotypes and their response to selenium nanoparticles application. *Nanomaterials, 13*(6), 998.

Prasad, P. V. V., & Staggenborg, S. A. (2008). Impacts of drought and/or heat stress on physiological, developmental, growth, and yield processes of crop plants. Response of crops to limited water: Understanding and modeling water stress effects on plant growth processes. *Advances in Agricultural Systems Modeling Series, 1*, 301–355.

Press Information Bureau. (2023). *Agriculture has the highest estimated percentage distribution of female workers followed by manufacturing as per the annual periodic labor force survey (PLFS)* Report 2021–22 27th March, 2023.

Razavizadeh, R., Adabavazeh, F., & Mosayebi, Z. (2023). Titanium dioxide nanoparticles improve element uptake antioxidant properties and essential oil productivity of Melissa officinalis L. seedlings under in vitro drought stress. *Environmental Science and Pollution Research, 30*(43), 98020–98033. https://doi.org/10.1007/s11356-023-29384-x

Reddy, R. S. (2019). *Principles of agronomy* (6th ed.). Kalyani Publishers.

Sabagh, E. (2020). Maize adaptability to heat stress under changing climate. In *Plant stress physiology*. IntechOpen.

Samreen, T., Rasool, S., Kanwal, S., Riaz, S., & Nazir, M. Z. (2022). Role of nanotechnology in precision agriculture. *Environmental Sciences Proceedings, 23*(1), 17.

Sanghera, G. S., Malhotra, P. K., Singh, H., & Bhatt, R. (2019). Climate change impact in sugarcane agriculture and mitigation strategies. *Harnessing Plant Biotechnology and Physiology to Stimulate Agricultural Growth, 1*(1), 99–115.

Semida, W. M., Abdelkhalik, A., Mohamed, G. F., Abd El-Mageed, T. A., Abd El-Mageed, S. A., Rady, M. M., & Ali, E. F. (2021). Foliar application of zinc oxide nanoparticles promotes drought stress tolerance in eggplant (Solanum melongena L.). *Plants, 10*(2), 421.

Sharma, I., Kaur, M., & Pati, P. K. (2017). Brassinosteroids: A promising option in deciphering remedial strategies for abiotic stress tolerance in rice. *Frontiers in Plant Science, 8*, 2151.

Singh, A., Tiwari, S., Pandey, J., Lata, C., & Singh, I. K. (2021). Role of nanoparticles in crop improvement and abiotic stress management. *Journal of Biotechnology, 337*, 57–70.

Thakur, S., Asthir, B., Kaur, G., Kalia, A., & Sharma, A. (2021). Zinc oxide and titanium dioxide nanoparticles influence heat stress tolerance mediated by antioxidant defense system in wheat. *Cereal Research Communications, 50*, 1–12.

Tiwari, R., Sheoran, S., & Rane, J. (2014). Wheat improvement for drought and heat tolerance. In R. S. Shukla, P. C. Mishra, R. Chatrath, R. K. Gupta, S. S. Tomar, & I. Sharma (Eds.), *Recent trends on production strategies of wheat in India*. Directorate of Wheat Research.

Tyagi, M., & Pandey, G. C. (2022). Physiology of heat and drought tolerance in wheat: An overview. *Journal of Cereal Research, 14*(1), 13–25

Upadhyaya, A. M., Srivastava, M. C., Sharan, P., & Roy, S. K. (2021). Silicon nanostructure-based photonic MEMS sensor for biosensing application. *Journal of Nanophotonics, 15*(2), 026001-026001.

WMO. (2020). State of the global climate, WMO No. 1264, 2020.

Yang, X., Lu, M., Wang, Y., Wang, Y., Liu, Z., & Chen, S. (2021). Response mechanism of plants to drought stress. *Horticulturae, 7*(3), 50.

Younis, A. A., Khattab, H., & Emam, M. M. (2020). Impacts of silicon and silicon nanoparticles on leaf ultrastructure and TaPIP1 and TaNIP2 gene expressions in heat stressed wheat seedlings. *Biologia Plantarum, 64*(1), 343.

Zaghdoud, C., & Nagaz, K. (2022). Zinc oxide nanoparticles enhances germination performance of wheat (Triticum durum L.) seeds under individual and combined salinity and copper treatments. *Journal of Oasis Agriculture and Sustainable Development, 4*(1), 7–17.

Zhao, L., Peng, B., Hernandez-Viezcas, J. A., Rico, C., Sun, Y., Peralta-Videa, J. R., &Gardea-Torresdey, J. L. (2012). Stress response and tolerance of Zea mays to CeO2 nanoparticles: Cross talk among H2O2, heat shock protein, and lipid peroxidation. *ACS Nano, 6*(11), 9615–9622.

Zheng, Y., Yang, Z., Xu, C., Wang, L., Huang, H., & Yang, S. (2020). The interactive effects of daytime high temperature and humidity on growth and endogenous hormone concentration of tomato seedlings. *Hort Science, 55*, 1575–1583.

New-Generation Mitigation Practices for Drought Management in Rice

14

Kavita Kumari, Manimala Mahato, V. V. S. Jaya Krishna, and Debarati Bhaduri

Abstract

Drought, or insufficient soil moisture hampering crop growth, is a common abiotic stress for many field crops. This is also reported for rice-cultivated areas in many countries across the world. In this chapter, a detailed discussion was made covering the tolerance mechanisms in rice under drought, agronomic management along with genetic and physiological approaches for drought mitigation, in addition to application of simulation models for predicting the impact of drought stress. Availability of soil water, itself a life-saving input for rice cultivation, can also regulate the nutrient transport by mass flow within the soil and may cause both stresses simultaneously. Response to drought stress may vary with the crop growth stages it faces, i.e., early vegetative stage (before tillering), mid-season or reproductive stage (flowering time), terminal drought (at maturity–/grain-filling time). The prominent effects of drought stress like reduced transpiration limits growth and biomass development, reduced nutrient availability and uptake, spikelet sterility, and reduced sink size may limit achievable yields. Overcoming the genetic barriers, several breeding approaches like pedigree selection, recurrent selection, back cross-breeding, etc., have been followed to improve the yield potential, and improved varieties like Sahbhagi Dhan, Anjali, Vandana, CR Dhan 40 emerged. Further, specific genomic region, a region associated with a specific phenotype, i.e., Quantitative Trait Loci (QTLs) related to drought tolerance in rice that influence both overall yield and its contributing traits was also identified across the world. Predicting the yield loss by use of artificial intelligence or simulation models are recent developments in realizing the drought impact.

K. Kumari · M. Mahato · D. Bhaduri (✉)
Crop Production Division, ICAR-Central Rice Research Institute, Cuttack, Odisha, India

V. V. S. Jaya Krishna
Department of Agronomy, Bidhan Chandra Krishi Viswavidyalaya, Nadia, West Bengal, India

R. Dubey et al. (eds.), *Drought and Heat Stress in Agriculture*,
https://doi.org/10.1007/978-981-96-5735-3_14

Overall, this chapter provides a detailed representation of the cause, impact, and mitigation strategies of drought with a specific case of rice.

Keywords

Drought · Agro-management · Rice · Drip irrigation · Mulching · Breeding approaches · Anti-transpirants

14.1 Introduction

Rice is a nationally as well as globally important cereal crop. Hence, any yield-limiting stress, be it biotic (pest, diseases, etc.) or abiotic (submergence, heat, drought), is of prime significance to manage and control to the best possible extent. As per the estimate, worldwide approximately 20–25 m ha area of rainfed rice has faced drought stress. The eastern Indo-Gangetic plain, accounted >20 m ha of rain-fed rice, is often considered the largest and most frequently visited drought-prone area in Asia. Drought stress showed importance regarding the yield losses in severe form in the states of Chhattisgarh, Madhya Pradesh, Bihar, Jharkhand, Orissa, and Uttar Pradesh, often considered as rice bowl (IRRI, 2006).

Water shortage commonly happens in rice fields with light-textured soils that do not form a hardpan, more specifically in upper toposequence fields, but more frequent in the lowest, heavy-textured fields of the area or regions, annually receiving less than 1000 mm rainfall. In rainfed cultivation, drought stress can occur at any time of the season, but direct damage may occur either intermittently at several times during the season or during reproductive development and early grain-filling. Categorically, drought stress can be of three types—(i) early drought (between seed sowing/transplanting to tillering), (ii) mid-season drought (drought at flowering time), and (iii) terminal drought (drought at maturity/grain-filling) (Swain et al., 2014).

It is, therefore, crucial that both agronomic and genetic management strategies focus on the best use of available soil moisture for all crop phenophases, i.e., crop establishment, active growth, and yield. Rice yields in drought-prone rainfed systems remained low and unstable and limited between 1.0 and 2.5 t ha^{-1}, where traditional and non-tolerant rice varieties were cultivated. Hence, introduction of drought mitigation packages through drought-resistant rice varieties and improved management practices, ensuring water conservation, yield sustenance, and financial security to rice farmers (Pandey & Bhandari, 2009).

To a rice farmer, the word “drought” signifies not only physical water shortage that affects plant growth and development, but a water deficit in the complete field operations starting from land preparation, transplanting, fertilizer application, and weed control operations. As reported earlier, rice yield is linearly correlated with the number of days in any growing season when the cultivated soil is saturated (Haefele et al., 2009). It has been observed that intermittent soil drying majorly hampers biomass production and, finally, reflects in total yield potential. The ability to maintain biomass accumulation in relatively dry soils is, therefore, a key feature required in drought-resistant varieties. An improved germplasm surely promises in reducing risk and enhancing yield in drought-prone rice-based lowlands. However, introducing new germplasm or a drought-tolerant variety comes with the optimization of improved agro-management techniques. Management strategies should be either alleviating stress by escaping drought (increased water input or avoiding dry periods) or to moderate the effect of drought by reducing non-transpirational losses (Serraj et al., 2011).

Apart from water deficit and associated effects, drought is also linked to crop nutrition. The uptake of important nutrients like N, Ca, Mg, Na is carried on by mass flow with the transpiration stream, while under drought stress lower transpiration of plants means lower nutrient uptake. In other dominant nutrient transport processes, lower soil moisture also restricts nutrient diffusion to the roots (mainly P and K) because the diffusion coefficient decreases. Moreover, due to redox potential gradients (transition from anaerobic /waterlogged soil to aerobic/unsaturated soil due to reduces the availability of N, P, Fe, Si, Mn, and Mo, but increases the availability of S, Zn, and Cu (Ponamperuma, 1978).

In this chapter, a detailed discussion was made covering the tolerance mechanisms in rice under drought, agronomic management along with genetic and physiological approaches for drought mitigation, in addition to application of simulation models for predicting the impact of drought stress.

14.2 Rice Tolerance Mechanisms to Drought

Drought is a major abiotic stress that significantly restricts global rice production and increases the vulnerability of food security (Rasheed et al., 2020). Drought adversely affects crop growth parameters and ultimately leads to yield decline, with the extent of damage influenced by factors such as the severity and duration of the stress and the plant's growth stage (Panda et al., 2021a, 2021b). Rice plants exhibit a range of adaptations to endure and thrive amidst drought conditions, essential for cultivation in water-scarce regions. These adaptations encompass various physiological and morphological strategies to conserve water and sustain essential metabolic functions (Salgotra & Chauhan, 2023). These strategies include either drought escape, avoidance, or tolerance. Drought escape is characterized by a plant's ability to complete its life cycle before significant soil water deficits develop, while drought avoidance involves maintaining relatively high tissue water potential despite limited soil moisture. Drought tolerance, on the other hand, describes a plant's ability to survive despite low tissue water content (Kumar et al., 2016; Rollins et al., 2013). The plant's morphological, physiological, biochemical, and molecular responses to drought have been illustrated in Fig. 14.1.

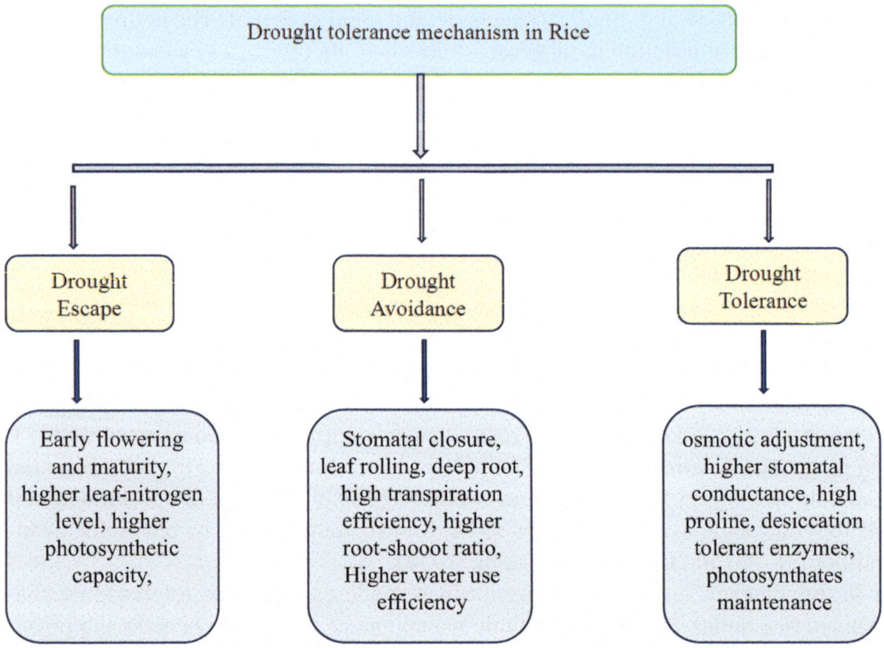

Fig. 14.1 Different responses and mechanisms of drought tolerance in rice plants (Modified from Rasheed et al., 2020; Panda et al., 2021a, 2021b)

14.3 Agronomic Management for Drought Mitigation in Rice

14.3.1 Crop Establishment Technique

14.3.1.1 Dry Direct Seeding of Rice

Dry direct seeding rice (DDSR) involves establishing the crop from seeds in non-puddled and unsaturated soil, contrasting with the traditional method of transplanting seedlings into flooded fields, thus addressing water and labor shortages in sustainable rice production while boosting productivity (Liang et al., 2023). Studies show that DDSR requires 30–51% less total water for land preparation and can result in 32–88% higher crop productivity, saving 49–55% on labor compared to traditional methods (Sandhu et al., 2019). However, DDSR often yields lower compared to transplanted rice despite alleviating some water and labor issues. Challenges such as poor crop establishment, weed control, panicle sterility, lodging, and the lack of suitable varieties and agronomic practices contribute to these lower yields. Among these, the lack of stable, high-yielding varieties adapted to the dry direct seeding system is a significant limitation. Developing such varieties is crucial to maximize yield potential. Moreover, dry direct seeding rice offers several advantages over traditional methods, including efficient water use, suitability for mechanization, improved yield and quality, better stand establishment, and earlier maturity, leading to higher economic returns (Farooq et al., 2011; Ishfaq et al., 2020). DDS rice typically performs better in drought conditions due to its deeper, finer, and more extensive root system (Ingram et al., 1990). Additionally, it enhances the sustainability of rice-wheat cropping systems by enabling timely sowing of succeeding crops and avoiding soil conflicts (Anjum, 2019). Weed management in dry direct seeding rice can be effectively done through pre-emergence and post-emergence herbicide applications or manual hand weeding, ensuring successful crop establishment.

14.3.1.2 Aerobic Rice

Aerobic rice, sown directly with a drill seeder in well-drained, non-puddled soils, represents a novel approach to cultivation (Bouman et al., 2005). Aerobic rice combines the drought tolerance of upland rice with the yield potential of lowland rice, leading to nearly 50% water saving compared to traditional lowland rice cultivation (Joshi et al., 2018). This method, still in developmental stages, holds promise for utilization in drought-prone rainfed lowlands (Kato & Katsura, 2014). The primary hurdle facing aerobic rice is the identification of suitable varieties possessing high yield potential, desired grain quality, and drought tolerance, as many traditional upland rice varieties lack these attributes (Atlin et al., 2009; Haefele et al., 2016). Through traditional upland farming techniques and improved irrigated cultivars, aerobic rice cultivars have been developed with high yield potential and moderate drought tolerance (Ramulu et al., 2016). These cultivars demonstrate high yields under appropriate management practices, facilitated by deep root systems and tolerance to water stress during both vegetative and reproductive stages.

14.3.2 Crop Variety Selection

14.3.2.1 Early Maturing Varieties

Planting early maturing rice varieties can help mitigate the effects of drought by allowing the crop to complete its lifecycle before severe water stress occurs. For the drought escape mechanism, the first priority is placed on early maturity because of its simplicity and high heritability. Agarwal and Khan (2002) also reported that early maturing upland rice genotypes partially escaped terminal drought. Apart from high root density and root weight, the duration of the crop was very important for selecting drought-tolerance genotypes. Moreover, medium-duration varieties were more affected by moisture stress at flowering and milk stages, whereas short-duration varieties escaped the stress at these stages (Sheela & Alexander, 1995; Ray et al., 2016).

14.3.2.2 Drought-Tolerant Varieties

Planting drought-tolerant rice varieties stands out as the most effective strategy for enduring water scarcity while sustaining productivity. These varieties typically possess deeper root systems, enabling them to access water from deeper soil layers. The utilization of improved varieties can boost upland rice productivity by up to 40%

(Bujarbaruah et al., 2002). Drought-resistant upland rice varieties feature fewer but thicker roots that penetrate deeper soil layers during dry spells (Namuco et al., 1993). Traditional varieties exhibit greater drought tolerance at the reproductive stage compared to high-yielding ones, attributed to their higher seedling survival and vigor, lower transpiration index, consistently higher water potential in plant tissues, lower chlorophyll stability index, higher root-to-shoot ratio, and efficient allocation of dry matter into reproductive sinks even under drought conditions (Ray et al., 2016). In recent years, several varieties tailored for drought-prone regions emerged, including Sahbhagi dhan, Shushk Samrat, CR dhan 40, NDR 97, and Indira Barani dhan (Table 14.1).

14.3.3 Tillage

Conservation tillage methods, like minimal or no-till practices, retain residue on the soil surface, which reduces evaporation and enhances soil structure. This helps in mitigating high temperatures, stress, and drought while also decreasing water losses, leading to improved water use efficiency (Sisti et al., 2004). Additionally, conservation tillage reduces soil compaction and crusting, and increases organic matter content, which enhances infiltration rate and water-holding capacity (Tyagi et al., 2020; Wallander et al., 2013). Implementing this practice not only saves time and energy but also aids in early sowing and minimizes terminal drought (Das et al., 2020). Studies have shown significant water savings under zero tillage, while efficient residue management enhances soil quality, leading to improved plant growth and stress tolerance (Bhan & Behera, 2014; Das et al., 2014). Moreover, research indicates that conservation tillage methods improve plant water stress, resulting in better photosynthetic capacity and, ultimately, higher crop yields (Mukherjee et al., 2022; Sher et al., 2021).

14.3.4 Seed/Seedling and Sowing/Planting

Early maturing, drought-tolerant variety seeds having agronomic traits such as yield potential, disease resistance, and grain quality are opted.

14.3.4.1 Seed Priming/Hardening

Seed hardening or seed priming for combating occasional moisture stress requires urgent attention When dry seeds are soaked in water or chemical solutions, quiescent cells are hydrated, initiating germination. However, shade drying before sowing halts the triggered germination. Physiological preconditioning through imbibition-drying-reimbibition upon sowing prepares plants to resist adverse weather conditions (Ray et al., 2016). It has been seen that drought tolerance can be induced by seed hardening with potassium salt, and among them, seed hardening with 4% KCl salt alone or combined with higher K fertilizer application and anti-transpirant spraying at tillering stage increased tiller number, root volume, water

Table 14.1 Drought tolerant varieties in India and their specifications

Variety	Developed by	Recommended areas	Drought specification	Yield potential	Resistance/ Tolerance
Sahbhagi Dhan	CRURRS, Hazaribag; IRRI, Philippines	Jharkhand, Bihar, UP, Odisha	Highly drought-tolerant, early maturing	2.0–2.5 t/ha (with stress), 3.8–4.5 t/ha (without stress)	Highly resistant to leaf blast, moderately resistant to brown spot and sheath blight
Shusk Samrat	Narendra Dev University, Masoda	Eastern Uttar Pradesh, Bihar, Chhattisgarh	Drought-prone rainfed upland and lowland areas Semi-dwarf, early maturity, high yielding	3.0–3.5 t/ha	Moderately resistant to major insects and pests, tolerant to sheath rot and brown spot
CR Dhan 40	CRURRS, Hazaribag	Jharkhand, Bihar, Maharashtra	Drought-prone upland areas, Medium-tall, early-maturing	3.0 t/ha (direct seeded), 4.0 t/ha (transplanted)	Moderately resistant to brown spot and leaf blast
Vandana	CRRI, Cuttack	Jharkhand, Bihar, Odisha, Chhattisgarh	Drought-prone rainfed upland areas, Tall, early maturing	2.5–3.0 t/ha (direct seeded), 3.5–4.5 t/ha (transplanted)	Moderately resistant to brown spot and blast, weed competitive
Anjali	CRURRS, Hazaribag	Bihar, Jharkhand, Odisha, Assam, Chhattisgarh	Drought-prone upland areas, Semi-tall, early maturing	3.0 t/ha (direct seeded), 4.0 t/ha (transplanted)	Highly resistant to brown spot & gall midge biotype 5 and 1, moderately resistant to leaf blast
Indira Barani dhan-1	Indira Gandhi Krishi Viswavidyalaya	Chhattisgarh	Rainfed shallow lowland ecosystem, Medium, slender grain, early maturing	4.0–4.5 t/ha	Highly resistant to bacterial leaf blight, brown spot, and stem borer

economy, water use efficiency, and yield, offering drought protection (Khatua, 2002; Thakuria & Choudhry, 1995). Further, seed treatment with 1% $CaCl_2$ solution and 385 ppm sodium phosphate improved germination, crop establishment, and increased grain yield because of better survival under drought (Ananda & Reddy, 2002; Ray et al., 2016).

14.3.4.2 Time of Sowing/Planting

Rice planting at proper timing can avoid peak drought periods, mitigate water scarcity effects, ensuring critical growth stages receive sufficient water (Ghosh, 2002). Suboptimal plant population due to untimely sowing limits upland rice yields. Early sowing enhances rainfall availability during seedling to grain filling stages, aiding in drought escape (Bodner et al., 2015). Further, strong root development from rapid and healthy crop establishment due to early sowing minimizes future drought effects (Vance et al., 2014). Early sown crops exhibit higher water use efficiency, whereas delayed sowing beyond normal dates adversely affects rice seedling growth, plant stand, and yield due to heavy rains, inadequate weed control, and terminal drought (Ray et al., 2016).

14.3.4.3 Method of Sowing

Choosing the appropriate seedling method is crucial for achieving an optimal plant stand, reducing seed wastage, minimizing seedling mortality from drought, and facilitating weed control. Ananda and Reddy (2002) found higher grain yields with seed drilling compared to broadcasting, as broadcasting led to increased lodging risk. Seeds dibbled or continuously sown behind the plow, typically placed at 5–6 cm depth, showed better results than broadcasting or drilling at 2–3 cm depth. Deeper seed placement behind the plow resulted in a uniform plant stand, improved drought tolerance, and higher yields compared to other methods (Chandra et al., 1991; Ray et al., 2016). This might be attributed to better root growth and moisture extraction from deeper soil layers, as well as increased nutrient uptake by crops sown in lines or behind the plow, compared to broadcast sowing (Khatua, 2002).

14.3.5 Nutrient Management

Diverse and occasionally conflicting studies exist regarding the impact of nutrient supplementation on enhancing plant growth during drought stress (Ashraf et al., 2011). Typically, the external provision of nutrients can lead to improvements, reductions, or no discernible effects on plant growth under water-deficient conditions. The outcome is contingent on factors such as the severity of drought, soil nutrient levels, and other environmental conditions (Alam, 1999; Hu & Schmidhalter, 2005). However, many studies proved that external application of nutrients through fertilizers under drought situation might eliminate the negative effects of stress on plant by improving the nutrient concentration in plant (Alsafar & AI-Hassan, 2009; Percival et al., 2003). In recent times, there has been significant interest in the foliar feeding of nutrients, as this method allows for the direct delivery of nutrients to the foliage and is potentially a faster means of nutrient supply compared to uptake from the soil when root system either not fully developed or prevailing of aberrant situation. Thus, foliar application is effective way for correction of nutrient deficiencies that cannot be effectively remedied through soil amendments (Cakmak & Kirkby, 2008; Sarkar et al., 2007). A brief discussion regarding various role of major, minor, and beneficial nutrients under moisture stress is presented below.

14.3.5.1 Major Nutrients

Importance of nitrogen as a major nutrient is widen not only as a structural, genetical, and metabolic compound (Tisdale & Nelson, 1975) but also as stress mitigator through efficient management (Mahpara et al., 2019; Shangguan et al., 2000). Nitrogen (N) holds paramount importance for plant structural integrity, growth, physiology, and stress mitigation (Ata-Ul-Karim et al., 2016; Song et al., 2010). It is intricately linked to heightened chlorophyll biosynthesis, increased photosynthetic activity, and efficient utilization of solar radiation (Hassan et al., 2005; Waraich et al., 2011). In terms of agricultural output during moisture stress, the absorption and utilization of nitrogen by plants are crucial (Binghua et al., 2012). Nitrogen fertilizer application under lower water potential in crops facilitates cellular membrane stability, carbon partitioning, carbohydrate accumulation, osmoregulation, and antioxidative defense mechanisms, contributing to overall improvement in development (Saneoka et al., 2004; Saud et al., 2017). Increased levels of nitrogen under moisture stress found to be improved the total root length in rice, which in turn increased shoot dry weight by improving the stomatal conductance and photosynthetic rate (Thiem & Akira, 2015). Similarly, Menge et al., 2019 examined how regulation of N doses in upland rice under induced drought conditions manipulates the root morphology and productivity of the crop. Major role of phosphorus under drought stress lies upon development of root growth and proliferation, which helps to enhance the moisture uptake and utilization (Jin et al., 2015; Waraich et al., 2011). Phosphorus plays a crucial role in maintaining cell turgidity and cell membrane stability by enhancing stomatal conductance and the rate of net photosynthesis, contributing to optimal leaf water potential and stress tolerance (Sato et al., 2010; Tariq et al., 2018). Additionally, the incorporation of phosphorus under various abiotic stresses has significant effects on tissue formation, as well as carbohydrate and lipid metabolisms. It influences chlorophyll biosynthesis, photosynthesis, and the production of essential compounds like adenosine triphosphate (ATP) and reduced nicotinamide adenine dinucleotide phosphate (NADPH). Phosphorus also affects enzymatic functions, including activities related to ribulose-1,5-bisphosphate (RuBisCo) and ATPase. Furthermore, it plays a role in phytohormonal functions such as abscisic acid (ABA) and indole acetic acid (IAA), as well as energy transfer and storage, reproductive development, and scavenging activities against reactive oxygen species (ROSs) like hydrogen peroxide (H_2O_2), malondialdehyde (MDA), singlet oxygen ($^1O_2$), superoxide anions (O_2), and hydroxyl radicals (OH), etc. Phosphorus also influences the accumulation of osmolytes, such as proline, soluble sugars, and proteins (Begum et al., 2020; Yang et al., 2019). Several studies have illustrated the role of potassium in imparting resilience against various abiotic stresses. Potassium involves in major plant physiological and biochemical mechanisms viz. osmotic adjustment, stomata opening, regulation of transpiration, root architecture, enzymatic upregulations, optimization of nutrient uptake and utilization, increased water use efficiency, and prevention of the accumulation of reactive oxygen species (ROSs) induced by drought, etc. (Wang et al. 2013a, b; Zamani et al., 2020). In essence, potassium emerges as a key element in maintaining the optimal energy status of plants and bolstering their ability to withstand various environmental challenges. Increased doses of potassium (@ 120 kg ha^{-1}) in lowland rice under drought stress during the reproduction stage slowed the leaf rolling, improved yield attributes and yield along with lowered the spikelet sterility (Faustino et al., 2019). Zain and Ismail (2016) reported from the experiment based on different sources of potassium with various doses that increased potassium dose improved the transpiration rate, which in turn improved the nutrient uptake that helped to reduce the oxidative stress in rice plants under stress. They illustrated a positive correlation among transpiration rate and antioxidant activity. Further, the experiment also concluded that the application of potassium through KCl-enhanced NAR that potassium applied through K_2SO_4 due to greater absorption of potassium from KCl than K_2SO_4.

14.3.5.2 Secondary and Minor Nutrients

Calcium is recognized for its involvement in facilitating the stress response in drought situations and aiding plant acclimation to stress conditions (Palta, 2000). In the context of drought recovery, calcium is essential as it activates the plasma membrane enzyme ATPase. This enzyme is crucial for pumping back nutrients lost during cell damage caused by drought stress (Palta, 2000). Calcium is also known to mitigate the effects of drought stress and mediate the efficiency of post-stress recovery in plants (Devi & Kar, 2013). Manganese plays a crucial role in activating various metabolic enzymes within the tricarboxylic cycle and is an integral component of photosystem II. Additionally, it contributes to ATP synthesis and supports the activity of RuBP carboxylase (Upadhyaya et al., 2011). Besides that, manganese is recognized for its significant involvement in the detoxification of reactive oxygen species (ROSs) such as superoxide and hydrogen peroxide (Millaleo et al., 2010). Conversely, a deficiency in manganese results in oxidative stress in plants, leading to chlorophyll damage and hindered photosynthetic activity (Hajiboland, 2012). Zinc, on the other hand, plays a significant role in elevating auxin levels, leading to enhanced root growth and subsequently improving drought tolerance in plants (Bennett & Skoog, 2002). Major involvement of zinc in physiological processes that is activation of catalytic enzymes such as carboxypeptidase, superoxide dismutase, RNA polymerase, and alkaline phosphatase (Hajiboland, 2012; Yuvaraj & Subramanian, 2020). Extended periods of drought stress in plants can adversely affect various cellular metabolic components, including NADPH activity. The application of zinc mitigates photooxidative damage, diminishes the generation of reactive oxygen species (ROS), and enhances the concentration of osmolytes like superoxide dismutase (SOD) (Waraich et al., 2011). In plants, Fe assists in chlorophyll synthesis and is essential for the maintenance of structural integrity as well as functional aspects of the chloroplast. The Fe-S cluster functions as a prosthetic group for various Fe-S proteins and assumes regulatory roles under conditions of oxidative stress (Balk & Pilon, 2011). Utilizing iron (Fe) as a plant nutrient has demonstrated notable ameliorative effects in countering stress by promoting the production of antioxidative enzymes, including catalase (CAT), peroxidase (POD), and superoxide dismutase (SOD), which play a crucial role as primary scavengers of reactive oxygen species (ROSs) (Khan et al., 2009). Boron (B) is imparting drought tolerance responses by aiding seed germination, sugar transport, flower, and pollen development (Waraich et al., 2011). Copper, a redox-active transition element, holds a crucial role in various plant processes, including photosynthesis, carbon and nitrogen metabolism, respiration, and defense against oxidative stress (Syuhada et al., 2014). Meanwhile, Cobalt contributes to drought tolerance in plants by enhancing water utilization efficiency and reducing transpiration rates (Arif et al., 2016; DalCorso et al., 2014). Lala et al. (2019) investigated the effect of exogenous application nutrients through sole or combined application of P, K, Ca, Zn, and Fe in different rice varieties under drought and well-watered situations and revealed that increased growth attributes, dry matter partitioning, chlorophyll concentration, carbohydrate, antioxidant activities. Ultimately, the combined application of nutrients resulted in 53, 53, and 49% higher yields over P or K application.

14.3.5.3 Beneficial Nutrients

Though silicon requirement in crops is low compare to other nutrients, its beneficial role in stress mitigation was proven by many studies. Several studies have investigated the positive impact of silicon (Si) on plants experiencing drought stress in not only triggering the defense mechanisms but also regulates parameters such as relative water content (RWC), net photosynthetic ratio, intercellular CO_2 levels, stomatal conductance, and transpiration ratio (Gong et al., 2008; Hussain et al., 2021; Romero-Aranda et al., 2006). Furthermore, silica plays a crucial role in enhancing physiological activities and boosting cellular metabolic rates in plants when exposed to drought stress. This, in turn, improves water use efficiency, promotes growth, and increases biomass (Gong et al., 2003; Li et al., 2018). Chen et al. (2011) reported from their study that exogenous application of Si with different concentration improved root traits, photosynthetic rate, and transpiration rate rice plants under drought stress. The beneficial effects of silicon (Si) on plants experiencing drought conditions were linked to a decrease in transpiration (Agarie et al., 1998; Gao et al., 2006). On the other hand, El-Okkiah et al., 2022 demonstrated silicon application through foliar application of potassium silicate during vegetative growth stage accelerating the physiological and biochemical parameters to alleviate the stress effect and improve the plant growth. Ghouri et al. (2021) evaluated the role of the combined application of silicon and selenium in minimize the harmful effect of drought in rice through improving chlorophyll content and reducing water loss and enhanced the growth and productivity of rice. As, combined application of silicon and selenium reduced the development of reactive oxygen species (Verma et al., 2021), and elevated selenium (Se) concentration within the cell serves as a protective mechanism for chloroplasts, leading to the accumulation of starch within them (Mozafariyan et al., 2017); which improves crop yield as well as nutritional quality (Ahmad et al., 2021).

14.3.5.4 Antioxidants, PGPR, and Others

Apart from major or minor micronutrients, application of antioxidants to rice was assessed as alleviation of drought effect. Urmi et al. (2023) reported that, application of salicylic acid or proline or combine application of both to rice crop proved to be drought mitigator through improving photosynthetic rate, stomatal conductance, transpiration rate, relative water content, uptake of nutrient, growth, yield attributes, and yield of the crop. They also reported the enhanced activity of antioxidant enzymes such as catalase (CAT), ascorbate peroxidase (APX), glutathione peroxidase (GPX), and superoxide dismutase (SOD). This resulted in a reduction in the levels of hydrogen peroxide (H_2O_2) and malondialdehyde (MDA) in plants subjected to water stress. As external application of salicylic acid (SA) and/or proline has the potential to boost the water potential gradient and enhance the absorption of mineral elements, thereby mitigating the adverse effects of drought stress on crop growth (Tayyab et al., 2020) also stimulating cell expansion and elongation (Kibria et al., 2016). Further, salicylic acid (SA) and proline play a crucial role in osmotic adjustment, enhancing the root's ability to absorb water (Blum, 2009). This adjustment results in optimized soil water retention, leading to increased stomatal transpiration and subsequent higher carbon dioxide (CO_2) fixation. This, in turn, may stimulate greater biomass production and contribute to higher crop yields. SA and proline contribute to elevated biomass and crop yield by promoting the enlargement of leaf size and increasing the number of leaves, thereby enhancing photosynthetic capabilities (Merwad et al., 2018). Additionally, studies suggest that SA and proline significantly boost cytokinin levels in leaves and grains, facilitating improved nitrogen absorption and utilization, heightened photosynthetic capacity, and grain yield (Talla et al., 2016). Many studies have reported the role of plant growth-promoting rhizobacteria (PGPR) in the alleviation of drought stress in different crops. PGPR improves nutrient uptake by inducing or producing plant growth regulators (Vejan et al., 2016) and several processes like nitrogen fixation, siderophore production, sequestration, and mobilization of nutrients (Hardoim et al., 2015). Increment of stomatal conductance, photosynthesis, and concentration of stress-induced metabolites through application of *Trichoderma harzianum* was reported by Shukla et al., 2012 and shoots fresh weight, plant vigor, stomata conductance, and photosynthetic status through application of *Azospirillum Brasilense* interacted with arbuscular mycorrhizal fungi (AMF) was reported by Ruíz-Sánchez et al. (2011). Hazman et al., 2023 exemplified the beneficial role of biochar—compost mixture in alleviating drought stress in rice. The results of their experiment revealed better shoot biomass, root anatomical and architectural traits with the application of biochar compost mixture.

14.3.6 Water Conservation Practices & Management Strategies

14.3.6.1 Alternate Wetting and Drying

Various water-saving irrigation techniques have been extensively studied nationwide, including intermittent irrigation, alternating wetting and drying irrigation, and midseason flooding and drainage with intermittent irrigation. Among these, alternating wetting and drying (AWD) has emerged as a prevalent and effective method for conserving water. AWD involves subjecting fields to intermittent flooding, where irrigation is halted and water is allowed to subside until the soil reaches a specified moisture level. Then, the field is reflooded. This approach has been shown to reduce water usage by about 23% compared to continuously flooded rice systems (Bouman & Tuong, 2001; Lampayan et al., 2015).

The determination of AWD thresholds varies and can be based on parameters such as volumetric water content, gravimetric water content, days after ponded water disappears, qualitative indicators like "hair cracking," soil water potential in the rooting zone (SWP), and field water level (FWL). Among these, SWP and FWL are the most commonly used and quantifiable. They are categorized into two groups for AWD threshold: (1) Severe AWD, where SWP in the rooting zone drops below −20 kPa (SWP < -20 kPa); and (2) Mild AWD, where SWP in the rooting zone does not drop below −20 kPa (SWP ≥ -20 kPa) or, if FWL is measured, it does not drop more than 15 cm below the soil surface (FWL $\leq$ 15 cm), also referred to as Safe AWD (Bouman et al., 2005).

14.3.6.2 Drip Irrigation Technique

In regions grappling with water scarcity, the adoption of micro irrigation methods such as sprinkler and drip irrigation is being encouraged to mitigate the risk of yield reduction. While drip irrigation remains relatively underutilized in rice cultivation, it offers significant advantages in enhancing water-use efficiency by delivering water directly to the root zone, thereby minimizing evaporative and runoff losses (Kumar et al., 2016). Recent research has highlighted the superiority of subsurface drip irrigation for irrigating aerobic rice or varieties suitable for drip-based cultivation, particularly when combined with fertigation. These integrated approaches have been found to augment root volume, root oxidizing power, canopy photosynthesis, and translocation efficiency, thereby fostering the growth of aboveground plant parts. The utilization of drip irrigation technology in rice farming represents a pivotal strategy for boosting productivity while conserving water resources (Arouna et al., 2023; Ashoka et al., 2015; Meher et al., 2020; Ramamurthy et al., 2009).

According to Parthasarathi et al. (2018), drip irrigation has resulted in a substantial 29% increase in aerobic rice yield, accompanied by a 50% improvement in water-saving efficiency and subsequent enhancement in water productivity. Furthermore, studies focusing on alleviating pressure on underground water reserves due to the depletion of aquifers, in the face of projected climate changes, have underscored the tangible benefits offered by drip irrigation systems. These benefits include significant savings in irrigation water and energy, alongside improvements in nitrogen use efficiency and net income (Sidhu et al., 2019) (Table 14.2).

Table 14.2 Effect of drip irrigation on water use and rice yield

Location	Impact on irrigation water use and rice yield	References
Nuapada, India	Drip irrigation technology can reduce irrigation water costs by 2.0 to 5.6 times, lowering overall water consumption significantly	Meher et al. (2020)
Coimbatore, India	Yield of aerobic rice and water savings increased by 29% and 50%, respectively, using drip irrigation compared to surface irrigation	Parthasarathi et al. (2018)
Punjab, India	Subsurface drip fertigation reduced irrigation water use for rice by 48–53% and improved energy consumption and nutrient utilization efficiency	Sidhu et al. (2019)
Bhopal, India	SRI under drip irrigation system provided the greatest plant height, root length, yield, and yield-contributing factors compared to conventional irrigation	Rao et al. (2017)
Maharashtra, India	Water productivity increased from 0.46 to 0.67 kg/m^3 using drip fertigation, with rice yields 13–28% higher across varieties and regions	Padmanabhan (2019)
Tamil Nadu, India	Grain yield significantly increased with subsurface drip irrigation approach, saving 27.0% water without reducing yield compared to traditional methods.	Parthasarathi et al. (2018)

Source: Arouna et al. (2023)

Rao et al. (2017) also demonstrated the superior efficiency of drip irrigation systems for rice production compared to conventional paddy cultivation under continuous flooding, particularly in terms of enhancing water productivity and conserving water and energy resources.

Moreover, the adoption of drip irrigation for rice production presents multifaceted advantages, encompassing agronomic, economic, and environmental realms. These include reductions in fertilizer use, leaching of fertilizers, diseases, pests, and weeds, as well as benefits such as power saving and reduced labor requirements. Environmentally, drip irrigation contributes to reductions in water usage, greenhouse gas emissions, arsenic uptake, and improvements in rice quality (Ragaglini et al., 2014).

14.3.6.3 Supplemental Irrigation

In severe drought conditions, supplemental irrigation becomes crucial for sustaining rice production. This involves utilizing alternative water sources such as groundwater, reservoirs, or harvested rainwater to supplement natural rainfall. The application of water during critical growth stages, known as life-saving or supplemental irrigation, is essential to prevent total crop failure. These critical stages, including active tillering, panicle initiation, booting, heading, and flowering, are particularly sensitive to moisture deficits, which can lead to significant yield losses. Implementing one or two need-based life-saving irrigations during critical stages can optimize yield under drought conditions (Oweis & Hachum, 2006; Praharaj et al., 2017). It's advisable to apply a limited amount of water during these stages, preferably through drip or sprinkler irrigation, to ensure optimal moisture levels and prevent crop loss.

14.3.7 Mulching

Applying mulch to the soil surface plays a vital role in conserving soil moisture by reducing evaporation and moderating soil temperature. Organic mulches, such as straw or crop residues, are commonly used in rice cultivation to maintain soil moisture, particularly during drought conditions. This practice can significantly reduce unproductive water loss, which accounts for 30–60% of total applied water, by mitigating soil evaporation. Mulching improves water-use efficiency by 10–20% through enhancing infiltration rates reducing evaporation loss, runoff, and soil temperature fluctuations (Ossom et al., 2001; Ranjan et al., 2017; Ahmad et al., 2015). Studies have shown that mulch treatments result in higher water-use efficiency, increased relative water content of leaves, and reduced weed populations under water stress conditions compared to no mulch (Ahmad et al., 2015). Additionally, mulches suppress weed growth by limiting light penetration into the soil, thereby enhancing water availability to crop plants during drought situations (Tyagi et al., 2020).

14.3.8 Laser Land Leveling

Implementing technologies like laser leveling to ensure uniform water distribution and precision irrigation scheduling can optimize water use efficiency. Studies have demonstrated that laser land leveling can reduce water consumption by up to 30% compared to traditional land leveling methods.

14.3.9 Hydrogel

Hydrogel or Superabsorbent polymer (SAP) amendments in soils mitigate moisture loss during drought, thereby reducing evaporation and percolation, and aiding crops to withstand drought more effectively. According to Reddy et al. (2021), sandy loam soil treated with hydrogel exhibited a 20% improvement in water retention capacity compared to untreated soil. Hydrogels can absorb more than 400 times their weight in water through this mechanism. They respond to moisture stress by retaining water when surroundings dry out and rehydrate upon moisture application (Attri et al., 2022). Hydrogel apart from retaining large amounts of water, also serves as a reservoir for plants' essential nutrients (Wang & Gregg, 1990). Johnson (1984) found that adding hydrogel at 2 g/kg increased sand's water-holding capacity from 171% to 402%. Hydrogel application reduces irrigation needs by enhancing water retention, delaying wilting onset (Taylor & Halfacre, 1986). Rehman et al. (1984) found that sowing rice on beds with hydrogel amendment was the most effective technique. This method not only improved the performance of aerobic rice but also enhanced its growth and yield more than other sowing techniques.

14.3.10 Use of Antitranspirants

Antitranspirants are chemical compounds that aid in enhancing drought resistance by stabilizing the structure of plant cells (Ouda et al., 2007). Antitranspirants, used to reduce water loss in plants, are less commonly applied to rice but can help manage water stress. Examples include film-forming substances like polyvinyl acetate, stomatal closure agents such as abscisic acid, reflective particles like kaolin clay, and plant extracts like seaweed. Careful consideration of growth stage, environmental factors, and regulations is crucial. Consulting agricultural experts' aids in determining suitability and efficacy in specific conditions, ensuring effective water conservation and drought mitigation in rice cultivation. Reflective antitranspirants like Kaolin and Green Miracle spray reduce leaf temperature through reflectance mechanisms (Nakano & Uehara, 1996; Rosati et al., 2006). Application of long-chain fatty alcohol, an organic source, under water-stressed conditions improves crop growth and yield. Antitranspirant positively affect rice's growth parameters and dry matter production. Patel et al. (2019) reported higher dry matter production in plants treated with long chain fatty alcohol and kaolin compared to untreated plants.

14.3.11 Crop Diversification

Integrating rice with other crops in a rotation system can enhance soil health and mitigate drought-related losses. Crop diversification offers alternative income sources for farmers during drought periods. Substituting rice with direct-seeded, low-water, drought-tolerant crops, or intercropping legumes can stabilize upland yields, particularly for traditional rice farmers. In rice-dominant areas, partial substitution of rice with intercrops in light-textured upland rice soils provides yield certainty against total rice failure due to drought. Intercropping rice with pigeon pea, groundnut, and black gram increases rice equivalent yield compared to sole cropping (Kar et al., 2004). Diversification and substituting rice with low-water rainfed crops like maize, black gram, green gram, cowpea, groundnut, and pigeon pea through sole cropping or intercropping are viable options for mitigating drought and enhancing upland rice productivity with assured and early returns (Kar & Verma, 2002).

14.4 Genetic and Physiological Approaches for Drought Mitigation in Rice

Continuous endeavors are underway to develop rice varieties that can withstand drought conditions, thereby minimizing reliance on water resources while concurrently improving nutritional content and increasing grain yield during periods of water scarcity (Selvaraj et al., 2017). Numerous initiatives, including traditional breeding methods, molecular techniques, and genomic approaches, are currently being implemented to cultivate plants with increased resilience to drought conditions (Oliveira et al., 2020). Important genetic and physiological approaches associated developing drought resistance in rice are discussed below.

14.4.1 Conventional Breeding Strategies

Another better option to improve rice production in sustainable and economically viable way is by breeding high yielding varieties with drought resistant cultivars. The main criteria for selection in conventional breeding technique is selection superior grain yielding cultivars in drought-prone areas; this, approach has proven to be ineffective because of low heritability and the significant influence of genotype-environment interaction (Chourasia, 2017; Upadhyaya & Panda, 2019). In conventional breeding, the emphasis on selection criteria has progressively shifted toward physiological traits, as they exhibit faster growth and rely on genetic diversity (Dixit et al. 2014a, b; Atkinson & Urwin, 2012; Chourasia, 2017). These strategies are designed to produce high-yielding varieties under water surplus conditions, and these may continue to produce moderate to high yield during drought conditions (Dixit et al. 2014a, b; Madhu et al., 2022). Traditional breeding techniques are mainly helpful in germplasm preservation and development of novel genetic characteristics through sexually different parent hybridization. Many renowned rice research institutes like IRRI, Philippines and IIRR Hyderabad, Telangana performing research on traditional breeding techniques and developed elite cultivars that are exhibiting resilience to numerous diseases and abiotic stresses have been observed in the last three decades (Madhu et al., 2022; Panda et al., 2021a, b). Some traditional breeding techniques that have been used in recent years are pedigree selection, recurrent selection, backcross breeding, and induced mutation.

14.4.2 Pedigree Selection

Pedigree selection, a traditional and extensively utilized breeding technique in rice enhancement, proves particularly effective for instilling resistance, particularly when traits are controlled by major genes (Chengqi et al., 2024; Oladosu et al., 2019). A notable advantage of pedigree selection lies in its ability to combine numerous genes governing both biotic and abiotic factors. However, a significant drawback of pedigree selection is its time-intensive nature, demanding regular evaluations of numerous lines throughout planting seasons and the maintenance of meticulous records on selection criteria. Proficiency in breeding materials, coupled with a comprehensive understanding of genotype–environment interactions, is paramount for successful pedigree selection. For traits influenced by numerous genes, pedigree selection may not be appropriate; in such instances, the diallel mating design would be more suitable for selection (Khush, 1984; Oladosu et al., 2019). Generally, in self-pollinating crops such as rice, recurrent selection is the preferred choice among plant breeders over pedigree selection. The comprehensive selection process for developing drought-tolerant rice is illustrated in Fig. 14.2.

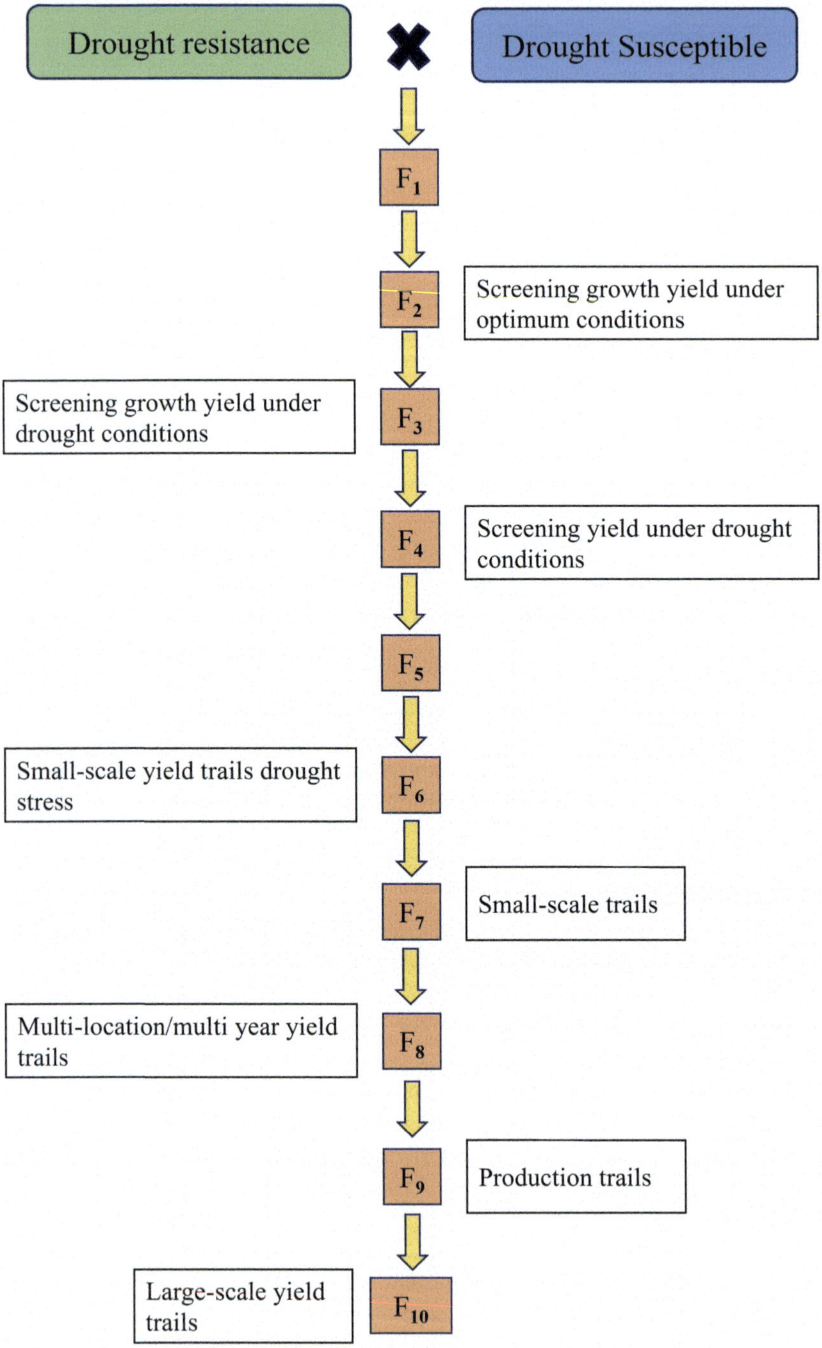

Fig. 14.2 Flowchart of Selection procedure for drought tolerance in rice. (Source: Madhu et al., 2022)

14.4.3 Recurrent Selection

Varietal improvement employs recurrent selection, a technique that combines favourable alleles through multiple crosses while preserving genetic diversity. This approach yields shorter and well-defined breeding cycles, more accurate genetic gains, and the creation of highly diverse breeding lines. Extensive research has been conducted on this method in maize, rice, and soybean (Oladosu et al., 2019). In wheat, it has proven successful in enhancing grain yield (Rebetzke et al., 2002) kernel weight, and the percentage of grain protein (Delzer & Hareland, 1995). In summary, the superior effectiveness of this method in improving agronomic traits and enhancing drought tolerance establishes its superiority over the pedigree selection method.

14.4.4 Back Cross Breeding

This technique assists in the development of drought-resistant cultivars (Lafitte et al., 2006; Madhu et al., 2022; Oladosu et al., 2014). The backcrossing technique is commonly utilized in rice breeding to transfer specific genes controlling a desired trait from a donor parent to a recipient parent. This approach aims to decrease the genome of the donor parent while enhancing the genetic makeup of the recipient parent, ultimately increasing the likelihood of obtaining a high recovery of the desired traits (Madhu et al., 2022). This method provides a precise and effective means of generating multiple superior breeding lines with the desired genetic characteristics.

14.4.5 Induced Mutation

The technique of induced mutation is employed in plant mutation breeding to generate novel varieties with desired traits. In Myanmar, the Manawthukha rice variety underwent exposure to ^{300}Gy of gamma radiation from a ^{60}Co source, aiming to assess its drought resistance by refraining from irrigation from 90 days posttransplantation until harvest. Through six generations of evaluation and selection, two mutant lines, MK-D-2 and MK-D-3, were identified as drought-resistant using physiological screening techniques (Soe et al., 2016). Similarly, in Iran, 11 lines exhibiting drought-tolerant characteristics were selected from the indigenous rice landrace "Tarom Mahalli" after exposure to an optimal gamma radiation dose of ^{230} Gy (Hallajian et al., 2014). Indonesia utilized induced mutations to create a super green rice mutant that demonstrates resilience to drought, high productivity, and water efficiency (Efendi et al., 2017). In Malaysia, enhanced lines, MR219-9 and MR219-4, displaying both high production potential and drought tolerance, were derived from the widely cultivated MR219 rice variety (Rahim et al., 2012).

14.4.6 Modern Techniques for Drought Mitigation with Marker-Assisted Breeding

Currently, substantial initiatives are underway to advance the development of rice cultivars that exhibit resilience to drought conditions. The modern breeding methodology involves the identification of genes or quantitative trait loci (QTLs) associated with specific phenotypic traits. Following the mapping process, these identified genetic elements are incorporated into an elite gene pool. Subsequently, the process involves the application of marker-associated selection (MAS) to facilitate the development of drought-resistant cultivars. Marker-associated selection is an indirect approach in which desired traits are chosen based on markers, including morphological, biochemical, or DNA/RNA variations linked to specific characteristics such as high yield, drought resistance, or tolerance (Hassan et al., 2023; Lamont et al., 2014). Utilizing biomarkers and contemporary breeding methods, such as molecular breeding and gene editing tools, is strongly advised to expedite the process of developing new varieties (Gouda et al., 2020). The ability to tolerate drought involves a complex mechanism regulated by numerous genes and quantitative trait loci (QTLs) (Suh et al., 2015).

To understand the methodology of marker-associated modern breeding techniques better, a simplified illustration of the marker-assisted selection method is depicted in Fig. 14.3. Marker-assisted selection (MAS) is a powerful breeding strategy employed in crop improvement, particularly for developing stress-tolerant varieties such as drought-tolerant rice. The process begins with the identification of drought-responsive quantitative trait loci (QTLs) (some drought-related QTLs are listed in Table 14.3) through genetic studies. Subsequently, molecular markers, such as single nucleotide polymorphisms (SNPs) or simple sequence repeats (SSRs), are developed and linked to these QTLs (Suh et al., 2015). In the genotyping phase, a diverse population of rice cultivars undergoes DNA analysis using these molecular markers, enabling the identification of individuals carrying the desired QTLs associated with drought tolerance. Phenotyping follows, subjecting the genotyped population to drought stress conditions to evaluate and record the phenotypic responses. The selection of elite genotypes involves choosing individuals exhibiting both favorable genotypes and phenotypes. These elite genotypes become donors in the backcrossing and breeding stage, where they are crossed with popular rice varieties to introgress the desired QTLs while retaining essential agronomic traits. Marker-assisted selection is then iteratively applied in successive generations, integrating genotyping, phenotyping, and backcrossing, to refine and enhance the drought tolerance traits. The process culminates in field testing, validation, and the eventual release of drought-tolerant rice cultivars for widespread adoption in agricultural practices.

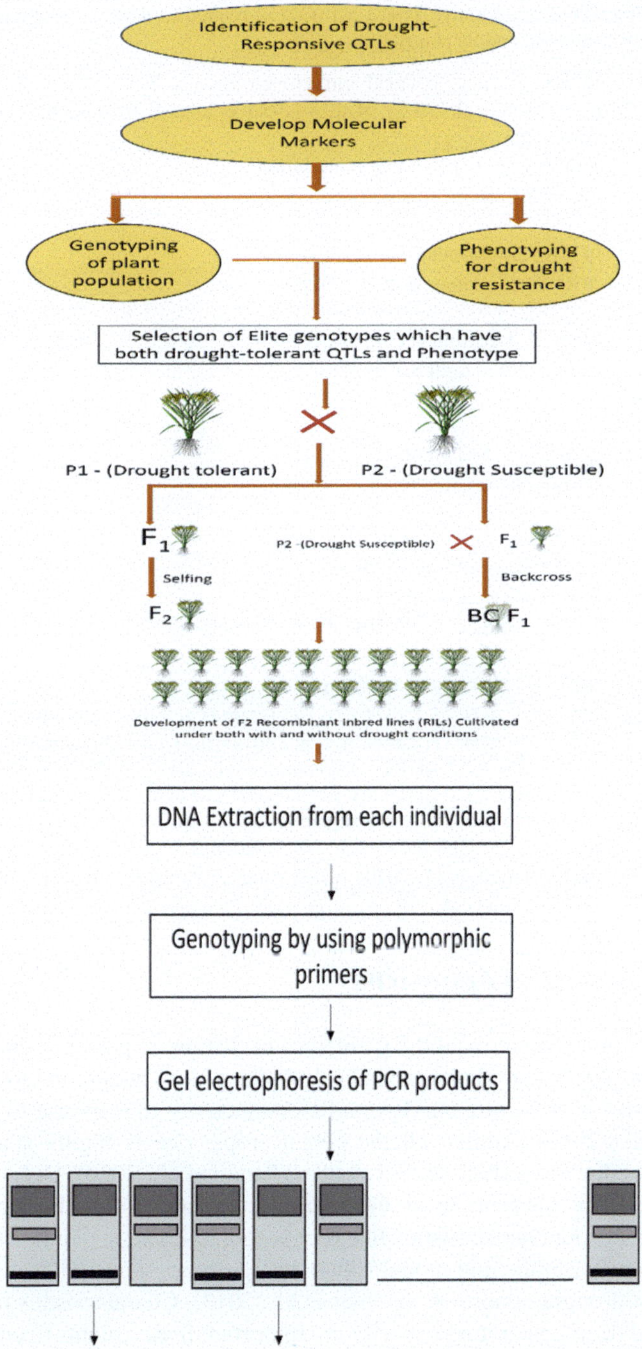

Fig. 14.3 Flowchart showing marker-assisted selection. (Source: Chengqi et al., 2024; Hasan et al., 2021)

Table 14.3 Identifying Quantitative Trait Loci (QTLs) related to drought tolerance in rice that influence both overall yield and its contributing traits

Traits	QTLs
Leaf rolling	qlr8.1 (Lin et al., 2007), qLR9.1 (Barik et al., 2019), qDLR8.1 (Dixit et al. 2014a, b)
Leaf drying	qLD9.1, qLD12.1 (Barik et al., 2019)
Relative water content	qRWC9.1 (Barik et al., 2019)
Transpiration	qDTR8 (Ramchander et al., 2016)
Total dry matter yield	qHGW1 (Sellamuthu et al., 2015)
Plant height	qPH1.1 (Trijatmiko et al., 2014)
Days to heading, grain/ panicle	qPSS8.1 (Trijatmiko et al., 2014)
Flowering day	qHGW2.2 (Sellamuthu et al., 2015)
Panicle number, grain weight	qGy7 (Zhao et al., 2008)
Panicle number	qgy3.1 (Lanceras et al., 2004)
Panicle length	qPL-9 (Zhao et al., 2008)
Spikelet fertility	qSF9.1 (Barik et al., 2019)
Grain number	qDTY8.1 (Vikram et al., 2016)
Grain Yield	qDTY1.1, qDTY3.2, qDTY10.1 (Vikram et al., 2016), (Ghimire et al., 2012), qDTY1.2, qDTY1.3(Sandhu et al., 2014), qDTY2.1 (Venuprasad et al., 2012), qDTY2.2, qDTY2.3 (Palanog et al., 2014), qDTHI2.3 (Mishra et al., 2019), qDTY3.1, qDTY6.1, qDTY6.2 (Dixit et al., 2014a, b), qDTY4.1, qDTY9.1, qDTY10.2 (Swamy et al., 2013), qDTY9.1A (Dixit et al. 2014), qDTY12.1 (Sellamuthu et al., 2015), qPDL1.2, qPNF3.1 (Lin et al., 2007)
Harvest index	qHI9.1 (Barik et al., 2019), qSf6 (Wang et al. 2013a, b), qPNF3.1 (Sellamuthu et al., 2015)

14.4.7 Physiological Approaches

Roots are the first structures in the plant system that are exposed to drought stress (Niones et al., 2012) and play a crucial role in the plant's adaptation to drought by adjusting both root structure and hydraulic conductivity in response to such stress (Maurel et al., 2010). Among all the cereal crops, rice is highly susceptible to drought stress due to its shallow root system (Pandit et al., 2020). Deep-rooted rice cultivars are more resistant to drought stress compared to shallow-rooted ones. Modification of root characteristics like root length, thickness, depth, root-to-shoot ratio, root density, branching, nodal roots, etc., are important in developing resistance against drought conditions (Comas et al., 2013). Characteristics like densely branched and deep penetrating capacity are important in rice plants to enhance their water uptake and maintain their water reserves during severe drought conditions (Gowda et al., 2011).

Aquaporins (AQPs) are intrinsic proteins that play a central role in the formation of permeable channels facilitating the intracellular transport of water and various solutes. This function is vital for maintaining hydraulic conductivity and meeting the transpiration demands of the plant (Pawłowicz & Masajada, 2018). These proteins facilitate passive exchange of gaseous molecules like CO_2, O_2, and water molecules through the membrane. Genes related to AQP family are reported very less. To date, only a limited number of AQP genes in rice have been thoroughly studied regarding their functional responses to drought stress. Lian et al. (2004) found that the rice water channel 3 (RWC3/OsPIP1;3) protein, regulated by a stress-inducible promoter, is instrumental in mitigating drought effects in genetically modified lowland rice. This suggests that the manipulation of OsPIP1;3 expressions can significantly contribute to enhancing drought tolerance in rice plants.

A crucial physiological approach to alleviate drought stress involves phytohormone-mediated drought tolerance. This strategy focuses on the regulation of plant cell water loss and water status under severe drought conditions, primarily through the maintenance of abscisic acid (ABA) levels. This hormonal control is highlighted in studies by de Ollas and Dodd (2016), Kuromori et al. (2018), and Zhang et al. (2020). Abscisic acid, a plant hormone, is released in response to drought stress. It serves a pivotal function in facilitating stomatal closure, thereby assisting in the mitigation of water loss from the plant via transpiration (Kim et al., 2010). In rice, this hormone enhances root growth and promotes water uptake, and also increases the soluble sugar accumulation to maintain standard water content in leaf and osmotic balance during drought conditions (Pattanagul, 2011). Several genes that are responsive to ABA significantly mitigate drought stress and promote optimal plant growth even in drought conditions. For example, ASR gene (OsASR5, OsASR 1, etc.) which contains ABA/WDS domain plays vital role in ABA-mediated stress responses in rice (Gonzalez & Iusem, 2014).

14.5 Application of Artificial Intelligence (AI) and Simulation Models in Drought Monitoring

Assessment, monitoring, and forecasting of drought provide prior advantages to manage drought stress efficiently for overall production system irrespective of crop in long run under climate change scenarios. Recent research has witnessed widespread utilization of AI in analysing drought across different regions, indicating a burgeoning impact and a trajectory for increased future influence. To comprehend diverse drought concepts, establishing distinct drought management systems becomes essential. These systems should monitor various factors influencing drought and implement measures to mitigate its adverse effects. Modeling and probabilistic analysis of climatic data to predict drought vulnerability mapping is becoming the future research (Kikon & Deka, 2022). Agricultural drought is notably responsive to the variation in climatic factors across space and time, especially during the rainfed crop season unlike hydrological, meteorological, and socioeconomic droughts (Houmma et al., 2022). Globally, the collective efforts in various approaches have led to the development of approximately 150 indices both including single and multivariate composite indices for the assessment, classification, and monitoring of drought (Svoboda & Fuchs, 2016). Recent advancements in geospatial technologies, artificial intelligence, and the utilization of cloud computing platforms like Google Earth Engine have significantly propelled the field of drought studies. In this context, the adoption of sophisticated machine learning techniques, as data-driven approaches, is increasingly recognized as an essential alternative to traditional methods relying on statistical analysis or domain expertise (Jung et al., 2021; Shen et al., 2019). Numerous studies focusing on crop modeling, exploring diverse scenarios concerning the impact of climate change on rice production, have been extensively documented. Many of these studies delve into specific countries, examining various management approaches, while others concentrate on adaptation methods and the dynamics of rice cropping systems (Ahmad et al., 2012; Yoshida et al., 2015). Here, some enlisted simulation models (Table 14.4) given specifically used for rice production to study the effect of climate change (Kao et al., 2013; Vaghefi et al., 2013; Zhang et al., 2016).

Table 14.4 Simulation models applied for rice production

Name of the model	Description	Application
CERES_RICE	A model designed for knowledge transfer in agro-technology, focusing on a process-based and management-oriented approach	Modeling crop growth and yield reactions to alterations in management practices Conducting assessments of regional climate change impacts, and examining the environmental influence on different genotypes
RICESYS	A demographic model that explores the growth and development of rice under the influence of temperature and solar radiation	Evaluating the impacts of climate change, elucidating the dynamics of rice growth and development, analysing yield and its components concerning transplanting date and planting density Assessing the influence of climate change on rice plants, their weeds, and herbivores
RICEMOD	An ecophysiological model for irrigated rice production based on Fortran and Basic programming languages	Growth and yield components predictions, water stress
MACROS	Assessment of different scenarios of climate change impact on rice production	Evaluations of the impact of climate change, planning optimal planting dates, and estimating grain yield
SIMRIW	A model that simulates growth physiological processes	Assessment of various cultivation technologies and climatic productivity.
ORYZA2000	A follow-up to a sequence of rice growth models, serving as a comprehensive update and integration of the ORYZA1 model for potential production, ORYZA-W for water-limited production, and ORYZA-N for nitrogen-limited production	Thorough physiological examination of field experiments, the assessment of crop performance, and optimization of crop management tailored to a specific biophysical environment (climate and soil). This includes considerations for expected climate change, as well as activities related to breeding and the evaluation of germplasm

14.6 Conclusion

Rice-growing nations across the world have invested their scientific manpower for improving drought management in rice. Drought stress, that varies with climate and soil types, can be havoc depending on crop stages and severity. The stress may be even irrecoverable sometimes leading to significant reduction in crop biomass, complete dryness, spikelet sterility, incomplete grain formation, and complete loss of yield. As a future research prospective, breeding of improved varieties with drought tolerance ability shall be a prioritized area. Large sets of breeding lines can be evaluated in an artificial screening facility where stress is imposed intermittently from transplanting (lowland) or maximum tillering (upland) up to harvesting stage, only life-saving irrigation may be provided when soil below field capacity in the root zone. It has been observed that yields in the stress trial hence be reduced to a range of 20–40% of yields under non-stress conditions. Optimization of agro-management is another approach that supports the drought tolerance (mostly by conserving the adequate soil moisture) to ensure better yield and yield attributing parameters even also the drought hits moderately for traditional or non-tolerant cultivars. Both the breeding approach and improved agro-management may elevate the drought stress tolerance for rice crops addressing the climate-resilient agriculture. With the advancement of science, simulation models were developed that predict the drought occurrence and future crop growth scenario near perfectly for rice crops, adapted by agronomists, physiologists, and soil scientists across the world. We hope this chapter comprehensively enriches the knowledge base for new generation drought mitigation strategies for rice crops across a wide range of science readers.

References

Agarie, S., Agata, W., Kubota, F., et al. (1998). Effects of silicon on transpiration and leaf conductance in rice plants (*Oryza sativa* L.). *Plant Production Science, 2*, 89–95.

Agarwal, A. P., & Khan, R. U. (2002). Performance of upland rice genotypes under terminal drought. In *National symposium abstract on upland rice production systems 26–28 September, Hazaribag, Jharkhand* (p. 33).

Ahmad, S., Ahmad, A., Soler, C. M. T., Ali, H., Zia-Ul-Haq, M., Anothai, J., Hussain, A., Hoogenboom, G., & Hasanuzzaman, M. (2012). Application of the CSM-CERES-Rice model for evaluation of plant density and nitrogen management of fine transplanted rice for an irrigated semiarid environment. *Precision Agriculture, 13*, 200–218.

Ahmad, S., Raza, M. A. S., Saleem, M. F., Zahra, S. S., Khan, I. H., Ali, M., Shahid, A. M., Iqbal, R., & Zaheer, M. S. (2015). Mulching strategies for weeds control and water conservation in cotton. *Journal of Agricultural and Biological Sciences, 10*(8), 299–306.

Ahmad, Z., Anjum, S., Skalicky, M., Waraich, E. A., Muhammad Sabir Tariq, R., Ayub, M. A., Hossain, A., Hassan, M. M., Brestic, M., Sohidul Islam, M., & Habib-Ur-Rahman, M. (2021). Selenium alleviates the adverse effect of drought in oilseed crops camelina (*Camelina sativa* L.) and Canola (*Brassica napus* L.). *Molecules, 26*(6), 1699–1716.

Alam, S. M. (1999). Nutrient uptake by plants under stress conditions. In M. Pessarakli (Ed.), *Plant and crop stress* (pp. 285–313). Marcel Dekker.

Alsafar, M. S., & Al-Hassan, Y. M. (2009). Effect of nitrogen and phosphorus fertilizers on growth and oil yield of indigenous mint (*Mentha longifolia* L.). *Biotechnology, 8*, 380–384.

Ananda, M. G., & Reddy, V. C. (2002). Effect of sowing methodsand seed treatments on grain yield of rainfed low land paddy varieties. *Mysore Journal of Agricultural Sciences, 36*, 231–236.

Anjum, S. A. (2019). Interactive effect of rice produciton systems and tillage systems in rice-wheat cropping system. *Pakistan Journal of Science, 71*(2), 21.

Arif, N., Yadav, V., Singh, S., Singh, S., Ahmad, P., Mishra, R. K., Sharma, S., Tripathi, D. K., Dubey, N. K., & Chauhan, D. K. (2016). Influence of high and low levels of plant-beneficial heavy metal ions on plant growth and development. *Frontiers in Environmental Science, 4*, 69.

Arouna, A., Dzomeku, I. K., Shaibu, A. G., & Nurudeen, A. R. (2023). Water management for sustainable irrigation in rice (Oryza sativa L.). *Production: A Review, 13*(6), 1522.

Ashoka, P., Kadasiddappa, M. M., & Sanjey, M. T. (2015). Enhancing water productivity through microirrigation technologies in Indian agriculture. *Annals of Plant and Soil Research, 17*(Special Issue), 601–605.

Ashraf, M., Akram, N. A., Al-Qurainy, F., & Foolad, M. R. (2011). Drought tolerance: Roles of organic osmolytes, growth regulators, and mineral nutrients. *Advances in Agronomy, 111*, 249–296. https://doi.org/10.1016/B978-0-12-387689-8.00002-3

Ata-Ul-Karim, S. T., Liu, X., Lu, Z., Yuan, Z., Zhu, Y., & Cao, W. (2016). In-season estimation of rice grain yield using critical nitrogen dilution curve. *Field Crops Research, 195*, 1–8. https://doi.org/10.1016/j.fcr.2016.04.027

Atkinson, N. J., & Urwin, P. E. (2012). The interaction of plant biotic and abiotic stresses: From genes to the field. *Journal of Experimental Botany, 63*, 3523–3543.

Atlin, G. N., Venuprasad, R., Bernier, J., Zhao, D., Virk, P., & Kumar, A. (2009). Rice germplasm development for drought-prone environments: progress made in breeding and genetic analysis at the International Rice Research Institute. In *Drought frontiers in rice: Crop improvement for increased rainfed production* (pp. 35–59).

Attri, M., Bharti, V., Ahmad Nesar, N., Mehta, S., Bochalya, R. S., Kumar Bansal, K., & Sandhu, R. (2022). Improved irrigation practices for higher agricultural productivity: A review. *International Journal of Environment and Climate Change, 12*, 51–61.

Balk, J., & Pilon, M. (2011). Ancient and essential: the assembly of iron–sulfur clusters in plants. *Trends in Plant Science, 16*, 218–226.

Barik, S. R., Pandit, E., Pradhan, S. K., Mohanty, S. P., & Mohapatara, T. (2019). Genetic mapping of morpho-physiological traits involved during reproductive stage drought tolerance in rice. *PLoS One, 14*(12), e0214979.

Begum, N., Ahanger, M. A., & Zhang, L. (2020). AMF inoculation and phosphorus supplementation alleviates drought induced growth and photosynthetic decline in *Nicotiana tabacum* by up-regulating antioxidant metabolism and osmolyte accumulation. *Environmental and Experimental Botany, 176*, 104088. https://doi.org/10.1016/j.envexpbot.2020.104088

Bennett, J. P., & Skoog, F. (2002). Preliminary experiments on the relation of growth-promoting substances to the rest period in fruit trees. *Plant Physiology, 13*, 219–225. https://doi.org/10.1104/pp.13.2.219

Bhan, S., & Behera, U. K. (2014). Conservation agriculture in India–Problems, prospects and policy issues. *International Soil and Water Conservation Research, 2*(4), 1–2.

Binghua, L., Liang, C., Mingjun, L., Dong, L., Yangjun, Z., & Fengwang, M. (2012). Interactive effects of water and nitrogen supply on growth, biomass partitioning, and water-use efficiency of young apple trees. *African Journal of Agricultural Research, 7*, 978–985.

Blum, A. (2009). Effective use of water (EUW) and not water-use efficiency (WUE) is the target of crop yield improvement under drought stress. *Field Crops Research, 112*, 119–123.

Bodner, G., Nakhforoosh, A., & Kaul, H. P. (2015). Management of crop water under drought: a review. *Agronomy for Sustainable Development, 35*, 401–442.

Bouman, B. A., & Tuong, T. P. (2001). Field water management to save water and increase its productivity in irrigated lowland rice. *Agricultural Water Management, 49*(1), 11–30.

Bouman, B. A., Peng, S., Castaneda, A. R., & Visperas, R. M. (2005). Yield and water use of irrigated tropical aerobic rice systems. *Agricultural Water Management, 74*(2), 87–105.

Bujarbaruah, K. M., Sharma, B. K., Pattanayak, A., Annaduraland, A., & Das, A. (2002). Problems and prospects of upland rice in North eastern Hill region of India. In *National symposium abstract on upland rice production, 26–28 September; Hazaribag, Jharkhand* (pp. 16–17).

Cakmak, I., & Kirkby, E. A. (2008). Role of magnesium in carbon partitioning and alleviating photooxidative damage. *Physiologia Plantarum, 133*, 692–704.

Chandra, D., Moorthy, B. T. S., Jha, K. P., & Manna, G. B. (1991). Agronomic practices for augmenting rice production in rainfed upland ecosystem of Orissa. *Indian Journal of Agronomy, 24*, 199–202.

Chen, W., Yao, X., Cai, K., & Chen, J. (2011). Silicon alleviates drought stress of rice plants by improving plant water status, photosynthesis, and mineral nutrient absorption. *Biological Trace Element Research, 142*, 67–76.

Chengqi, Z., Yuxuan, Y., Yafan, H., Jifeng, Y., & Zhicheng, S. (2024). Drought-tolerant rice at molecular breeding Eras: An emerging reality. *Rice Science, 31*(2), 179–189.

Chourasia, K. N. (2017). Resistance tolerance mechanism under water deficit (drought) condition in plants. *International Journal of Current Microbiology and Applied Sciences, 6*, 66–78.

Comas, L. H., Becker, S. R., Cruz, V. M., Byrne, P. F., & Dierig, D. A. (2013). Root traits contributing to plant productivity under drought. *Frontiers in Plant Science, 4*, 442.

DalCorso, G., Manara, A., Piasentin, S., & Furini, A. (2014). Nutrient metal elements in plants. *Metallomics, 6*, 1770–1788.

Das, A., Lal, R., Patel, D. P., Idapuganti, R. G., Layek, J., Ngachan, S. V., Ghosh, P. K., Bordoloi, J., & Kumar, M. (2014). Effects of tillage and biomass on soil quality and productivity of lowland rice cultivation by small scale farmers in North Eastern India. *Soil and Tillage Research, 143*, 50–58.

Das, A., Layek, J., Idapuganti, R. G., Basavaraj, S., Lal, R., Rangappa, K., Yadav, G. S., Babu, S., & Ngachan, S. (2020). Conservation tillage and residue management improves soil properties under a upland rice–rapeseed system in the subtropical eastern Himalayas. *Land Degradation & Development, 31*(14), 1775–1791.

de Ollas, C., & Dodd, I. C. (2016). Physiological impacts of ABA-JA interactions under water-limitation. *Plant Molecular Biology, 91*, 641–650.

Delzer, B. W., & Hareland, G. A. (1995). Recurrent selection for grain protein in hard red spring wheat. *Crop Science, 35*, 730–735.

Devi, S. H., & Kar, M. (2013). Amelioration of moisture stress effect by $CaCl_2$ pre-treatment in upland rice. *Indian Journal of Plant Physiology, 18*, 384–387. https://doi.org/10.1007/S40502-014-0058-Y

Dixit, S., Singh, A., & Kumar, A. (2014a). Rice breeding for high grain yield under drought: A strategic solution to a complex problem. *International Journal of Agronomy*. https://doi.org/10.1155/2014/863683

Dixit, S., Singh, A., Cruz, M. T. S., Maturan, P. T., Amante, M., & Kumar, A. (2014b). Multiple major QTL led to stable yield performance of rice cultivars across varying drought intensities. *BMC Genetics, 15*, 16. https://doi.org/10.1186/1471-2156-15-16

Efendi, B., Sabaruddin, Z., & Lukman, H. (2017). Mutation with gamma raysirradiation to assemble green super rice tolerant to drought stress and high yield rice (*Oryza sativa* L.). *International Journal of Advanced Science, Engineering and Technology, 5*, 1–5.

El-Okkiah, S. A. F., El-Afry, M. M., Shehab Eldeen, S. A., El-Tahan, A. M., Ibrahim, O. M., Negm, M. M., Alnafissa, M., El-Saadony, M. T., Almazrouei, H. M. R. S., AbuQamar, S. F., El-Tarabily, K. A., & Selim, D. A. (2022). Foliar spray of silica improved water stress tolerance in rice (*Oryza sativa* L.) cultivars. *Frontiers in Plant Science, 13*, 935090. https://doi.org/10.3389/fpls.2022.935090

Farooq, M. K., Siddique, K. H., Rehman, H., Aziz, T., Lee, D. J., & Wahid, A. (2011). Rice direct seeding: Experiences, challenges and opportunities. *Soil and Tillage Research, 111*(2), 87–98.

Faustino, G. P., Hernandez, J. E., Desamero, N. V., & Cruz, R. T. (2019). Field application of potassium fertilizer mitigates impact of reproductive stage drought stress on rice yield. *Rice-Based Biosystems Journal, 5*, 1–17.

Gao, X., Zou, C., Wang, L., et al. (2006). Silicon decreases transpiration rate and conductance from stomata of maize plants. *Journal of Plant Nutrition, 29*, 1637–1647.

Ghimire, K. H., Quiatchon, L. A., Vikram, P., Swamy, B. P. M., Dixit, S., Ahmed, H., et al. (2012). Identification and mapping of a QTL (qDTY1.1) with a consistent effect on grain yield under drought. *Field Crops Research, 131*, 88–96.

Ghosh, A. (2002). Growth and yield of upland rice genotypes under Normal and late sown conditions. In *National symposium abstract on upland rice production systems, 26–28 September, Hazaribag, Jharkhand* (pp. 43–44).

Ghouri, F., Ali, Z., Naeem, M., Ul-Allah, S., Babar, M., Baloch, F. S., Chattah, W. S., & Shahid, M. Q. (2021). Effects of silicon and selenium in alleviation of drought stress in rice. *Silicon*. https://doi.org/10.1007/s12633-021-01277-z

Gong, H. J., Chen, K. M., Chen, G. C., Wang, S. M., & Zhang, C. L. (2003). Effects of silicon on growth of wheat under drought. *Journal of Plant Nutrition, 26*, 1055–1063. https://doi.org/10.1081/PLN-120020075

Gong, H. J., Chen, K. M., Zhao, Z. G., Chen, G. C., & Zhou, W. J. (2008). Effects of silicon on defense of wheat against oxidative stress under drought at different developmental stages. *Biologia Plantarum, 52*, 592–596. https://doi.org/10.1007/s10535-0080118-0

Gonzalez, R. M., & Iusem, N. D. (2014). Twenty years of research on Asr (ABA-stress-ripening) genes and proteins. *Planta, 239*, 941–949.

Gouda, G., Gupta, M. K., Donde, R., Mohapatra, T., Vadde, R., & Behera, L. (2020). Marker-assisted selection for grain number and yield-related traits of rice (*Oryza sativa* L.). *Physiology and Molecular Biology of Plants, 26*, 885–898. https://doi.org/10.1007/S12298-020-00773-7

Gowda, V. R. P., Henry, A., Yamauchi, A., Shashidhar, H. E., & Serraj, R. (2011). Root biology and genetic improvement for drought avoidance in rice. *Field Crops Research, 122*, 1–13.

Haefele, S. M., Siopongco, J. D. L. C., Boling, A. A., Bouman, B. A. M., & Tuong, T. P. (2009). Transpiration efficiency of rice (*Oryza sativa* L.). *Field Crops Research, 111*, 1–10.

Haefele, S. M., Kato, Y., & Singh, S. (2016). Climate ready rice: Augmenting drought tolerance with best management practices. *Field Crops Research, 190*, 60–69.

Hajiboland, R. (2012). Effect of micronutrient deficiencies on plants stress responses. In P. Ahmad & M. N. V. Prasad (Eds.), *Abiotic stress responses in plants* (pp. 283–329). Springer.

Hallajian, M. T., Ebadi, A. A., Mohammadi, M., Muminjanov, H., Jamali, S. S., & Aghamirzaei, M. (2014). Integration of mutation and conventional breeding approaches to develop new superior drought-tolerant plants in rice (*Oryza sativa*). *Annual Research & Review in Biology, 4*, 1173–1186.

Hardoim, P. R., van Overbeek, L. S., Berg, G., Pirttilä, A. M., Compant, S., Campisano, A., et al. (2015). The hidden world within plants: ecological and evolutionary consider ations for defining functioning of microbial endophytes. *Microbiology and Molecular Biology Reviews, 79*(3), 293–320.

Hasan, N., Choudhary, S., Naaz, N., Sharma, N., & Laskar, R. A. (2021). Recent advancements in molecular marker-assisted selection and applications in plant breeding programmes. *Journal of Genetic Engineering and Biotechnology, 19*, 128.

Hassan, M. J., Wang, F., Ali, S., & Zhang, G. (2005). Toxic effects of cadmium on rice as affected by nitrogen fertilizer form. *Plant and Soil, 277*, 359–365.

Hassan, M. A., Dahu, N., Hongning, T., Qian, Z., Yueming, Y., Yiru, L., & Shimei, W. (2023). Drought stress in rice: morphophysiological and molecular responses and marker-assisted breeding. *Frontiers in Plant Science, 14*, 1215371. https://doi.org/10.3389/fpls.2023.1215371

Hazman, M., Fawzy, S., Hamdy, A., Khaled, A., Mahmoud, A., Khalid, E., Ibrahim, H. M., Gamal, M., Elyazeed, N. A., Saber, N., Ehab, M., & Kabil, F. (2023). Enhancing rice resilience to drought by applying biochar–compost mixture in low-fertile sandy soil. *Beni-Suef University Journal of Basic and Applied Sciences, 12*, 74. https://doi.org/10.1186/s43088-023-00411-7

Houmma, I. H., El Mansouri, L., Gadal, S., Garba, M., & Hadria, R. (2022). Modelling agricultural drought: A review of latest advances in big data technologies. *Geomatics, Natural Hazards and Risk, 13*(1), 2737–2776. https://doi.org/10.1080/19475705.2022.2131471

https://www.jains.com/Company/news/blog/Crop_gasping_for_some_air.htm

Hu, Y., & Schmidhalter, U. (2005). Drought and salinity: A comparison of their effects on the mineral nutrition of plants. *Journal of Plant Nutrition and Soil Science, 168*, 541–549.

Hussain, S., Mumtaz, M., Manzoor, S., Shuxian, L., Ahmed, I., Skalicky, M., et al. (2021). Foliar application of silicon improves growth of soybean by enhancing carbon metabolism under shading conditions. *Plant Physiology and Biochemistry, 159*, 43–52. https://doi.org/10.1016/j.plaphy.2020.11.053

Ingram, K. T., Real, J. G., Maguling, M. A., Obien, M. A., & Loresto, G. C. (1990). Comparison of selection indices to screen lowland rice for drought resistance. *Euphytica, 48*, 253–260.

IRRI. (2006). *Drought-prone environments -Rice Knowledge Bank*. International Rice Research Institute. http://www.knowledgebank.irri.org/ricebreedingcourse/Incorporating_drought_tolerance_as_a_rice_breeding_objective.htm

Ishfaq, M., Akbar, N., Anjum, S. A., & Anwar-Ijl-Haq, M. (2020). Growth, yield and water productivity of dry direct seeded rice and transplanted aromatic rice under different irrigation management regimes. *Journal of Integrative Agriculture, 19*(11), 2656–2673.

Jin, J., Lauricella, D., Armstrong, R., Sale, P., & Tang, C. (2015). Phosphorus application and elevated CO_2 enhance drought tolerance in field pea grown in a phosphorus-deficient vertisol. *Annals of Botany, 116*, 975–985.

Johnson, M. S. (1984). The effects of gel-forming polyacrylamides on moisture storage in sandy soils. *Journal of the Science of Food and Agriculture, 35*(11), 1196–1200.

Joshi, R., Singh, B., & Shukla, A. (2018). Evaluation of elite rice genotypes for physiological and yield attributes under aerobic and irrigated conditions in Tarai areas of western Himalayan region. *Current Plant Biology, 13*, 45–52.

Jung, J., Maeda, M., Chang, A., Bhandari, M., Ashapure, A., & Landivar-Bowles, J. (2021). The potential of remote sensing and artificial intelligence as tools to improve the resilience of agriculture production systems. *Current Opinion in Biotechnology, 70*, 15–22.

Kao, C. M., Tu, Y. T., Cheng, S. F., Surampalli, R. Y., & Zhang, T. C. (2013). Modelling the impact of climate change on agriculture and food production. In R. Y. Surampalli, T. C. Zhang, C. S. P. Ojha, B. Gurjar, R. D. Tyagi, & C. M. Kao (Eds.), *Americam Society of Civil Engineers* (pp. 454–470). Preston.

Kar, G., & Verma, H. N. (2002). Improved sustainability of rainfed upland sub-humid ecosystem of watershed based on probabilistic rainfall. *Hydrology and Watershed, 1*, 318–327.

Kar, G., Singh, R. and Verma, H. N. (2004). Alternative cropping strategies for assured and efficient crop production in upland rain fed rice areas of Eastern India based on rainfall analysis. *Agricultural Water Management, 67*, 47–62.

Kato, Y., & Katsura, K. (2014). Rice adaptation to aerobic soils: physiological considerations and implications for agronomy. *Plant Production Science, 17*(1), 1–2.

Khan, M. A., Ansari, R., Ali, H., Gul, B., & Nielsen, B. L. (2009). Panicum turgidum, a potentially sustainable cattle feed alternative to maize for saline areas. *Agriculture, Ecosystems and Environment, 129*, 542–546.

Khatua, S. (2002). *Studies on drought management in upland rice*. M.Sc. (Ag) thesis, Department of Agronomy, OUAT, Bhubaneswar.

Khush, G. S. (1984). IRRI breeding program and its worldwide impact on increasing rice production. In *Gene manipulation in plant improvement* (pp. 61–94). Springer.

Kibria, M. G., Farzana, K., Matin, M. A., & Hoque, M. A. (2016). Mitigating water stress in wheat (BARI Gom-26) by exogenous application of proline. *Fundamental and Applied Agriculture, 1*, 118–123.

Kikon, A., & Deka, P. C. (2022). Artificial intelligence application in drought assessment, monitoring and forecasting: A review. *Stochastic Environmental Research and Risk Assessment, 36*, 1197–1214. https://doi.org/10.1007/s00477-021-02129-3

Kim, T. H., Böhmer, M., Hu, H., Nishimura, N., & Schroeder, J. I. (2010). Guard cell signal transduction network: advances in understanding abscisic acid, CO_2 and Ca^{2+} signalling. *Annual Review of Plant Biology, 61*, 561–591.

Kumar, R., Trivedi, H., Yadav, R., Das, B., & Bist, A. S. (2016). Effect of drip irrigation on yield and water use efficiency on brinjal (Solanum melongena) cv. Pant samrat. *International Journal of Engineering Sciences & Research Technology, 5*(10), 7–17.

Kuromori, T., Seo, M., & Shinozaki, K. (2018). ABA transport and plant water stress responses. *Trends in Plant Science, 23*, 513–522.

Lafitte, H. R., Li, Z. K., Vijayakumar, C. H. M., Gao, Y. M., Shi, Y., Xu, J. L., et al. (2006). Improvement of rice drought tolerance through backcross breeding: Evaluation of donors and selection in drought nurseries. *Field Crops Research, 97*, 77–86.

Lala, B., Gautama, P., Nayaka, A. K., Maharana, S., Tripathi, R., Shahid, M., Baiga, M. J., Raja, R., Kato, Y., Srivastava, A. K., & Singh, S. (2019). Tolerant varieties and exogenous application of nutrients can effectively manage drought stress in rice. *Archives of Agronomy and Soil Science*. https://doi.org/10.1080/03650340.2019.1587749

Lamont, S. J., Dekkers, J. C. M., & Zhou, H. (2014). Immunogenetics and the mapping of immunological functions. In *Avian immunology* (2nd ed., pp. 205–221). https://doi.org/10.1016/B978-0-12-396965-1.00011-X

Lampayan, R. M., Rejesus, R. M., Singleton, G. R., & Bouman, B. A. (2015). Adoption and economics of alternate wetting and drying water management for irrigated lowland rice. *Field Crops Research, 170*, 95–108.

Lanceras, J. C., Pantuwan, G., Jongdee, B., & Toojinda, T. (2004). Quantitative trait loci associated with drought tolerance at reproductive stage in rice. *Plant Physiology, 135*(1), 384–399.

Li, Z., Song, Z., Yan, Z., Hao, Q., Song, A., Liu, L., et al. (2018). Silicon enhancement of estimated plant biomass carbon accumulation under abiotic and biotic stresses. a meta-analysis. *Agronomy for Sustainable Development, 38*, 26. https://doi.org/10.1007/s13593-018-0496-4

Lian, H. L., Yu, X., Ye, Q., Ding, X. S., Kitagawa, Y., Kwak, S. S., et al. (2004). The role of aquaporin RWC3 in drought avoidance in rice. *Plant & Cell Physiology, 45*, 481–489.

Liang, C., Li, Y., Zhang, K., Wu, Z., Liu, J., Liu, J., Zhou, C., Wang, S., Li, F., & Sui, G. (2023). Selection and yield formation characteristics of dry direct seeding rice in Northeast China. *Plants, 12*(19), 3496.

Lin, M. H., Lin, C. W., Chen, J. C., Lin, Y. C., Cheng, S. Y., Liu, T. H., et al. (2007). Tagging rice drought- related QTL with SSR DNA markers. *Crop, Environment & Bioinformatics, 4*(1), 65–76.

Madhu, B., Sukrutha, B., Singh, N. U., & Venkateswarao, G. (2022). Breeding strategies for improvement of drought tolerance in rice: recent approaches, and future outlooks. In *Sustainable rice production-challenges, strategies and opportunities*. IntechOpen. https://doi.org/10.5772/intechopen.107313

Mahpara, S., Shahnawaz, M., Rehman, K., Ahmad, R., & Khan, F. U. (2019). Nitrogen fertilization induced drought tolerance in sunflower: a review. *Pure and Applied Biology, 8*(2), 1675–1683. https://doi.org/10.19045/bspab.2019.80110

Maurel, C., Simonneau, T., & Sutka, M. (2010). The significance of roots as hydraulic rheostats. *Journal of Experimental Botany, 61*, 3191–3198.

Meher, W., Sagar, P., Kumar, K., Meher, G. S., & Priyabhavana, G. (2020). Integration of on-farm drip irrigation system in rice cultivation (more crop- per drop). *International Journal of Creative Research Thoughts, 8*, 3555–3561.

Menge, D. M., Onyango, J. C., Yamauchi, A., Kano-Nakata, M., Asanuma, S., Thi, T. T., Inukai, Y., Kikuta, M., & Makihara, D. (2019). Effect of nitrogen application on the expression of drought-induced root plasticity of upland NERICA rice. *Plant Production Science, 22*(2), 180–191.

Merwad, A. R. M., Desoky, E. S. M., & Rady, M. M. (2018). Response of water deficit-stressed *Vigna unguiculata* performances to silicon, proline, or methionine foliar application. *Scientia Horticulturae, 228*, 132–144.

Millaleo, R., Reyes-Díaz, M., Ivanov, A. G., Mora, M. L., & Alberdi, M. (2010). Manganese as essential and toxic element for plants: Transport, accumulation, and resistance mechanisms. *Journal of Soil Science and Plant Nutrition, 10*(4), 470–481. https://doi.org/10.4067/S0718-95162010000200008

Mishra, S. S., Behera, P. K., & Panda, D. (2019). Genotypic variability for drought tolerance-related morpho-physiological traits among indigenous rice landraces of Jeypore tract of Odisha, India. *Journal of Crop Improvement, 33*, 254–278.

Mozafariyan, M., Pessarakli, M., & Saghafi, K. (2017). Effects of selenium on some morphological and physiological traits of tomato plants grown under hydroponic condition. *Journal of Plant Nutrition, 40*(2), 139–144.

Mukherjee, D., Maji, A., Mahapatra, S., & Mandal, B. (2022). Response of various sowing date and tillage options on the performance of wheat under late sown condition. *International Journal of Bioresource Science, 02*, 115–121.

Nakano, A., & Uehara, Y. (1996). The effects of kaolin clay on cuticle transpiration in tomato. In *International symposium on plant production in closed ecosystems* (Vol. 440, pp. 233–238).

Namuco, O. S., Ingram, K. T., & Fuentes, L. T. (1993). Root characteristics of rice genotypes with different drought responses. *IRRN, 18*, 38–39.

Niones, J. M., Suralta, R. R., Inukai, Y., & Yamauchi, A. (2012). Field evaluation on functional roles of root plastic responses on dry matter production and grain yield of rice under cycles of transient soil moisture stresses using chromosome segment substitution lines. *Plant and Soil, 359*, 107–120.

Oladosu, Y., Rafii, M. Y., Abdullah, N., Abdul Malek, M., Rahim, H. A., Hussin, G., et al. (2014). Genetic variability and selection criteria in rice mutant lines as revealed by quantitative traits. *The Scientific World Journal*. https://doi.org/10.1155/2014/190531

Oladosu, Y., Rafii, M. Y., Samuel, C., Fatai, A., Magaji, U., Kareem, I., Kamarudin, Z. S., Muhammad, I., & Kolapo, K. (2019). Drought resistance in rice from conventional to molecular breeding: A review. *International Journal of Molecular Sciences, 20*(14), 3519. https://doi.org/10.3390/ijms20143519

Oliveira, I. C. M., Guilhen, J. H. S., de Oliveira Ribeiro, P. C., Gezan, S. A., Schaffert, R. E., Simeone, M. L. F., & Pastina, M. M. (2020). Genotype-by environment interaction and yield stability analysis of biomass sorghum hybrids using factor analytic models and environmental covariates. *Field Crops Research, 257*, 107929.

Ossom, E. M., Pace, P. F., Rhykerd, R. L., & Rhykerd, C. L. (2001). Effect of mulch on weed infestation, soil temperature, nutrient concentration, and tuber yield in Ipomoea batatas (L.) Lam. in Papua New Guinea. (144). *Tropical Agriculture, 78*(3).

Ouda, S. A., Mesiry, T. E., Adballah, E. F., & Gaballah, M. S. (2007). Effect of water stress on the yield of soybean and maize grown under different intercropping patterns. *Australian Journal of Basic and Applied Sciences, 1*, 578–585.

Oweis, T., & Hachum, A. (2006). Water harvesting and supplemental irrigation for improved water productivity of dry farming systems in West Asia and North Africa. *Agricultural Water Management, 80*(1–3), 57–73.

Padmanabhan, S. (2019). Drip irrigation technology for rice cultivation for enhancing rice productivity and reduction water consumption. In *Proceedings of the 3rd World Irrigation Forum (WIF3), Bali, Indonesia*.

Palanog, A. D., Swamy, B. M., Shamsudin, N. A. A., Dixit, S., Hernandez, J. E., Boromeo, T. H., et al. (2014). Grain yield QTLs with consistent-effect under reproductive stage drought stress in rice. *Field Crops Research, 161*, 46–54.

Palta, J. P. (2000). Stress interactions at the cellular and membrane levels. *Horticultural Science, 25*, 1377–1381.

Panda, D., Mishra, S. S., & Behera, P. K. (2021a). Drought tolerance in rice: focus on recent mechanisms and approaches. *Rice Science, 28*(2), 119–132.

Panda, D., Mishra, S. S., & Behera, P. K. (2021b). Drought tolerance in rice: Focus on recent mechanisms and approaches. *Rice Science, 28*(2), 119–132.

Pandey, S., & Bhandari, H. (2009). Drought: Economic costs and research implications. In J. Serraj, J. Bennet, & B. Hardy (Eds.), *Drought frontiers in rice: Crop improvement for increased rainfed production* (pp. 3–17). World Scientific Publishing.

Pandit, E., Panda, R. K., Sahoo, A., Pani, D. R., & Pradhan, S. K. (2020). Genetic relationship and structure analysis of root growth angle for improvement of drought avoidance in early and mid-early maturing rice genotypes. *Rice Science, 27*, 124–132.

Parthasarathi, T., Vanitha, K., Mohandass, S., & Vered, E. (2018). Evaluation of drip irrigation system for water productivity and yield of rice. *Agronomy Journal, 110*(6), 2378–2389.

Patel, R., Raha, P., & Kundu, P. S. (2019). Impact of antitranspirant application on nutrient uptake and yield of rice (Oryza sativa) grown under water stress condition in inceptisol of Varanasi, Uttar Pradesh. *International Journal of Communication Systems, 7*(6), 1327–1331.

Pattanagul, W. (2011). Exogenous abscisic acid enhances sugar accumulation in rice. *Asian Journal of Plant Sciences, 10*, 212–219.

Pawłowicz, I., & Masajada, K. (2018). Aquaporins as a link between water relations and photosynthetic pathway in abiotic stress tolerance in plants. *Gene, 687*, 166–172.

Percival, D. C., Janes, D. E., Stevens, D. E., & Sanderson, K. (2003). Impact of multiple fertilizer applications on plant growth, development, and yield of wild low bush blueberry (*Vaccinium Angustifolium* Aiton). *Acta Horticulturae, 626*, 415–421.

Ponamperuma, F. N. (1978). Electrochemical changes in submerged soils and the growth of rice. In *Soils and rice* (pp. 421–444). International Rice Research Institute.

Praharaj, C. S., Singh, U., Singh, S. S., & Kumar, N. (2017). Micro-irrigation in rainfed pigeonpea–Upscaling productivity under Eastern Gangetic Plains with suitable land configuration, population management and supplementary fertigation at critical stages. *Current Science*, 95–107.

Ragaglini, G., Triana Jimeno, F., Tozzini, C., Taccini, F., Mantino, A., Puggioni, A., Vered, E., & Bonari, E. (2014). Water saving and reduced arsenic uptake in aerobic rice (Oryza sativa L.): Feasibility of drip irrigation under Mediterranean climate. In *Atti del XLIII Convengo della Società Italiana di Agronomia. Società Italiana di Agronomia.*

Rahim, H. A., Zarifth, S. K., Bhuiyan, M. A. R., Narimah, M. K., Wickneswari, R., & Abdullah, M. Z. (2012). Evaluation and characterization of advanced rice mutant line of rice (*Oryza sativa*), MR219-4 and MR219-9 under drought condition. In *Proceedings of research development seminar, Bangi (Malaysia), 26–28 September, 2012.*

Ramamurthy, V., Patil, N. G., Venugopalan, M. V., & Challa, O. (2009). Effect of drip irrigation on productivity and water-use efficiency of hybrid cotton (Gossypium hirsutum) in Typic Haplusterts. *Indian Journal of Agricultural Sciences, 79*(2), 118–121.

Ramchander, S., Raveendran, M., & Robin, S. (2016). Mapping QTLs for physiological traits associated with drought tolerance in rice (*Oryza sativa* L.). *Journal of Investigative Genomics, 3*(3), 56–61.

Ramulu, V., Rao, V. P., Devi, M. U., Kumar, K. A., & Radhika, K. (2016). Evaluation of drip irrigation and fertigation levels in aerobic rice for higher water productivity. In *Anais do World Irrigation Forum, Chiang Ma* (pp. 1–10).

Ranjan, P., Patle, G. T., Prem, M., & Solanke, K. R. (2017). Organic mulching-a water saving technique to increase the production of fruits and vegetables. *Current Agriculture Research Journal, 5*(3), 371.

Rao, K. V., Gangwar, S., Keshri, R., Chourasia, L., Bajpai, A., & Soni, K. (2017). Effects of drip irrigation system for enhancing rice (Oryza sativa L.) yield under system of rice intensification management. *Applied Ecology and Environmental Research, 15*, 487–495.

Rasheed, A., Hassan, M. U., Aamer, M., Batool, M., Sheng, F. A., Ziming, W. U., & Huijie, L. I. (2020). A critical review on the improvement of drought stress tolerance in rice (Oryza sativa L.). *Notulae Botanicae Horti Agrobotanici Cluj-Napoca, 48*(4), 1756–1788.

Ray, M., Sahoo, K. C., Patro, H., & Biswasi, S. (2016). Drought management in upland rice-a review. *Plant Archives, 16*(2), 497–504.

Rebetzke, G. J., Condon, A. G., Richards, R. A., & Farquhar, G. D. (2002). Selection for reduced carbon isotope discrimination increases aerial biomass and grain yield of rainfed bread wheat. *Crop Science, 42*, 739–745. https://doi.org/10.2135/cropsci2002.7390

Reddy, B. S., Vasudevan, S., Mani, S. K., Uppaluri, S., & Sudakar, M. (2021). Drought mitigation through hydrogel application in rice (Oryza sativa L.) cultivation. *Journal of Experimental Biology and Agricultural Sciences, 9*(6), 727–733.

Rehman, A., Ahmad, R., & Safdar, M. (1984). Effect of hydrogel on the performance of aerobic rice sown under different techniques. *Journal of the Science of Food and Agriculture, 35*(11), 1196–1200.

Rollins, J. A., Habte, E., Templer, S. E., Colby, T., Schmidt, J., & Von Korff, M. (2013). Leaf proteome alterations in the context of physiological and morphological responses to drought and heat stress in barley (Hordeum vulgare L.). *Journal of Experimental Botany, 64*(11), 3201–3212.

Romero-Aranda, M. R., Jurado, O., & Cuartero, J. (2006). Silicon alleviates the deleterious salt effect on tomato plant growth by improving plant water status. *Journal of Plant Physiology, 163*, 847–855. https://doi.org/10.1016/j.jplph.2005.05.010

Rosati, A., Metcalf, S. G., Buchner, R. P., Fulton, A. E., & Lampinen, B. D. (2006). Physiological effects of kaolin applications in well-irrigated and water-stressed walnut and almond trees. *Annals of Botany, 98*(1), 267–275.

Ruíz-Sánchez, M., Armada, E., Muñoz, Y., De Salamone, G., Aroca, I. E., Ruíz-Lozano, R., et al. (2011). *Azospirillum* and arbuscular mycorrhizal colonization enhance rice growth and physiological traits under well-watered and drought conditions. *Journal of Plant Physiology, 168*, 1031–1037.

Salgotra, R. K., & Chauhan, B. S. (2023). Ecophysiological responses of rice (Oryza sativa L.) to drought and high temperature. *Agronomy, 13*(7), 1877.

Sandhu, N., Singh, A., Dixit, S., Cruz, M. T. S., Maturan, P. C., Jain, R. K., et al. (2014). Identification and mapping of stable QTL with main and epistasis effect on rice grain yield under upland drought stress. *BMC Genetics, 15*, 63.

Sandhu, N., Yadaw, R. B., Chaudhary, B., Prasai, H., Iftekharuddaula, K., Venkateshwarlu, C., Annamalai, A., Xangsayasane, P., Battan, K. R., Ram, M., & Cruz, M. T. (2019). Evaluating the performance of rice genotypes for improving yield and adaptability under direct seeded aerobic cultivation conditions. *Frontiers in Plant Science, 10*, 159.

Saneoka, H., Moghaieb, R. E. A., Premachandra, G. S., & Fujita, K. (2004). Nitrogen nutrition and water stress effects on cell membrane stability and leaf water relations in Agrostis palustris Huds. *Environmental and Experimental Botany, 52*, 131–138.

Sarkar, D., Mandal, B., & Kundu, M. C. (2007). Increasing use efficiency of boron fertilisers by rescheduling the time and methods of application for crops in India. *Plant and Soil, 301*, 77–85.

Sato, A. M., Catuchi, T. A., Ribeiro, R. V., & Souza, G. M. (2010). The use of network analysis to uncover homeostatic responses of a drought-tolerant sugarcane cultivar under severe water deficit and phosphorus supply. *Acta Physiologiae Plantarum, 32*, 1145–1151.

Saud, S., Fahad, S., Yajun, C., Ihsan, M. Z., Hammad, H. M., Nasim, W., Amanullah, J., Arif, M., & Alharby, H. (2017). Effects of nitrogen supply on water stress and recovery mechanisms in kentucky bluegrass plants. *Frontiers in Plant Science, 8*, 983.

Sellamuthu, R., Chandrababu, R., & Serraj, R. (2015). Mapping QTLs for reproductive stage drought resistance traits using an advanced backcross population in upland rice. *Crop Science, 55*, 1–12.

Selvaraj, M. G., Ishizaki, T., Valencia, M., Ogawa, S., Dedicova, B., Ogata, T., et al. (2017). Overexpression of an Arabidopsis thaliana galactinol synthase gene improves drought tolerance in transgenic rice and increased grain yield in the field. *Plant Biotechnology Journal, 15*, 1465–1477.

Serraj, R., McNally, K. L., Slamet-Loedin, I., Kohli, A., Haefele, S. M., Atlin, G., & Kumar, A. (2011). Drought resistance improvement in rice: an integrated genetic and resource management strategy. *Plant Production Science, 14*(1), 1–14.

Shangguan, Z. P., Shao, M. A., & Dyckmans, J. (2000). Nitrogen nutrition and water stress effects on leaf photosynthetic gas exchange and water use efficiency in winter wheat. *Environmental and Experimental Botany, 44*(2), 141–149. https://doi.org/10.1016/S0098-8472(00)00064-2

Sheela, K. R., & Alexander, V. T. (1995). Performance of rainfed rice (Oryza sativa) as influenced by varieties and nutrient levels. *Indian Journal of Agronomy, 40*(3), 407–411.

Shen, R., Huang, A., Li, B., & Guo, J. (2019). Construction of a drought monitoring model using deep learning based on multi-source remote sensing data. *International Journal of Applied Earth Observation and Geoinformation, 79*, 48–57.

Sher, A., Arfat, M. Y., Ul-Allah, S., Sattar, A., Ijaz, M., Manaf, A., Qayyum, A., Zuan, A. T., Nasif, O., & Gasparovic, K. (2021). Conservation tillage improves productivity of sunflower (Helianthus annuus L.) under reduced irrigation on sandy loam soil. *PLoS One, 16*(12), e0260673.

Shukla, N., Awasthi, R. P., Rawat, L., & Kumar, J. (2012). Biochemical and physiological responses of rice (*Oryza sativa* L.) as influenced by *Trichoderma harzianum* under drought stress. *Plant Physiology and Biochemistry, 54*, 78–88.

Sidhu, H., Jat, M., Singh, Y., Sidhu, R. K., Gupta, N., Singh, P., Singh, P., Jat, H., & Gerard, B. (2019). Sub-surface drip fertigation with conservation agriculture in a rice-wheat system: A breakthrough for addressing water and nitrogen use efficiency. *Agricultural Water Management, 216*, 273–283.

Sisti, C. P., dos Santos, H. P., Kohhann, R., Alves, B. J., Urquiaga, S., & Boddey, R. M. (2004). Change in carbon and nitrogen stocks in soil under 13 years of conventional or zero tillage in southern Brazil. *Soil and Tillage Research, 76*(1), 39–58.

Soe, H. M., Myat, M., Khaing, Z. L., Nyo, N. M., & Phyu, P. T. (2016). Development of drought tolerant Mutant from Rice var. Manawthukha through mutation breeding technique using 60Co gamma source. *The International Journal of Innovative Research in Science, Engineering and Technology, 4*, 11205–11212.

Song, C. J., Ma, K. M., Qu, L. Y., Liu, Y., Xu, X. L., Fu, B. J., & Zhong, J. F. (2010). Interactive effects of water, nitrogen and phosphorus on the growth, biomass partitioning and water-use efficiency of Bauhinia faberi seedlings. *Journal of Arid Environments, 74*, 1003–1012.

Suh, J. P., Cho, Y. C., Won, Y. J., Ahn, E. K., Baek, M. K., Kim, M. K., et al. (2015). Development of resistant gene-pyramided japonica rice for multiple biotic stresses using molecular marker-assisted selection. *Plant Breeding and Biotechnology, 3*, 333–345. https://doi.org/10.9787/PBB.2015.3.4.333

Svoboda, M., & Fuchs, B. (2016). *Handbook of drought indicators and indices*. World Meteorological Organization (WMO).

Swain, P., Mohanty, S., Bhattacharyya, P., Baig, M. J., Chattopadhyay, K., Singh, O. N., & Sarkar, R. K. (2014). *Rice cultivation in rainfed upland and drought prone areas under changed climate* (CRRI Technology Bulletin-101). CRRI, ICAR.

Swamy, B. P. M., Ahmed, H. U., Henry, A., Mauleon, R., Dixit, S., Vikram, P., et al. (2013). Genetic, physiological, and gene expression analyses reveal that multiple QTL enhance yield of rice mega variety IR64 under drought. *PLoS One, 8*(5), e62795.

Syuhada, N., Jahan, M. S., Khandaker, M. M., Nashriyah, M., Khairi, M., Nozulaidi, M., & Razali, M. B. (2014). Application of copper increased corn yield through enhancing physiological functions. *Australian Journal of Basic and Applied Sciences, 8*, 282–286.

Talla, S. K., Panigrahy, M., Kappara, S., Nirosha, P., Neelamraju, S., & Ramanan, R. (2016). Cytokinin delays dark-induced senescence in rice by maintaining the chlorophyll cycle and photosynthetic complexes. *Journal of Experimental Botany, 67*, 1839–1851.

Tariq, A., Pan, K., Olatunji, O. A., Graciano, C., Li, Z., Sun, F., Zhang, L., Wu, X., Chen, W., Song, D., et al. (2018). Phosphorous fertilization alleviates drought effects on *Alnus cremastogyne* by regulating its antioxidant and osmotic potential. *Scientific Reports, 8*, 5644.

Taylor, K. C., & Halfacre, R. G. (1986). The effect of hydrophilic polymer on media water retention and nutrient availability to Ligustrum lucidum. *HortScience, 21*(5), 1159–1161.

Tayyab, N., Naz, R., Yasmin, H., Nosheen, A., Keyani, R., Sajjad, M., Hassan, M. N., & Roberts, T. H. (2020). Combined seed and foliar pre-treatments with exogenous methyl jasmonate and salicylic acid mitigate drought-induced stress in maize. *PLoS One, 15*, e0232269.

Thakuria, R. K., & Choudhry, J. K. (1995). Effect of seed priming or dry seeded ahu rice. *Indian Journal of Agronomy, 40*(3), 412–414.

Thiem, T. T., & Akira, Y. (2015). Effect of nitrogen application levels on growth of rice under drought stress conditions. *Journal of Scientific Development, 13*(8), 1388–1396.

Tisdale, S. L., & Nelson, W. L. (1975). Elements required in plant nutrition. In S. L. Tisdale & W. L. Nelson (Eds.), *Soil fertility and fertilizers* (pp. 66–104). Macmillan Publishing, Inc.

Trijatmiko, K. R., Prasetiyono, J., Thomson, M. J., Cruz, C. M. V., Moeljopawiro, S., & Pereira, A. (2014). Meta analysis of quantitative trait loci for grain yield and component traits under reproductive-stage drought stress in an upland rice population. *Molecular Breeding, 34*, 283–295.

Tyagi, V., Nagargade, M., & Singh, R. K. (2020). Agronomic interventions for drought management in crops. In *New frontiers in stress management for durable agriculture* (pp. 461–476).

Upadhyaya, H., & Panda, S. K. (2019). Drought stress responses and its management in rice. In M. Hasanuzzaman, M. Fujita, K. Nahar, & J. K. Biswas (Eds.), *Advances in rice research for abiotic stress tolerance* (pp. 177–200). Elsevier.

Upadhyaya, H., Panda, S. K., & Dutta, B. K. (2011). $CaCl_2$ improves post drought recovery potential in *Camellia sinensis* (L) O. Kuntze. *Plant Cell Reports, 30*(4), 495–503. https://doi.org/10.1007/s00299-010-0958-x

Urmi, T. A., Islam, M. M., Zumur, K. N., Abedin, M. A., Haque, M. M., Siddiqui, M. H., Murata, Y., & Hoque, M. A. (2023). Combined effect of salicylic acid and proline mitigates drought stress in rice (*Oryza sativa* L.) through the modulation of physiological attributes and antioxidant enzymes. *Antioxidants, 12*, 1438. https://doi.org/10.3390/antiox12071438

Vaghefi, N., Nasir Shamsudin, M., Radam, A., & Rahim, K. A. (2013). Modelling the impact of climate change on rice production. An overview. *Journal of Applied Sciences, 13*(5649), 5660.

Vance, W. H., Bell, R. W., Johansen, C., Haque, M. E., Musa, A. M., Shahidullah, A. K. M., & Mia, M. N. N. (2014). Optimum time of sowing for rainfed winter chickpea with one-pass mechanised row-sowing: an example for small-holder farms in north-west Bangladesh. *Crop & Pasture Science, 65*, 602–613.

Vejan, P., Abdullah, R., Khadiran, T., Ismail, S., & Nasrulhaq Boyce, A. (2016). Role of plant growth promoting rhizobacteria in agricultural sustainability: A review. *Molecules, 21*(5), 573.

Venuprasad, R., Bool, M. E., Quiatchon, L., & Atlin, G. N. (2012). A QTL for rice grain yield in aerobic environments with large effects in three genetic backgrounds. *Theoretical and Applied Genetics, 124*, 323–332.

Verma, K. K., Song, X. P., Lin, B., Guo, D. J., Singh, M., Rajput, V. D., Singh, R. K., Singh, P., Sharma, A., Malviya, M. K., Chen, G. L., & Li, Y. R. (2021). Silicon induced drought tolerance in crop plants: physiological adaptation strategies. *Silicon, 22*, 1–15. https://doi.org/10.1007/s12633-021-01071-x

Vikram, P., Swamy, B. P. M., Dixit, S., Trinidad, J., Cruz, M. T. S., Maturan, P. C., et al. (2016). Linkages and interactions analysis of major effect drought grain yield QTLs in rice. *PLoS One, 11*(3), e0151532.

Wallander, S., Aillery, M., Hellerstein, D., & Hand, M. (2013). *The role of conservation programs in drought risk adaptation*. Economic Research Report Number 148. (p. 34). United States Department of Agriculture.

Wang, Y. T., & Gregg, L. L. (1990). Hydrophilic polymers—Their response to soil amendments and effect on properties of a soilless potting mix. *Journal of the American Society for Horticultural Science., 115*(6), 943–948.

Wang, M., Zheng, Q., Shen, Q., & Guo, S. (2013a). The critical role of potassium in plant stress response. *International Journal of Molecular Sciences, 14*(4), 7370–7390. https://doi.org/10.3390/ijms14047370

Wang, Y., Zang, J. P., Sun, Y., Ali, J., Xu, J. L., & Li, Z. K. (2013b). Background-independent quantitative trait loci for drought tolerance identified using advanced backcross introgression lines in rice. *Crop Science, 53*(2), 430–441.

Waraich, E. A., Ahmad, R., Ashraf, M. Y., & Saifullah Ahmad, M. (2011). Improving agricultural water use efficiency by nutrient management in crop plants. *Acta Agriculturae Scandinavica. Section B, Soil and Plant Science, 61*, 291–304.

Yang, J., Fan, W., & Zheng, S. (2019). Mechanisms, and regulation of Aluminium induced secretion of organic acid anions from plant roots. *Journal of Zhejiang University. Science. B, 20*, 513–527.

Yoshida, R., Fukui, S., Shimada, T., Hasegawa, T., Ishigooka, Y., Takayabu, I., & Iwasaki, T. (2015). Adaptation of rice to climate change through a cultivar-based simulation: a possible cultivar shift in eastern Japan. *Climate Research, 64*, 275–290.

Yuvaraj, M., & Subramanian, K. S. (2020). Significance of zinc in plant nutrition. *Biotica Research Today, 2*(8), 823–825.

Zain, N. A. M., & Ismail, M. R. (2016). Effects of potassium rates and types on growth, leaf gas exchange and biochemical changes in rice (*Oryza sativa*) planted under cyclic water stress. *Agricultural Water Management, 164*, 83–90.

Zamani, S., Naderi, M. R., Soleymani, A., & Nasiri, B. M. (2020). Sunflower (*Helianthus annuus* L.) biochemical properties and seed components affected by potassium fertilization under drought conditions. *Ecotoxicology and Environmental Safety, 190*, 110017. https://doi.org/10.1016/j.ecoenv.2019.110017

Zhang, T., Li, T., Yang, X., & Simelton, E. (2016). Model biases in rice phenology under warmer climates. *Scientific Reports, 6*, 27355. https://doi.org/10.1038/srep27355

Zhang, Y., Wang, X., Luo, Y., Zhang, L., Yao, Y., Han, L., et al. (2020). OsABA8ox2, an ABA catabolic gene, suppresses root elongation of rice seedlings and contributes to drought response. *Crop Journal, 8*, 480–491.

Zhao, X. Q., Xu, J. L., Zhao, M., Lafitte, R., Zhu, L. H., Fu, B. Y., et al. (2008). QTLs affecting morpho-physiological traits related to drought tolerance detected in overlapping introgression lines of rice (*Oryza sativa*). *Plant Science, 174*(6), 618–625.

Navigating Agricultural Drought: Building Resilience Through Monitoring and Forecasting

15

Alka Rani, Vinay Kumar Sehgal, Rajkumar Dhakar, Abhilash, and Kavita Kumari

Abstract

Agricultural drought poses significant challenges to agricultural productivity, food security, and economic stability, particularly in the face of escalating climate change. Remote sensing technology plays a pivotal role in monitoring drought conditions by providing near-real-time data on crucial indicators such as rainfall, soil moisture, evapotranspiration, and vegetation health. While traditional single-parameter indices like the Standardized Precipitation Index (SPI), Normalized Difference Vegetation Index (NDVI), etc., have been used, their limitations have prompted the development of composite indices that offer comprehensive understanding by integrating multiple data sources through advanced weighting techniques. Additionally, forecasting agricultural drought is evolving, with hybrid models showing promise in accurately predicting drought. Several global agricultural drought monitoring and early warning systems are functioning across various parts of the world such as the United States Drought Monitor, South Asia Drought Monitoring System, etc. India has also implemented its own drought monitoring and declaration system making use of science-based parameters and decision-tree-based approach. However, India still lacks a robust agricultural drought forecasting and prognosis system tailored to its specific needs

A. Rani (✉) · V. K. Sehgal · R. Dhakar
Division of Agricultural Physics, ICAR – Indian Agricultural Research Institute, New Delhi, India
e-mail: Alka.Rani1@icar.gov.in

Abhilash
Division of Natural Resource and Management, ICAR – Central Island Agricultural Research Institute, Port Blair, Andaman and Nicobar Islands, India

K. Kumari
Crop Production Division, ICAR – National Rice Research Institute, Cuttack, Odisha, India

R. Dubey et al. (eds.), *Drought and Heat Stress in Agriculture*,
https://doi.org/10.1007/978-981-96-5735-3_15

for different regions. Establishing such a system is crucial for providing timely alerts and supporting proactive drought management strategies, ultimately enhancing resilience in the face of agricultural drought.

Keywords

Agricultural drought · composite drought index · drought indices · monitoring · forecasting

Abbreviations

AMSR-E	Advanced Microwave Scanning Radiometer for Earth Observing System
ANN	Artificial Neural Network
ARIMA	Autoregressive Integrated Moving Average
AVHRR	Advanced Very High-Resolution Radiometer
CART	Classification and Regression Tree
CHIRPS	Climate Hazards Group InfraRed Precipitation with Station data
CMI	Crop Moisture Index
ECMWF	European Centre for Medium-Range Weather Forecasts
EDI	Evaporative Drought Index
ENSO	El Nino Southern Oscillation
ESI	Evaporative Stress Index
ET	Evapotranspiration
fAPAR	fraction of Absorbed Photosynthetically Active Radiation
GIMMS	Global Inventory Modeling and Mapping Studies
GLDAS	Global Land Data Assimilation System
GPCC	Global Precipitation Climatology Centre
GWDI	Groundwater Drought Index
IMD	India Meteorological Department
LST	Land Surface Temperature
MAI	Moisture Adequacy Index
MERRA	Modern-Era Retrospective Analysis for Research and Applications
ML	Machine Learning
MNCFC	Mahalanobis National Crop Forecast Centre
MODIS	Moderate Resolution Imaging Spectroradiometer
NADAMS	National Agricultural Drought Assessment and Monitoring System
NDDI	Normalized Difference Drought Index
NDVI	Normalized Difference Vegetation Index
NDWI	Normalized Difference Water Index
NMDI	Normalized Multiband Drought Index
NOAA	National Oceanic and Atmospheric Administration

PASM	Percentage Available Soil Moisture
PCA	Principle Component Analysis
PCI	Precipitation Condition Index
PDSI	Palmer Drought Severity Index
RF	Random Forest
SARIMA	Seasonal Autoregressive Integrated Moving Average
SMA	Soil Moisture Anomaly
SMAPI	Soil Moisture Anomaly Percentage Index
SMCI	Soil Moisture Condition Index
SMDI	Soil Moisture Deficit Index
SPEI	Standardized Precipitation Evapotranspiration Index
SPI	Standardized Precipitation Index
SSI	Standardized Soil Moisture Index
SVI	Standardized Vegetation Index
SVM	Support Vector Machine
SWDI	Soil Water Deficit Index
TCI	Temperature Condition Index
TRMM	Tropical Rainfall Measuring Mission
USDM	United States Drought Monitor
VCI	Vegetation Condition Index
VHI	Vegetation Health Index
VTCI	Vegetation Temperature Condition Index
VWSI	Vegetation Water Supply Index

15.1 Introduction

Drought is a complex natural hazard with no universal definition and numerous adverse effects on agriculture, economy, industries, ecology, life, and livelihoods of humans (Wilhite et al., 2007). Drought has been categorized into five types from a disciplinary perspective: meteorological, agricultural, hydrological, socio-economical, and ecological. Agriculture is the most important sector for feeding the populace, guaranteeing food security, and sustaining the livelihoods of millions of farmers. Agricultural drought is referred to as a condition initiated by failure in precipitation leading to a decline in soil moisture and consequent crop failure without any reference to surface water resources. Droughts can have disastrous effects on agriculture, resulting in crop failure and financial losses. Concerns about the effects of drought events on agricultural production and sustainability have grown because of drought events' increased frequency and severity in recent decades. With its extensive agricultural landscapes and reliance on rainfed agriculture, India is particularly vulnerable to the ill effects of drought. Out of 3.29 million km^2 of geographical area in India, about 1.07 million km^2 of land is subjected to different degrees of water stress and drought conditions (Gupta et al., 2011; Mishra & Desai, 2005). In India, drought areas are mainly confined to the southern and western parts of the country (Gupta et al., 2011). The risk of agricultural drought is further anticipated to increase due to an increase in the frequency and intensity of heatwaves as well as erratic monsoonal rainfall patterns induced by climate change (Tripathy et al., 2023). Thus, to reduce the adverse effects of agricultural drought, it is necessary to understand the agricultural drought phenomenon and its progression in both qualitative and quantitative terms. The success of agricultural drought preparedness and mitigation depends upon timely information on its onset, progress, intensity, and areal extent. The agricultural drought monitoring and outlook system may assist in alleviating the negative consequences of agricultural drought by providing timely information on contingency and mitigation measures to all the farmers, bureaucrats, policymakers, and concerned stakeholders, as well as for undertaking the appropriate relief operations in the affected areas.

Advances in remote sensing technology and the availability of meteorological datasets have revolutionized the area of drought monitoring and early warning in recent years (Liu et al., 2017). Remote sensing provides comprehensive and near-real-time data on vegetation health, soil moisture, and other crucial indicators of drought conditions over large areas, which makes it feasible to monitor and anticipate drought occurrences with better precision and spatial resolution (Hazaymeh & Hassan, 2016). A more thorough knowledge of the complicated interactions between climatic factors and agricultural management practices can be accomplished by merging remote sensing data with meteorological information. The monitoring of agricultural drought is normally performed using indices based on various indicators such as precipitation, soil moisture, evapotranspiration (ET), temperature, vegetation health, etc. Numerous drought indices based on remote sensing and meteorological datasets are available in the literature with its own strengths and weaknesses. Some of these commonly used drought indices have been discussed in this chapter. However, the presence of range of indices and indicators makes the decision-making process complicated resulting in responses not being triggered on time or in appropriate way. This can have considerable economic consequences as well as a loss of trust in the authorities responsible for triggering agricultural drought preparedness and responses. Furthermore, it is likely that the main users of agricultural drought indices are policymakers and stakeholders who do not necessarily have the required knowledge of the strengths, weaknesses and correct interpretation of each index. Therefore, now the research is mainly focused on integration of various indices in a single composite index, which can incorporate more information about various factors and impacts of agricultural drought. The single combined drought indicator that makes use of meteorological, soil moisture, and remote-sensing-derived information enables the efficient monitoring of the onset of agricultural drought and its evolution in time and space. Such combined indicator also makes it easier to compare drought intensity across space and time scales. The integrated or composite drought index has been developed by several researchers across various parts of the world. Some of the most prominent composite indices have also been discussed in this chapter.

The presence of an agricultural drought outlook or forecasting system provides numerous benefits for farmers, policymakers, and communities. By combining data on soil moisture levels, precipitation forecasts, crop conditions, and water reservoir levels, etc., the system can forecast potential drought impacts on agriculture weeks or months in advance. This allows farmers to proactively make decisions to mitigate crop losses, such as adjusting planting dates, applying irrigation, or purchasing crop insurance. Policymakers can use the forecasts to set up assistance programs for impacted farmers or to anticipate potential impacts on food prices and food security. Overall, an effective agricultural drought monitoring and forecasting system enables advanced planning and risk reduction across the food system to improve resilience for farmers and consumers alike. The system provides valuable lead time for key decisions in the face of climatic uncertainty. This enables the policymakers or bureaucrats to take appropriate measures like irrigation water management, cultivation of drought-tolerant varieties, supplemental feeding programs, etc., in order to prevent or mitigate the effects of drought on crop production and thus minimize economic losses (Wilhite, 2000). The forecasting of agricultural drought can be done by various statistical methods such as regression or time-series analysis, predictive or dynamic modeling, machine learning models, etc. These forecasting methods have been briefly discussed in this chapter. Furthermore, the discussion expands to global operational agricultural drought monitoring and early warning systems, shedding light on the intricate network of tools and methodologies utilized across the globe. Overall, this chapter equips you with a comprehensive understanding of the indices and approaches used worldwide to monitor and forecast agricultural drought, empowering proactive decision-making for a more resilient future.

15.2 Drought

A drought is defined as "a period of abnormally dry weather sufficiently prolonged for the lack of water to cause serious hydrologic imbalance in the affected area." (Huschke, 1959). Wilhite et al. (2007) defined drought as a severe moisture deficit below expected levels that restricts some type of activity. In simple words, drought is a period of drier than normal conditions that lead to water-related impacts like crop damage, water supply shortage, etc. It primarily originates from a shortage of precipitation over an extended period of time. Drought should not be confused with aridity, which is a permanent climatic feature, whereas drought is a temporary aberration. Drought is a normal and recurrent feature of climate that may occur in all climatic zones ranging from dry to humid. However, the characteristics of drought may vary from one region to another based on intensity, duration, or spatial extent. There is no precise and universally accepted definition of drought due to its complex nature. Its definition depends upon the regions, needs, and disciplinary perspectives.

15.2.1 Major Drought Events in India

There were 25 major drought years identified in India from 1871 to 2015 when the rainfall deficit in the all India summer monsoon rainfall was more than 10%. These drought years are 1873, 1877, 1899, 1901, 1904, 1905, 1911, 1918, 1920, 1941, 1951, 1965, 1966, 1968, 1972, 1974, 1979, 1982, 1985, 1986, 1987, 2002, 2009, 2014 and 2015 (Manual for Drought Management, 2016).

15.2.2 States and Regions Affected by Drought

Figure 15.1 depicts the frequency of drought occurrences in Indian districts from the year 2000 to 2019 (20 years) as declared by state governments. It is clear from the map that certain parts of Rajasthan, Maharashtra, Karnataka, and Andhra Pradesh are most prone to droughts. In India, 68% of the cropped area is vulnerable to droughts as these areas receive less than 1125 mm of mean annual rainfall. The areas where mean annual rainfall is less than 750 mm are classified as "chronically drought prone," whereas the areas with mean annual rainfall of 750–1125 mm are "drought prone." These drought-prone areas are mainly present in the western and peninsular parts of India having arid, semi-arid to sub-humid climate (Manual for Drought Management, 2016).

15.2.3 Causes of Drought

Drought begins with insufficient rainfall, leading to reduced soil moisture and a decline in both surface and groundwater levels. This, in turn, adversely impacts the growth and yield of crops if irrigation resources are limited. Extended periods of low rainfall can result in socio-economic problems such as a scarcity of drinking water, the loss of human and animal life, migration, and other related issues. Uneven distribution of rainfall, less annual average rainfall with high variability, lack of irrigation facilities, and decline in groundwater table increases the susceptibility to droughts. Poor rainfall in subsequent years intensifies the impact of drought due to less chance of recharging of groundwater or surface water reservoirs, and replenishment of soil moisture (Manual for Drought Management, 2016). Drought in agriculture can be caused by short-term precipitation shortages resulting in decreased soil moisture content, and/or increase in temperature causing higher ET levels above water supply (Hazaymeh & Hassan, 2016).

15.2.4 Types of Droughts

There are mainly five types of droughts based on physical aspects:

(i) *Meteorological drought*: When the rainfall of an area is below 25% of the normal rainfall, then this condition is known as meteorological drought. According to the India Meteorological Department (IMD), when the deficiency in rainfall at a particular meteorological subdivision is between 26% to 50% of its long-term average, then it is known as 'moderate drought'. If rainfall deficiency is more than 50%, then it is known as 'severe drought'.

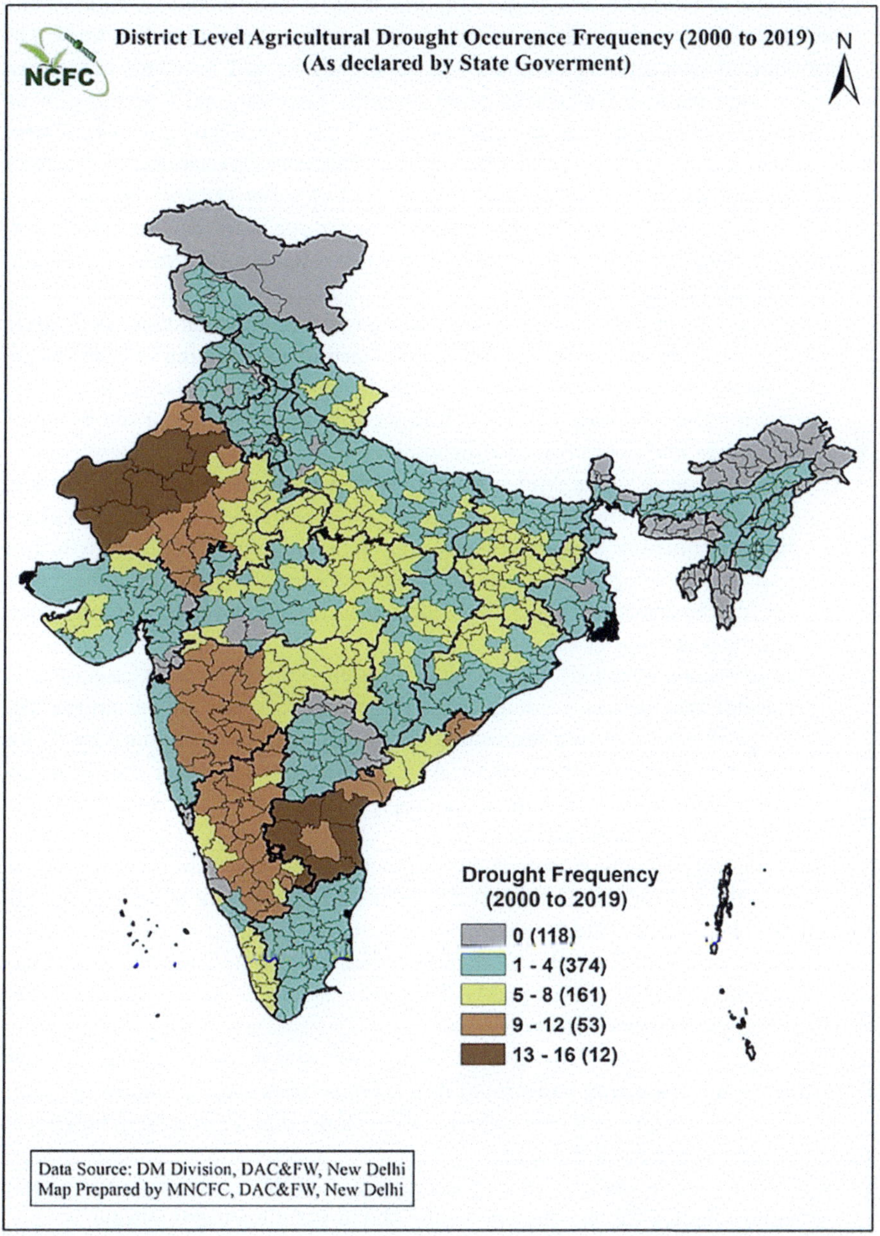

Fig. 15.1 District-level frequency of agricultural drought occurrence as declared by the state governments from 2000 to 2019. (Source: Manual for Drought Management—updated upto 2020)

(ii) *Agricultural drought*: When the soil moisture is insufficient to support the healthy crop growth resulting in crop failure or less yield, then it is known as agricultural drought. is caused by a deficiency in soil moisture content or higher than usual ET affecting plant growth. The National Commission on Agriculture (1976) defined agricultural drought as a period of four consecutive weeks of severe meteorological drought or with a weekly rainfall of 5 mm or less during the period from mid-May to mid-October (*kharif* season) when 80% of the country's total crop is planted, or six such consecutive weeks during the rest of the year (Ray et al., 2015). There are four categories of agricultural drought discussed below:

(a) *Early season drought*: Early season drought occurs when there is delay in the commencement of rainfall at the starting of cropping season during crop sowing. Sometimes, early rainfall appeals to the farmers for crop sowing, but the long dry spell followed by it results in the withering of seedlings and poor crop establishment.

(b) *Mid-season drought*: Mid-season drought occurs when there is a break in the southwest monsoon. It coincides with the vegetative crop growth phase when deficiency in soil moisture may lead to stunted plant growth, low leaf area development or reduced plant population. However, some crops are able to cope with water stress situations by regulating leaf area to intercept less amount of solar radiation.

(c) *Late-season drought*: Late-season drought occurs by early cessation of rainfall which may coincide with reproductive phase of the crops. The moisture stress during reproductive phase leads to forced maturity of the crop due to rise in temperature. As the moisture availability during reproductive phase has huge impact on crop yield, therefore, late-season drought can reduce the crop yields.

(d) *Apparent drought*: Apparent drought occurs when the rainfall in the region is sufficient for one crop but not for others. For example, if there is rainfall deficiency in the paddy-grown areas, then the rainfall may be sufficient for other crops like sorghum, maize, finger-millet, etc., having less water requirement, but not for paddy. Therefore, apparent drought occurs by the wrong selection of crops.

(iii) *Hydrological drought*: The reduction in water supply from surface or subsurface water resources like stream flows, lakes, reservoirs, groundwater, etc., on a watershed or river basin scale due to prolonged precipitation deficiency is known as hydrological drought.

(iv) *Socioeconomic drought*: The socioeconomic drought results from the human activities affected by one or more of the above-mentioned drought types.

(v) *Ecological drought*: Ecological drought is an episodic deficit in water availability that drives ecosystems beyond thresholds of vulnerability, impacts ecosystem services, and triggers feedback in natural and/or human systems.

15.2.5 Impacts of Droughts

Droughts have environmental, economic, and social impacts. A severe drought event can decrease the availability of drinking water to humans and animals, which may cause deaths (Chand & Biradar, 2017). Prolonged drought can reduce the water levels in both surface water bodies like reservoirs, lakes, ponds, rivers, etc., and in groundwater. The decline in water availability can lead to loss of biodiversity. The frequency and severity of forest fires also increase (Littell et al., 2016). The crop yields reduced, which affects the food availability resulting in inflation of food prices. There is increased vulnerability of insect–pest attack to the crops. The productivity of livestock is reduced. The industries which are dependent on agriculture encounter economic losses (Edwards et al., 2019). The unemployment rate in the drought-affected areas also increases. People residing in drought-affected areas tend to migrate to other areas. Thus, droughts impose multifaceted impacts across various sectors, profoundly affecting ecosystems, agriculture, water resources, economies, and communities.

15.3 Agricultural Drought Monitoring System in India

The impacts of agricultural drought depend upon timing, intensity, spatial extent, and duration (Mishra & Singh, 2010). For instance, occurrence of low-intensity drought over a longer time period may not severely affect the crop, but a short period drought during flowering or grain formation stage of the crop may acutely affect productivity as plants are more sensitive to water stress in reproductive stage. The monitoring of agricultural droughts is important in the planning and management of food production. In India, the incremental progress in monitoring of agricultural drought was done as described below:

15.3.1 Anomaly of Rainfall and Aridity Index by IMD

India Meteorological Department (IMD) monitors the incidence, spread, intensification and cessation of drought on a weekly time scale over the country based on Aridity Anomaly Index. Aridity Index is computed by using Thornthwaite's (1948) water balance technique. Aridity Index is the percentage of annual water deficit to annual water need or annual potential ET. Aridity Anomaly Index is the departure of aridity index values from normal, expressed in percentage. Positive anomalies are used for agricultural drought intensity classification. IMD also monitors meteorological drought using Standardized Precipitation Index (SPI) at a monthly time scale.

15.3.2 National Agricultural Drought Assessment and Monitoring System (NADAMS)

National Agricultural Drought Assessment and Monitoring System (NADAMS) project (1986), developed by National Remote Sensing Centre (NRSC) and used by MNCFC (2012), generated agricultural drought reports for the *kharif* season considering prevalence, severity level, and persistence of agricultural drought at state/district/sub-district level at monthly time scale for 14 states (Andhra Pradesh, Bihar, Chhattisgarh, Gujarat, Haryana, Jharkhand, Karnataka, Madhya Pradesh, Maharashtra, Odisha, Rajasthan, Tamil Nadu, Telangana, and Uttar Pradesh) of India at district level, out of which, 5 states (Andhra Pradesh, Karnataka, Haryana, Maharashtra and Telangana) are being monitored at sub-district level. They take into account the anomaly in NDVI/NDWI, ground information about irrigation percentage, rainfall deviation, and sown area deviation, also additional information like change in crop calendar and lag between rainfall and NDVI/NDWI to provide drought warning (in June, July and August) and drought declaration (September, October) (Sivakumar et al., 2011).

15.3.3 Methodology of Manual for Drought Management (Updated Upto 2020)

The first attempt of scientifically monitoring and declaration of agricultural drought was made in the "Manual for Drought Management—2009". A major paradigm shift came with the revised "Manual for Drought Management—2016" in which both "causative" and "impact" technological indicators were proposed for use, how they should be collected, agencies responsible, their weightages and timelines for different activities were formulated for main monsoonal *"kharif"* agricultural season. Expanding on the adoption by States and success of these protocols, the drought monitoring and declaration protocol for *"rabi"* crop season was also proposed in 2018. Since 2017, states are monitoring agricultural drought conditions in India using the methodology given in Manual for Drought Management (2016). Currently about 17 states of India (Andhra Pradesh, Assam, Bihar, Chhattisgarh, Gujarat, Haryana, Jharkhand, Karnataka, Madhya Pradesh, Maharashtra, Odisha, Punjab, Rajasthan, Tamil Nadu, Telangana, Uttar Pradesh, and West Bengal) are monitoring agricultural drought at state/district/sub-district level at fortnightly or monthly time scale. They take into account the anomaly in NDVI/NDWI derived from MODIS or AWiFS, Percentage Available Soil Moisture (PASM), or Moisture Adequacy Index (MAI) computed using soil moisture model, dry spell derived from IMD rainfall data, Standardized Precipitation Index (SPI) computed from NOAA CPC rainfall data, ground information about irrigation percentage, sown area deviation, hydrological indices like Reservoir Storage Index (RSI), Groundwater Drought Index (GWDI), Stream-flow Drought Index (SFDI), change in crop calendar and lag between rainfall, along with ground truth information. For declaration of drought, mandatory indicators, i.e., SPI and dry spell, are used. If these mandatory indicators indicate the presence of drought, then any three impact indicators like deviation in NDVI or NDWI, crop area sown, MAI or PASM, hydrological indicators like SFDI, RSI, or GWDI are analyzed. If two of the selected 3 impact indicators are in severe category and one is in moderate category, then it is classified as severe drought. If two of the selected three impact indicators are in the moderate category, then it is classified as moderate drought. The condition is normal for all other cases. Minimal ground-truth verification is done to validate the drought severity. The drought declaration for *kharif* season is done by 30th October and for *rabi* season by 31st March.

However, the current agricultural drought monitoring system India several challenges like contamination of satellite data by clouds, lack of long-term and high-resolution satellite data, low spatial resolution of soil moisture data, and difficulty in assessing the effects of consecutive droughts. The vegetation indices used for drought assessment are also by crop phenology, changes in cropping patterns or poor vegetation health caused by factors other than water stress. These problems need to be resolved to strengthen the agricultural drought monitoring system India. operational agricultural drought forecasting or prognosis system is still lacking in India.

15.4 Indices for Agricultural Drought Monitoring or Forecasting

For monitoring or forecasting of agricultural drought, drought indicators like precipitation, temperature, ET, soil moisture, vegetation status, etc., are collected in near-real time, and desired indices are computed for evaluation. Accurate, quantitative, and objective assessment of these indices determines the timing, spatial extent, severity, and duration of the agricultural drought (Hazaymeh & Hassan, 2016). The information obtained monitoring and forecasting provides guidance to the policy makers, bureaucrats, and farmers to take actions. The drought preparedness actions focus on the adoption of during non-drought period to reduce the impact of drought events, if they occur. The drought preparedness measures include selection of appropriate crop management practices, contingency planning, suitable policy formulation, development of irrigation facilities, crop or livestock insurance, watershed management, agro-advisories, agro-climatic regional planning, etc. O drought response actions are during drought event to reduce its impact. Therefore, accurate and timely monitoring and forecasting of agricultural drought is essential for reducing the potential impacts of drought events. Numerous indices are available in the literature for monitoring or forecasting agricultural drought. Some of the most commonly used indices are tabulated below:

15.4.1 Drought Indices Based on Meteorological Data

The most commonly used drought indices computed from the meteorological data are briefly discussed in Table 15.1.

15.4.2 Drought Indices Based on Vegetation Data

The most commonly used drought indices computed from the vegetation data is briefly discussed in Table 15.2.

15.4.3 Drought Indices Based on Soil Moisture Data

The most commonly used drought indices computed from the soil moisture data has been briefly discussed in Table 15.3.

15.4.4 Integrated or Composite Drought Indices

The analysis of agricultural drought using a single indicator may not be sufficient as agricultural drought is influenced by various factors like rainfall, soil moisture, temperature, ET, etc. Therefore, several indices are integrated by various researchers using different methods to accurately represent the agricultural drought conditions. An ideal agricultural drought index should integrate various parameters like rainfall, temperature, ET, runoff, and other water supply indicators that impact crop growth and development during drought episodes into a single number (Sivakumar et al., 2011). An integrated drought index would help the policymakers and stakeholders to take appropriate and timely decisions as they might not be aware of the strengths, weaknesses, and correct interpretation of each drought index. Therefore, now the research is mainly focused on integration of various indices into a single index to incorporate more information about various factors and their impact on agricultural drought. Some of the integrated indices for monitoring or assessment of agricultural drought are briefly discussed further.

15.4.4.1 Multivariate Standardized Drought Index (MSDI)

Multivariate Standardized Drought Index (MSDI) was developed by Hao and AghaKouchak (2013) for determination of drought onset and termination in California and North Carolina states of United States using the concept of copula by combining SPI and SSI. This MSDI can be computed at different time scales just like SPI. It indicated drought onset as early as SPI, whereas drought persistence similar to SSI.

Table 15.1 Common drought indices based on meteorological data

S. No.	Index	Description	Reference	Advantages	Limitations
1	Percent of normal	Percent of normal = $\left(\frac{P_i}{\overline{P}}\right)\times 100$ where P_i: actual precipitation $\overline{P}$: normal precipitation (mean precipitation of past 30 years) The index value <80% represents drought condition.	Hayes (2006)	It can be computed at various time scales like daily, weekly, monthly, seasonal or annual. Simple, transparent, easy, and quick to calculate.	It cannot be used for comparing droughts across different seasons and regions. It is limited only to the precipitation following normal distribution. It lacks robustness.
2	Standardized Precipitation Index (SPI)	The long-term precipitation record is first fitted to an incomplete gamma probability distribution, which is then transformed into a normal distribution such that the mean of SPI for the location and desired period is zero. SPI for period i is calculated as: $\mathrm{SPI}_i = \frac{x_i - x_{\mathrm{mean}}}{\sigma}$ where, x_i is standardized rainfall of a grid for period i; x_{mean} and σ are long-term mean and standard deviation of standardized rainfall for the same period The negative SPI value indicates drier conditions whereas positive values indicate wetter conditions.	McKee et al. (1993)	It can be computed at various time scales. SPI computed for 6 months or less represent agricultural impacts and for 12 months or longer represent hydrological impacts. It can be used for comparing drought across different agro-climatic regions. It represents both wet and dry conditions.	As it lacks temperature component, therefore, it doesn't represent ET and overall water balance of a region. Thus, it is difficult to compare drought events having similar SPI values but different temperature scenarios.

(continued)

Table 15.1 (continued)

S. No.	Index	Description	Reference	Advantages	Limitations
3	Deciles	The long-term record of precipitation of an area is arranged in ascending order, and then divided into 10 equal parts, where each part is known as decile. The lowest 10% indicates serious rainfall deficiency.	Gibbs and Maher (1967)	Simple and easy to calculate. It can be computed at various time scales like daily, weekly, monthly, seasonal or annual.	Long period of precipitation data is required for accurate results. Impact of other variables for determination of drought are not considered.
4	Palmer Drought Severity Index (PDSI)	It is based on water balance approach in which precipitation, temperature, and available soil water content data is used to compute ET, runoff, soil recharge, and moisture loss from the surface soil. Potential ET is computed using Thornthwaite method. It is computed on weekly or monthly scale. Negative PDSI value indicates drought conditions.	Palmer (1965)	More comprehensive index as it considers precipitation, temperature, ET, and soil moisture. Effective in determining long-term drought.	Sophisticated computation process. It neglects the water input from irrigation, snowfall, frozen soil, and time lag between rainfall and run off.
5	Standardized Precipitation Evapotranspiration Index (SPEI)	Methodology of its computation is similar to that of SPI, but the difference between precipitation and potential ET (P-PET) is used as the input variable instead of precipitation only. Negative SPEI value indicates drought conditions.	Vicente-Serrano et al. (2010)	It is more comprehensive in drought monitoring than SPI as it accounts for both precipitation and temperature. It can be used to compare drought across different regions.	Longer period of data (>30 years) is required for its computation. It is sensitive to the method used for computing potential ET.

6	Precipitation Condition Index (PCI)	$\text{PCI} = \left(\frac{P_i - P_{\min}}{P_{\max} - P_{\min}} \right)$ where, P_i is the precipitation of current time period; $P_{\max}$ and $P_{\min}$ are maximum and minimum precipitation, respectively for the same time period in the historical dataset The PCI < 0.5 indicates drought conditions.	Du et al. (2013)	Simple to compute. Comparable across time and regions.	It doesn't consider the temporal distribution of precipitation in a region.
7	Evaporative Stress Index (ESI)	Standardized anomaly or z-score of relative ET fraction ($f_{\text{RET}} = \text{ET}_a/\text{ET}_{\text{ref}}$), where ET_a is actual ET retrieved from ALEXI model and ET_{ref} is the reference ET derived from Penman-Monteith (FAO-56) equation. The negative ESI value indicates drought conditions.	Anderson et al. (2011)	It can capture the early signals of agricultural drought caused by extended periods of hot, dry, and windy conditions resulting in rapid soil moisture depletion.	ALEXI model requires LST data for predicting actual ET which is contaminated by cloud cover.
8	Temperature Condition Index (TCI)	$\text{TCI} = \left(\frac{\text{LST}_{\max} - \text{LST}_i}{\text{LST}_{\max} - \text{LST}_{\min}} \right) \times 100$ where, LST_i is the LST of current time period; $\text{LST}_{\max}$ and $\text{LST}_{\min}$ are maximum and minimum LST, respectively, for the same time period in the historical dataset. TCI < 40% indicates drought conditions.	Kogan (1995)	Easy to get the required input data. It determines the stress on vegetation due to temperature and excessive wetness.	Requires clear-sky conditions as LST data can be contaminated by the presence of clouds. Requires long record period for effective results.

(continued)

Table 15.1 (continued)

S. No.	Index	Description	Reference	Advantages	Limitations
9	Evaporative Drought Index (EDI)	$\text{EDI} = 1 - \frac{\text{ET}}{\text{PET}}$ where, ET and PET are actual and potential ET, respectively. EDI > 0.6 indicate drought condition.	Yao et al. (2010)	It can be computed at various time scales. It can reflect drought conditions, not reflected by rainfall deficits. No need for local calibration.	For comparison of moisture stress using EDI over different regions, the EDI should be standardized.
10	Evaporative Demand Drought Index (EDDI)	It is the departure of reference ET, aggregated across a time window of interest relative to historical conditions in percentile. $\text{EDDI} = W - \frac{C_0 + C_1 W + C_2 W^2}{1 + d_1 W + d_2 W^2 + d_3 W^3}$ where, $C_0 = 2.515517$, $C_1 = 0.802853$, $C_2 = 0.010328$, $d_1 = 1.432788$, $d_2 = 0.189269$, and $d_3 = 0.001308$. For $P(E_{0i}) \leq 0.5$, $W = \sqrt{-2\ln\left[P(E_{0i})\right]}$, and for $P(E_{0i}) \geq 0.5$, replace $P(E_{0i})$ with $[1 - P(E_{0i})]$ and reverse the sign of EDDI. Here, $P(E_{0i})$ is the empirical probability of atmospheric evaporative demand (E_0) aggregated across the period of interest.	Hobbins et al. (2016)	It is not sensitive to land-surface type, therefore, it is valid for all ecological regions. It provides early warning of drought. It is multi-scalar or can be computed at different timescales.	Sophisticated computation process.

Table 15.2 Common drought indices based on vegetation data

S. No.	Index	Description	Reference	Advantages	Limitations
1	Normalized Difference Vegetation Index (NDVI)	$\text{NDVI} = \frac{\rho_{\text{NIR}} - \rho_{\text{Red}}}{\rho_{\text{NIR}} + \rho_{\text{Red}}}$ where, ρ_{NIR} and ρ_{Red} are surface reflectance in near-infrared and red band wavelengths, respectively.	Rouse et al. (1974), Weier and Herring (2000)	Indicate vegetation greenness or health conditions. Most widely used index. Serve as an important indicator for computation of other vegetation-based indices. Very high resolution and great spatial coverage.	Sensitive to cloud contamination as low NDVI could be due to atmospheric conditions rather than vegetation condition. Cannot compare drought conditions over different regions.
2	Normalized Difference Water Index (NDWI)	$\text{NDWI} = \frac{\rho_{\text{NIR}} - \rho_{\text{SWIR}}}{\rho_{\text{NIR}} + \rho_{\text{SWIR}}}$, where, ρ_{NIR} and ρ_{SWIR} are surface reflectance in near-infrared and short-wave infrared band wavelengths, respectively.	Gao (1996)	Indicate changes in the water content in the vegetation canopy.	Other factors apart from drought like change in land cover or insect-pests attack can decrease the NDWI. So, this index should not be used alone for assessing drought condition.
3	Standardized Vegetation Index (SVI)	$\text{SVI} = \frac{\text{NDVI}_i - \text{NDVI}}{\sigma_{\text{NDVI}}}$, where, NDVI_i is the NDVI of current time period. NDVI is the mean NDVI of that time period, and σ_{NDVI} is the standard deviation of NDVI of that time period	Peters et al. (2002)	It compares the vegetation status in current year with that of other years for the same time period. Easy to calculate.	The poor vegetation condition could be due to factors other than drought like floods, change in crop cover or crop type, insect–pest attack, etc. Therefore, it should be used along with some other drought indicators.

(continued)

Table 15.2 (continued)

S. No.	Index	Description	Reference	Advantages	Limitations
4	Vegetation Condition Index (VCI)	$VCI = 100 \times (\frac{NDVI_i - NDVI_{min}}{NDVI_{max} - NDVI_{min}})$, where, $NDVI_i$ is the NDVI of current time period; $NDVI_{max}$ and $NDVI_{min}$ are maximum and minimum NDVI, respectively, for the same time period in the historical dataset.	Kogan (1990)	Compares the NDVI of current period with the range of values in historical data for determination of drought condition. It can be used to compare drought across different time periods and ecological regions.	Requires NDVI data over a longer period for accurate results. Sensitive to cloud contamination as low NDVI could be due to atmospheric conditions rather than vegetation conditions, which may produce erroneous results. Produce erroneous results when NDVI data has trend due to improvement in technology.
5	Vegetation Temperature Condition Index (VTCI)	$VTCI = \frac{LST_{NDVIi.max} - LST_{NDVIi}}{LST_{NDVIi.max} - LST_{NDVIi.min}}$ where: $LST_{NDVI\,i.\,max} = a + b \times NDVI_i$ $LST_{NDVI\,i.\,min} = a' + b' \times NDVI_i$ where, $LST_{NDVI\,i.\,max}$ and $LST_{NDVI\,i.\,min}$ are maximum and minimum land surface temperatures of pixels which have same $NDVI_i$ value in a study region, respectively. $LST_{NDVI\,i}$ denotes land surface temperature of one pixel whose NDVI value is $NDVI_i$. a, b, a', and b' are coefficients estimated on pixels from an area large enough to comprise the whole range of soil moisture content from field capacity to wilting point. It ranges from 0 to 1. Lower the value of VTCI, higher is the drought severity.	Wang et al. (2001)	It works better at regional scale. This index is site-specific and time-specific.	It cannot be calculated over a small study area, or area having variable topography.
6	Vegetation Health Index (VHI)	$VHI = \alpha VCI + (1 - \alpha)TCI$ It combines VCI and TCI with a weightage factor α which determines their respective contributions. Lesser VHI values denote more drought severity.	Kogan (1990, 2001)	It monitors drought more comprehensively as it combines both VCI and TCI.	It requires appropriate data fusion for VCI and TCI. It doesn't work well in the areas where the relationship between NDVI and LST is positive, especially at high latitudes.

7	Normalized Multiband Drought Index (NMDI)	$NMDI = \frac{\rho_{0.86\mu m} - (\rho_{1.64?m} - \rho_{2.13?m})}{\rho_{0.86\mu m} + (\rho_{1.64?m} - \rho_{2.13?m})}$ where, $\rho_{0.86\ \mu m}$, $\rho_{1.64\ \mu m}$, and $\rho_{2.13\ \mu m}$ are surface reflectance in the wavelengths 0.86, 1.64, and 2.13 µm, respectively.	Wang and Qu (2007)	It estimates moisture content in both soil and vegetation, therefore, it has enhanced sensitivity toward drought. In bare soil or in very less vegetation cover, it acts as soil drought indicator, while at higher vegetation cover, it acts as vegetation drought indicator.	NMDI saturates with vegetation having LAI more than 5. In moderate vegetation condition, i.e., LAI between 0.5 to 1, the role of NMDI as soil or vegetation drought indicator blurs.
8	Crop Moisture Index (CMI)	It is based on the concept of abnormal ET deficit, calculated as the difference between computed actual ET and computed potential ET. Actual ET is based on the temperature and precipitation that occurs during the week and computed soil moisture in both the topsoil and subsoil layers.	Palmer (1968)	It responds quickly to short-term crop water stress conditions. It can be used to compare drought conditions across various regions. It is easy to compute from precipitation, temperature, and soil moisture data.	It is not suitable for the long-term agricultural drought.
9	Vegetation Water Supply Index (VWSI)	$VWSI = \frac{NDVI}{LST}$ Lower value of VWSI indicates insufficient water supply to the crops resulting in drought.	Cai et al. (2010)	This index combines both NDVI and LST to measure drought on agricultural crops as the NDVI decreases and LST increases under drought conditions.	VSWI slopes may vary from one place to another. The difference in the range of NDVI and LST is large, therefore, very small value of VWSI is generated.
10	Normalized Difference Drought Index (NDDI)	$NDDI = \frac{NDVI - NDWI}{NDVI + NDWI}$ NDDI >1 indicates drought conditions.	Gu et al. (2007)	It combines both vegetation greenness and wetness conditions.	It is not applicable for short-term drought monitoring.

Table 15.3 Common drought indices based on soil moisture data

S. No.	Index	Description	Reference	Advantages	Limitations
1	Soil Moisture Anomaly (SMA)	SMA is calculated using the daily soil moisture output of JRC's LISFLOOD hydrological model. First of all, Soil Moisture Index (SMI) is computed using following formula, and then the z-score of SMI across the available historical time period is taken as SMA $SMI = 1 - \frac{1}{1+\left(\frac{\theta}{\theta_{50}}\right)^{6}}$ where, Θ is the daily soil moisture and Θ_{50} is the mean between the wilting point and the field capacity.	Sepulcre-Canto et al. (2012)	The update of SMA with the weather forecast data in the model can provide the risk of drought along with its duration and severity.	LISFLOOD model need to be calibrated and validated for the region using ground-truth data.
2	Soil Moisture Deficit Index (SMDI)	SMDI is calculated by using the available soil moisture output of SWAT model from which soil water deficit (SD) in percentage on weekly basis is computed with respect to the long-term soil moisture content. SMDI for any given week (*j*) is computed using following formula: $SMDI_j = \frac{\sum_{t=1}^{j} SD_t}{25t+25}$ where, '*t*' is the time in weeks. The SMDI for further weeks is computed on incremental basis. Lower the value of SMDI, higher is the drought severity.	Narasimhan and Srinivasan (2005)	As it takes into account the different depths of the soil profile, therefore, it could be used for monitoring drought in different crops.	As it uses the available soil moisture content output from the SWAT model, therefore, the SWAT model needs to be calibrated and validated according to the local conditions. It is sophisticated to compute.

3	Percent Available Soil Moisture (PASM)	$\text{PASM} = \left(\frac{\text{SM} - \text{PWP}}{\text{FC} - \text{PWP}}\right) \times 100$ where, SM is the current volumetric soil moisture content, FC and PWP is the field capacity and permanent wilting point of the soil, respectively. The agricultural drought occurs when the PASM <75%.	Manual for Drought Management (2016)	Simple and easy to compute.	Site-specific data of soil moisture, field capacity, and permanent wilting point is required for its computation, which is not easily available for a larger area.
4	Moisture Adequacy Index (MAI)	$\text{MAI} = \left(\frac{\text{AET}}{\text{PET}}\right) \times 100$ where, AET and PET are actual and potential ET, respectively, calculated using weekly soil water balance. MAI < 75% indicate drought condition.	Manual for Drought Management (2016)	It is a better indicator for assessing the degree of adequacy of rainfall and soil moisture to meet the potential water requirement of the crops.	Complex computation as much data is needed for the soil water balance approach.
5	Standardized Soil moisture Index (SSI)	$\text{SSI} = \frac{\text{SM}_i - \overline{SM}}{\sigma_{\text{SM}}}$ where, SM_i is the volumetric soil moisture content of current time period, $\overline{\text{SM}}$ is the mean soil moisture of that time period, and σ_{SM} is the standard deviation of soil moisture of that time period. SSI < −1 denotes drought conditions.	Xu et al. (2018)	It compares the soil moisture deviations with the historical mean to determine the drought condition. Simple and easy to calculate. It can be used to compare drought across regions.	Long-term soil moisture data is required for accurate results.

(continued)

Table 15.3 (continued)

S. No.	Index	Description	Reference	Advantages	Limitations
6	Soil Moisture Condition Index (SMCI)	$\text{SMCI} = 100 \times \left(\frac{\text{SM}_i - \text{SM}_{\text{min}}}{\text{SM}_{\text{max}} - \text{SM}_{\text{min}}} \right)$, where, SM_i is the soil moisture content of current time period; SM_{max} and SM_{min} are maximum and minimum soil moisture content, respectively for the same time period in the historical dataset.	Zhang and Jia (2013)	Compares the soil moisture status of current period with the range of values in historical data for determination of drought conditions. It can be used to compare drought across regions.	Long-term soil moisture data is required for accurate results.
7	Empirical Standardized Soil Moisture Index (ESSMI)	It is computed by fitting a nonparametric empirical probability density function to historical time series of soil moisture observations, and then transforming it into a normal distribution. The z-score of this normally distributed soil moisture data is taken as ESSMI, where negative values denote dry soil conditions.	Carrão et al. (2016)	ESSMI has better ability to assess dry and wet soil conditions due to flexible nonparametric empirical probability density function. It can be computed at different time scales and can be used to compare drought situations across different regions.	It is challenging to apply the appropriate nonparametric empirical probability density function to the soil moisture data over a large area which makes its computation sophisticated.
8	High-resolution Soil Moisture Drought Index (HSMDI)	AMSR-E soil moisture product is downscaled to 1 km using MODIS and TRMM data through random forest technique, and HSMDI is computed using following formula: $\text{HSMDI} = \left(\frac{\text{SM}_i - \text{SM}_{\text{min}}}{\text{SM}_{\text{max}} - \text{SM}_{\text{min}}} \right)$, where, SM_i is the downscaled soil moisture content of current time period; SM_{max} and SM_{min} are maximum and minimum downscaled soil moisture content, respectively, for the same time period in the historical dataset.	Park et al. (2017)	It is good for monitoring meteorological and agricultural drought due to higher spatial resolution.	Cannot monitor hydrological drought. Its performance is dependent upon the MODIS data as the data quality get affected by several factors like clouds, aerosols, etc.

9	Soil Water Deficit Index (SWDI)	$SWDI = \left(\frac{\theta - \theta_{FC}}{\theta_{AWC}}\right) \times 10$, where, Θ is the current volumetric soil moisture content, Θ_{FC} is the field capacity and Θ_{AWC} is the available water content, i.e., the difference between field capacity and permanent wilting point. Negative values of SWDI indicate drought conditions.	Martínez-Fernández et al. (2015)	Good indicator for monitoring agricultural drought.	Need the data of field capacity and permanent wilting point of the soil, which makes it challenging for applying over a large area.
10	Soil Moisture Anomaly Percentage Index (SMAPI)	The soil moisture output from Variable Infiltration Capacity (VIC) model is used to compute SMAPI $SMAPI = \left(\frac{SM_i - \overline{SM}}{\overline{SM}}\right) \times 100$, where, SM_i is the daily VIC-simulated soil moisture content and $\overline{SM}$ is its corresponding multi-year average soil moisture content. SMAPI ≤ −5% denotes drought.	Mao et al. (2017)	Simple and effective index for spatio-temporal analysis of regional drought.	VIC model needs to be calibrated and validated for a particular region.

15.4.4.2 Combined Deficit Index (CDI)

Combined Deficit Index (CDI) was developed by Vyas et al. (2015) for agricultural drought assessment over semiarid tract (Gujarat, Maharashtra and Karnataka) of India. This index was computed using the following formula.

$$\mathrm{CDI} = ?\mathrm{RAIN}_d + (1\text{-}?)\,\mathrm{NDVI}_d,$$

where, $NDVI_d$ and $RAIN_d$ are the percent deficits of monthly NDVI composite and sum of antecedent tri-monthly rain, respectively, from their means over the normal years (2010, 2011, and 2013). α and $(1-\alpha)$ are the weights assigned to $RAIN_d$ and $NDVI_d$, respectively through simple ranking method. The NDVI data at the lag of 1 month of rainfall data were taken for computing CDI considering the time lag in the NDVI response due to rainfall deficit. The CDI was evaluated with SPI and negative percent anomaly of LST, Gross Primary Productivity, ET, and agricultural yield in terms of deficit from normal mean. This CDI was able to assess late-season agricultural drought only.

15.4.4.3 Blended Drought Index (BDI)

Blended Drought Index (BDI) was developed by Luetkemeier et al. (2017) for estimating the impact of drought in the semiarid Cuvelai-Basin of Angola and Namibia. In BDI, standardized drought indicators of precipitation (SPI), vegetation (VHI) computed from the combination of VCI and TCI with relative weights of 70/30, ET (SPEI), and soil moisture (SSI) were combined using a copula function. Gaussian copula function was found most suitable to combine the individual indicators. The drought characteristics like frequency of drought occurrence, severity of the impact of drought events and the duration of droughts as per each indicator were used for developing final BDI. The BDI was validated using millet/sorghum crop yields and local water consumption data.

15.4.4.4 Vegetation Drought Response Index (VegDRI)

Vegetation Drought Response Index (VegDRI) was developed by Brown et al. (2008) for monitoring agricultural drought in the United States. VegDRI combines NDVI datasets as derived from NOAA–AVHRR with climate-based SPI and PDSI drought indices as derived from observations from selected stations of the synoptic network along with static biophysical information such as elevation, land use, soil water capacity, or percentage of irrigated agriculture. In VegDRI, Percent of Average Seasonal Greenness and Start of Seasonal Anomaly were derived from the NDVI datasets. These indices were combined using Cubist model to generate three seasonal, rule-based, piecewise linear regression VegDRI models. The self-calibrated PDSI was used for training the Cubist model. This index was designed for operational near-real time use in the United States with a spatial resolution of 1 km and a temporal update of every 14 days. Later VegDRI was developed for Canada (Tadesse et al., 2017) and South Korea (Nam et al., 2018). Based on VegDRI concept, Wu et al. (2015) developed Integrated Surface Drought Index (ISDI) for China.

15.4.4.5 Process-Based Accumulated Drought Index (PADI)

Zhang et al. (2017) linked drought development phases with crop phenology to quantify the accumulative impact of drought on crops using Process-based Accumulated Drought Index (PADI) developed through Evolution Process-based Multi-sensor Collaboration (EPMC) framework in three different provinces of China with varied agro-climatic conditions. EPMC combines precipitation data from Global Precipitation Climatology Centre (GPCC), root zone soil moisture data from GLDAS 2.0, and vegetation condition data from AVHRR in the form of PCI, SMCI, and VCI, respectively, to categorize agricultural drought into four phases, i.e., latency, onset, development and recovery. The "Latency" phase starts when precipitation deficit takes place, i.e., PCI < 0.5. When precipitation deficit results in root-zone soil moisture deficit, i.e., SMCI <0.5, which starts affecting crops, then the drought "Onset" phase begins. When this continued soil moisture stress starts affecting crop conditions, then the drought "Development" phase begins, which is indicated by VCI below 0.6. The "Recovery" phase begins when the precipitation occurs after third phase such that both PCI and SMCI indicate normal conditions. These drought evolution phases were further integrated with the crop phenology and water-deficit sensitivity coefficients to produce weekly PADI. The PADI was found to be well correlated with wheat yield loss and SPI. This index doesn't provide "snapshot" of the drought condition unlike most of the drought indices, but takes into account the evolution of drought process along with the crop phenology to assess the impact of agricultural drought on crops.

15.4.4.6 Combined Drought Indicator (CDI)

Sepulcre-Canto et al. (2012) proposed a Combined Drought Indicator (CDI) to detect drought in Europe. The concept behind the CDI is informed by an idealized cause–effect relationship for agricultural drought. This cause–effect relationship assumes that a shortage of precipitation (the cause) leads to a soil moisture deficit, which results in a reduction of vegetation production (the effect). Such a relationship gives an opportunity to provide an early warning system for agricultural drought by identifying which stage of the cause–effect relationship is current. The CDI is presented as a first step towards such a system. In its simplest form, the CDI is able to identify which stage of the cause–effect relationship has been reached by an agricultural drought event (Sepulcre-Canto et al., 2012). They took SPI, the anomalies of soil moisture, and the anomalies of the fraction of Absorbed Photosynthetically Active Radiation (fAPAR). The Combined Drought Indicator (CDI) gives a synthetic and synoptic overview of the drought situation using a classification scheme composed of three warning levels: "watch" when a relevant precipitation shortage is observed, "warning" when this precipitation shortage translates into a soil moisture anomaly and "alert" when either these two conditions are accompanied by an anomaly in the vegetation condition or precipitation deficit results directly in vegetation stress. This index was further revised by Cammalleri et al. (2021) in which two additional classes, i.e., "partial recovery" and "full recovery" were introduced to denote the recovery of precipitation and vegetation condition, respectively, after the drought event. This system is currently being implemented in the European Commission's European Drought Observatory (EDO) for the monitoring and early warning of agricultural drought in Europe.

15.4.4.7 Scaled Drought Condition Index (SDCI)

Scaled Drought Condition Index (SDCI) was formulated by Rhee et al. (2010) for monitoring the agricultural drought at both arid and humid regions of the United States using LST and NDVI from MODIS, and precipitation data from TRMM. These variables were scaled from 0 to 1 and then combined using weights derived through regression analyses to compute SDCI. This index was found to have good correlations with *in-situ* drought indices, i.e., PDSI, Z-index, SPI-3, and SPI-6 as well as agreed well with the United States Drought Monitor (USDM) maps. However, its performance can be further improved by inclusion of long-term dataset, which may provide reliable scaling for discriminating the drought conditions from normal conditions.

15.4.4.8 Synthesized Drought Index (SDI)

Synthesized Drought Index (SDI) was proposed by Du et al. (2013) for the Shandong province of China by taking the first principal component of VCI, TCI, and PCI. VCI and TCI were computed from MODIS NDVI and LST products, respectively, whereas PCI was computed from TRMM precipitation product. SDI was found to be well correlated with SPI-3, crop yield variation, and drought-affected areas. However, its performance can be further improved by integration of soil moisture variation.

15.4.4.9 Geographically Independent Integrated Drought Index (GIIDI)

Jiao et al. (2019) proposed a Geographically Independent Integrated Drought Index (GIIDI) for drought assessment in the United States. For formulation of GIIDI, TCI, and VCI computed from MODIS data, SMCI computed from AMSR-E data, and PCI computed from TRMM data were combined using local ordered weighted averaging model. This GIIDI was found to perform better than individual PCI, TCI, SMCI, VCI, and SDCI with respect to their correlation with PDSI, SPI-1, SPI-3, and SPEI-6. It could be used to monitor both short-term and long-term droughts across most of the climatic divisions of the United States.

15.4.4.10 Integrated Agricultural Drought Index (IADI)

Integrated Agricultural Drought Index (IADI) was formulated by Liu et al. (2020) for agricultural drought monitoring in the North China Plain. He used standardized NDVI, LST, and precipitation along with soil water capacity and elevation to develop IADI at 1 km spatial resolution and 1-month temporal interval using Back-propagation Neural Network in which *in-situ* soil moisture content unto 10 cm soil depth was used as the dependent variable. IADI was found to be highly correlated with 10 cm depth soil moisture, as well as SPI-3 and SPEI-3.

15.4.4.11 Combined Drought Monitoring Index (CDMI)

Combined Drought Monitoring Index (CDMI) was developed by Han et al. (2021) for Shaanxi Province of China. He used NDVI, LST, and ET from MODIS, and precipitation from TRMM as inputs which were standardized to the scale of 0 to 1. These scaled inputs were subjected as independent variables in the random forest model with relative soil moisture as the dependent variable. The variable importance derived from the random forest model was then homogenized as factor coefficient of each variable to develop a linear model. This linear model was termed as CDMI. Thus, he used variable importance of random forest model for assigning weightage to each variable. He validated the effectiveness of CDMI with SPI at 1, 3, 6, 9, and 12 months' time scale, and relative soil moisture. He found that the performance of CDMI is superior to other drought indices like NDDI, NMDI, VCI, TCI, NDWI, etc., in monitoring drought.

15.4.4.12 Integrated Remote Sensing Drought Monitoring Index (IRSDI)

Integrated Remote Sensing Drought Monitoring Index (IRSDI) was proposed by Sun et al. (2017) for drought monitoring in the Huai River Basin in China. He computed 16 remote-sensing-derived drought index available in literature and used variance inflation factor to evaluate multi-collinearity among these indices. He selected four indices which had no multi-collinearity and highest correlation with the soil moisture, i.e., TCI, Anomaly Vegetation Index, Visible and Shortwave Infrared Drought Index, and Global Vegetation Moisture Index. These indices were integrated using multiple linear regression with soil moisture as dependent variable to develop equations for computing IRSDI for each month. The developed IRSDI used to monitor agricultural drought with 90% confidence level.

15.4.4.13 Combined Drought Indicator for Ethiopia (CDI-E)

Combined Drought Indicator for Ethiopia (CDI-E) was formulated by Bayissa et al. (2019) to monitor the spatial and temporal extents of historic agricultural drought events in Ethiopia. In this index, SPI, LST anomaly, standardized NDVI, and soil moisture anomaly computed from CHIRPS, Terra MODIS, e-MODIS, NOAH soil moisture datasets, respectively, were combined using weights assigned through principal component analysis (PCA). The squared value of eigen-vectors of the first principal component was used as weights at 1 km grid for each month. Thus, CDI-E was developed at spatial resolution of 1 km at monthly time interval. The classification of drought severity was done using classification scheme of SPI given by McKee et al. (1993). This index was found to be well correlated with cereal crop yields and 3-month SPI.

15.4.4.14 Integrated Drought Monitoring Index (IDMI)

Integrated Drought Monitoring Index (IDMI) was proposed by Kumar et al. (2021) to monitor agricultural drought over Tamil Nadu state of India by integrating the PCI, TCI, SMCI, and VCI through weights assigned by first principal component in the PCA technique. The IDMI was showing good correlation with 3-month SPI and crop yields of major crops, i.e., maize, pearl millet and sorghum. The sensitivity analysis showed that PCI is the most influencing parameter followed by VCI and TCI. IDMI was found to have good potential in assessing and monitoring spatio-temporal variation of agricultural drought.

15.4.4.15 Drought Severity Index (DSI)

Drought Severity Index (DSI) was given by Mu et al. (2013) for assessing global terrestrial drought severity using remote sensing data. He has used MODIS NDVI and ET data for its computation. The sum of the z-score of NDVI and ratio of actual to potential ET is taken, and further standardized to formulate DSI. The negative value of DSI indicates drought conditions. The DSI was produced at spatial resolution of 1 km and at 8-day, monthly, and annual temporal intervals. The authors evaluated the DSI using data from several major drought events, including the 2003 European drought, the 2005 Amazon drought, and the 2010 Russian drought. The DSI was able to capture the spatial and temporal patterns of these drought events, and it was also correlated with other drought indices, such as the Palmer Drought Severity Index (PDSI).

15.4.4.16 Integrated Drought Index (IDI)

Integrated Drought Index (IDI) was developed by Shah and Mishra (2020) for drought monitoring in India by integrating the response to meteorological, agricultural, and hydrological droughts along with ground water storage in terms of 12-month SPI, 1-month SSI, 4-month Standardized Runoff Index (SRI), and 1-month Standardized Groundwater Index (SGI), respectively. These indices were integrated using a Gaussian copula to develop IDI. The input data of runoff, soil moisture, and ground water required for the computation of SRI, SSI, and SGI, respectively, were simulated using a hydrological model known as Variable Infiltration Capacity with SIMple Groundwater Model (VIC-SIMGM), and rainfall data for SPI was taken from IMD. The authors evaluated the IDI using data from several major drought events in India, including the 2002 drought in Gujarat and the 2010 drought in Maharashtra. The IDI was able to capture the spatial and temporal patterns of these drought events, and it was also correlated with other drought indices, such as the Palmer Drought Severity Index (PDSI).

15.4.4.17 Multivariate Phenology-Based Agricultural Drought Index (MADI)

The Multivariate phenology-based Agricultural Drought Index (MADI) was proposed by Das et al. (2020) for characterizing agricultural drought over India using long-term crop phenological parameters like length of growing period and integrated NDVI during *kharif* season using Savitzky—Golay filtering technique. The GIMMS NDVI data was taken for this purpose. The data of long-term length of growing period and integrated NDVI was de-trended and standardized to eliminate the effect of improvement in agricultural technologies. The standardized length of growing period and standardized integrated NDVI were fitted using joint probability function to constitute MADI. The MADI varies from −4 to +4, where negative value represents drought condition. The performance of MADI was evaluated with SPI, SSI, and MSDI, which showed good agreement in terms of drought intensity and spatial extent.

15.4.4.18 Combined Drought Index (CDI)

Various researchers have developed different types of CDI for monitoring agricultural drought. Chattopadhyay et al. (2020) developed CDI for monitoring agricultural drought by assigning suitable weightages to parameters such as soil moisture at 60 cm depth, rainfall, SWDI, CMI, ESI, NDVI, VCI, TCI, and VHI. This CDI was computed for *kharif* crop season over Andhra Pradesh, Chhattisgarh, Haryana, Maharashtra, and Telangana states of India. Kulkarni et al. (2020) have developed Combined Drought Indicator to monitor agricultural drought in the Marathwada region of India by assigning weightage to 3-month SPI, LST anomaly, SNDVI and soil moisture anomaly using PCA, which showed good correlation with cereal crop yields. Prajapati et al. (2022) combined 3-month SPI, 5-month VCI, and 5-month Streamflow Drought Index (SDI) through PCA to develop CDI for Marathwada region of India. Danodia et al. (2021) developed CDI based on Analytical Hierarchical Process method for assigning weightage to VCI, TCI, PCI, and SMCI for agricultural drought assessment in Bundelkhand region of India, which showed significant correlation with crop yield anomaly.

15.4.4.19 Microwave Integrated Drought Index (MIDI)

Microwave Integrated Drought Index (MIDI) was proposed by Zhang and Jia (2013) based on microwave remote sensing data to monitor short-term drought over semi-arid regions of China. MIDI was calculated at a spatial resolution of 0.25° using the combination of three microwave remote sensing data products, i.e., Tropical Rainfall Measuring Mission (TRMM) Precipitation Radar data, the Advanced Microwave Scanning Radiometer for Earth Observing System (AMSR-E) land surface temperature data, and the AMSR-E soil moisture data, where each variable is scaled from 0 to 1, and proper weightages were assigned to them based on the region. MIDI showed good correlation with 1-month and 3-month SPI.

15.4.4.20 Multivariate Drought Composite Index (MDCI)

Hanadé Houmma et al. (2023) formulated Multivariate Drought Composite Index (MDCI) for monitoring agricultural drought over Sahelian agro-systems. For development of MDCI, variables pertaining to vegetation stress like VCI, proportion of vegetation, and smoothed NDVI; water stress like PCI, Rainfall Deficit Index, Normalized Rainfall Deficit Index, and Normalized Rainfall Efficiency Index; and heat stress like TCI, smoothed brightness temperature and scaled TCI were taken. The weightage of each variable was assigned through the relative importance computed using random forest algorithm with inter-annual fluctuation in the yield of cereal crops as dependent variable. The MDCI was able to capture the spatial and temporal patterns of drought in the Sahel region. The MDCI was showing good correlation with other drought indices, such as the PDSI, SPI, VHI, and climatic water deficit index.

15.5 Approaches for Forecasting Agricultural Drought

The major approaches used for agricultural drought forecasting are regression analysis, time-series analysis or stochastic modeling, probabilistic modeling, machine learning models, dynamic models, and hybrid models (Fung et al., 2020) which are further discussed below:

15.5.1 Regression Analysis

Regression analysis is a statistical approach in which the relationship of dependent or response variable is explored with respect to one or more independent variables or predictors by finding a best-fitting line or curve (Sykes, 1993). The most common types of regression analysis include simple linear regression, multiple linear regression, logistic regression, polynomial regression, and log-linear regression. Kumar and Panu (1997) developed a multiple linear regression model to predict the grain yield of pearl millet as an indicator of the agricultural crop in the Jodhpur district of Rajasthan, India by using amount of rainfall and number of rainy days in the month of July, August, and September, as well as days for delayed sowing as predictors. Using this technique, they were able to predict the severity level of agricultural drought several months in advance, based on the percent deviation in crop yield. Liu and Juárez (2001) used multiple linear regression techniques to predict drought onset in the northeast Brazil by establishing the relationship between NDVI anomalies and El Nino Southern Oscillation (ENSO) indices viz., Nino 3.4, Southern Oscillation Index (SOI), North Atlantic Sea Surface Temperature (NATL), South Atlantic Sea Surface Temperature (SATL) and Dipole 2. The developed models were able to predict agricultural drought onset prior to 4 months with 68% accuracy. Zambrano et al. (2018) predicted inter-annual variability in agricultural productivity in Chile using optimal linear regression with seasonal cumulative NDVI as response variable and SPI, Pacific Decadal Oscillation, and Multivariate ENSO oscillation indices, and length of the growing season as predictor variables. The developed optimal linear regression model showed good accuracy for smaller lead time with mean R^2 value of 0.95 for one-month lead time, which can contribute to the early warning system of agricultural drought. However, the regression analysis has some limitations in its usage, such as assumptions of linearity, independence, and normality of the observations; problem of multi-collinearity, under-fitting, and over-fitting; and sensitivity to outliers, which makes it poor for long-lead forecasting.

15.5.2 Time-Series Analysis

Stochastic or time-series models are statistical models created especially for analysing and predicting data that changes over time. These models account for the data's innate randomness and time dependence. The foundation of time-series models is the idea that a variable's future values depend on its past values and sometimes on additional variables. In doing so, they enable forecasting of future values by capturing the temporal dependencies and patterns in the data. Time series models have a number of significant benefits, including their capacity to conduct systematic searches for model-building purposes such as identification, estimation, and diagnostic checks (Mishra & Desai, 2005). The two most commonly used and important stochastic models are Autoregressive Integrated Moving Average (ARIMA) and Seasonal Autoregressive Integrated Moving Average (SARIMA) (Mishra et al., 2007). The serial linear correlation between observations is effectively taken into account by ARIMA, but SARIMA is better able to describe time series that exhibit nonstationarity both within and across seasons. Various researchers have used ARIMA or SARIMA models for forecasting drought indices like Palmer's Drought Index by Rao and Padmanabhan (1984); SPI by Mishra and Desai (2005), Han et al. (2013) and Alam et al. (2014); VTCI by Han et al. (2010); NDVI by Fernández-Manso et al. (2011); SPEI by Mossad and Alazba (2015), and many more. However, the applicability of these time series models for drought forecasting has limitations such as their inability to capture nonlinear characteristics, assumption of time series stationarity, high computational cost, and sensitivity to outliers.

15.5.3 Probabilistic Modeling

Probabilistic modelling is used for drought forecasting as it is able to deal with complex distributions (Fung et al., 2020). These models estimate the likelihood or probability of future drought events based on historical data of pertinent variables. Probabilistic modelling can be done by various techniques like Bayesian network, Markov chain, and copula. Salakpi et al. (2022) reported that the Hierarchical Bayesian model was more accurate in forecasting agricultural drought at 4–10 weeks lead time in different agro-ecological regions and vegetation land cover as compared to a regular Bayesian auto-regression distributed lags model. Zarekarizi et al. (2021) proposed a new probabilistic drought forecasting framework that combines dynamic land surface modeling, statistical Bayesian multivariate analysis, satellite soil moisture data assimilation, and high-performance computing to generate deterministic and probabilistic drought forecasts at various lead times, including 1, 2, and 3 months. Wu et al. (2023) integrated vine copulas with Bayesian model averaging to build agricultural drought prediction model for 1–6 months lead times. Paulo and Pereira (2007) used Markov chain model on monthly SPI data of 67 years in Alentejo region of southern Portugal to predict drought class up to 3 months ahead. Khadr (2016) forecasted SPI for the Blue Nile river basin in Egypt using homogeneous hidden Markov chain model. Madadgar and Moradkhani (2013) developed a Bayesian framework for probabilistic seasonal drought forecasting that uses copula functions to model the joint distribution of precipitation and streamflow. The model was able to provide probabilistic forecasts of drought conditions for the Aral Sea basin in Central Asia. Zhang et al. (2021) forecasted Standardized Soil moisture Index (SSI) as an indicator of agricultural drought using meta Gaussian copula model, based on GLDAS-derived initial soil moisture conditions, SPI, and large-scale drivers ENSO, which performed well in semiarid and sub-humid regions during spring but showed poor performance in humid regions during summer.

15.5.4 Machine Learning Models

The data-driven machine learning algorithms have been widely used for the agricultural drought forecasting by several researchers due to their non-parametric nature and ability to capture complex, non-linear relationships. Liu et al. (2017) forecasted agricultural drought at 1–2 weeks lead time based on SWDI using Support Vector Machine (SVM) with Leaf Area Index and soil moisture retrievals from AMSR-E and SMAP as inputs. Tadesse et al. (2010) developed Vegetation Outlook (VegOut) which predicts a standardized seasonal greenness (SSG) measure at multiple time steps (2–6 weeks into the future) based on the analysis of "historical patterns" (i.e., patterns at each 1 km grid cell and time of the year) of satellite, climate, and oceanic data over an 18-year period (1989–2006) using Cubist model. Tian et al. (2018) used support vector regression to predict SPEI over Xiangjiang River basin of China. Nafii et al. (2022) used Artificial Neural Network (ANN) model to predict SPEI in Morocco. Sardar et al. (2021) used Convolution Neural Network to predict drought or no drought condition based on NDVI with accuracy of up to 96% in the Kolar district of Karnataka. Marj and Meijerink (2011) used ANN to predict NDVI which is an indicator of agricultural drought. In order to predict the SPEI for 17 climatology stations in Pakistan's Northern Area and Khyber Pakhtunkhwa, Ali et al. (2017) used the multilayer perceptron–ANN model. According to their findings, the model was able to accurately capture the variance in a few selected drought indices across a one-month time scale. Brust et al. (2021) formulated "DroughtCast" which is a machine learning framework for forecasting the USDM. It takes daily meteorological data from local weather stations along with satellite observed soil moisture, Gross Primary Productivity, Land Cover, ET, and multivariate ENSO index as inputs into a recurrent neural network to accurately forecast the USDM between 1 and 12 weeks into the future. Dikshit et al. (2020) used random forest (RF) model to predict SPEI at New South Wales region of Australia. Feng et al. (2019) reported that the bias-corrected RF model outperformed SVM and multilayer perceptron neural network in predicting SPEI. Chen et al. (2023) found that the RF model was able to predict spring soil drought based on winter climate in the southern Songnen Plain of China. Kan et al. (2023) reported that the RF model performed better than SVM, CART, multivariate linear regression, ANN and ARIMA in predicting soil moisture as an indicator of agricultural drought in Sweden. Park et al. (2020) developed a drought prediction model to overcome the challenges of short-term forecasting remote sensing-based drought indices (SDCI, SPI) with numerical model outputs. Thus, 8-day drought forecasts were created by combining the RF algorithm and convolutional long-term memory (ConvLSTM). The method was proven to be helpful for operationally providing high-resolution predicted drought conditions. Houmma et al. (2022) found in his review of drought forecasting techniques that the random forest model has better performance than other ML algorithms like SVM, ANN, CART, or Cubist in most of the studies. However, the performance of ML algorithms is highly dependent on the quantity and diversity of data, as well as optimization of hyper-parameters to achieve higher accuracy (Houmma et al., 2022).

15.5.5 Dynamic Models

Dynamic modeling is a type of drought forecasting that uses mathematical models to simulate the behaviour of the climate system. These models can be used to predict how droughts will develop and evolve over time, and they can also be used to assess the impact of droughts on water resources and ecosystems (Feller & Vaseva, 2014). General circulation models (GCMs) serve as the foundation for dynamic (physical) models, which contain links between the land, ocean, and atmosphere (Thober et al., 2015). However, these models are sensitive to the prediction interval and the spatial extent taken into consideration. Another drawback of climate models is that their prediction ability is dependent on the climate context. In this context, Yuan and Wood (2013) emphasized that, in contrast to extratropical regions, North-American Multi-Model Ensemble (NMME) model performs better in tropical regions. According to Thober et al. (2015), the NMME dynamically based seasonal drought forecasting system outperformed forecasts based on the Streamflow Prediction Set (ESP) model in terms of accuracy. But when it comes to identifying the seasonality of hydrological and meteorological dryness, the ESP model continues to be the most widely used model. Another approach of dynamic modeling is ensemble forecasting in which multiple forecasts are generated using different model configurations or initial conditions to capture the uncertainty inherent in drought prediction. For instance, the European Centre for Medium-Range Weather Forecasts (ECMWF) utilizes ensemble forecasting to estimate the probability of drought occurrence.

15.5.6 Hybrid Models

In hybrid models, multiple techniques are used for forecasting drought in which the remarkable abilities of individual techniques are harnessed to produce better accuracy and higher lead time as compared to individual models. The hybrid models are produced by coupling two or more machine learning (ML) models or forecasting techniques and by incorporating novel data processing techniques or parameter optimization techniques in machine learning models. Xu et al. (2020) proposed a hybrid ARIMA-Support Vector Regression (SVR) model for prediction of SPI whose prediction accuracy was higher than the single ARIMA model and predicted results of 1–2 months ahead showed good agreement with the actual data. Belayneh and Adamowski (2013) used ANN, SVR and coupled wavelet-ANN to forecast SPI, out of which, coupled wavelet-ANN performed best. Aghelpour et al. (2021) reported that the optimization of hyper-parameters in SVM using dragonfly algorithm can predict PDSI better than the ARIMA and ANN. Danandeh Mehr et al. (2022) reported that the integration of genetic algorithm with RF can improve the accuracy of the RF model by 30–40% in the prediction of SPEI. Das et al. (2020) reported that the preprocessing of data by wavelet packet transform before incorporating into ML algorithms of ANN and SVR, outperformed the individual ANN and SVR models in prediction of SPI. Altunkaynak and Jalilzadnezamabad (2021) combined the PDSI and the Discrete Wavelet Transform (DWT) tools with the k-Nearest Neighbour (kNN), Fuzzy, and SVM approach to improve the accuracy of drought predictions in the Marmara region of Turkey. This study shows that these hybrid models were performing better than stand-alone models in terms of prediction accuracy, out of which W-Fuzzy model was better than the W-KNN and W-SVM models. Başakın et al. (2021) reported that the performance of an adaptive neuro-fuzzy inference system (ANFIS) with empirical mode decomposition (EMD) to forecast the self-calibrated PDSI in Turkey was better than the ANFIS model alone. Adnan et al. (2021) combined random vector functional link (RVFL) with the particle swarm optimization (PSO), genetic algorithm (GA), grey wolf optimization (GWO), social spider optimization (SSO), salp swarm algorithm (SSA), and hunger games search algorithm (HGS) to predict 3-, 6-, 9- and 12-month SPI in Bangladesh, and found that the HGS algorithm performed better than the other algorithms and significantly increased the RVFL method's accuracy in drought forecasting. Khan et al. (2020) reported that the hybrid model of Wavelet-ARIMA–ANN performed better than Wavelet-ANN and ANN only in forecasting SPI and Standard Index of Annual Precipitation in Langat River Basin of Malaysia. Wu et al. (2021) used wavelet-ARIMA long-short-term memory (WT-ARIMA-LSTM) model to predict precipitation, which was further used to compute the China Z-Index drought episodes. The results indicate that the WT-ARIMA-LSTM model performed better than the ARIMA and LSTM models individually. Houmma et al. (2022), in their review of several drought forecasting methodologies, emphasized that hybrid models superior performance over single models, as they incorporate both pre-processing and metaheuristic algorithm techniques. Furthermore, despite significant recent in hybrid model strategies, no as the most accurate forecasting approach. Since drought prediction a challenge, there is still space for researchers to improve hybrid approaches for applications (Alawsi et al., 2022).

15.6 Global Drought Monitoring and Early Warning Systems

The operational drought monitoring and early warning systems present in different parts of the world briefly discussed in this section.

15.6.1 Global Integrated Drought Monitoring and Prediction System (GIDMaPS)

The Global Integrated Drought Monitoring and Prediction System (GIDMaPS) is developed by Hao et al. (2014) at the University of California, United States. It can be accessed from https://drought.eng.uci.edu/. GIDMaPS integrates precipitation and soil moisture data to compute SPI, SSI, and MSDI derived from Modern-Era Retrospective analysis for Research and Applications (MERRA-Land), NLDAS, GLDAS, and Global Drought Climate Data Record (GDCDR) to monitor and predict meteorological and agricultural drought information across the globe. It monitors drought based on historical observations, and predicts drought by extending the persistence-based drought prediction method, using the concept of Ensemble Streamflow Prediction, to a multi-index framework. provides short-term forecasts for SPI, SSI, and MSDI. Its prediction component provides seasonal forecasts of the probability of drought occurrence with different severity levels. It was found that GIDMaPS was able to capture several major drought events across the globe. However, the grid sizes of the input data and their coarser spatial resolution limit their applicability at the local scale.

15.6.2 US Drought Monitor (USDM)

The United States Drought Monitor (USDM) is produced by the National Drought Mitigation Center (NDMC) at the University of Nebraska-Lincoln, the United States Department of Agriculture (USDA), and the National Oceanic and Atmospheric Administration (NOAA) to depict the location and intensity of drought across the United States. The concept of USDM was introduced by Svoboda et al. (2002) based on a combined drought index approach. The USDM integrates multiple indices such as SPI and PDSI as well as indicators such as vegetation and hydrologic conditions such as CMI, percentile of CPC soil moisture, topsoil moisture, percentile of USGS daily streamflow, percent of normal precipitation, and remotely sensed VHI, into a weekly map of drought classified on a scale from zero (abnormally dry) to four (exceptional drought). This information is later subjected to expert interpretation for refinement. Because of its composite nature, USDM could respond to the needs of various water users including water planners and the agriculture industry. The USDM is currently widely used the organizational level, for research, and by the media. The USDM can be accessed from https://droughtmonitor.unl.edu.

15.6.3 South Asia Drought Monitoring System (SADMS)

The South Asia Drought Monitoring System (SADMS) is a drought monitoring and early warning system that was developed by the International Water Management Institute (IWMI) in 2014. It can be accessed from https://dmsdemo.iwmi.org/

drought-monitor. SADMS uses a variety of data sources, including satellite imagery, precipitation data, soil moisture data, and ground truth observations, to track drought conditions in South Asia. The drought indices integrated into SADMS include TCI from MODIS LST, VCI from MODIS surface reflectance, PCI from TRMM daily rainfall, and SMCI from ESACCI soil moisture, all combined through PCA (Giriraj, 2017). The SADMS provides weekly map of drought conditions at 500 m spatial resolution.

15.6.4 African Flood and Drought Monitor (AFDM)

The African Flood and Drought Monitor (AFDM) was developed in 2008 by the Princeton Climate Institute (PCI), in collaboration with the University of Southampton and Princeton University, early warning of flood and drought conditions (Sheffield et al., 2014). AFDM can be accessed through https://hydrology.soton.ac.uk/apps/afdm/. The AFDM system is comprised of three parts. The first part includes simulations of the VIC land surface model forced by a hybrid reanalysis–observational meteorological dataset. The second part involves real-time monitoring component that updates the model run to 2–3 days from real time forced by bias-corrected and downscaled TMPA (Tropical Rainfall Measurement Mission Multi-satellite Precipitation Analysis) satellite precipitation and NOAA NCEP Global Forecast System (GFS) analysis fields of temperature and wind speed. The third part constitutes seasonal hydrological forecast component that uses bias-corrected and downscaled CFSv2 climate forecasts of precipitation and temperature to drive the model and provide ensemble predictions of drought conditions for precipitation, soil moisture, and streamflow. The system is built on a collection of ground-based, satellite-based, and model-based information that work together to produce weekly maps of the hydrological conditions in near real-time as well as forecasts for floods and droughts up to 7 days and 6 months, respectively, at 5 km spatial resolution.

15.6.5 Global Drought Information System (GDIS)

Global Drought Information System (GDIS) is an international initiative to compile the finest non-prescriptive drought data from regional sources and to give a direct comparison of drought conditions throughout the world. The portal of Global Drought Monitor is hosted by NOAA and can be accessed on https://gdis-noaa.hub.arcgis.com/pages/drought-monitoring. The "bottom up" approach is used to develop GDIS, in which drought conditions each continent provided by the respective nations merged. The global and regional drought indicators available on this portal are North American Drought Monitor, European Combined Drought Indicator, CMORPH Daily Standardized Precipitation Index, GPCC Global Drought Index, GPCC Standardized Precipitation Index, GPCC Standardized Precipitation

Evapotranspiration Index, MERRA2 Evaporative Demand Drought Index, VIIRS Vegetation Health Index, MODIS Evaporative Stress Index, GRACE-Based Root Zone Soil Moisture, Shallow Groundwater, and Surface Soil Moisture.

15.7 Conclusions

Agricultural drought, characterized by prolonged deficits in water availability for crop growth, has serious adverse impacts on agricultural productivity, food and nutritional security, economic losses farmers, and socioeconomic repercussions inflation and limited food supply. In the context of escalating climate change, its adverse impacts are projected to intensify, emphasizing the critical importance of robust monitoring and forecasting systems. The remote-sensing technology can play a crucial role in monitoring and forecasting of agricultural drought, as it provides near-real-time information on arious drought indicators such as rainfall, soil moisture, ET, temperature, and vegetation health. Leveraging these insights, researchers and policymakers can undertake timely interventions to mitigate the impacts of agricultural drought. Traditionally, drought monitoring relied on single-parameter indices such as SPI based on precipitation; VCI based on vegetation status; TCI based on Land Surface Temperature; ESI based on ET, or SMCI based on soil moisture. However, these approaches often fell short in capturing the complex dynamics of agricultural drought comprehensively. Recognizing this limitation, composite or integrated drought indices have been developed. These indices amalgamate data from multiple sources and employ sophisticated weighting techniques such as PCA, Analytical Hierarchy Process, linear regression, machine learning, copula, etc. to provide a more holistic assessment. Examples include the Vegetation Drought Response Index, Combined Drought Monitoring Index, Blended Drought Index, Combined Drought Indicator, Drought Severity Index, Process-based Accumulated Drought Index, etc. Furthermore, forecasting or early warning of agricultural drought is a dynamic and evolving field, encompassing a range of methodologies from regression analysis to machine learning models. Hybrid models, which integrate multiple approaches, have shown promising performance in accurately predicting drought onset, severity, and duration.

Globally, several agricultural drought monitoring and early warning systems exist, such as the USDM, GDIS, SADMS, AFDM, and GIDMaPS. However, India lacks a robust and near-real-time early warning system tailored to its specific needs. Establishing such a system is imperative to provide timely alerts and support proactive agricultural drought management strategies. Additionally, enhancing the accessibility and usability of drought information through user-friendly platforms can empower farmers, policymakers, and other stakeholders to make informed decisions and implement adaptive measures effectively. In conclusion, addressing the challenges posed by agricultural drought requires a multifaceted approach that integrates cutting-edge technology, interdisciplinary research, and stakeholder engagement. By leveraging remote sensing technology, developing integrated drought indices, refining forecasting methodologies, and establishing robust early warning systems, India can enhance its resilience to agricultural drought and safeguard the livelihoods of millions dependent on agriculture. Moreover, fostering collaboration between government agencies, research institutions, and local communities is essential to effectively mitigate the impacts of agricultural drought and build a sustainable future for Indian agriculture.

References

Adnan, R. M., Mostafa, R. R., Islam, A. R. M. T., Gorgij, A. D., Kuriqi, A., & Kisi, O. (2021). Improving drought modeling using hybrid random vector functional link methods. *Water, 13*(23), 3379.

Aghelpour, P., Mohammadi, B., Mehdizadeh, S., Bahrami-Pichaghchi, H., & Duan, Z. (2021). A novel hybrid dragonfly optimization algorithm for agricultural drought prediction. *Stochastic Environmental Research and Risk Assessment, 35*(12), 2459–2477.

Alam, N. M., Mishra, P. K., Jana, C., & Adhikary, P. P. (2014). Stochastic model for drought forecasting for Bundelkhand region in Central India. *Indian Journal of Agricultural Science, 84*, 79–84.

Alawsi, M. A., Zubaidi, S. L., Al-Bdairi, N. S. S., Al-Ansari, N., & Hashim, K. (2022). Drought forecasting: A review and assessment of the hybrid techniques and data pre-processing. *Hydrology, 9*(7), 115.

Ali, Z., Hussain, I., Faisal, M., Nazir, H. M., Hussain, T., Shad, M. Y., Mohamd Shoukry, A., & Hussain Gani, S. (2017). Forecasting drought using multilayer perceptron artificial neural network model. *Advances in Meteorology, 2017*, 1–19.

Altunkaynak, A., & Jalilzadnezamabad, A. (2021). Extended lead time accurate forecasting of palmer drought severity index using hybrid wavelet-fuzzy and machine learning techniques. *Journal of Hydrology, 601*, 126619.

Anderson, M. C., Hain, C., Wardlow, B., Pimstein, A., Mecikalski, J. R., & Kustas, W. P. (2011). Evaluation of drought indices based on thermal remote sensing of evapotranspiration over the continental United States. *Journal of Climate, 24*(8), 2025–2044.

Başakın, E. E., Ekmekcioğlu, Ö., & Özger, M. (2021). Drought prediction using hybrid soft-computing methods for semi-arid region. *Modeling Earth Systems and Environment, 7*, 2363–2371.

Bayissa, Y. A., Tadesse, T., Svoboda, M., Wardlow, B., Poulsen, C., Swigart, J., & Van Andel, S. J. (2019). Developing a satellite-based combined drought indicator to monitor agricultural drought: A case study for Ethiopia. *GIScience and Remote Sensing, 56*(5), 718–748.

Belayneh, A., & Adamowski, J. (2013). Drought forecasting using new machine learning methods. *Journal of Water And Land Development, 18*(9), 3–12.

Brown, J. F., Wardlow, B. D., Tadesse, T., Hayes, M. J., & Reed, B. C. (2008). The vegetation drought response index (VegDRI): A new integrated approach for monitoring drought stress in vegetation. *GIScience and Remote Sensing, 45*(1), 16–46.

Brust, C., Kimball, J. S., Maneta, M. P., Jencso, K., & Reichle, R. H. (2021). DroughtCast: A machine learning forecast of the United States drought monitor. *Frontiers in Big Data, 4*, 773478.

Cai, G., Du, M., & Liu, Y. (2010, October). Regional drought monitoring and analyzing using MODIS data—A case study in Yunnan Province. In *International conference on computer and computing technologies in agriculture* (pp. 243–251). Springer.

Cammalleri, C., Arias-Muñoz, C., Barbosa, P., de Jager, A., Magni, D., Masante, D., Mazzeschi, M., McCormick, N., Naumann, G., Spinoni, J., & Vogt, J. (2021). A revision of the combined drought indicator (CDI) used in the European drought observatory (EDO). *Natural Hazards and Earth System Sciences, 21*(2), 481–495.

Carrão, H., Russo, S., Sepulcre-Canto, G., & Barbosa, P. (2016). An empirical standardized soil moisture index for agricultural drought assessment from remotely sensed data. *International Journal of Applied Earth Observation and Geoinformation, 48*, 74–84.

Chand, K., & Biradar, N. (2017). Socio-economic impacts of drought in India. In *Drought mitigation and management* (pp. 245–263). Scientific Publishers.

Chattopadhyay, N., Malathi, K., Tidke, N., Attri, S. D., & Ray, K. (2020). Monitoring agricultural drought using combined drought index in India. *Journal of Earth System Science, 129*, 1 16.

Chen, X., Li, X., Jiang, B., Su, J., Zheng, X., & Wang, G. (2023). Prediction of spring agricultural drought using machine learning algorithms in the Southern Songnen plain, China. *Land Degradation and Development*. https://doi.org/10.1002/ldr.4720

Danandeh Mehr, A., Torabi Haghighi, A., Jabarnejad, M., Safari, M. J. S., & Nourani, V. (2022). A new evolutionary hybrid random forest model for SPEI forecasting. *Water, 14*(5), 755.

Danodia, A., Kushwaha, A., & Patel, N. R. (2021). Remote sensing-derived combined index for agricultural drought assessment of rabi pulse crops in Bundelkhand region, India. *Environment, Development and Sustainability, 23*, 15432–15449.

Das, P. K., Das, R., Das, D. K., Midya, S. K., Bandyopadhyay, S., & Raj, U. (2020). Quantification of agricultural drought over Indian region: A multivariate phenology-based approach. *Natural Hazards, 101*, 255–274.

Dikshit, A., Pradhan, B., & Alamri, A. M. (2020). Short-term spatio-temporal drought forecasting using random forests model at New South Wales, Australia. *Applied Sciences, 10*(12), 4254.

Du, L., Tian, Q., Yu, T., Meng, Q., Jancso, T., Udvardy, P., & Huang, Y. (2013). A comprehensive drought monitoring method integrating MODIS and TRMM data. *International Journal of Applied Earth Observation and Geoinformation, 23*, 245–253.

Edwards, B., Gray, M., & Hunter, B. (2019). The social and economic impacts of drought. *Australian Journal of Social Issues, 54*(1), 22–31.

Feller, U., & Vaseva, I. I. (2014). Extreme climatic events: Impacts of drought and high temperature on physiological processes in agronomically important plants. *Frontiers in Environmental Science, 2*, 39.

Feng, P., Wang, B., Li Liu, D., & Yu, Q. (2019). Machine learning-based integration of remotely-sensed drought factors can improve the estimation of agricultural drought in south-eastern Australia. *Agricultural Systems, 173*, 303–316.

Fernández-Manso, A., Quintano, C., & Fernández-Manso, O. (2011). Forecast of NDVI in coniferous areas using temporal ARIMA analysis and climatic data at a regional scale. *International Journal of Remote Sensing, 32*(6), 1595–1617.

Fung, K. F., Huang, Y. F., Koo, C. H., & Soh, Y. W. (2020). Drought forecasting: A review of modelling approaches 2007–2017. *Journal of Water and Climate Change, 11*(3), 771–799.

Gao, B. C. (1996). NDWI—A normalized difference water index for remote sensing of vegetation liquid water from space. *Remote Sensing of Environment, 58*, 257–266. https://doi.org/10.1016/S0034-4257(96)00067-3

Gibbs, W. J., & Maher, J. V. (1967). *Rainfall deciles drought indicators* (Bureau of Meteorology Bulletin, 48). Commonwealth of Australia.

Giriraj, A. I. (2017). South Asia Drought Monitoring System (SADMS)-Giriraj Amarnath-International Water Management Institute (IWMI), Sri Lanka.

Gu, Y., Brown, J. F., Verdin, J. P., Wardlow, B., Gu, Y., Brown, J. F., Verdin, J. P., & Wardlow, B. (2007). A five-year analysis of MODIS NDVI and NDWI for grassland drought assessment over the central Great Plains of the United States. *Geophysical Research Letters, 34*, L06407. https://doi.org/10.1029/2006GL029127

Gupta, A. K., Tyagi, P., & Sehgal, V. K. (2011). Drought disaster challenges and mitigation in India: Strategic appraisal. *Current Science*, 1795–1806.

Han, P., Wang, P. X., Zhang, S. Y., & Zhu, D. H. (2010). Drought forecasting based on the remote sensing data using ARIMA models. *Mathematical and Computer Modelling, 51*(11), 1398–1403.

Han, P., Wang, P., Tian, M., Zhang, S., Liu, J., & Zhu, D. (2013). Application of the ARIMA models in drought forecasting using the standardized precipitation index. In *Computer and comput-*

ing technologies in agriculture VI: 6th IFIP WG 5.14 international conference, CCTA 2012, Zhangjiajie, China, October 19–21, 2012, Revised selected papers, part I 6 (pp. 352–358). Springer.

Han, H., Bai, J., Yan, J., Yang, H., & Ma, G. (2021). A combined drought monitoring index based on multi-sensor remote sensing data and machine learning. *Geocarto International, 36*(10), 1161–1177.

Hanadé Houmma, I., Gadal, S., El Mansouri, L., Garba, M., Gbetkom, P. G., Mamane Barkawi, M. B., & Hadria, R. (2023). A new multivariate agricultural drought composite index based on random forest algorithm and remote sensing data developed for Sahelian agrosystems. *Geomatics, Natural Hazards and Risk, 14*(1), 2223384.

Hao, Z., & AghaKouchak, A. (2013). Multivariate standardized drought index: A parametric multi-index model. *Advances in Water Resources, 57*, 12–18.

Hao, Z., AghaKouchak, A., Nakhjiri, N., & Farahmand, A. (2014). Global integrated drought monitoring and prediction system. *Scientific Data, 1*(1), 1–10.

Hayes, M. J. (2006). *Drought indices. Van Nostrand's Scientific Encyclopedia*. Wiley. https://doi.org/10.1002/0471743984.vse8593

Hazaymeh, K., & Hassan, Q. K. (2016). Remote sensing of agricultural drought monitoring: A state of art review. *AIMS Environmental Science, 3*(4), 604–630.

Hobbins, M. T., Wood, A., McEvoy, D. J., Huntington, J. L., Morton, C., Anderson, M., & Hain, C. (2016). The evaporative demand drought index. Part I: Linking drought evolution to variations in evaporative demand. *Journal of Hydrometeorology, 17*(6), 1745–1761.

Houmma, I. H., El Mansouri, L., Gadal, S., Garba, M., & Hadria, R. (2022). Modelling agricultural drought: A review of latest advances in big data technologies. *Geomatics, Natural Hazards and Risk, 13*(1), 2737–2776.

Huschke, R. E. (1959). *Glossary of meteorology*. American Meteorological Society.

Jiao, W., Tian, C., Chang, Q., Novick, K. A., & Wang, L. (2019). A new multi-sensor integrated index for drought monitoring. *Agricultural and Forest Meteorology, 268*, 74–85.

Kan, J. C., Ferreira, C. S., Destouni, G., Haozhi, P., Passos, M. V., Barquet, K., & Kalantari, Z. (2023). Predicting agricultural drought indicators: ML approaches across wide-ranging climate and land use conditions. *Ecological Indicators, 154*, 110524.

Khadr, M. (2016). Forecasting of meteorological drought using Hidden Markov Model (case study: The upper Blue Nile river basin, Ethiopia). *Ain Shams Engineering Journal, 7*(1), 47–56.

Khan, M. M. H., Muhammad, N. S., & El-Shafie, A. (2020). Wavelet based hybrid ANN-ARIMA models for meteorological drought forecasting. *Journal of Hydrology, 590*, 125380.

Kogan, F. N. (1990). Remote sensing of weather impacts on vegetation in non-homogeneous areas. *International Journal of Remote Sensing, 11*(8), 1405–1419.

Kogan, F. N. (1995). Application of vegetation index and brightness temperature for drought detection. *Advances in Space Research, 11*, 91–100.

Kogan, F. N. (2001). Operational space technology for global vegetation assessments. *Bulletin of the American Meteorological Society, 82*(9), 1949–1964.

Kulkarni, S. S., Wardlow, B. D., Bayissa, Y. A., Tadesse, T., Svoboda, M. D., & Gedam, S. S. (2020). Developing a remote sensing-based combined drought indicator approach for agricultural drought monitoring over Marathwada, India. *Remote Sensing, 12*(13), 2091.

Kumar, V., & Panu, U. S. (1997). Predictive assessment of severity of agricultural droughts based on agro-climatic factors 1. *JAWRA Journal of the American Water Resources Association, 33*(6), 1255–1264.

Kumar, K. A., Reddy, G. O., Masilamani, P., Turkar, S. Y., & Sandeep, P. (2021). Integrated drought monitoring index: A tool to monitor agricultural drought by using time-series datasets of space-based earth observation satellites. *Advances in Space Research, 67*(1), 298–315.

Littell, J. S., Peterson, D. L., Riley, K. L., Liu, Y., & Luce, C. H. (2016). A review of the relationships between drought and forest fire in the United States. *Global Change Biology, 22*(7), 2353–2369.

Liu, W. T., & Juárez, R. N. (2001). ENSO drought onset prediction in Northeast Brazil using NDVI. *International Journal of Remote Sensing, 22*(17), 3483–3501.

Liu, D., Mishra, A. K., Yu, Z., Yang, C., Konapala, G., & Vu, T. (2017). Performance of SMAP, AMSR-E and LAI for weekly agricultural drought forecasting over continental United States. *Journal of Hydrology, 553*, 88–104.

Liu, X., Zhu, X., Zhang, Q., Yang, T., Pan, Y., & Sun, P. (2020). A remote sensing and artificial neural network-based integrated agricultural drought index: Index development and applications. *Catena, 186*, 104394.

Luetkemeier, R., Stein, L., Drees, L., & Liehr, S. (2017). Blended drought index: Integrated drought hazard assessment in the Cuvelai-Basin. *Climate, 5*(3), 51.

Madadgar, S., & Moradkhani, H. (2013). A Bayesian framework for probabilistic seasonal drought forecasting. *Journal of Hydrometeorology, 14*(7), 1685–1705.

Manual for Drought Management. (2016). Department of Agriculture, Cooperation and Farmers Welfare, Ministry of Agriculture and Farmers Welfare, Government of India, New Delhi.

Mao, Y., Wu, Z., He, H., Lu, G., Xu, H., & Lin, Q. (2017). Spatio-temporal analysis of drought in a typical plain region based on the soil moisture anomaly percentage index. *Science of the Total Environment, 576*, 752–765.

Marj, A. F., & Meijerink, A. M. (2011). Agricultural drought forecasting using satellite images, climate indices and artificial neural network. *International Journal of Remote Sensing, 32*(24), 9707–9719.

Martínez-Fernández, J., González-Zamora, A., Sánchez, N., & Gumuzzio, A. (2015). A soil water based index as a suitable agricultural drought indicator. *Journal of Hydrology, 522*, 265–273.

McKee, T. B., Doesken, N. J., & Kleist, J. (1993). The relationship of drought frequency and duration to time scales. In *Eighth conference on applied climatology* (Vol. 17 (No. 22), pp. 179–183). American Meteorological Society.

Mishra, A. K., & Desai, V. R. (2005). Drought forecasting using stochastic models. *Stochastic Environmental Research and Risk Assessment, 19*, 326–339.

Mishra, A. K., & Singh, V. P. (2010). A review of drought concepts. *Journal of Hydrology, 391*, 202–216.

Mishra, A. K., Desai, V. R., & Singh, V. P. (2007). Drought forecasting using a hybrid stochastic and neural network model. *Journal of Hydrologic Engineering, 12*(6), 626–638.

Mossad, A., & Alazba, A. A. (2015). Drought forecasting using stochastic models in a hyper-arid climate. *Atmosphere, 6*(4), 410–430.

Mu, Q., Zhao, M., Kimball, J. S., McDowell, N. G., & Running, S. W. (2013). A remotely sensed global terrestrial drought severity index. *Bulletin of the American Meteorological Society, 94*(1), 83–98.

Nafii, A., Taleb, A., El Mesbahi, M., Ezzaouini, M. A., & El Bilali, A. (2022). Early forecasting hydrological and agricultural droughts in the Bouregreg Basin using a machine learning approach. *Water, 15*(1), 122.

Nam, W. H., Tadesse, T., Wardlow, B. D., Hayes, M. J., Svoboda, M. D., Hong, E. M., Pachepsky, Y. A., & Jang, M. W. (2018). Developing the vegetation drought response index for South Korea (VegDRI-SKorea) to assess the vegetation condition during drought events. *International Journal of Remote Sensing, 39*(5), 1548–1574.

Narasimhan, B., & Srinivasan, R. (2005). Development and evaluation of Soil Moisture Deficit Index (SMDI) and Evapotranspiration Deficit Index (ETDI) for agricultural drought monitoring. *Agricultural and Forest Meteorology, 133*(1–4), 69–88.

Palmer, W. C. (1965). *Meteorological drought* (Weather Bureau Research Paper No. 45) (Vol. 30, p. 58). US Department of Commerce.

Palmer, W. C. (1968). Keeping track of crop moisture conditions, nationwide: The crop moisture index. *Weatherwise, 21*, 156–161.

Park, S., Im, J., Park, S., & Rhee, J. (2017). Drought monitoring using high resolution soil moisture through multi-sensor satellite data fusion over the Korean peninsula. *Agricultural and Forest Meteorology, 237*, 257–269.

Park, S., Im, J., Han, D., & Rhee, J. (2020). Short-term forecasting of satellite-based drought indices using their temporal patterns and numerical model output. *Remote Sensing, 12*(21), 3499.

Paulo, A. A., & Pereira, L. S. (2007). Prediction of SPI drought class transitions using Markov chains. *Water Resources Management, 21*, 1813–1827.

Peters, A. J., Walter-Shea, E. A., Ji, L., Vina, A., Hayes, M. J., & Svoboda, M. D. (2002). Drought monitoring with NDVI-based standardized vegetation index. *Photogrammetric Engineering and Remote Sensing, 68*, 71–75.

Prajapati, V. K., Khanna, M., Singh, M., Kaur, R., Sahoo, R. N., & Singh, D. K. (2022). PCA–based composite drought index for drought assessment in Marathwada region of Maharashtra state, India. *Theoretical and Applied Climatology, 149*(1–2), 207–220.

Rao, A. R., & Padmanabhan, G. (1984). Analysis and modeling of Palmer's drought index series. *Journal of Hydrology, 68*(1–4), 211–229.

Ray, S. S., Sesha Sai, M. V. R., & Chattopadhyay, N. (2015). Agricultural drought assessment: Operational approaches in India with special emphasis on 2012. In *High-impact weather events over the SAARC region* (pp. 349–364). Springer.

Rhee, J., Im, J., & Carbone, G. J. (2010). Monitoring agricultural drought for arid and humid regions using multi-sensor remote sensing data. *Remote Sensing of Environment, 114*(12), 2875–2887.

Rouse, J. W., Haas, R. H., Schell, J. A., & Deering, D. W. (1974). Monitoring vegetation systems in the Great Plains with ERTS. In *Proceedings of the third earth resources technology Satellite-1 symposium* (NASA SP-351) (pp. 301–317).

Salakpi, E. E., Hurley, P. D., Muthoka, J. M., Bowell, A., Oliver, S., & Rowhani, P. (2022). A dynamic hierarchical Bayesian approach for forecasting vegetation condition. *Natural Hazards and Earth System Sciences, 22*(8), 2725–2749.

Sardar, V. S., Yindumathi, K. M., Chaudhari, S. S., & Ghosh, P. (2021, October). Convolution neural network-based agriculture drought prediction using satellite images. In *2021 IEEE Mysore sub section international conference (MysuruCon)* (pp. 601–607). IEEE.

Sepulcre-Canto, G., Horion, S., Singleton, A., Carrao, H., & Vogt, J. (2012). Development of a combined drought indicator to detect agricultural drought in Europe. *Natural Hazards and Earth System Sciences, 12*, 3519–3531.

Shah, D., & Mishra, V. (2020). Integrated Drought Index (IDI) for drought monitoring and assessment in India. *Water Resources Research, 56*(2), e2019WR026284.

Sheffield, J., Wood, E. F., Chaney, N., Guan, K., Sadri, S., Yuan, X., Olang, L., Amani, A., Ali, A., Demuth, S., & Ogallo, L. (2014). A drought monitoring and forecasting system for sub-Sahara African water resources and food security. *Bulletin of the American Meteorological Society, 95*(6), 861–882.

Sivakumar, Mannava, V. K., Motha, R. P., Wilhite, D. A., & Wood, D. A. (2011). Agricultural drought indices. In *Proceedings of the WMO/UNISDR expert group meeting on agricultural drought indices, 2–4 June 2010, Murcia, Spain* (AGM-11, WMO/TD No. 1572; WAOB-2011) (p. 197). World Meteorological Organization.

Sun, P., Zhang, Q., Wen, Q., Singh, V. P., & Shi, P. (2017). Multisource data-based integrated agricultural drought monitoring in the Huai River basin, China. *Journal of Geophysical Research: Atmospheres, 122*(20), 10–751.

Svoboda, M., LeCompte, D., Hayes, M., Heim, R., Gleason, K., Angel, J., & Rippey, B. (2002). The drought monitor. *Bulletin of the American Meteorological Society, 83*(8), 1181–1190.

Sykes, A. O. (1993). *An introduction to regression analysis* (Coase-Sandor working paper series in law and economics). https://doi.org/10.1198/tas.2007.s74

Tadesse, T., Wardlow, B. D., Hayes, M. J., Svoboda, M. D., & Brown, J. F. (2010). The Vegetation Outlook (VegOut): A new method for predicting vegetation seasonal greenness. *GIScience and Remote Sensing, 47*(1), 25–52.

Tadesse, T., Champagne, C., Wardlow, B. D., Hadwen, T. A., Brown, J. F., Demisse, G. B., Bayissa, Y. A., & Davidson, A. M. (2017). Building the vegetation drought response index for Canada (VegDRI-Canada) to monitor agricultural drought: First results. *GIScience and Remote Sensing, 54*(2), 230–257.

Thober, S., Kumar, R., Sheffield, J., Mai, J., Schäfer, D., & Samaniego, L. (2015). Seasonal soil moisture drought prediction over Europe using the North American Multi-Model Ensemble (NMME). *Journal of Hydrometeorology, 16*(6), 2329–2344.

Thornthwaite, C. W. (1948). An approach toward a rational classification of climate. *Geographical Review, 38*, 55–94.

Tian, Y., Xu, Y. P., & Wang, G. (2018). Agricultural drought prediction using climate indices based on Support Vector Regression in Xiangjiang River basin. *Science of the Total Environment, 622*, 710–720.

Tripathy, K. P., Mukherjee, S., Mishra, A. K., Mann, M. E., & Williams, A. P. (2023). Climate change will accelerate the high-end risk of compound drought and heatwave events. *Proceedings of the National Academy of Sciences, 120*(28). https://doi.org/10.1073/pnas.2219825120

Vicente-Serrano, S. M., Beguería, S., & López-Moreno, J. I. (2010). A multiscalar drought index sensitive to global warming: The standardized precipitation evapotranspiration index. *Journal of Climate, 23*(7), 1696–1718.

Vyas, S. S., Bhattacharya, B. K., Nigam, R., Guhathakurta, P., Ghosh, K., Chattopadhyay, N., & Gairola, R. M. (2015). A combined deficit index for regional agricultural drought assessment over semi-arid tract of India using geostationary meteorological satellite data. *International Journal of Applied Earth Observation and Geoinformation, 39*, 28–39.

Wang, L., & Qu, J. J. (2007). NMDI: A normalized multi-band drought index for monitoring soil and vegetation moisture with satellite remote sensing. *Geophysical Research Letters, 34*(20), L20405. https://doi.org/10.1029/2007GL031021

Wang, P. X., Li, X. W., Gong, J. Y., & Song, C. (2001). Vegetation temperature condition index and its application for drought monitoring. In *IGARSS 2001. Scanning the present and resolving the future. Proceedings. IEEE 2001 international geoscience and remote sensing symposium* (Cat. No. 01CH37217) (Vol. 1, pp. 141–143). IEEE.

Weier, J., & Herring, D. (2000). *Normalized difference vegetation index (NDVI)*. NASA Earth Observatory.

Wilhite, D. A., Svoboda, M. D., & Hayes, M. J. (2007). Understanding the complex impacts of drought: A key to enhancing drought mitigation and preparedness. *Water Resources Management, 21*, 763–774.

Wilhite, D.A. (2000) Drought as a natural hazard: Concepts and definitions. In Handbook of Drought and Water Scarcity, pp. 3-18, CRC Press

Wu, J., Zhou, L., Mo, X., Zhou, H., Zhang, J., & Jia, R. (2015). Drought monitoring and analysis in China based on the Integrated Surface Drought Index (ISDI). *International Journal of Applied Earth Observation and Geoinformation, 41*, 23–33.

Wu, X., Zhou, J., Yu, H., Liu, D., Xie, K., Chen, Y., Hu, J., Sun, H., & Xing, F. (2021). The development of a hybrid wavelet-ARIMA-LSTM model for precipitation amounts and drought analysis. *Atmosphere, 12*(1), 74.

Wu, H., Su, X., Singh, V. P., AghaKouchak, A., & Liu, Z. (2023). Bayesian vine copulas improve agricultural drought prediction for long lead times. *Agricultural and Forest Meteorology, 331*, 109326.

Xu, Y., Wang, L., Ross, K. W., Liu, C., & Berry, K. (2018). Standardized soil moisture index for drought monitoring based on soil moisture active passive observations and 36 years of north American land data assimilation system data: A case study in the Southeast United States. *Remote Sensing, 10*(2), 301.

Xu, D., Zhang, Q., Ding, Y., & Huang, H. (2020). Application of a hybrid ARIMA–SVR model based on the SPI for the forecast of drought—A case study in Henan Province, China. *Journal of Applied Meteorology and Climatology, 59*(7), 1239–1259.

Yao, Y., Liang, S., Qin, Q., & Wang, K. (2010). Monitoring drought over the conterminous United States using MODIS and NCEP Reanalysis-2 data. *Journal of Applied Meteorology and Climatology, 49*(8), 1665–1680.

Yuan, X., & Wood, E. F. (2013). Multimodel seasonal forecasting of global drought onset. *Geophysical Research Letters, 40*(18), 4900–4905.

Zambrano, F., Vrieling, A., Nelson, A., Meroni, M., & Tadesse, T. (2018). Prediction of drought-induced reduction of agricultural productivity in Chile from MODIS, rainfall estimates, and climate oscillation indices. *Remote Sensing of Environment, 219*, 15–30.

Zarekarizi, M., Yan, H., Ahmadalipour, A., & Moradkhani, H. (2021). A probabilistic framework for agricultural drought forecasting using the ensemble data assimilation and Bayesian multivariate modeling. In *Global drought and flood: Observation, modeling, and prediction* (pp. 147–164). https://doi.org/10.1002/9781119427339.ch8

Zhang, A., & Jia, G. (2013). Monitoring meteorological drought in semiarid regions using multi-sensor microwave remote sensing data. *Remote Sensing of Environment, 134*, 12–23.

Zhang, H., Li, J., Hu, G., Liu, X., & Su, Z. (2017). Evaluating satellite-based indices for agricultural drought monitoring in Northeast China. *Remote Sensing, 9*(1), 46.

Zhang, R., Li, L., Zhang, Y., Huang, F., Li, J., Liu, W., Mao, T., Xiong, Z., & Shangguan, W. (2021). Assessment of agricultural drought using soil water deficit index based on ERA5-land soil moisture data in four southern provinces of China. *Agriculture, 11*(5), 411.

Machine Learning-Led Solutions in Indian Agriculture: Case of Drought and Heat Stresses

16

Anurag Ajay, Aranab Chakraborty, and Rachana Dubey

Abstract

The effect of climate change, through rising temperature, incidents of drought, and unusually high-intensive rains at times on crop development is well known. However, the conventional methods of mitigating these abiotic stresses in the Indian context largely include post-facto interventions such as proportionate compensation to farmers in the form of weather-based insurance, infrastructure-intensive goals such as improvement of irrigation systems, or promotion of tolerant varieties and resilient practices. In the given context of increasingly changing weather patterns, the unpredictability is quite high making post-facto solutions less promising and sustainable. In this chapter, we propose that machine learning applications can play a bigger role in mitigation and adaptation to abnormal weather patterns supplementing traditional methods. Going forward, we highlight how a policy framework for this can be developed to effectively tackle the emerging challenge. Finally we use case studies of existing practices to show where such a policy framework can fit in.

A. Ajay
International Maize and Wheat Improvement Centre (CIMMYT), Patna, India

Leibniz Institute for Agricultural Development in Transition Economies (IAMO), Halle (Saale), Germany

A. Chakraborty (✉)
International Maize and Wheat Improvement Centre (CIMMYT), Patna, India

Sozial-nachhaltige Transformation, Institute for Advanced Studies, Vienna, Austria
e-mail: arnab.chakraborty19_dev@apu.edu.in

R. Dubey
Division of Crop Research, ICAR Research Complex for Eastern Region, Patna, Bihar, India

R. Dubey et al. (eds.), *Drought and Heat Stress in Agriculture*,
https://doi.org/10.1007/978-981-96-5735-3_16

Keywords

Machine learning · Soil · Crop management · Sensor-based technologies · ML Policy framework · Policy monitoring and evaluation

16.1 Machine Learning Applications in Indian Agriculture

Machine learning (ML) application and research within agriculture can be classified into 4 broad categories (a) livestock management, (b) soil management, (c) water management, and (d) crop management based on the target group it serves (Benos et al., 2021). In India, a review on the Web of Science database reveals that the literature looking at the application of ML in the context of agriculture is mostly focused on precision agriculture, Internet of Things (IoT), and other sensor-based technologies for monitoring and modeling climate footprint of different production practices. Specific use of various forms of ML was employed in yield prediction, weed detection, disease detection, water management, soil management, and livestock production. The advanced version of ML, such as Deep Learning where images of plants can be used for detection of diseases is now coming up. During the years between 2020 and 2024, there were 37 publications on these subject areas with the highest number in 2022.

16.1.1 Water Resource Management

ML has become increasingly relevant in water resources management (WRM) across the world (Drogkoula et al., 2023), especially given the challenges posed by climate change and the need for sustainable water management practices. Monitoring, prediction, and sustainable use of water resources can be done effectively with the use of ML algorithms. The volume and type of water-related data are increasing due to large-scale sensor networks. Coupled with the growing availability of computational power, particularly ML, these data are now being transformed into actionable advice to manage water resources.

Applications of ML in WRM: The ML technologies can be applied to WRM under the following theoretical categories. The operationalization of these algorithms will involve a combination of all these technologies.

(a) Prediction: ML models can forecast water availability, demand, and quality, aiding decision-making.
(b) Clustering: Grouping similar hydrological patterns helps identify trends and anomalies.
(c) Reinforcement learning: Optimizing water allocation strategies through continuous learning, using big data from sensors and remotely sensed data.

Smart Irrigation Systems: It uses weather data that interacts with soil moisture data to determine the irrigation requirement of the crop. These algorithms can be operationalized to create smart irrigation systems with localized sensors and IoT by combining temperature and humidity data. These systems help conserve freshwater resources while ensuring optimal crop growth. These products also maximize irrigation efficiency by reducing water waste, while maintaining plant health and quality. Integration of ML in irrigation systems has helped in reducing water usage by up to 25% while crop yields were increased by 20–30%. As water availability is impacted greatly by weather shocks, it has become relevant to use smart water technology to cope with the challenges of scarcity.

The types of irrigation systems that can potentially use or are using ML for better efficiency are:

- Sprinkler—ML technology has transformed the traditional sprinkler system by collecting real-time weather data from rain sensors and soil conditions to plan the next irrigation. The system sends notifications to sprinklers to avoid overwatering.
- Drip—Sensors are placed in the soil to monitor soil moisture levels, which helps ML algorithms precisely control the water flow.
- Center pivot—For large-scale farming, in-field sensors are installed on moving irrigation equipment that collect data on soil and crop health. Sensors provide this information to control circle irrigation sprinklers and adjust the pivot's speed, stream, and angle of water flow.

16.1.2 Crop Management

Crop models use quantitative measurements of biophysiological processes to predict plant growth and development based on environmental conditions and crop management inputs. Results of crop-related models can help farmers in deciding what to grow, how to grow and even when to grow. Few models allow users to understand the possible impact of the various climatic factors on crop production. These can be crucial tools for understanding the effect of abiotic stresses such as heat and drought stress on various production systems. In India, ICAR has identified the potential of these technologies, and has initiated efforts for building the capacity of scientists to use these technologies effectively.[1]

16.1.2.1 Crop Yield Prediction

It is the process of accurately predicting the potential yield of a specific crop, during a particular season in a region using the ML algorithm. Although, the prediction accuracy of a model depends upon various factors including the quality of input data, various crop yield prediction models have been developed by Indian researchers to understand the effect of abiotic stressors as well as other factors on crop yield (Narawade et al., 2024). There are various types of ML algorithms that can be used for crop yield prediction, including regression, decision trees, and artificial neural

[1] See how the ICAR has been involved in training students towards greater sustainability https://icar.gov.in/application-machine-learning-agricultural-meteorology (Accessed on 21 May 2024)

networks. However, there is a dearth of models that focus on estimating the risks of drought and heat stress in Indian conditions.

16.1.2.2 Land Use Planning

ML when combined with remote sensing data generates a landscape view for crucial long-term planning. It can assess land-use changes in agriculture including forest cover. Climate models also rely much on remotely sensed data. One of the key factors for the rise of climate models in the sphere of agriculture has been in conjunction with the advances in remote sensing technologies. The combination of yield simulation models and remote sensing technologies has opened up scope for advanced decision support systems, crucial for policy-making in a vast country like India (Pokhariyal et al., 2023). Remote sensing can help the crop in better mitigation and adaptation to climate change as it can predict upcoming weather patterns.

16.1.3 Development of Resilient Varieties

A significant number of abiotic stress-resistant crop varieties have been developed by the national agricultural research system in India. However, there are specific limitations to the development of more appropriate varieties in the face of climate change and adoption by farmers (Boriah et al., 2023). ML-based deep learning can simplify the task of crop breeding. With the development of high-throughput omics technologies, large amounts of data have been generated on the genetic architecture of plants. These can be exploited for manipulating the desired characteristics of plants that are important for varietal development.

Most of the progress in the development of stress-tolerant varieties has been based on conventional breeding methods that rely on heritable, adaptive phenological traits. However, there is a need to understand the complex genetic, molecular, as well as physiological mechanisms that go into stress tolerance (Saint Cast et al., 2022). The combination of phenomics, climate science, and ML modeling can lead to more effective efforts in developing climate-smart varieties.

Despite the large numbers of stress-tolerant varieties being developed by researchers, there is no comprehensive database in India on the available varieties (apart from isolated efforts by ICAR–NIASM and ICAR–CRIDA). Moreover, these databases are not integrated with real-time data streams monitoring agricultural landscapes. The creation and integration of these databases can allow for effective DSSs for extension workers and farmers, increasing the scope of adoption of these practices.

16.2 Leveraging ML for Abiotic Stress Mitigation

Based on the existing research and capacity within the agricultural research system in India, we have proposed a few basic elements of a coherent and outcome-driven policy for ML in the Indian agricultural sector in the context of heat and drought stress mitigation.

The following are the key components of the proposed ML policy framework.

(a) Data collection and management: Establish a centralized database to collect and manage agricultural data across different regions of India. This database should include information on soil properties, weather patterns, crop types, and historical yield data. There are several public datasets that can be integrated for this purpose. Moreover, promotion of open data platforms where different stakeholders can share data, along with ensuring interoperability and standardization, can be a crucial step. Innovative models of data management and aggregation through farmers' participation and crowd-sourcing are also emerging.[2] These different methods of data collection and management would strengthen the foundation for advancing ML approaches and are the most crucial step.
(b) ML model development: Encourage and fund the development of ML models that can process the collected data to identify patterns and predict abiotic stress occurrences. These models should be tailored to the specific needs of Indian agriculture, considering the local environmental conditions and crop varieties. These should also be scalable and adaptable within different regions of India, given the diversity of agroclimatic conditions and cultivars across regions.
(c) Institutionalizing and capacity building: To truly integrate the ML tools within the agricultural system of the country, there is a need for creating institutional mechanisms for the use of these tools across different levels of government, and multiple stakeholders. This might be facilitated by creating user-friendly platforms. Subsequently, investment in capacity building for agricultural scientists, extension workers, and farmers in the use of ML tools is important. This will empower them to make data-driven decisions to combat abiotic stresses effectively.

For the policy framework to be effectively implemented, we would need to critically assess the possible pathways for implementing such a policy framework for integrating ML within Indian agriculture through different sectors. The following are the key components for implementation and scaling.

(a) Phased roll-out: Implementing the policy in phases, starting with high-priority regions and gradually expanding to cover the entire country can be an effective approach for the government. This phased approach allows for iterative learning and adjustments within the institutional structure. It can also allow for subsequent skill development, in parallel with the implementation process. However, this approach may have considerable limitations in terms of skilled manpower availability with the government, as well as a long-drawn timeline.
(b) Incentivizing the private sector: Private sector actors can be incentivized to engage in R&D in the domain of ML through subsidies or grants to create

[2] See Gramhal's data cooperative model for using data generated by farmers in India here: - https://www.gramhal.org/perspective (Accessed on 23 May 2024)

value-added agribusiness products that have significant benefits for users, as well as businesses. For example, ML technologies can be bundled with crop insurance schemes to create incentives for accurate crop protection for the insurance industry. While this approach can be useful in creating fast-paced product-prone torment, relying solely on the private sector can lead to exclusion of communities that have limited business prospects and are most prone to climatic risks, such as small and marginal farmers and rainfed regions.

(c) Innovation platform/sandbox approach: Multi-stakeholder platforms bringing together several viewpoints and interests from different actors within the agricultural system can be a good way of implementing any innovation. This can not only allow for all sectors to collaborate through traditional mechanisms such as public–private partnerships but also create room for unconventional models of collaboration through networks of community-based organizations, academic institutions, philanthropies, agri-startups, etc. These approaches help in creating collaborative processes of experimentation and learning at different scales. Research for development organizations such as the Consultative Group on International Agriculture Research (CGIAR) can play a crucial role in these domains (Cramer, 2024). An emerging initiative in the state of Telangana is a case in point. Initiated by the World Economic Forum and implemented with the support from Gates Foundation, and Digital Green, it brought together multiple Government departments, and academia to implement state-of-the art ML technology in the state for servicing chili and turmeric farmers.[3]

For the successful implementation of any policy, it is crucial to define specific and measurable indicators of success, so that they can be monitored and evaluated, at regular intervals. Continuous monitoring and evaluation of policy outcomes can act as a feedback mechanism for planning and strategizing, ensuring the policy is on track.

The modalities and agency for monitoring and evaluation will need to be chosen based on the pathways finalized for implementation. While a formal audit approach is suitable for private-sector-led pathway of scaling, a theory-based framework is suitable for a government led phased roll out model. For an innovation sandbox approach, an iterative and participatory evaluation framework can be more suitable.[4]

Undoubtedly, the widespread application of ML will help improve the way Indian agriculture performs. But there are several challenges and still a long way to go. The foremost challenge is immature data collection infrastructure. Under such circumstances, developing a robust machine-learning model that will work with a high and acceptable degree of accuracy becomes a challenge. A precise ML model

[3] Find the details for the 'Saagu Baagu' initiative in Telangana at https://indiaai.gov.in/news/telangana-govt-conceptualises-saagu-baagu-project-to-transform-agri-with-emerging-technologies (Accessed on 22 May 2024)

[4] See https://nethope.org/toolkits/the-artificial-intelligence-ai-suitability-toolkit-for-nonprofits/ (Accessed on 22 May 2024) for a participatory framework for evaluating the suitability of a AI/ML in any developmental context.

requires a lot of effort in the collection, standardization, and annotation of raw data to derive meaningful inferences. Additionally, the process comes with a cost. The government has been focused on developing a digital ecosystem that focuses on long-term aspects such as interoperability, data governance, data quality, and data protection in the agriculture sector. The development of AgriStack is a commendable step in this direction which will digitize farmers' data and land records. Indian private sector has already advanced in this domain. So, a public–private partnership seems a better proposition to take this endeavor forward at a faster pace.

16.3 Case Studies and Best Practices

Despite the wide-ranging support for integrating ML into the agricultural system in India, there have been very few successful cases of this. In this section, we have highlighted some pilots by tech companies and research institutes. We further present two cases that bring together various ML and ICT tools to empower the stakeholders in the agricultural system to reduce the impact of abiotic stresses. The first case from Maharashtra talks about an early drought warning system, and the second from Punjab shows how an ML-based decision support system has been used for mitigating the risk of heat stress. These cases not only highlight the impact of these interventions but also help us identify lessons on how to create buy-in from the end-users and other stakeholders involved in the dissemination and operation of the system.

Some tools under piloting in India are as follows:

(i) IBM Watson Decision Platform for Agriculture: IBM's Watson Decision Platform for Agriculture is used extensively in India to address issues related to drought and heat stress. The platform employs machine learning algorithms to analyze weather data, soil moisture levels, and crop health indicators. By providing real-time insights and predictive analytics, it helps farmers make informed decisions about irrigation, crop selection, and pest management. The technology has been particularly beneficial in states like Maharashtra and Karnataka, which are prone to drought conditions.

(ii) Microsoft AI for Earth Initiative: Microsoft's AI for Earth initiative has collaborated with Indian agricultural institutes to develop machine learning models that predict drought conditions and provide early warnings. These models analyze large datasets from remote sensing satellites, weather forecasts, and ground sensors to monitor soil moisture and temperature. One successful implementation is in Andhra Pradesh, where farmers receive alerts and recommendations on water management, helping them to mitigate the effects of drought and heat stress.

(iii) International Crops Research Institute for the Semi-Arid Tropics (ICRISAT): ICRISAT has developed a machine learning-based tool called "Agriculture Intelligence" that assists farmers in semi-arid regions of India. The tool uses data from local weather stations, remote sensing, and field sensors to predict crop yields under various stress conditions, including drought and heat stress. This predictive capability enables farmers to adjust their practices in advance, improving resilience and productivity.

(iv) Google AI for Social Good: Google's AI for Social Good program has worked on projects in India to address agricultural challenges through machine learning. One notable project involves using TensorFlow to analyze satellite images and predict areas at risk of drought. These predictions help local authorities and farmers implement water-saving techniques and plan crop cycles better. This initiative has shown positive results in the states of Rajasthan and Gujarat, which frequently face severe water scarcity.

(v) Tata Consultancy Services (TCS) Digital Farming Initiative: TCS has implemented its Digital Farming Initiative in various parts of India, leveraging machine learning to support farmers. The initiative includes the development of a mobile app that provides farmers with personalized advice based on ML analysis of weather data, soil health, and crop conditions. This advice helps farmers in regions like Telangana and Tamil Nadu to optimize their irrigation practices and reduce the impact of heat stress on crops.

While these initiatives show promise for future application, there are some successful experiences of implementing ML tools in India's local agricultural systems at scale:

(i) Predictive Analytics for Expectation maximization in Maharashtra: Maharashtra, a state in western India, frequently faces severe drought conditions affecting agricultural productivity and water availability. These droughts have significant socio-economic impacts, including crop failure, water scarcity, and economic distress among farmers. The state meteorological department has turned to predictive analytics to develop a predictive model to anticipate and mitigate these impacts effectively (Yashashree et.al., 2024).

    They are using a predictive model that utilizes hydrological, agricultural, and meteorological covariates to generate predicted indices of standardized precipitation and vegetation conditions, assessing drought severity. This ML technology is coupled with other ICT tools to create a custom-made alert system in the state to provide actionable advice to farmers on water management, crop selection, and alternative livelihoods. They are using SMS, mobile apps, and community radio, to disseminate warnings to farmers and local authorities. Added with comprehensive training of extension agents and farmers on using these recommendations, this can potentially reduce the drought risk of farmers significantly.

(ii) Precision agriculture in Punjab for heat stress resilience: Punjab, known as the "Granary of India," is a major agricultural state in northern India. However, the region faces significant challenges due to changing climate patterns, including increased temperatures and heat stresses that affect crop yields (Raza et al., 2025). To address these issues, precision agriculture technologies are being implemented to enhance productivity and sustainability. The precision agriculture models use data from remote sensing data from satellites and drones on crop health and field conditions, along with data from automated weather stations and soil sensors on temperature, humidity, soil moisture, acidity, etc. ML algorithms can predict crop growth patterns, pest infestations, and optimal planting times based on historical and real-time data.

Precision agriculture technologies that use automated machinery (like GPS-guided tractors and automatic irrigation systems) and variable rate technologies can benefit from the outcomes of the ML algorithms. But outcomes can be significantly improved by integration of the recommendations with the extension systems.

These cases point out best practices that should be considered while implementing ML in agriculture:

(a) Data integration: Successful case studies emphasize the importance of integrating diverse data sources, including weather data, satellite imagery, and soil health records, to train machine learning models effectively.
(b) Farmer training and engagement: Continuous education and support for farmers in using these technologies are crucial. This involves not only training on how to use the tools but also understanding the insights provided by machine learning models.
(c) Localized solutions: Tailoring machine learning solutions to local conditions, such as specific crop types and regional climate patterns, enhances the effectiveness of interventions.
(d) Collaborative efforts: Collaboration between tech companies, agricultural institutions, and government bodies ensures that machine learning solutions are scalable and sustainable.
(e) Real-time monitoring and feedback: Implementing real-time monitoring systems and providing instant feedback to farmers helps in making timely decisions, thereby mitigating the effects of drought and heat stress more effectively.

16.4 Considerations and Way Forward

Based on the learnings from the case studies, we can see that the integration of ML into the agricultural research and extension system can have significant economic and social benefits for the farmers and the society at large, with the rising risk of climate change. However, these cases demonstrate that ML technologies will have to be adapted according to local requirements and challenges, like drought in Maharashtra, and heat stress or varietal selection in Telangana.

It is also clear that there are only isolated initiatives by various actors in the country. In the absence of an integrative policy, scaling of the ML based technologies cannot be directed towards a common goal. A policy identifying the priority geographies and the appropriate pathway for scaling is required.

The policy should lay down the process for creating the required data infrastructure for these systems to work effectively. These protocols should take into consideration the ethical dimensions of data storage and utilizing personal data, borrowing from the global standards of public data management such as the GDPR in the EU.

In conclusion, as the first step, the Department of Agriculture should encourage experts from civil society organizations, research institutes, start-ups, and other stakeholders to come together for designing this policy with clearly defined goals. The policy framework should also define the goals and objectives in a participatory manner for it to act as an innovation platform to work towards the implementation of the policy in the future.

References

Benos, L., Tagarakis, A. C., Dolias, G., Berruto, R., Kateris, D., & Bochtis, D. (2021). Machine learning in agriculture: A comprehensive updated review. *Sensors, 21*(11), Article 11. https://doi.org/10.3390/s21113758

Boriah, K., Basavaraj, P., Kakade, V. D., Harisha, C., Khapte, P., Halagundegowda, G. D. K., Kulshreshtha, N., Vijayakumar, H., Naik, B., Jagadish Rane Sammi Reddy, K., & Pathak, H. (2023). Abiotic stress-tolerant crop varieties in India: Status and a way forward. In *Recent trends in plant breeding and genetic improvement*. IntechOpen. https://doi.org/10.5772/intechopen.1001916

Cramer, L. (2024). Linking science with policy: The importance of incorporating clear roles for knowledge brokers into research-for-development organizations. *Frontiers in Sustainable Food Systems, 8*, 1359190. https://doi.org/10.3389/fsufs.2024.1359190

Drogkoula, M., Kokkinos, K., & Samaras, N. (2023). A comprehensive survey of machine learning methodologies with emphasis in water resources management. *Applied Sciences, 13*(22), Article 22. https://doi.org/10.3390/app132212147

Narawade, V., Chaudhari, A., Mohammad, M. A., Dubey, T., & Jadhav, B. (2024). Agricultural crop yield prediction for Indian farmers using machine learning. In H. Sharma, A. Chakravorty, S. Hussain, & R. Kumari (Eds.), *Artificial intelligence: Theory and applications* (Vol. 843, pp. 75–86). Springer Nature Singapore. https://doi.org/10.1007/978-981-99-8476-3_7

Pokhariyal, S., Patel, N. R., & Govind, A. (2023). Machine learning-driven remote sensing applications for agriculture in India—A systematic review. *Agronomy, 13*(9), 2302. https://doi.org/10.3390/agronomy13092302

Raza, A., Ashraf, E., Sadaf, S., Khan, N. A., Shah, A. A., Alotaib, B. A., & Muzamil, R. M. (2025). The potential role of precision agriculture in building sustainable livelihoods and farm resilience amid climate change: a stakeholders' perspective from southern Punjab Pakistan. *Land, 14*(4), 770. https://doi.org/10.3390/land14040770

Saint Cast, C., Lobet, G., Cabrera-Bosquet, L., Couvreur, V., Pradal, C., Tardieu, F., & Draye, X. (2022). Connecting plant phenotyping and modelling communities: Lessons from science mapping and operational perspectives. *In Silico Plants, 4*(1), diac005. https://doi.org/10.1093/insilicoplants/diac005

Tiwari, G. S., Pareek, N., & Raverkar, K. P. (2018). Increased heat and drought stress under climate change and their impact on physiological growth and development of crops: A review. *Indian Journal of Agricultural Sciences, 88*(12), 1818. https://doi.org/10.56093/ijas.v88i12.85374

Yashashree, Mahale Nida, Khan Kunal, Kulkarni Shivali Amit, Wagle Preksha, Pareek Ketan, Kotecha Tanupriya, Choudhury Ashutosh, Sharma (2024) Crop recommendation and forecasting system for Maharashtra using machine learning with LSTM: a novel expectation-maximization technique. *Abstract Discover Sustainability 5*(1). https://doi.org/10.1007/s43621-024-00292-5

Part III

Policies and Future Prospects

Cultivating Resilience: A Policy Framework for Climate-Smart Agriculture

17

Sanjit Maiti and R. Roy Burman

Abstract

Policies serve as a crucial roadmap, guided by recommendations, to undertake a multitude of projects aimed at uplifting the vulnerable and marginalised segments of society. Good policies ensure the culmination of equality through the processes that promote equity. Thus, these policies try to bring vicissitudes of social strata on the same plane in terms of opportunities, economic sustainability, and upliftment in terms of better livelihood. Concerns regarding agricultural sustainability are to be seriously taken into consideration while framing the policies in the backdrop of ever accelerating rate of climate change since the last decade. Sustainability is an umbrella term with numerous issues under it which need to be addressed but most of the policies fail to integrate or harmonise the various sustainability issues (Streimikis J, Baležentis T. Sustain Dev 28(6):1702–1712. https://doi.org/10.1002/sd.2118, 2020). Policymakers focus on one or few issues overlooking the cruciality of the other issues. This makes the validation of the strength of the policies to channel the resources at the right target even more questionable. Food production should be operated in a planet-friendly manner on the side while taking care to not overstep into the environmental arena, which may lead to accelerated climate change. Thus, a delicate balance between ecological preservation and food production needs to be maintained. A robust policymaking process may be required to culminate this ideology. In this article, an attempt has been made not only to comprehend the importance but also to devise a method for the integration and mainstreaming of the concept of sustainability

S. Maiti
Veterinary Extension Education, ICAR-National Dairy Research Institute, Karnal, India

R. R. Burman (✉)
Division of Agricultural Extension, Indian Council of Agricultural Research, New Delhi, India
e-mail: rajarshi.burman@icar.gov.in

R. Dubey et al. (eds.), *Drought and Heat Stress in Agriculture*,
https://doi.org/10.1007/978-981-96-5735-3_17

and climate change into agricultural policymaking. It further attempts to answer the question of the importance of sustainable agricultural practices in combating climate change and how this can be achieved in the Indian context through devising a novel policymaking approach.

Keywords

Sustainability · Policy · Policymaking · Climate-centred policy

17.1 Climate, Agriculture, and Society: Interrelationship and Interdependency

It is established through the prior arguments that climate-smart agricultural practices (CSAPs) need to be included holistically in agricultural enterprises. To attain that, the policy recommendations should be devised in a climate-centric manner. For this, two prerequisites should be considered.

17.1.1 Farmer at the Heart of the Process

Since the ultimate consequence bearer of the policy recommendations are the farmers lying at the grassroots level, the policies must be developed by keeping them at the heart of the process. A top-down approach in policymaking implies the absence of consensus on the part of farmers and hence a participatory bottom-up approach may be required. Through this, the numerous stakeholders of agriculture can collaborate with farmers at the local level (who are major stakeholders) and thereby understand through the close association the needs of farmers, challenges in the adoption of climate-centred recommendations, and opportunities at hand for the inclusion of climate smartness. Farmers work in the field in actual reality and have profound knowledge of the agricultural processes. Hence, their inclusion can help in gaining their insights on the practicality of the issues.

A study in Ethiopia on farmers' participation in land use policies suggested that increasing the farmers' participation in policymaking can lead to increased ownership and lead to even social unrest and better implementation of policies (Ariti et al., 2018). A social action research in Germany that attempted to qualitatively assess the benefits of the participation of local stakeholders in the decision-making (for inclusion of agri-environmental schemes) revealed improved feedback of information, improved communication, building multidisciplinary networks among numerous participants, and increased acceptance of the agri-environmental schemes (Prager & Freese, 2009). Research on the inclusion of public organisations in decisions suggested that the participation of the farmers in policymaking may lead to broadened benefits in terms of improvement in gender equality, and eradication of poverty (Nkiaka et al., 2019). So, the inclusion of farmers not only as a mere provider of information but as important stakeholders in policymaking is requisite.

17.1.2 Awareness of Policymakers Towards Climate Change

Further, another requisite would be to bring awareness among the policymakers regarding the importance of the inclusion of climate-based aspects in the development of recommendations. The policymakers should be well acquainted with the impacts of climate change on the various agricultural enterprises pertaining to the productivity of crops and livestock, changes in pests and disease incidence dynamics due to erraticism of weather. Moreover, the overall effect on the socio-economic conditions of the farmers with whom they are supposed to work and for whom they must provide recommendations needs to be thoroughly assessed. Since climate change leads to change in the dynamics, it is crucial on the part of policymakers that they are well-versed in the knowledge of these effects. To attain this, numerous methods can be devised:

(i) *Inclusion of climate experts in the team*
(ii) *Awareness programmes about climate change*
(iii) *Preliminary survey for analysis of climatic situations*
(iv) *Assessing the livelihood vulnerability of the core farming community for whom the policies are being recommended*
(v) *Developing insights on the suitability of numerous climate-smart practices in the target area*
(vi) *Inclusion of KVK officials and SMS as providers of climatic as well as socio-demographic information on target area in the process*

17.2 Developing a Model Climate-Centred Policy for Agricultural Sustainability

A framework has been developed as shown in Fig. 17.1. The steps are as follows:

17.2.1 Developing a Team Involving a Multidisciplinary Approach

A multidisciplinary approach is incredibly important in policymaking for the inclusion of all the possible facets. Since one or more facets may be rendered less important or ignored if the team includes members from a single discipline. A multidisciplinary approach helps to attain the presentation of various facets of issues as perceived by different disciplines. These can be later merged to an optimal level as the policymaking progresses. This can be done by including the following:

17.2.2 Inclusion of Environmentalists

Numerous novel concepts like climate debt, agroecology, and environmental justice are gaining popularity among research fraternities. Not only limited to gaining

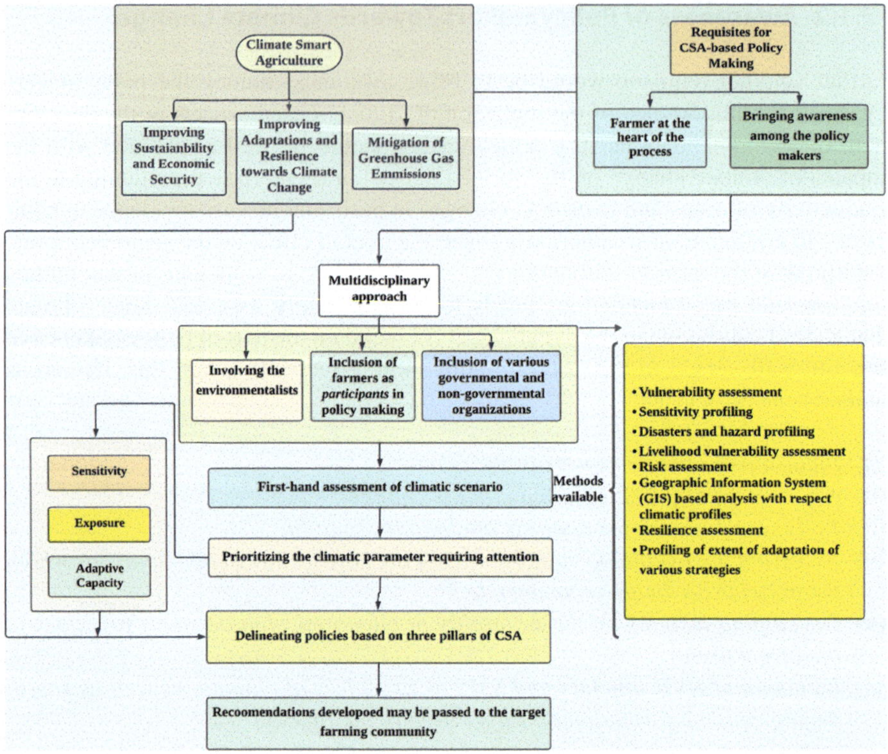

Fig. 17.1 Framework for developing a model climate-centred policy for agricultural sustainability

popularity but also influencing significantly the decision-making of the governance spaces of various nations by entering into debates (Bjork-James et al., 2022). Further, these ongoing debates focus on sustainability and mitigating climate change. These are in line with our basic aim to promote CSA practices, hence, it is imperative to have an insight on the numerous environmental aspects. For this purpose, environmentalist(s) can be included in the process of policymaking. They can provide valuable insights into the positive consequences as well as ramifications that may be lying underneath any CSA practice to be promoted. Further, they can be an asset for carrying out environmental assessments at the target site, which can help in better formulation of plan for recommendations. Moreover, environmentalists focus on soil, water, climate as well as biodiversity, the information on these aspects can lead to a development of a sustainable plan.

17.2.3 Inclusion of Farmers in the Decision-Making

Farmers are the final receptors of the process of policymaking. They have the final decision to undertake the recommendations developed through policymaking. The major barrier is the cultural compatibility of the practices which are being recommended (Schramm et al., 2020). A top-down approach that ignores the importance of farmers in policymaking, may lead to the formulation of recommendations which are either incompatible with the economic dynamics of the target area or the cultural belief. This may lead to the failure of the entire goal of the policymaking. Herein, the inclusion of the farmers at the initial point can help us achieve a major goal. It can lead to gaining quick feedback on the part of the farmers and also their perceptions towards the practices being promoted. Further, their ground knowledge can lead to gaining insights into the practical feasibility of the CSA in the target area and the possible barriers may be highlighted. Since the farmers will be involved in the various steps of the process, the rate of acceptance of the recommendations may become high owed to the consensus of the farmers with the numerous stakeholders. Evidently from a case in Kenya wherein farmers were actively involved in the Kenya Agricultural Carbon Project (KACP) in decision-making. The results suggest that the approach helped in generating improved knowledge through sharing and also strengthened the social capital. The results also suggested that high coordination among the major stakeholders increases the likelihood of acceptance of the project policies. Thus, increased participation led to improved socio-economic conditions along with increased acceptance of the CSA on the part of the farmers (Lee, 2017).

17.2.4 Inclusion of Governmental as Well as Non-governmental Organisations

Numerous governmental as well as non-governmental organisations are working actively towards mitigating the effects of climate change. The potential of these institutions may as well be tapped. They may be rendered as major stakeholders in the process with a major say in the decision-making process (Christensen & Lægreid, 2020; Sabatier, 2019). Numerous NGOs which are thriving on the principles of environmental activism may be approached for involvement in the process.

17.3 First-Hand Assessment of Climate Scenario

To develop a climate-centric policy, it must that before formulation, a prior survey of the climatic scenario of the target area should be carried out. Here, the scope climatic scenario includes various parameters like vulnerability to climate change, risk associated with climate change, area under exposure to climate change, sensitivity of production systems to climate change, resilience of the social unit, adaptive capacity, etc. It can be carried out using econometrics or an index development approach (Jha & Gundimeda, 2019; Maiti et al., 2014b). The multitude of indicators need to be chosen to accurately represent the construct to be measured. Indicator based approach helps in understanding the contribution of various factors in aggravating or alleviating the effects of climate change (Sahani et al., 2023). Further, the dynamics of major components can be understood. For example, in vulnerability, namely, exposure, sensitivity, and adaptive capacity are major components. After the analysis of the data, it can be understood whether the vulnerability is aggravated by a higher magnitude of all three or any one or two of the components. Suppose the sensitivity in an area is more. In that case, the focus can be given to the improvement of production systems which can done by use of climate-resilient crop varieties, soil testing, water conservation techniques, use of resilient livestock breeds, etc. On the other hand, if exposure is bounteous, the focus may be given to subjugating the ramifications of the effects of climate change by the development of a sound early warning system, providing timely climate-based agro-advisory services, bringing awareness regarding climate change among the farmers, conducting farmers field schools for climate change, etc. If the adaptive capacity is lower, the resources may be diverted in the direction of improving livelihood opportunities for the farmers, which can be done through vocational training, alternative livelihood options, organisations, transition in cropping systems, undertaking integrated farming systems, etc. Thus, the first-hand assessment can aid in better understanding the needs, challenges and opportunities pertinent to the goal of mitigating the effects of climate change.

- Methods at disposal for assessment of climatic scenarios (Table 17.1):

Table 17.1 Elaboration of methods at disposal to assess the climatic scenario

Methods for climatic assessment	References
Vulnerability assessment	Balaganesh et al. (2020), Chen et al. (2013), Maiti et al. (2014a) and Maiti et al. (2015)
Sensitivity profiling	Luers et al. (2003) and Shukla et al. (2011)
Disasters and hazard profiling	Jeganathan et al. (2021)
Livelihood vulnerability assessment	Jha and Gundimeda (2019) and Venus et al. (2022)
Risk assessment	Rama Rao et al. (2020)
Geographic information system (GIS) based analysis with respect climatic profiles	Feyissa et al. (2018) and Shukla et al. (2016a)
Resilience assessment	Mondal et al. (2023)
Profiling of extent of adaptation of various strategies	Goswami et al. (2018) and Maiti et al. (2014)

17.4 Prioritising the Climatic Parameter Requiring Attention

Based on the preliminary survey, the extent of climate change and the ramifications of it can be determined. This leads to the next step of the process wherein based on the magnitude of the parameters being used in climate change, the severity of the parameters can be judged, based on which the discussion may be conducted with various experts in the team. Based on the discussion, a decision may be taken concerning how much the climate-centric approach is required. If, in a case, the extent of vulnerability or risk due to climate change is less, the climate-centric approach may be altogether dismissed or vice versa, and the focus on climate-centric approach may be increased. Further, the resources may be diverted to a motive based on the severity of sub-component of the risk or vulnerability. A decision may be taken regarding whether the focus should be on increasing the adaptive capacity, reducing the impact of the hazards, or nullifying the sensitivity. Based on the decision, a tailored plan may be further developed to accommodate the requirements of the situation and resources may allocated accordingly.

17.4.1 Climate-Smart Agriculture

Climate-smart agriculture (CSA) is a holistic approach to farming that aims to address the challenges posed by climate change while ensuring food security and sustainability. At its core, CSA integrates three main objectives: increasing agricultural productivity and incomes, adapting, and building resilience to climate change, and reducing greenhouse gas emissions. One of the key principles of CSA is the adoption of practices that enhance resilience to climate variability and extremes. This includes techniques such as conservation agriculture, agroforestry, crop diversification, and water management strategies that help farmers cope with changing weather patterns.

The concept of CSA cannot just be limited to a set of new practices, but its scope reveals that it is a holistic and integrated approach to the implementation of agricultural development policies for enhancing food security resilience and livelihood opportunities under the reality of climatic variability being faced indiscriminately by the society.

Elaborating further following Aryal et al. (2020), the CSA can be classified into the following types of practices (Table 17.2):

In action research undertaken in Indo-Gangetic plains of India wherein the economic benefits of following CSA practices were determined. Most farmers undertook practices, namely, improved crop varieties, diversification of cropping, zero tillage and laser land levelling. The results showed that investing a few thousand in adoption led to benefits ranging from 6951 to 15,712 per year per hectare (Khatri-Chhetri et al., 2016). Operational land holdings, household profile, access to market and other infrastructure were found to significantly affect the adoption of CSA in Indo Gangetic Plains, and hence it was suggested that engagement of multiple stakeholders was imperative for broad-scale undertaking of CSA which in turn

Table 17.2 Typology of CSA technologies following Aryal et al. (2020)

Type of CSA technology	Practices recommended
Knowledge-smart technologies	Enhancement of social safety nets, improving the farmer's information network, integration of ICT-based knowledge centres at rural levels
Energy smart technologies	Zero tillage, residue management, precision nutrients management, machinery usage right to scale
Nitrogen smart technologies	Cultivation of leguminous crops, site-specific nutrition management
Carbon smart technologies	Management of livestock, improving livestock productivity, livestock nutrition management
Water smart technologies	Mulching, using raised beds for cultivation, laser land levelling
Weather smart technologies	Use of climatic agro-advisory services, crop insurance, early warning system

requires reform in the agricultural policies (Aryal et al., 2018). Sustainability and productivity (in terms of yield per hectare) of the cotton crop among the farmers of Punjab, Pakistan, were found to be enhanced due to CSA constituting improved irrigation practice, soil health management and crop management as against the prevailing climate change scenario, water shortage and depletion of the water table (Jamil et al., 2021). CSA was found to have a significantly positive impact on productivity of crops and income. Hence, the policymaker should scale the adoption of CSA by undertaking efficient agricultural extension services (Tanti et al., 2024; Tanti & Jena, 2023).

17.5 Delineating Policies Based on Three Pillars of Climate-Smart Agriculture (CSA)

Delineation of policy recommendations based on the three pillars of CSA. CSA is based on three pillars, namely, improving sustainability, increasing adaptation and resilience towards climate change and enhancement of mitigation of GHG emissions. Hence, the policymaking should be focussed on these three aspects.

17.5.1 Policies to Improve Sustainability

Sustainability implies holistic usage of resources in such a manner that the livelihood of future generations is not compromised for the present needs. For the attainment of this ideology, policies should be framed appropriately. To promote sustainability, numerous practices such as shifting in cropping patterns (Abbas et al., 2022), undertaking an integrated farming systems approach (Dutta et al., 2021) and tapping the potential of alternative livelihood practices may be promoted. The stakeholders should be actively involved in understanding the ground-level reality and feasibility of undertaking such ventures. In case of economic barriers, the development of social safety nets (Ngesa et al., 2021; Omerkhil et al., 2020), enhancement of infrastructural services (Chen et al., 2013; Zia & Wagner, 2015), and organisation of smallholder farmers and resource-poor farmers into SHGs (Shukla et al., 2016b) should be promoted. These activities may altogether lead to improved economic conditions, which may aid in adoption of novel agricultural practices. In case of cultural incompatibility factors, the stakeholders should facilitate working with farmers at the ground level and understand their cognition towards the novel practices. Further, the awareness campaigns followed by field schools to demonstrate the benefits of new practices along with myth-busting regarding false ideas prevailing among the farmers should be carried out.

17.5.2 Improving Adaptation Towards Climate Change

Undertaking of adaptation strategies requires on the part of the farmers that they are aware of the effects of climate change, or they can perceive the losses in crops and livestock amidst the changing climate. Adaptation refers to adjustment carried out by individuals in the face of increasing climate in order to build resilience towards the effects of the climatic events and carry out the agricultural venture profitably. The policies should take into consideration the prevailing cropping systems, farming patterns and livelihood options being availed by the farmers in the target area. This should be followed by the testing their feasibility in the current changing climatic scenario. Further wherever a potential discrepancy with respect to future sustainability and profitability of any aspect is observed, appropriate adaptation measures need to be devised. This information can be gained by discussion with experts, field visits in the area where the adaptations have been undertaken successfully, etc. The most appropriate measures may be chosen. Based on this, an unfailing set of recommendations may be developed which can be used by the policymaker. This awareness is a prerequisite to increase the acceptance of recommendations pertinent to adaptation among the farmers. Numerous studies indicate that the gender of the household head, economic conditions, household profile, access to the market, etc. are significant determinants of the extent of adoption of adaptation strategies among farmers (Aryal et al., 2018; Di Falco & Sharma-Khushal, 2019; Maiti et al., 2016). Hence, this information on these factors too should be available to the policymaker for efficient decision-making.

17.5.3 Improving Mitigation of GHG Emissions

GHG emissions lead to increased temperatures across the globe. Of the full GHG outflows, 10–12% is owed to the agrarian sector. Livestock account for more than 70% of the methane gas outflow (Pathak, 2023; Rojas-Downing et al., 2017). Since, Methane is one of the major greenhouse gases, an increment within the concentration can lead to quickened worldwide warming. Major emissions are concentrated within the tropical and sub-tropical regions of the globe(Srinivasa Rao et al., 2017).

Since the bovine and little ruminant populaces are rising yearly, the rate of rise in methane gas emanations may be quickened. Methane emanations are anticipated to rise within the coming decades with an unpredictably higher impact on Uttar Pradesh, Madhya Pradesh, and Andhra Pradesh (Kumari et al., 2016). Subsequently, it is imperative to plan different pathways and techniques to limit the emanations for relieving the impacts of climate alteration which may be due to the quickened rate of worldwide warming. Since mitigation pathways do not have a direct economic significance to the farmers, hence it is difficult to convince farmers to undertake the mitigation strategies. Interestingly, many adaptation strategies also enhance the mitigation of greenhouse gases like use of improved cultivars, residue management, use of direct seeded rice, improved tillage practices, improved irrigation practices, and use of climate resilient cultivars (Kragt et al., 2012; Singh et al., 2022). These factors reduce the direct and indirect emissions per kg of yield of the concerned crops. Moreover, livestock in agriculture is one of the major contributors to greenhouse gas emissions, so numerous methods have been devised to reduce it. The techniques include more inclusion of concentrated feed in the diet, use of rumen bacteria, biotechnological interventions, and improving the milk productivity of dairy animals (Sahoo et al., 2021; Wall et al., 2010). Further, most mitigation strategies demand a significant role to be played by the governance and require a transformative approach hence the policymakers should focus on the enhancement of infrastructures along with the engagement of various research organisations, government organisations and non-governmental organisations in the mitigation process. Hence, the recommendations should be framed accordingly in a manner that governance can be actively involved in the process.

17.6 Case Studies on Building Resilience Through CSA-Based Interventions

In the process of bringing about full-fledged changes in the status of agricultural communities' resilience, the role of extension agencies and other stakeholders working concurrently cannot be disregarded. Furthermore, understanding the features of change that such interventions bring, can lead to improved insight for future climate-centric projects that engage directly with small and commercial farms. This can provide a solid foundation for sound policy advocacy for improving the resilience of agricultural commodities comprehensively. As a result, this section delves into the various therapies targeted at strengthening resilience.

Garard et al. (2022) implemented an intervention in which professionals and villagers were encouraged to interact and work together to create a community resilient to the effects of climate change. In this attempt, ICT tools, specifically information systems, were implemented in multiple communities, generating information acceptable for both genders who play unique agricultural tasks. The information system supplied timely rainfall forecasts, as well as temperature data, which were then used by farmers to make good decisions about a variety of agricultural techniques. It focused on three components of rural agriculture: efforts for mitigating the effects of climate change, activities for efficiently managing resources, and finally, replacing maladaptation methods with climate-smart ones. Garg et al. (2020) conducted research in the central Bundelkhand region of India, incorporating climate-smart methods for water management into villages through the employment of extension agents. The approach resulted in a 2–5 metre increase in groundwater table level, alleviating the community's water scarcity. This resulted in a larger area being placed under cultivation. The 20% of land that was previously fallow was brought under intense cultivation. Furthermore, crop yields improved by 11–60%, resulting in higher household incomes.

Šūmane et al. (2018) found that the RETHINK project's varied approaches to addressing resilience-building difficulties were useful. An extension method in Austria highlighted the value of farmer-led networks in leveraging the work of vocational training institutions and field schools in agriculture. These networks were focused on informal knowledge building and knowledge transfer. Furthermore, a case study conducted in Denmark focused on a watershed in which farmers acquired local landscape management through enhanced learning at both the individual and community levels. Furthermore, this approach resulted in the formation of social capital.

Furthermore, in Ireland, farmers served as intermediaries for the dissemination of revolutionary organic manure application practices. The case was mostly about novel nutrient management approaches. Farmers used decision support systems and collaborated in the form of a network. This resulted in enhanced nutrient utilisation. Saran et al. (2022) developed PROTOCOL, a strategy that included a variety of agricultural treatments targeted exclusively at women farmers and aimed to improve climate resilience. It emphasised the importance of a variety of measures, including information distribution, finance-based approaches, natural resource management approaches, and social networking approaches, in developing women farmers' adaptive capacity. These resulted in beneficial effects in terms of monetary growth, socio-demographic relief, and improved mitigation and decision-making. Furthermore, in Ireland, farmers served as intermediaries for the dissemination of revolutionary organic manure application practices. The case was mostly about novel nutrient management approaches. Farmers used decision support systems and collaborated in the form of a network. This resulted in enhanced nutrient utilisation.

The case studies unequivocally show that the extension agents are crucial in educating and showing to farmers the range of climate-smart practices. Efforts to create knowledge networks result in improved decision-making and more cooperation among farmers. Thus, the extension workers must put in adequate effort to execute climate-smart solutions at the village level. Utilising ICT can also enhance the way interventions are implemented by enabling farmers to become self-sufficient in terms of the knowledge they need to do their jobs and keep up with changes in agricultural methods, financing, and the supply and demand of different farm commodities.

Extension agents can contribute to improving cooperation amongst the many parties. Additionally, extension agents have the power to persuade a variety of stakeholders to foster a sense of unity within farming communities, which in turn fosters the growth of organisations that can improve the profitability, productivity, and marketing of agricultural enterprises—whether they be farmer-producer companies, farmers-producer organisations, or specific public–private partnerships.

17.6.1 What Can the Policies Help Us Achieve?

Numerous studies suggest that the undertaking of CSA on farms have led to a significant increase in income through the productivity of the concerned crops (Khatri-Chhetri et al., 2016; Tanti et al., 2024). Thus, the number of resources needed per unit yield has reduced. Hence, incorporating the CSA in policymaking will assure sustainability. CSA-based policies will focus on the generation and accumulation of assets, which will help develop the farmers' adaptive capacity. Since most farmers in India are smallholders, fewer assets are a constraint in expanding the farm business or utilising innovative technologies like precision agriculture. Hence, improved adaptive capacity will ensure the upscaling and outscaling of business. Integration of various stakeholders operating at different can be achieved through CSA policies. This may further aid in gaining support from these stakeholders' as and when needed.

Gender equity may be ensured by improving the participation of women farmers not as a mere information provider but as active decision-makers in the climate smartness narrative. Insurance and development of social safety nets can be achieved which will help the farmers to adapt better in the face of extreme climatic vagaries. Social safety nets ensure economic and food security in case of extreme losses. Moreover, higher community cohesiveness is observed in communities where social safety nets are strengthened. Further, low emissions are the major goal that all organisations at the global level tend to achieve. Since mitigating activities can be better integrated if CSA-based policies are being implemented, it can assure a reduction in global emissions of greenhouse gases through appropriate measures.

References

Abbas, Q., Han, J., Bakhsh, K., Ullah, R., Kousar, R., Adeel, A., & Akhtar, A. (2022). Adaptation to climate change risks among dairy farmers in Punjab, Pakistan. *Land Use Policy, 119*. https://doi.org/10.1016/j.landusepol.2022.106184

Ariti, A. T., van Vliet, J., & Verburg, P. H. (2018). Farmers' participation in the development of land use policies for the Central Rift Valley of Ethiopia. *Land Use Policy, 71*, 129–137. https://doi.org/10.1016/j.landusepol.2017.11.051

Aryal, J. P., Rahut, D. B., Maharjan, S., & Erenstein, O. (2018). Factors affecting the adoption of multiple climate-smart agricultural practices in the Indo-Gangetic Plains of India. *Natural Resources Forum, 42*(3), 141–158. https://doi.org/10.1111/1477-8947.12152

Aryal, J. P., Jat, M. L., Sapkota, T. B., Rahut, D. B., Rai, M., Jat, H. S., Sharma, P. C., & Stirling, C. (2020). Learning adaptation to climate change from past climate extremes: Evidence from recent climate extremes in Haryana, India. *International Journal of Climate Change Strategies and Management, 12*(1), 128–146. https://doi.org/10.1108/IJCCSM-09-2018-0065

Balaganesh, G., Malhotra, R., Sendhil, R., Sirohi, S., Maiti, S., Ponnusamy, K., & Sharma, A. K. (2020). Development of composite vulnerability index and district level mapping of climate change induced drought in Tamil Nadu, India. *Ecological Indicators, 113*, 106197. https://doi.org/10.1016/j.ecolind.2020.106197

Bjork-James, C., Checker, M., & Edelman, M. (2022). Transnational social movements: Environmentalist, indigenous, and agrarian visions for planetary futures. *Annual Review of Environment and Resources, 47*, 583–608. https://doi.org/10.1146/ANNUREV-ENVIRON-112320-084822

Chen, Y., Shu, L., & Burbey, T. J. (2013). Composite subsidence vulnerability assessment based on an index model and index decomposition method. *Human and Ecological Risk Assessment, 19*(3), 674–698. https://doi.org/10.1080/10807039.2012.691405

Christensen, T., & Lægreid, P. (2020). Balancing governance capacity and legitimacy: How the Norwegian government handled the COVID-19 crisis as a high performer. *Public Administration Review, 80*(5), 774–779. https://doi.org/10.1111/puar.13241

Di Falco, S., & Sharma-Khushal, S. (2019). Cognitive drivers, and the effect of information on climate change adaptive behaviour in Fiji Islands. *Environmental Science and Policy, 92*, 245–254. https://doi.org/10.1016/j.envsci.2018.11.019

Dutta, S., Garai, S., Mandal, S., & Maiti, S. (2021). Farmers-led adaptation strategies to climate change for sustaining coastal agriculture in Sundarbans region of India. In *Climate Change and Resilient Food Systems: Issues, Challenges, and Way Forwards* (pp. 261–285). https://doi.org/10.1007/978-981-33-4538-6_10

Feyissa, G., Zeleke, G., Gebremariam, E., & Bewket, W. (2018). GIS based quantification and mapping of climate change vulnerability hotspots in Addis Ababa. *Geoenvironmental Disasters, 5*(1), 1–17. https://doi.org/10.1186/s40677-018-0106-4

Garard, J., Wood, S. L. R., Sabet-Kassouf, N., Ventimiglia, A., Matthews, H. D., Ubalijoro, É., Chaudhari, K., Ivanova, M., & Luers, A. L. (2022). Moderate support for the use of digital tracking to support climate-mitigation strategies. *One Earth, 5*(9), 1030–1041. https://doi.org/10.1016/j.oneear.2022.08.005

Garg, K. K., Singh, R., Anantha, K. H., Singh, A. K., Akuraju, V. R., Barron, J., Dev, I., Tewari, R. K., Wani, S. P., Dhyani, S. K., & Dixit, S. (2020). Building climate resilience in degraded agricultural landscapes through water management: A case study of Bundelkhand region, Central India. *Journal of Hydrology, 591*, 125592. https://doi.org/10.1016/j.jhydrol.2020.125592

Goswami, R. K., Maiti, S., Garai, S., Jha, S. K., Bhakat, M., Chandel, B. S., & Kadian, K. S. (2018). Coping mechanisms adopted by the livestock dependents of drought prone districts of Bihar, India. *Indian Journal of Animal Sciences, 88*(3), 356–364. https://doi.org/10.56093/ijans.v88i3.78378

Jamil, I., Jun, W., Mughal, B., Raza, M. H., Imran, M. A., & Waheed, A. (2021). Does the adaptation of climate-smart agricultural practices increase farmers' resilience to climate change? *Environmental Science and Pollution Research, 28*(21), 27238–27249. https://doi.org/10.1007/s11356-021-12425-8

Jeganathan, A., Andimuthu, R., & Kandasamy, P. (2021). Climate risks and socio-economic vulnerability in Tamil Nadu, India. *Theoretical and Applied Climatology, 145*(1–2), 121–135. https://doi.org/10.1007/s00704-021-03595-z

Jha, R. K., & Gundimeda, H. (2019). An integrated assessment of vulnerability to floods using composite index – A district level analysis for Bihar, India. *International Journal of Disaster Risk Reduction, 35*, 101074. https://doi.org/10.1016/j.ijdrr.2019.101074

Khatri-Chhetri, A., Aryal, J. P., Sapkota, T. B., & Khurana, R. (2016). Economic benefits of climate-smart agricultural practices to smallholder farmers in the Indo-Gangetic Plains of India. *Current Science, 110*(7), 1251–1256. https://doi.org/10.18520/cs/v110/i7/1251-1256

Kragt, M. E., Pannell, D. J., Robertson, M. J., & Thamo, T. (2012). Assessing costs of soil carbon sequestration by crop-livestock farmers in Western Australia. *Agricultural Systems, 112*, 27–37. https://doi.org/10.1016/j.agsy.2012.06.005

Kumari, S., Dahiya, R. P., Naik, S. N., Hiloidhari, M., Thakur, I. S., Sharawat, I., & Kumari, N. (2016). Projection of methane emissions from livestock through enteric fermentation: A case study from India. *Environmental Development, 20*, 31–44. https://doi.org/10.1016/j.envdev.2016.08.001

Lee, J. (2017). Farmer participation in a climate-smart future: Evidence from the Kenya agricultural carbon market project. *Land Use Policy, 68*, 72–79. https://doi.org/10.1016/j.landusepol.2017.07.020

Luers, A. L., Lobell, D. B., Sklar, L. S., Addams, C. L., & Matson, P. A. (2003). A method for quantifying vulnerability, applied to the agricultural system of the Yaqui Valley, Mexico. *Global Environmental Change, 13*(4), 255–267. https://doi.org/10.1016/S0959-3780(03)00054-2

Maiti, S., Jha, S. K., Garai, S., Nag, A., Chakravarty, R., Kadian, K. S., & Meena, B. S. (2014a). Determinants to climate change adaptation among the Brokpa pastoral nomads of Western Arunachal Pradesh. *Indian Journal of Extension Education, 14*(2), 36–40.

Maiti, S., Jha, S. K., Garai, S., Nag, A., Chakravarty, R., Kadian, K. S., Chandel, B. S., Datta, K. K., & Upadhayay, R. C. (2014b). Adapting to climate change: Traditional coping mechanism followed by the Brokpa pastoral nomads of Arunachal Pradesh, India. *Indian Journal of Traditional Knowledge, 13*(4), 752–761.

Maiti, S., Jha, S. K., Garai, S., Nag, A., Chakravarty, R., Kadian, K. S., Chandel, B. S., Datta, K. K., & Upadhyay, R. C. (2015). Assessment of social vulnerability to climate change in the eastern coast of India. *Climatic Change, 131*(2), 287–306. https://doi.org/10.1007/s10584-015-1379-1

Maiti, S., Jha, S. K., Garai, S., Nag, A., Bera, A. K., Bhattacharya, D., Kale, R. B., & Deb, S. M. (2016). Climate change awareness among the livestock rearers of east coast of India. *Indian Journal of Animal Sciences, 86*(7), 799–809. https://doi.org/10.56093/ijans.v86i7.59950

Mondal, M., Biswas, A., Mandal, S., Bhattacharya, S., & Paul, S. (2023). Developing micro level resilience index for Indian Sundarban adopting resilience indicators for measurement and analysis (RIMA) methodology. *Geosystems and Geoenvironment, 2*(1), 100129. https://doi.org/10.1016/j.geogeo.2022.100129

Ngesa, O., Grey, S., Duveskog, D., & Atieno, I. (2021). Characteristics and determinants of the resilience of smallholder farmers: Lessons from application of the RIMA II methodology in Eastern Africa. In *Handbook of Climate Change Management: Research, Leadership, Transformation* (Vol. 2, pp. 1547–1569). https://doi.org/10.1007/978-3-030-57281-5_67

Nkiaka, E., Taylor, A., Dougill, A. J., Antwi-Agyei, P., Fournier, N., Bosire, E. N., Konte, O., Lawal, K. A., Mutai, B., Mwangi, E., Ticehurst, H., Toure, A., & Warnaars, T. (2019). Identifying user needs for weather and climate services to enhance resilience to climate shocks in sub-Saharan Africa. *Environmental Research Letters, 14*(12), 123003. https://doi.org/10.1088/1748-9326/ab4dfe

Omerkhil, N., Chand, T., Valente, D., Alatalo, J. M., & Pandey, R. (2020). Climate change vulnerability and adaptation strategies for smallholder farmers in Yangi Qala District, Takhar, Afghanistan. *Ecological Indicators, 110*, 105863. https://doi.org/10.1016/j.ecolind.2019.105863

Pathak, H. (2023). Impact, adaptation, and mitigation of climate change in Indian agriculture. *Environmental Monitoring and Assessment, 195*(1), 52. https://doi.org/10.1007/s10661-022-10537-3

Prager, K., & Freese, J. (2009). Stakeholder involvement in agri-environmental policy making – Learning from a local- and a state-level approach in Germany. *Journal of Environmental Management, 90*(2), 1154–1167. https://doi.org/10.1016/j.jenvman.2008.05.005

Rama Rao, C. A., Raju, B. M. K., Adlul Islam, A. V. M. S. R., Rao, K. V., Ravindra Chary, G., Nagarjuna Kumar, R., Prabhakar, M. Sammi Reddy, K., Bhaskar, S., & Chaudhari, S. K. (2020). *Risk and vulnerability assessment of indian agriculture to climate change*. Protocolo de Prevención y Organización Del Regreso a La Actividad Lectiva En Los Centros Educativos de Castilla y León Para El Curso Escolar 2020/2021. https://scholar.google.com/scholar?&q=RamaRaoCA%2CRaju%2CBMK%2CIslamA%2CSubbaRaoAVM%2CRaoKV%2CRavindraCharyG%2CNagarjunaKumarR%2CPrabhakarM%2CSammiReddyK%2CBhaskarS%2CChaudhariSK%282019%29RiskandvulnerabilityassessmentofIn

Rojas-Downing, M. M., Nejadhashemi, A. P., Harrigan, T., & Woznicki, S. A. (2017). Climate change and livestock: Impacts, adaptation, and mitigation. *Climate Risk Management, 16*, 145–163. https://doi.org/10.1016/j.crm.2017.02.001

Sabatier, P. A. (2019). Theories of the policy process. In *Theories of the policy process*. https://doi.org/10.4324/9780367274689.

Sahani, S., Zade, S., Panja, A., Garai, S., & Maiti, S. (2023). Framework for climate change risk assessment. In *Social science dimensions of climate resilient agriculture*. https://www.researchgate.net/profile/Sanjit-Maiti/publication/372909405_Social_Science_Dimensions_of_Climate_Resilient_Agriculture/links/64cd3c4191fb036ba6c6d311/Social-Science-Dimensions-of-Climate-Resilient-Agriculture.pdf#page=86

Sahoo, N., Pattnaik, S., Pattnaik, M., & Mohapatra, S. (2021). Methane emission and strategies for mitigation in livestock. In *Environmental and agricultural microbiology: Advances and applications* (pp. 257–274). Wiley. https://doi.org/10.1002/9781119525899.ch12

Saran, A., Singh, S., Gupta, N., Walke, S. C., Rao, R., Simiyu, C., Malhotra, S., Mishra, A., Puskur, R., Masset, E., White, H., & Sharma Waddington, H. (2022). PROTOCOL: Interventions promoting resilience through climate-smart agricultural practices for women farmers: A systematic review. *Campbell Systematic Reviews, 18*(3). https://doi.org/10.1002/cl2.1274

Schramm, P. J., Al Janabi, A. L., Campbell, L. W., Donatuto, J. L., & Gaughen, S. C. (2020). How indigenous communities are adapting to climate change: Insights from the climate-ready tribes initiative. *Health Affairs, 39*(12), 2153–2159. https://doi.org/10.1377/hlthaff.2020.00997

Shukla, R., Chakraborty, A., & Joshi, P. K. (2011). Vulnerability of agriculture to climate change: District level assessment in the Indo-Gangetic Plains. *Current Science, 101*(3), 429–453. https://doi.org/10.1016/j.gloplacha.2014.03.001%0A, https://doi.org/10.1016/j.ijdrr.2014.09.011%0A, https://doi.org/10.1016/j.ijdrr.2015.01.004%0A, https://doi.org/10.1016/j.ecolind.2016.06.041%0A, https://doi.org/10.1016/j.ecolind.2017.02.006%0A

Shukla, R., Sachdeva, K., & Joshi, P. K. (2016a). Inherent vulnerability of agricultural communities in Himalaya: A village-level hotspot analysis in the Uttarakhand state of India. *Applied Geography, 74*, 182–198. https://doi.org/10.1016/j.apgeog.2016.07.013

Shukla, R., Sachdeva, K., & Joshi, P. K. (2016b). An indicator-based approach to assess village-level social and biophysical vulnerability of agriculture communities in Uttarakhand, India. *Journal of Mountain Science, 13*(12), 2260–2271. https://doi.org/10.1007/s11629-016-4058-4

Singh, R., Singh, A., Sheoran, P., Fagodiya, R. K., Rai, A. K., Chandra, P., Rani, S., Yadav, R. K., & Sharma, P. C. (2022). Energy efficiency and carbon footprints of rice-wheat system under long-term tillage and residue management practices in western Indo-Gangetic Plains in India. *Energy, 244*. https://doi.org/10.1016/j.energy.2021.122655

Srinivasa Rao, C., Girija Veni, V., Prasad, J. V. N. S., Sharma, K. L., Chandrasekhar, C., Rohilla, P. P., & Singh, Y. V. (2017). Improving carbon balance with climate-resilient management practices in tropical agro-ecosystems of Western India. *Carbon Management, 8*(2), 175–190. https://doi.org/10.1080/17583004.2017.1309202

Streimikis, J., & Baležentis, T. (2020). Agricultural sustainability assessment framework integrating sustainable development goals and interlinked priorities of environmental, climate and agriculture policies. *Sustainable Development, 28*(6), 1702–1712. https://doi.org/10.1002/sd.2118

Šūmane, S., Kunda, I., Knickel, K., Strauss, A., Tisenkopfs, T., Rios, I. d. I., Rivera, M., Chebach, T., & Ashkenazy, A. (2018). Local and farmers' knowledge matters! How integrating informal and formal knowledge enhances sustainable and resilient agriculture. *Journal of Rural Studies, 59*, 232–241. https://doi.org/10.1016/j.jrurstud.2017.01.020

Tanti, P. C., & Jena, P. R. (2023). Perception on climate change, access to extension service and energy sources determining adoption of climate-smart practices: A multivariate approach. *Journal of Arid Environments, 212*, 104961. https://doi.org/10.1016/j.jaridenv.2023.104961

Tanti, P. C., Jena, P. R., Timilsina, R. R., & Rahut, D. B. (2024). Enhancing crop yields and farm income through climate-smart agricultural practices in Eastern India. *Mitigation and Adaptation Strategies for Global Change, 29*(5), 1–28. https://doi.org/10.1007/s11027-024-10122-8

Venus, T. E., Bilgram, S., Sauer, J., & Khatri-Chettri, A. (2022). Livelihood vulnerability and climate change: A comparative analysis of smallholders in the Indo-Gangetic plains. *Environment, Development and Sustainability, 24*(2), 1981–2009. https://doi.org/10.1007/s10668-021-01516-8

Wall, E., Simm, G., & Moran, D. (2010). Developing breeding schemes to assist mitigation of greenhouse gas emissions. *Animal, 4*(3), 366–376. https://doi.org/10.1017/S175173110999070X

Zia, A., & Wagner, C. H. (2015). Mainstreaming early warning systems in development and planning processes: Multilevel implementation of Sendai framework in Indus and Sahel. *International Journal of Disaster Risk Science, 6*(2), 189–199. https://doi.org/10.1007/s13753-015-0048-3

Food Security and Climate Change: Policy Implications

18

S. V. Bangararaju Tatipudi, Prabhat Kishore, Devesh Kumar Pant, and Neha Joshi

Abstract

The sustainable development goals (SDGs) target to end hunger, achieve food security, improve nutrition, and take urgent action to combat climate change and its impacts. For India, food security continues to be high on its list of development priorities because the country's reduction in hunger and malnutrition is not commensurate with relatively high rates of economic growth. In the last two decades, the undernourished population has declined marginally (~2%), and child wasting under the age of five has increased by 2%. According to the fifth National Family Health Survey (NFHS-5), the proportion of children under 5 years who are underweight is significantly high in the states of Bihar (41%), Chhattisgarh (31.3%), and Uttar Pradesh (32.1%). While significant portions of the Indian populace experience severe undernutrition, increasing incomes, and urbanization, there is evidence of a transition in the food landscape, switching from cereals toward high-value agricultural products like horticulture produce, poultry products, fish, and meat. Consequently, there will be an increase in total food demand driven by population growth and also an increasing indirect demand for animal feed and biofuel.

Keywords

Food grain production · Climate change · Sustainable development goals

S. V. B. Tatipudi (✉) · P. Kishore · D. K. Pant · N. Joshi
ICAR-National Institute of Agricultural Economics and Policy Research, New Delhi, India

R. Dubey et al. (eds.), *Drought and Heat Stress in Agriculture*,
https://doi.org/10.1007/978-981-96-5735-3_18

18.1 Agricultural Production and Net Availability of Food

The total food grain production in the country steadily increased from 52.0 million tonnes in 1951–52 to 315.7 million tonnes in 2021–22, as seen in Fig. 18.1. This manifold increase in food grain is mainly due to increased yields, as yield increased from 536 kg/ha in 1951–52 and rose to 2419 kg/ha in 2021–22. However, in Table 18.1, it can be observed that India's cereal yield is much lower than the world average, and many comparable countries like China and the United States indicate huge potential for improving the agricultural yield in India.

Figure 18.2 shows that the annual per capita net availability of food grains has increased by about 30% since 1951–52. The per capita availability of rice has moderately increased from 58 kg in 1951–52 to 69 kg in 2021–22. In contrast, there is a significant increase in per capita wheat availability from 24 to 68 kg in the same time period. However, the per capita pulse availability scenario is alarming as it decreased from 22.1 to 19.6 kg.

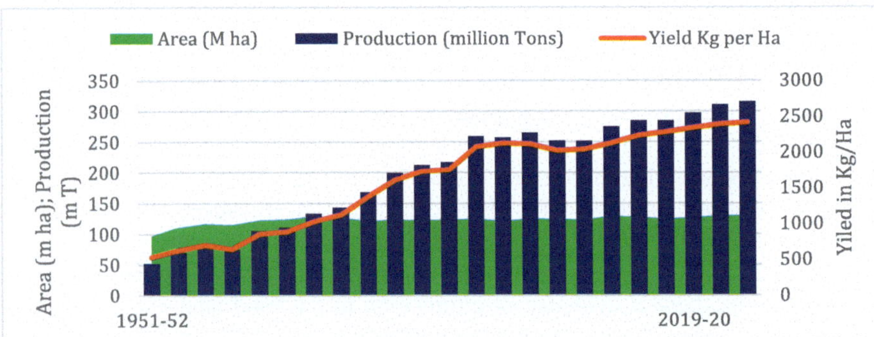

Fig. 18.1 Trends in food grains area, production, and yield. (Source: Government of India (2022). Agriculture Statistics at a glance)

Table 18.1 Cereal yield in different countries

Country	Yield (Kg per hectare)
India	3566.6
China	6379.6
Pakistan	3405.7
Bangladesh	5005.3
USA	8071.7
World average	4182.38

Source: World Bank Database

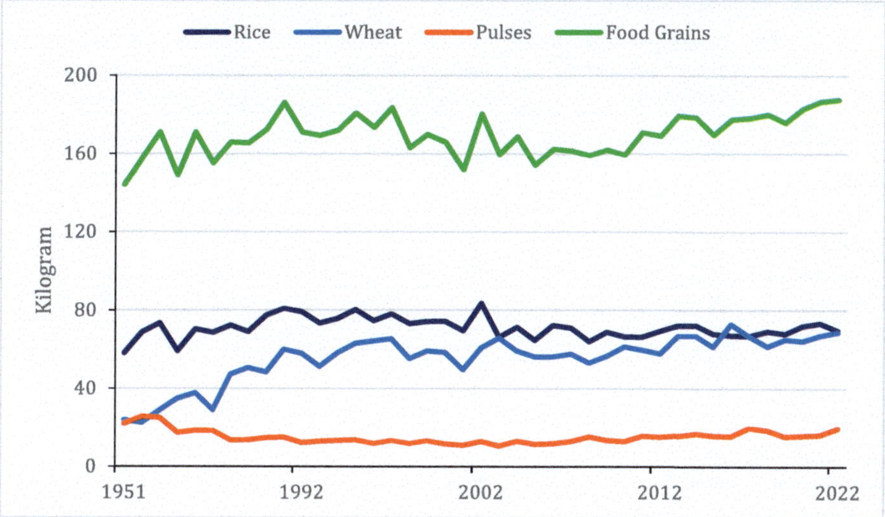

Fig. 18.2 Trends in per capita net availability of food grains. (Source: Government of India (2022). Agriculture Statistics at a glance)

18.2 Nutritional Status of the Country

According to the Global Food Security Index report, in 2022, India is ranked 68 out of 113 countries, and in the Global Hunger Index 2022, India is ranked 107 out of 136 countries, indicating a dire nutritional status. Within the country, there are widespread disparities. Table 18.2 shows that states like Punjab, Haryana, Kerala, Goa, and Sikkim have performed better, whereas the eastern states like Bihar, Jharkhand, West Bengal, and Chhattisgarh are poor performers in terms of child nutrition (NFHS-5).

18.3 Future Demand for Food Grains

India is the most populous country in the world, and by 2036, India's population will be 1518 million. Agricultural production should be commensurately increased to feed the increased population. Total demand for food grains is estimated at 326 million tonnes in 2030–31, gradually increasing to 402 million tonnes in 2047–48 in the Business-as-usual scenario, as seen in Table 18.3. Amongst food grains, the growth in demand for maize, pulses, and nutri-cereals will be significantly higher than the growth in demand for rice and wheat. The demand for eggs, meat, fish, and milk will increase almost fourfold by 2047–48. If this increased demand is not met by the production (Table 18.4), the prices are going to skyrocket, which adversely affects the people with low income. All these demand projections for food are in a Business-as-usual scenario (conservative estimates) based on the assumption that

Table 18.2 Nutritional status of children in India

State	Proportion of children under 5 who are stunted (%)	Proportion of children under 5 who are wasted (%)	Proportion of children under 5 who are underweight (%)
Northern region			
Himachal Pradesh	30.8	17.4	25.5
Punjab	24.5	10.6	16.9
Uttarakhand	27	13.2	21
Uttar Pradesh	39.7	17.3	32.1
Haryana	27.5	11.5	21.5
Eastern region			
Bihar	42.9	22.9	41
Chhattisgarh	34.6	18.9	31.3
Odisha	31	18.1	29.7
Jharkhand	39.6	22.4	39.4
West Bengal	33.8	20.3	32.2
Western region			
Goa	25.8	19.1	24
Gujarat	39	25.1	39.7
Maharashtra	35.2	25.6	36.1
Rajasthan	31.8	16.8	27.6
Central region			
Madhya Pradesh	35.7	19	33
Chhattisgarh	34.6	18.9	31.3
North eastern region			
Assam	35.3	21.7	32.8
Arunachal Pradesh	28	13.1	15.4
Sikkim	22.3	13.7	13.1
Nagaland	32.7	19.1	26.9
Meghalaya	46.5	12.1	26.6
Manipur	23.4	9.9	13.3
Mizoram	28.9	9.8	12.7
Tripura	32.3	18.2	25.6
Southern Region			
Andhra Pradesh	31.2	16.1	29.6
Karnataka	35.4	19.5	32.9
Kerala	23.4	15.8	19.7
Tamil Nadu	25	14.6	22
Telangana	33.1	21.7	31.8

Source: NFHS 5, 2019–2021

Table 18.3 Projected total food demand in India under the business-as-usual scenario

Crops	2019–20	2025–26	2030–31	2035–36	2040–41	2047–48	Compound growth rate
Food grains	277	303	326	353	371	402	1.39
Cereals	251	272	290	313	327	353	1.27
Rice	103	107	110	111	113	114	0.40
Wheat	100	106	111	115	117	119	0.65
Nutri-cereals	19	20	22	23	26	29	1.60
Maize	27	36	45	60	67	86	4.39
Pulses	26	31	35	40	44	49	2.38
Animal source food	24	32	40	49	44	49	2.38
Eggs	5	7	8	10	12	16	4.32
Meat	7	9	12	14	17	21	4.31
Fish	12	16	20	24	29	37	4.27
Milk	186	243	294	349	405	480	3.56
Vegetables	199	238	270	301	330	365	2.28
Fruits	108	136	160	184	206	233	2.90
Sugar and products	34	37	39	41	43	44	1.05
Edible oils	22	25	27	29	30	31	1.23

Source: NITI Aayog (2024)

Table 18.4 Forecast of agricultural production under business-as-usual scenario in India

Crops	2019–20 (Actual)	2025–26	2030–31	2035–36	2040–41	2047–48	Compound growth rate
Food grains	299	332	368	396	417	457	1.58
Cereals	276	305	337	361	378	409	1.47
Rice	119	133	145	150	153	154	0.97
Wheat	108	117	131	141	147	160	1.48
Nutri-cereals	19	17	16	16	15	14	−1.23
Maize	29	36	43	51	62	80	3.84
Pulses	23	27	31	35	40	47	2.70
Animal source food	28	38	46	55	63	74	3.62
Eggs	6	9	11	13	16	19	4.56
Meat	9	10	12	14	15	18	2.71
Fish	14	19	23	27	32	37	3.66
Milk	198	258	308	358	408	478	3.31
Vegetables	188	224	254	285	318	367	2.50
Fruits	102	122	141	161	182	214	2.78
Sugar and products	33.7	42	43	45	49	50	1.50
Edible oils	12	14	15	18	20	24	2.69

Source: NITI Aayog (2024)

Table 18.5 Projected household food demand (in Kg) in India under a business-as-usual scenario

Crops	2019–20	2025–26	2030–31	2035–36	2040–41	2047–48	Compound growth rate
Food grains	195	208	219	229	238	248	0.90
Cereals	178	188	195	202	208	214	0.69
Rice	93	97	100	102	104	106	0.46
Wheat	76	82	86	90	93	96	0.85
Nutri-cereals	6.6	7.0	7.5	8.2	9.2	10.8	1.84
Maize	1.6	1.6	1.7	1.7	1.7	1.7	0.26
Pulses	17	20	23	27	30	34	2.60
Animal source food	14	19	24	30	37	46	4.48
Eggs	3	4	5	7	8	10	4.48
Meat	5	7	9	11	13	16	4.48
Fish	6	8	11	13	16	20	4.48
Milk	104	136	166	198	230	276	3.67
Vegetables	138	166	190	213	236	263	2.43
Fruits	27	34	41	47	54	62	3.18
Sugar and products	14	16	17	18	20	21	1.45
Edible oils	14	16	17	18	20	21	1.45

Source: NITI Aayog (2024)

India's GDP growth is moderate. In a high GDP growth scenario, demand for food will increase much more.

As seen in Table 18.5, by 2047–48, the compound growth rate of household demand for eggs, meat, and fish is more than 4%. It is more than 3% for milk and fruits.

Millets are better adapted to diverse environmental conditions through nitrogen-use and water-use efficiencies, tolerant to insects, pests, and diseases, and resilient to environmental stresses induced by climate change. The climate-resilient characteristics and nutritional benefits of small millets have been extensively researched, showing that millets could be the preferred staple crop in areas affected by hunger. Compared to most other food crops, millet can better mitigate many negative effects of climate change. They are nutrient-dense and generally contain more iron, calcium, beta-carotene, and other nutrients than wheat and rice. Millets are also rich in dietary fiber, almost absent in rice. With a low glycemic index and being gluten-free, millet-based diets are excellent for people with diabetes and children with celiac disease (Tiwari et al., 2022).

The area under millets has decreased from 38.8 m ha in 1951–52 to 22.6 m ha in 2021–22. However, production increased from 16.1 million tonnes to 50.9 million tonnes in the same time period due to significant yield gains (Fig. 18.3). Realizing the overall importance of millets, the Indian government proposed to celebrate 2023 as the International Year of Millets. FAO and the United Nations have recognized

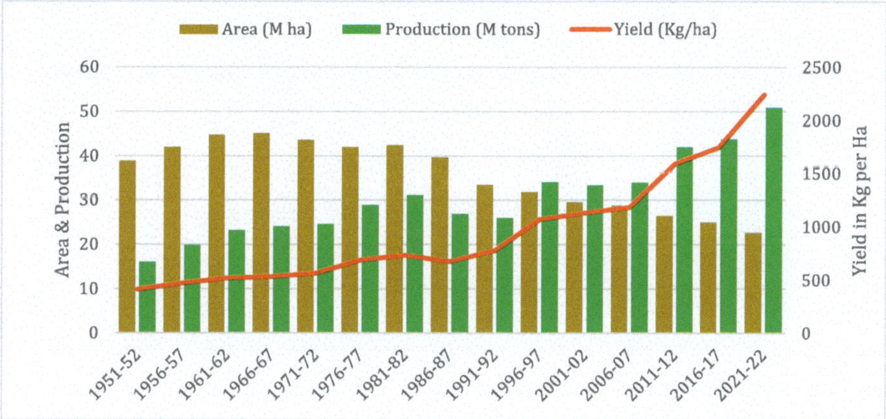

Fig. 18.3 Trends in area, production, and yield of nutri-cereals. (Source: Government of India (2022). Agriculture Statistics at a glance)

2023 as the International Year of Millets to raise awareness about their health and nutritional benefits.

Pulses are a crucial source of energy, protein, essential amino acids, dietary fiber, minerals, and vitamins and play a vital role in global nutritional security. In India, pulses are particularly significant, as the country has the world's largest population adhering to a vegetarian diet. For this population, pulses serve as the primary source of protein. They are also a key element of agricultural biodiversity, essential for combating climate change by enhancing biological nitrogen fixation in the soil, which reduces the need for synthetic fertilizers. Additionally, pulses contribute to higher yields in subsequent non-legume crops (Sandeep Kumar et al., 2023). Weather aberration adversely affects pulse production as the majority of pulses are grown on marginal land and primarily in rainfed regions.

The domestic production of pulses in India is insufficient to meet the country's demand, leading to imports from other countries. In 2015–16, pulse inflation reached an all-time high of 34.8%, and India imported 6.6 million tonnes of pulses in 2016–17 (Figs. 18.4 and 18.5). Subsequently, imports declined due to various government measures. This dependence on imports causes prices to fluctuate at the retail level. This high volatility affects poor people's protein intake, causing nutritional insecurity (Figs. 18.6 and 18.7).

18.4 Effect of Climate Change on Food Security

Food security exists when all people, at all times, have physical and economic access to sufficient, safe, and nutritious food that meets their dietary needs and food preferences for an active and healthy life (World Food Summit, 1996). The dimensions of food security are as follows:

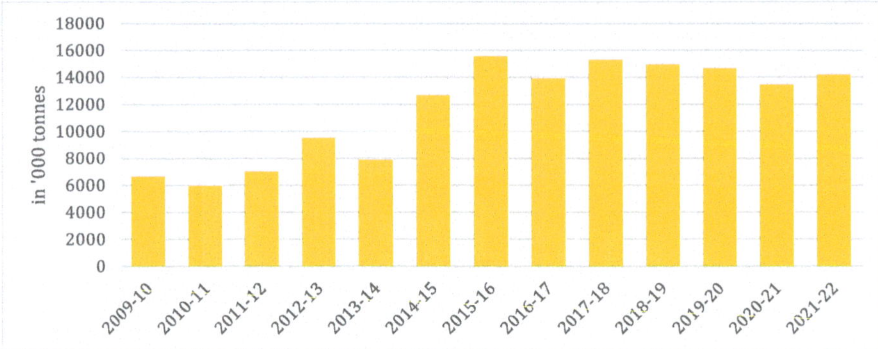

Fig. 18.4 Import of vegetable oils by India. (Source: Government of India (2022). Agriculture Statistics at a glance)

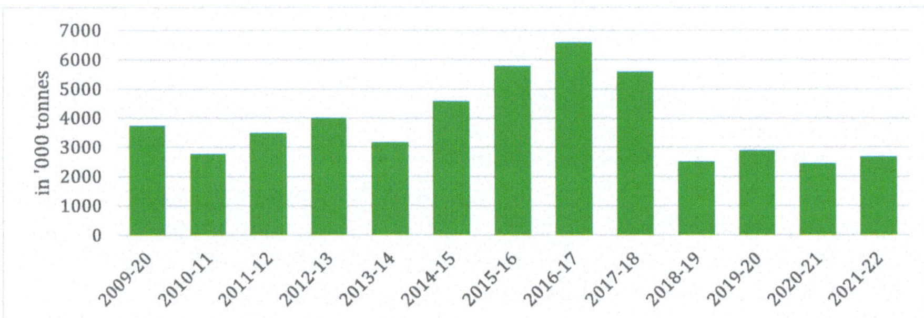

Fig. 18.5 India's edible oil inflation. (Source: Government of India (2022). Agriculture Statistics at a glance)

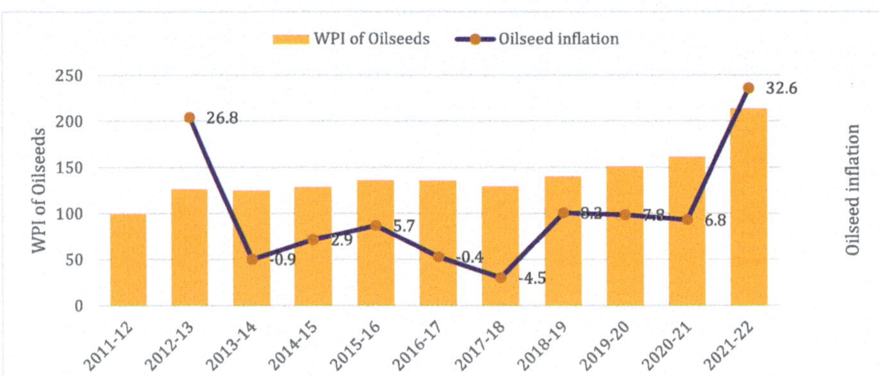

Fig. 18.6 Import of pulses in India. (Source: Government of India (2022). Agriculture Statistics at a glance)

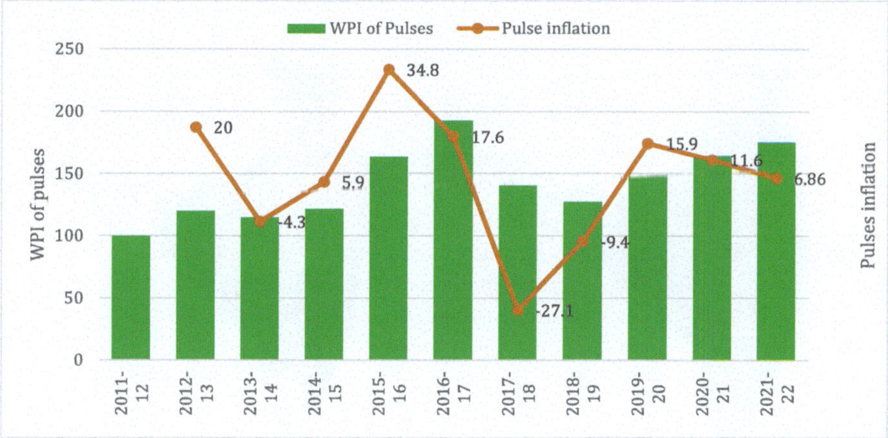

Fig. 18.7 India's pulses inflation. (Source: Government of India (2022). Agriculture Statistics at a glance)

(a) *Food availability*: Sufficient quantities of appropriate quality food, supplied through domestic production or imports.
(b) *Food access*: Access by individuals to adequate resources for acquiring appropriate foods for a nutritious diet.
(c) *Utilization*: The use of food through adequate diet, clean water, sanitation, and health care to reach a state of nutritional well-being where all physiological needs are met.
(d) *Stability*: To be food secure, a population, household, or individual must have access to adequate food at all times. They should not risk losing access to food due to sudden shocks or cyclical events.

Thus, sufficient food production alone cannot ensure a country's food security. Food security is one of the prominent concerns due to climate change. Climate change impacts food security in multifaceted ways, affecting crops, livestock, forestry, fisheries, and aquaculture and leading to significant social and economic repercussions, such as reduced incomes, compromised livelihoods, disrupted trade, and negative health effects. Importantly, the overall impact of climate change is influenced not just by the severity of the climatic events but also by the underlying vulnerabilities of the affected populations.

Climate change can impact food security through multiple pathways, such as precipitation, temperature, humidity, and solar radiation. One significant pathway is the effect of climate change on food production through precipitation, which affects the total water availability of any region. Water availability/scarcity directly affects crop yield & production and also indirectly influences disease & pest attacks, as about 78% of the total fresh water available is used in agriculture. Irrigation is considered a key catalyst in the food production system, besides other agricultural inputs and farm management practices anchored on household characteristics,

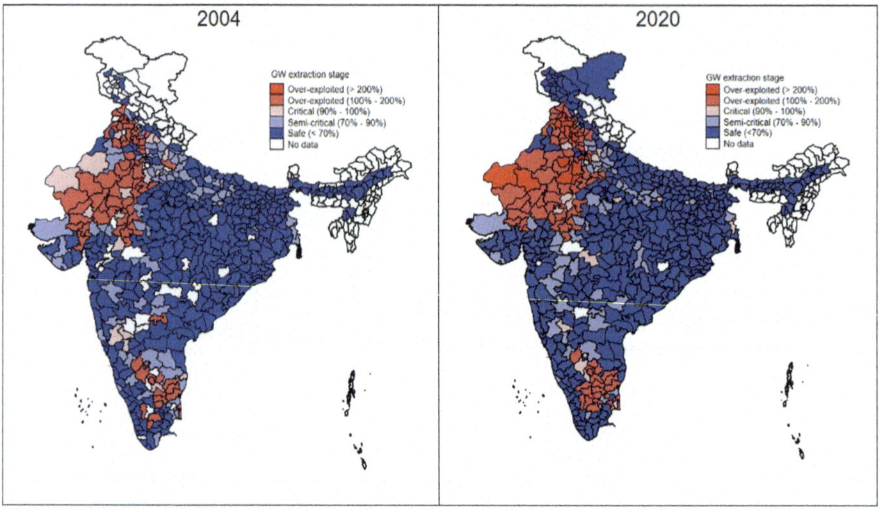

Fig. 18.8 Groundwater extraction across Indian districts. (Source: CGWB, 2020)

environmental factors, and institutional support. With the advent of the Green Revolution era, groundwater has emerged as a dominant source of irrigation and contributed substantially to improving cropping intensity and resilience to drought. In addition, access to groundwater improves irrigation water scheduling in terms of timing and quantity, resulting in relatively higher crop yield and water productivity over surface irrigation systems. However, increased dependency on groundwater irrigation led to its overextraction, resulting in groundwater depletion in many parts of the country.

In India, groundwater utilization almost doubled, and overexploited blocks almost tripled in the last two decades. There is wide spatial heterogeneity in groundwater utilization, and most of the areas of north-western states are witnessing overexploitation of groundwater (Fig. 18.8). The higher extraction of groundwater is rooted in the higher need for groundwater irrigation for higher food grain production or changing cropping patterns more biased toward water-intensive crops.

The country's rice and wheat are cultivated on nearly 40% of the gross cropped area, contributing almost 3/4th of total food grain production. However, the rice–wheat cropping system is considered a major contributor to groundwater depletion in the country (Devineni et al., 2022). Table 18.6 shows that groundwater depletion causes a reduction in food grain production and cropping intensity, which will have repercussions on India's food security.

Another pathway is through the alteration of atmospheric CO_2 levels, which affects both biomass and nutritional quality. Additionally, climate change can exacerbate food safety risks during transport and storage (Watts et al., 2015; Fanzo et al., 2018). Moreover, the direct impacts of changing weather can affect human health by exposing the agricultural workforce to extreme temperatures. Extreme temperatures can alter metabolic demands and cause physiological stress, potentially

Table 18.6 Estimated impact of groundwater depletion on food grain production

Authors	Outcome variables	Study area	Major findings
Sekhri (2013)	Food grain production	All India	With 1 meter decline in groundwater in a year reduces food-grain production by 8%, water-intensive crop production by 9%, and cash crops by 5%
Jain et al. (2021)	Cropping intensity	All India	With current groundwater depletion trends, cropping intensity may decrease by 20% nationwide and 68% in groundwater-depleted regions in 2025
Bhattarai et al. (2021)	Yield & food grain production	All India	With 1-meter decline in groundwater level, mean yield, and production decline by 1–3% and 4–8%, respectively, for winter crops

leading to a situation where people need more food to cope with the heat while being less able to produce it (Watts et al., 2015).

18.5 Initiatives by the Government and Policy Issues

18.5.1 The National Innovations on Climate Resilient Agriculture (NICRA)

The National Innovations on Climate Resilient Agriculture (NICRA) project, initiated by the Indian Council of Agricultural Research (ICAR) in February 2011, is a comprehensive initiative to bolster the resilience of Indian agriculture against the impacts of climate change. This network project focuses on strategic research and technology demonstration to enhance the adaptability and mitigation capabilities of Indian agriculture. The research scope encompasses crops, livestock, fisheries, and natural resource management, addressing the multifaceted challenges of climate change. NICRA is structured into four key components: strategic research, technology demonstration, capacity building, and sponsored/competitive grants, each designed to contribute to the overall goal of climate-resilient agriculture in India. The budget for the Climate Resilient Agriculture initiative declined since 2016–17 and remained stagnant for the past 7 years, as seen in Fig. 18.9. The NICRA project was initiated with much fanfare later, the project is surviving on with a minimal budget allocation.

18.5.2 Food Security Act and Public Distribution System

The government of India has been procuring grains through various procurement agencies to provide subsidized food grains to the citizens. There has been a significant increase in the procurement of rice and wheat over time (Fig. 18.10). To

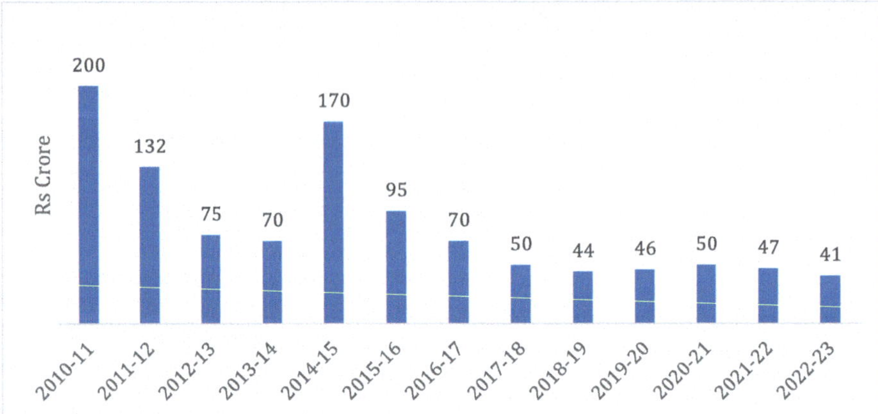

Fig. 18.9 The budget for climate resilient initiative budget of ICAR over the years. (Source: Government of India, various years)

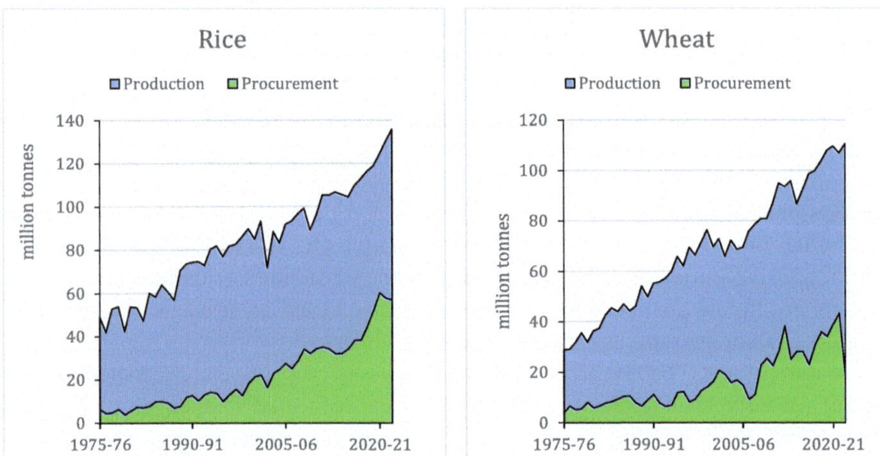

Fig. 18.10 Trends in the production and procurement of paddy and wheat. (Source: Government of India, 2022)

provide legal entitlement to the people for food, the Government of India enacted "The National Food Security Act (NFSA)" in the year 2013, also known as "Right to Food Act," to provide subsidized food grains to 2/3 of the population (to 75% in rural areas and 50% in urban areas. The act includes a scheme of Midday Meals, Integrated Child Development Services, a Public Distribution System, maternity entitlement, etc. Currently, the government is providing rice and wheat to about 80 crores of its citizens free of cost.

Table 18.7 India's Updated First Nationally Determined Contributions (NDCs) under Paris Agreement

Contribution number	Contribution description
1	Promote a sustainable lifestyle based on conservation and moderation through the 'Lifestyle for Environment' movement.
2	Adopt a climate-friendly, cleaner development path compared to others at similar economic stages.
3	Reduce GDP emissions intensity by 45% by 2030 from 2005 levels.
4	Achieve 50% electric power capacity from nonfossil fuels by 2030, with the help of the transfer of technology and low-cost international finance including from the Green Climate Fund (GCF).
5	Create an additional carbon sink of 2.5–3 billion tonnes CO_2 equivalent through forest and tree cover by 2030.
6	Enhance investments in climate-vulnerable sectors like agriculture, water, Himalayas, coasts, health, and disaster management.
7	Mobilize domestic and international funds for climate mitigation and adaptation actions.
8	Build capacities and frameworks for rapid diffusion of climate technology and joint R&D.

Source: UNFCCC (2022)

18.5.3 India's Updated First Nationally Determined Contribution

Pursuant to decision 1/CP.20, India submitted its intended nationally determined contribution (NDC) to the United Nations Framework Convention on Climate Change (UNFCCC) on 2 October 2015 (Table 18.7). In accordance with decision 1/CP.21, paragraph 22, India's intended NDC is now its first NDC under the Paris Agreement. Article 4, paragraph 9 of the Paris Agreement provides that each party shall communicate a nationally determined contribution every 5 years.

At the country level, the National Action Plan on Climate Change (NAPCC) outlines a national strategy that aims to enable the country to adapt to climate change and enhance the ecological sustainability of India's development path. It stresses that maintaining a high growth rate is essential for increasing the living standards of the vast majority of people in India and reducing their vulnerability to the impacts of climate change. Eight national missions form the core of the national action plan. They focus on promoting an understanding of climate change, adaptation and mitigation, energy efficiency, and natural resource conservation. Among the eight national missions, the national mission for sustainable agriculture aims at making agriculture more productive, sustainable, remunerative, and climate resilient by promoting location-specific integrated farming systems, soil and moisture conservation measures, comprehensive soil health management, efficient water management practices, and mainstreaming rain-fed technologies.

18.5.4 Need for a Coherent National Policy for Climate-Smart Agriculture

Transitioning to climate-smart agriculture necessitates coherence among the various policies designed to support the sustainable increase of agricultural productivity and incomes, as well as climate change adaptation and mitigation. However, aligning these diverse policy processes to address multiple objectives simultaneously is challenging. Developing ad hoc policies without a systematic assessment can lead to inefficiencies in achieving climate-smart agriculture goals.

National policy-making processes are driven by a variety of objectives. When policies are considered and formulated in isolation, they may lead to disconnected actions. Policies are often created in sectoral "silos," but even within a single sector, there is a need to harmonize policy objectives. Agricultural sustainability is increasingly recognized as a priority in India. However, most current policies focus on achieving productivity and profitability gains. The disconnect in policy actions also arises from the competition for natural and financial resources. The different spatial and temporal scales of the problems that policymakers must address, including climate change, create varied implementation incentives. Given the often limited human resources and current socio-political systems, it is understandable that policymakers prioritize urgent issues over long-term ones. For example, in India, governments incentivize groundwater use for agriculture through various schemes to increase food security. While this boosts agricultural production, it also creates significant trade-offs between current and future production possibilities.

18.5.4.1 Evaluating Policy Coherence to Enhance Policy Effectiveness

Enhancing policy coherence requires systematically evaluating existing policies and their impact on the country's prioritized objectives. New policies must address many goals, and regulatory frameworks need to become more holistic and intricate. Prevailing agricultural and nonagricultural policies can have both intended and unintended effects on objectives related to sustainable productivity growth, climate change adaptation, and mitigation. It is crucial to evaluate whether current policies provide consistent signals to farmers and if feedback loops exist to connect farmers' on-the-ground experiences with policymakers for improved policy formulation. By assessing the impacts of policies on multiple objectives, decisions can be made to harness synergies and achieve acceptable trade-offs. A key principle is to create a participatory policy-making process that includes all key stakeholders where possible.

18.5.4.2 Agricultural Planning and Public Expenditure for Climate-Resilient Agriculture

Effective coordination in planning and budgeting for climate-resilient agriculture among various national and local agencies is essential to ensure alignment with national objectives. Cross-sectoral climate-resilient agricultural policies must consider current legislation and regulations, incorporating them into existing legal and regulatory frameworks when necessary. There are numerous opportunities to create synergies, where a single policy instrument or regulation supporting climate-resilient agriculture can simultaneously achieve multiple goals, such as expanding social safety transfers, promoting sustainable energy development, enhancing land-use regulations, and reducing disaster risk.

Annual budgets translate national planning and expenditure frameworks into budget allocations for responsible sectoral agencies and are crucial for implementing climate-resilient agriculture programs or action plans. To achieve the objectives of climate-resilient agriculture, it is vital to align appropriate financing with the agencies tasked with carrying out these actions. This planning phase should include broader national objectives, such as those outlined in NDCs and NAPCC. In addition to investment budgets, it is important to consider budgets for operations and maintenance. Annual planning within a specific sector often relies on a sectoral medium-term plan, such as a time-bound agriculture sector development plan that sets medium-term goals and priorities. Beyond directing investments toward climate-resilient agriculture, it is advisable to climate-proof other budgetary decisions and investments in the agriculture sector. This entails examining investment plans through a "climate change lens" to evaluate how climate change may affect the attainment of investment goals.

18.5.5 Market-Based Instruments for Climate-Resilient Agriculture

18.5.5.1 Taxes and Subsidies on Inputs

Governments assist farmers and agri-businesses by offering various incentives for agricultural production, such as input or output subsidies. Additionally, governments may impose taxes on specific inputs or outputs to discourage environmentally harmful activities or to implicitly promote green alternatives. For example, in 2014, the government of India deregulated diesel prices, which led to the government lifting diesel subsidies. Generally, energy or water use subsidies can boost production but may also encourage unsustainable practices. For instance, energy subsidies have contributed to aquifer depletion in water-scarce countries like India (Ignaciuk, 2015). There is a free electricity supply to agriculture in states like Punjab, Andhra Pradesh, Tamil Nadu, and Telangana, which led to perilous levels of decline in the water table, especially in Punjab. Alternatively, the governments can replace the free supply of electricity to agriculture with direct benefit transfer (DBT), which can abate the problem of the "tragedy of commons."

18.5.5.2 Subsidized Finance to Climate-smart Technologies

Considering the capital-intensive nature of climate-smart agriculture (CSA) technologies, it is crucial to recognize how financial constraints can hinder their adoption. Generally, adoption rates are relatively high, with farmers predominantly renting these technologies from custom-hiring centers. There is significant potential for bolstering CSA technology rentals through government subsidies and cooperatives. Farmers do not need substantial credit or collateral since most CSA technologies are rented rather than owned. However, they require quick access to smaller amounts of credit to afford the rental of CSA technologies (Villalba et al., 2024).

18.5.5.3 Payments for Ecosystem Services

Payment for an ecosystem service is a policy instrument that offers incentives to encourage the adoption of climate-smart agriculture practices. Agricultural ecosystems offer both marketable and nonmarketable ecosystem services. The capacity to provide these services is constrained by the nature of production systems and methods. While input-intensive production systems yield short-term productivity gains, they often lead to long-term ecological and environmental damage. India's agriculture faces numerous ecological challenges, such as groundwater overexploitation, loss of agricultural biodiversity, and soil erosion. Policy measures to improve agriculture should shift from solely focusing on productivity enhancement to promoting ecologically sustainable agricultural production (Manjula et al., 2019).

Even though most of the ecosystem services are valued and recognized by people, they do not have markets, leading to a lack of incentives to provide them that come with their prices (Swinton et al., 2007). There needs to be a policy framework at the country level to incentivize farmers to produce non-marketable ecosystem services. In this aspect, a lot can be learned from the clean development mechanism (CDM)-based payment of money for carbon sequestration in agriculture, an ecosystem service.

18.6 Conditional Social Protection Measures

When designing social protection instruments to enhance the resilience of smallholders to climate-related risks, it is crucial to provide long-term support to facilitate necessary behavioral changes (Scognamillo & Sitko, 2021). One of the challenges in transitioning to climate-resilient agriculture is making sure that it benefits the most vulnerable households and communities. In some regions, climate change is expected to exacerbate the vulnerability of impoverished and food-insecure communities, heightening the importance and need for social protection measures to ensure food security. Channeling of MNERGA works toward works like desilting of ponds, building check dams, planting windbreaks, and construction of drainage channels can help mitigate agriculture from extreme weather events caused by climate change. The benefits of social protection interventions may take time to manifest and often become evident only during disasters.

References

Bhattarai, N., Pollack, A., Lobell, D. B., Fishman, R., Singh, B., Dar, A., & Jain, M. (2021). The impact of groundwater depletion on agricultural production in India. *Environment Research Letter, 16*, 085003. https://doi.org/10.1088/1748-9326/ac10de

CGWB. (2020). *National compilation on dynamics groundwater resources of India*. Central Ground Water Board. Ministry of Jal Shakti. Government of India.

Devineni, N., Perveen, S., & Lall, U. (2022). Solving groundwater depletion in India while achieving food security. *Nature Communications, 13*, 3374. https://doi.org/10.1038/s41467-022-31122-9

Fanzo, J., Davis, C., McLaren, R., & Choufani, J. (2018). The effect of climate change across food systems: Implications for nutrition outcomes. *Global Food Security., 18*, 12. https://doi.org/10.1016/j.gfs.2018.06.001

Government of India. (2022). *Agricultural Statistics at a glance*. Directorate of Economics and Statistics, Ministry of Agriculture and Farmers Welfare.

Government of India. (various years). *Department of Agricultural Research and Education. Expenditure budget. Ministry wise summary of budget provisions*. Ministry of Agriculture and Farmers welfare.

Ignaciuk, A. (2015). *Adapting agriculture to climate change: A role for public policies* (OECD Food, Agriculture and Fisheries Papers 85). OECD Publishing. https://doi.org/10.1787/5js08hwvfnr4-en

India's Updated First Nationally Determined Contribution Under Paris Agreement (2021–2030). Government of India. https://unfccc.int/sites/default/files/NDC/2022-08/India%20Updated%20First%20Nationally%20Determined%20Contrib.pdf

Jain, M., Fishman, R., Mondal, P., Galford, G. L., Bhattarai, N., Naeem, S., Lall, U., Singh, B., & Defries, R. S. (2021). Groundwater depletion will reduce cropping intensity in India. *Science Advances, 7*(9) https://www.science.org/doi/10.1126/sciadv.abd2849

Manjula, M., Venkatachalam, L., Mukhopadhyay, P., & Kumar, L. (2019). Ecosystem services approach for revitalizing agriculture in India. *Current Science*. https://doi.org/10.18520/cs/v116/i5/723-727

NFHS 5. (2021). *National Family Health Survey. Compendium of fact sheets- Key Indicators*. Ministry of Health & Family Welfare. Government of India.

Niti Aayog. (2024). *Working group report on crop husbandry, agriculture inputs, demand and supply*. https://www.niti.gov.in/sites/default/files/202402/Working%20Group%20Report%20on%20Demand%20%26%20Supply%20Final_V9.pdf

Sandeep Kumar, K. A., Gopinath, S. S., Meena, R. S., Srinivasarao, C., Bedwal, S., Jangir, C. K., Mrunalini, K., Jat, R., & Praharaj, C. S. (2023). Pulse-based cropping systems for soil health restoration, resources conservation, and nutritional and environmental security in rainfed agroecosystems. *Frontiers in Microbiology, 13*. https://doi.org/10.3389/fmicb.2022.1041124

Scognamillo, A., & Sitko, N. J. (2021). Leveraging social protection to advance climate-smart agriculture: An empirical analysis of the impacts of Malawi's Social Action Fund (MASAF) on farmers' adoption decisions and welfare outcomes. *World Development., 146*, 105618. https://doi.org/10.1016/j.worlddev.2021.105618

Sekhri, S. (2013). *Missing water: Agricultural stress and adaptation strategies in response to groundwater depletion in India* (Virginia Economics Online Paper 406). University of Virginia. https://econpapers.repec.org/paper/virvirpap/406.htm

Swinton, S. M., Lupi, F., Robertson, G. P., & Hamilton, S. K. (2007). Ecosystems services and agriculture: Cultivating agricultural ecosystems for diverse benefits. *Ecological Economics, 64*(2), 245. https://doi.org/10.1016/j.ecolecon.2007.09.020

Tiwari, H., Naresh, R. K., Kumar, L., Katari, S. K., Tewari, S., Saini, A., Yadav, R. K., & Asati, R. (2022). Millets for food and nutritional security for small and marginal farmers of North West India in context of climate change: A review. *International Journal of Plant & Soil Science*, 1694. https://doi.org/10.9734/ijpss/2022/v34i232594

UNFCCC. (2022). *India's Updated First Nationally Determined Contribution Under Paris Agreement (2021–2030).* Government of India. https://unfccc.int/sites/default/files/NDC/2022-08/India%20Updated%20First%20Nationally%20Determined%20Contrib.pdf

Villalba, R., Joshi, G., Daum, T., & Venus, T. E. (2024). Financing climate-smart agriculture: A case study from the Indo-Gangetic Plains. In *Mitigation and adaptation strategies for global change* (Vol. 29). Kluwer Academic Publishers. https://doi.org/10.1007/s11027-024-10127-3

Watts, N., Adger, W. N., Agnolucci, P., Blackstock, J., Byass, P., Cai, W., Chaytor, S., Colbourn, T., Collins, M., Cooper, A., Cox, P. M., Depledge, J., Drummond, P., Ekins, P., Galaz, V., Grace, D., Graham, H., Grubb, M., Haines, A., Hamilton, I., Hunter, A., Jiang, X., Li, M., Kelman, I., Liang, L., Lott, M., Lowe, R., Luo, Y., Mace, G., Maslin, M., Nilsson, M., Oreszczyn, T., Pye, S., Quinn, T., Svensdotter, M., Venevsky, S., Warner, K., Xu, B., Yang, J., Yin, Y., Yu, C., Zhang, Q., Gong, P., Montgomery, H., & Costello, A. (2015). Health and climate change: policy responses to protect public health. *Lancet, 386*, 1861. https://doi.org/10.1016/S0140-6736(15)60854-6

World food summit. (1996). *Report of the world food summit.* Food and Agriculture Organization of the United Nations. https://www.fao.org/4/w3548e/w3548e00.htm

Sustainable Agriculture and Climate Policy: Integrating Modern Technologies for Crop Stress Mitigation

19

Sayantan Sarkar and Abhijit Rai

Abstract

Agriculture, as we know it, is facing unprecedented challenges due to the impact of climate change. The increasing frequency of extreme weather events, shifting rainfall patterns, and rising temperatures pose a significant threat to food security and the sustainability of crop production on a global scale. This chapter delves into the critical intersection of modern technologies and climate policy as a response to these pressing challenges. It explores the integration of advanced tools, including remote sensing, machine learning, IoT, and artificial intelligence, to monitor, manage, and adapt to the physiological aspects of plant stress. Through a detailed exploration of innovative approaches, it highlights how these technologies can enhance agricultural resilience and align with climate policies. This chapter serves as a practical guide for policymakers, researchers, and practitioners seeking to navigate the evolving landscape of sustainable agriculture in an era of climate uncertainty. It emphasizes the importance of combining advanced technologies with effective policy measures to mitigate the impact of crop stress, ensuring that agriculture remains productive and resilient in the face of changing climatic conditions.

Keywords

Sustainable agriculture · Climate change · Food security · Machine learning

S. Sarkar
Texas A&M AgriLife Research and Extension Center, Corpus Christi, TX, USA

A. Rai (✉)
Department of Plant Sciences, College of Agriculture and Bioresources,
University of Saskatchewan, Saskatoon, SK, Canada
e-mail: abr922@mail.usask.ca

R. Dubey et al. (eds.), *Drought and Heat Stress in Agriculture*,
https://doi.org/10.1007/978-981-96-5735-3_19

19.1 Introduction

Sustainable agriculture is pivotal in addressing the pressing challenge of feeding a growing global population while mitigating environmental degradation. Climate change adds an additional layer of urgency, as shifts in temperature, precipitation patterns, and the frequency of extreme weather events disrupt traditional agricultural practices. According to the Intergovernmental Panel on Climate Change (IPCC), climate change is already affecting agriculture, impacting crop yields and food security (Shukla et al., 2019). It highlights that climate change poses risks to global food production and exacerbates existing challenges.

Agriculture faces multifaceted challenges stemming from climate change, including altered growing seasons, increased pest and disease prevalence, and more frequent extreme weather events. These challenges are compounded by population growth, urbanization, and changing dietary habits, straining global food systems (Calzadilla et al., 2013; Ruane & Rosenzweig, 2019). Studies indicate that climate change is responsible for a decline in global crop yields, with regions such as Africa and South Asia particularly vulnerable. Increased temperatures and altered precipitation patterns contribute to yield variations across different crops (Bello et al., 2012; Kumar et al., 2018).

Modern technologies offer innovative solutions to enhance agricultural resilience in the face of climate change. Tools such as remote sensing (RS), machine learning (ML), the Internet of Things (IoT), and artificial intelligence (AI) empower farmers and policymakers to make informed decisions for sustainable practices (Adhikari et al., 2023). Several studies demonstrate the efficacy of RS in monitoring crop conditions. Satellite imagery allows for the early detection of stress factors, enabling timely interventions to mitigate potential yield losses. ML plays a crucial role in precision agriculture (Sarkar, 2021a, b). Algorithms analyze vast datasets to identify patterns related to crop health, optimize resource allocation, and reduce environmental impact (Sarkar & Jha., 2020). The IoT facilitates real-time data collection through sensor networks in agriculture. This continuous monitoring enables adaptive management practices, responding promptly to changing environmental conditions. AI provides advanced decision-support systems. AI algorithms analyze complex interactions within agroecosystems, offering insights into optimal crop management strategies for stress mitigation (Chapu et al., 2022).

In this chapter, we will explore each of these technologies in greater detail, examining their applications, successes, and challenges. By synthesizing scientific findings, this chapter aims to provide a comprehensive guide for stakeholders seeking sustainable and climate-resilient agricultural solutions.

19.2 Climate Change and Its Impact on Agriculture

Climate change is an intricate web of long-term shifts in temperature, precipitation patterns, and other climatic parameters. Human activities, predominantly the burning of fossil fuels and deforestation, have significantly increased the concentration

Table 19.1 Different aspects of climate change and their stress symptoms on crops

Aspect	Impact	Call to action
Temperature changes	Warming temperatures affect crop maturation and growing seasons	Develop resilient crop varieties that can withstand temperature fluctuations. Implement precision agriculture for adaptive management
Extreme weather events	Increased frequency and intensity of events disrupt supply chains	Emphasize sustainable water management practices. Implement adaptive strategies, such as precision agriculture, to respond to changing conditions
Rainfall variability	Unpredictable patterns lead to crop failures and reduced yields	Invest in research for climate-resilient crops through breeding programs. Implement precision agriculture to optimize resource usage based on rainfall patterns
Heat stress	Affects photosynthesis and reduces grain yield in certain crops	Explore and implement adaptive strategies, such as breeding heat-tolerant crop varieties. Invest in precision agriculture for efficient resource management
Disease dynamics	Warmer temperatures favor pests and pathogens, impacting plant health	Develop strategies for disease-resistant crops. Implement precision agriculture for real-time monitoring and targeted interventions
Drought conditions	Limit water availability, affecting nutrient absorption and crop health	Implement sustainable water management practices. Invest in technologies that enhance water use efficiency and resilience to drought
Photosynthetic response to heat stress	Differential impact on C3 and C4 plants in terms of photosynthetic efficiency	Prioritize research on heat-tolerant varieties, especially for C3 plants. Promote precision agriculture practices for efficient resource use in varying temperature conditions
Overall agricultural resilience	Multifaceted challenges necessitate adaptive strategies	Advocate for the integration of advanced technologies in agriculture. Support breeding programs for climate-resilient crops. Emphasize precision agriculture for sustainable practices

of greenhouse gases in the atmosphere, contributing to the warming of the planet. Studies confirm that concentrations of carbon dioxide, methane, and nitrous oxide have reached unprecedented levels since the preindustrial era (Shukla et al., 2019). These greenhouse gases trap heat in the atmosphere, leading to global warming and alterations in climate patterns.

Agriculture, being intimately tied to weather patterns, is highly susceptible to the impacts of climate change (Table 19.1). Shifts in temperature, alterations in precipitation, and the increasing frequency of extreme weather events pose significant challenges to crop production and food security. Past studies highlight that climate change is affecting not only the average temperature but also the variability in temperature, which can have profound implications for crop development and yields (Thornton et al., 2014). Extreme weather events, such as hurricanes, droughts, and floods, are becoming more frequent and intense. The IPCC Special Report on

Managing the Risks of Extreme Events and Disasters to Advance Climate Change Adaptation (2012) underscores the vulnerability of agriculture to such events, leading to disruptions in supply chains and food production systems (Murray & Ebi, 2012). For example, changes in rainfall patterns disrupt the growth cycles of crops. Excessive rainfall can lead to waterlogged soils, affecting root health and nutrient absorption. Conversely, extended periods of drought stress plants, limiting their capacity to photosynthesize and produce essential nutrients. A study by Kyei-Mensah et al. (2019) observed that increased rainfall variability negatively affects crop yields in various regions, emphasizing the vulnerability of agriculture to unpredictable precipitation. Unpredictable rainfall directly influences crop yields. Erratic patterns can lead to crop failures, reduced yields, and economic losses for farmers (Högy et al., 2013; Lüttger et al., 2018; Liu et al., 2021). The studies also found that extreme precipitation events can result in significant reductions in wheat and barley yields, impacting food production and availability.

Elevated temperatures influence the physiological processes of plants, affecting growth and development. Higher temperatures can accelerate plant maturation, altering the timing of flowering and fruiting (Bita & Gerats, 2013). They can lead to changes in the duration of the growing season, affecting the timing of key developmental stages in crops such as corn and soybeans. Warming temperatures have a direct correlation with reduced crop yields (Jha et al., 2014). Heat stress during critical developmental stages can lead to diminished grain quality and quantity. Current research found that for every 1 °C increase in temperature, there is a significant decline in global crop yields, underscoring the vulnerability of agriculture to warming climates (Jagadish et al., 2020).

Climate change can alter the prevalence and distribution of plant diseases. Warmer temperatures and changes in humidity create favorable conditions for the spread of pests and pathogens. Studies indicate that increased temperatures can enhance the severity of diseases like rusts and blights in crops, affecting overall plant health (Luck et al., 2011). Elevated disease severity can lead to significant reductions in crop yields. Plants under stress from diseases are less productive and more susceptible to other environmental stressors (Newton et al., 2011). Climate-induced changes in disease dynamics contribute to substantial losses in crop yields, particularly in staple food crops like rice and wheat.

Drought situations have an impact on plant water uptake. Drought situations limit water availability, impacting plant water uptake and, consequently, nutrient absorption, and overall plant health (Rai et al., 2022). The increasing frequency of drought events globally and their adverse effects on crop production are prevalent, especially in regions reliant on rain-fed agriculture (Bennett et al., 2021). Prolonged drought conditions are associated with significant reductions in crop yields, affecting food security and economic stability (Rai et al., 2020). Drought-induced yield losses in major crops like maize, wheat, soybean, and rice are substantial, posing challenges for global food production (Sarkar et al., 2022).

Heat stress significantly affects plant physiology, particularly in the context of photosynthesis. C3 plants, like wheat and soybeans, experience challenges in the Calvin cycle—the central process of photosynthesis—due to heat stress. In contrast,

C4 plants, such as maize and sugarcane, exhibit more resilience by employing specific mechanisms to counter the impact of elevated temperatures (Hamilton et al., 2008; Wang et al., 2008). C3 plants struggle with increased heat. The Calvin cycle, a crucial part of photosynthesis, becomes less efficient under heat stress. This results in reduced photosynthetic activity, especially during reproductive stages (Balota & Sarkar, 2020; Kumar et al., 2017). The same study shows that C4 plants respond more robustly to heat stress. These plants maintain higher rates of carbon fixation even in elevated temperatures (Crafts-Brandner et al., 2002). Their specific adaptations help them navigate heat challenges more effectively. The influence of heat stress extends to crop yield. For C3 plants, compromised photosynthesis can lead to reduced grain yield. Conversely, C4 plants, with their adaptive mechanisms, often maintain a more stable yield under heat stress (Killi et al., 2017). Therefore, C3 plants, such as wheat, show yield implications that indicate reduced productivity, especially in staple crops, while C4 plants exhibit a more stable yield, even in the face of heat stress.

Therefore, the impacts of climate change on agriculture are multifaceted, affecting plant growth, yield, disease dynamics, and exacerbating drought situations. These examples underscore the urgency of implementing adaptive strategies and integrating advanced technologies to mitigate the adverse effects on global food security. Recognizing the inevitability of ongoing climate change, there is an urgent call for adaptive strategies in agriculture. These strategies involve a multifaceted approach, including the development of resilient crop varieties, sustainable water management practices, integration and domestication of newer and multipurpose crops, and the integration of advanced technologies to monitor and respond to changing conditions (Baath et al., 2023, 2024; Sarkar & Islam, 2016, 2017). They signify the importance of adaptive strategies in agriculture, including the development of climate-resilient crops through breeding programs and the implementation of precision agriculture practices (Patil et al., 2016; Sarkar, 2020). The need for adaptive strategies demonstrates that even with aggressive mitigation efforts, a certain degree of climate change is inevitable, requiring proactive adaptation in agricultural systems (Farber, 2006). As a result, we need to explore how modern technologies serve as integral components of these adaptive strategies, offering precision and efficiency in managing the complex challenges posed by a changing climate. This intersection is pivotal for developing holistic approaches that ensure the sustainability and resilience of agriculture in the face of evolving climatic conditions.

19.3 Advanced Tools for Monitoring and Management

In the pursuit of sustainable agriculture amid climate challenges, the integration of advanced tools for monitoring and management becomes a linchpin. The arsenal of modern technologies is a profound one, elucidating their roles in climate monitoring, adaptive management, and the enhancement of agricultural resilience (Table 19.2).

Table 19.2 Advanced tools, their types, and usage for crop stress management

Advanced tools	Types	Usage
Remote sensing	Satellite imagery Aerial multispectral, hyperspectral, and thermal sensors	Provide a comprehensive overview of large agricultural landscapes, detect changes in vegetation, soil moisture, and other environmental factors
Machine learning	Crop yield forecasting Disease prediction Proactive management	Analyze historical data on climate conditions, soil quality, and crop performance to predict future yields and enhance adaptive capacity in disease management
Internet of things	Connectivity across agricultural landscape Dynamic information gathering	Connect sensors, devices, and machinery to collect real-time data on climate conditions, enabling automated responses, and enhancing overall agricultural resilience
Artificial intelligence	Big data Processing vast datasets Real-time decision-making Variable rate application	Process extensive datasets, understand nuanced plant responses to stress, and enable real-time evolution of adaptive management strategies in agriculture

19.3.1 Remote Sensing Technologies

RS technologies stand at the forefront of modern agriculture, offering invaluable capabilities for climate monitoring. These tools, encompassing satellite imagery and unmanned aerial vehicles (UAVs), revolutionize our ability to observe, measure, and respond to dynamic environmental conditions (Sarkar et al., 2022).

Satellite Imagery: Satellite imagery provides a comprehensive overview of large agricultural landscapes, capturing data at different spectral bands. This technology enables the detection of subtle changes in vegetation health, soil moisture levels, and other environmental factors (Karthikeyan et al., 2020). Satellite-based monitoring can be harnessed to study the impact of unpredictable rainfall on crop yields. By analyzing satellite data, researchers can identify regions experiencing water stress, allowing for timely interventions such as irrigation adjustments or crop diversification.

Unmanned Aerial Vehicles (UAVs): UAVs, commonly known as drones, offer a more granular and agile perspective compared to satellites. They can be deployed at lower altitudes, capturing high-resolution imagery with greater frequency. This flexibility is particularly valuable for precision agriculture applications (Sarkar et al., 2020). In the context of climate monitoring, UAVs provide detailed insights into smaller agricultural plots, offering a localized view of temperature variations, plant stress, and pest infestations. Current research underscores the significance of monitoring drought conditions through UAV-based assessments, allowing for targeted responses to mitigate crop losses (Balota et al., 2021).

Real-Time Insights and Proactive Responses: One of the key strengths of RS technologies is their ability to provide real-time insights into climate patterns. Continuous monitoring allows farmers and policymakers to track changes in weather conditions, identify anomalies, and predict potential challenges. For instance, if satellite data indicates a deviation from normal rainfall patterns, farmers can implement water conservation measures or adjust planting schedules to optimize crop performance (Yang et al., 2013).

Connecting Climate Monitoring to Adaptive Strategies: Recent studies emphasize the role of RS as a sentinel in the face of climate uncertainties. Unpredictable rainfall and drought, identified through RS, directly impact crop yields (Levizzani & Cattani, 2019). By integrating these technologies into adaptive management strategies, farmers and stakeholders can take proactive measures to enhance resilience. Whether it is modifying irrigation practices, selecting drought-tolerant crop varieties, or adjusting planting dates, the insights derived from RS become actionable steps toward climate-resilient agriculture (Dedieu et al., 2014).

19.3.2 Applications of Machine Learning

ML, a powerful subset of AI, emerges as a transformative force in agriculture, leveraging predictive modeling and data analysis to enhance resilience in the face of climate challenges.

Crop Yield Forecasting: One of the primary applications of ML in agriculture is crop yield forecasting. By analyzing historical data on climate conditions, soil quality, and crop performance, ML models can predict future yields with remarkable accuracy (Sarkar et al., 2021). This predictive capability aids farmers in optimizing resource allocation, making informed decisions on planting, irrigation, and harvesting. The insights derived from these models contribute to adaptive strategies that mitigate the impact of climate stress on crop productivity.

Disease Prediction and Proactive Management: ML algorithms prove invaluable in addressing the challenges posed by increased disease severity under climate stress (Chapu et al., 2022). These algorithms excel in identifying intricate patterns in data, enabling early detection of diseases in crops. By analyzing environmental variables, such as temperature, humidity, and precipitation, alongside crop health indicators, ML models can predict the onset of diseases before visible symptoms appear. This foresight allows for proactive management strategies, such as targeted pesticide application or crop rotation, minimizing the impact of diseases on overall crop health and yield.

Enhanced Adaptive Capacity: The integration of ML into agriculture heightens the adaptive capacity of farming practices. Traditional approaches to disease management often rely on visible symptoms, leading to reactive responses. ML, however, enables a shift toward proactive and data-driven decision-making (Sarkar et al., 2021). As these algorithms continuously learn from new data inputs, they become adept at recognizing evolving patterns and trends, empowering farmers to stay ahead of potential challenges.

Connecting ML to Climate-Responsive Strategies: The insights derived from ML models create a bridge between climate stress and adaptive strategies. By understanding the patterns that precede adverse climate conditions or disease outbreaks, farmers can implement timely interventions (Sarkar et al., 2022). For example, if an ML model predicts a higher likelihood of disease based on prevailing weather conditions, farmers can implement targeted measures to protect crops (Chapu et al., 2022). This proactive approach aligns to build resilience in agriculture, ensuring that the adaptive strategies are not only responsive but also anticipatory.

19.3.3 IoT in Agriculture and Its Role in Climate Resilience

The IoT weaves a web of connectivity throughout the agricultural landscape, intertwining sensors, devices, and machinery to bolster climate resilience. The IoT transforms agriculture into a dynamic network, perpetually collecting and disseminating information for adaptive and responsive strategies (Zecca, 2019).

Connectivity Across the Agricultural Landscape: The essence of IoT lies in the interconnectedness it fosters among various components of agriculture. Sensors embedded in fields, on machinery, and within facilities collect a myriad of data points, ranging from soil moisture levels to machinery performance (Ahmed et al., 2022; Mehmood et al., 2022). This interconnected system forms a real-time network that communicates seamlessly, offering a holistic view of the agricultural environment.

Dynamic Information Gathering and Dissemination: In the context of climate resilience, IoT operates as an active and dynamic information hub. Studies have shown that the vulnerability of C3 plants like wheat to heat stress finds practical applications through IoT. Real-time data on temperature and humidity, collected by sensors deployed throughout the fields, are continuously relayed to a central system (Hazmy et al., 2023). This information is then disseminated to stakeholders, providing a comprehensive understanding of the current climate conditions.

Automated Responses for Climate Mitigation: One of the significant strengths of IoT lies in its ability to trigger automated responses based on real-time data. In the face of heat stress, for instance, the IoT system can initiate automated irrigation to maintain optimal soil moisture levels (Symeonaki et al., 2019). Similarly, shading mechanisms can be activated to mitigate the impact of excessive heat on vulnerable crops. These automated responses serve as proactive measures and contribute to climate-resilient agriculture.

Enhancing Crop Health and Productivity: The practical applications of IoT extend beyond mere data collection. By harnessing real-time information on climate conditions, farmers can implement precise and targeted interventions to enhance crop health and productivity (Shukla et al., 2021). IoT's ability to facilitate timely responses to changing environmental factors contributes to the overall resilience of agriculture in the face of climate uncertainties.

Connecting IoT to Adaptive Strategies: The role of IoT in climate resilience is intrinsically tied to adaptive strategies. The continuous flow of real-time data enables stakeholders to make informed decisions swiftly. If the IoT system detects conditions conducive to heat stress, it can trigger responses that mitigate potential damage (Jabbari et al., 2023). This connection between data-driven insights and adaptive strategies underscores the transformative impact of IoT on the agricultural landscape.

19.3.4 Integration of Artificial Intelligence for Adaptive Management

AI emerges as a cornerstone in the realm of adaptive management within agriculture, ushering in an era where decisions transcend mere data-driven analyses, becoming intelligent and context-aware. AI, through sophisticated algorithms, navigates the experiences of C3 and C4 plants under heat stress for adaptive management (Dadios et al., 2022).

AI in Adaptive Management: The infusion of AI into adaptive management heralds a paradigm shift. Decisions cease to be static responses to data points; instead, they become dynamic, evolving in real time based on intelligent algorithms. This adaptability is crucial in the context of agriculture, where the impacts of climate stress are intricate and ever-changing (Geli et al., 2019). The experiences of C3 and C4 plants under heat stress, previously explored, become focal points for nuanced considerations within the adaptive management framework (Chandrashekharaiah et al., 2021).

Processing Vast Datasets: AI algorithms possess the capability to process vast datasets, encompassing an array of variables crucial for adaptive management. These datasets may include plant physiological responses, climate data, historical performance metrics, and more. The intricate dance between various factors is analyzed in real time, allowing AI to discern patterns, trends, and anomalies that might elude traditional approaches (Garrett et al., 2022).

Considerations for Plant Physiological Responses: The adaptive approach facilitated by AI delves deep into plant physiological responses. Understanding how plants react to heat stress, how their metabolic processes adjust, and how these responses influence overall growth and yield becomes integral. The work of Sage and Kubien (2007), which highlights the resilience of C4 plants under heat stress, serves as a blueprint for AI algorithms to consider such nuanced physiological responses in real-time decision-making.

Context-Aware Decision-Making: AI's true prowess in adaptive management lies in its context-aware decision-making capabilities. The algorithms not only process data but interpret it in the context of the specific agricultural ecosystem. This means considering the unique characteristics of crops, the intricacies of local climates, and the historical performance of agricultural practices. The adaptive strategies, exemplified by Sage and Kubien's (2007) study on C4 plants, ensure that decisions align with the changing dynamics of climate stress.

Real-Time Evolution of Management Strategies: The crux of AI in adaptive management lies in its ability to enable real-time evolution of strategies. As climate stress patterns change, AI algorithms continually adjust management approaches (Sarkar and Jha, 2025). This ensures that interventions, be they related to irrigation, fertilization, or pest control, remain synchronized with the dynamic conditions of the agricultural environment (Tobal & Mokhtar, 2014).

The tools presented here are not mere innovations; they are practical manifestations of the symbiotic relationship between technological advancements and policy imperatives. As we scrutinize RS, ML, IoT, and AI in the context of agriculture, the overarching goal remains clear—to fortify agriculture against the impacts of climate change, ensuring a resilient and adaptive future for global food production.

19.4 The Intersection of Modern Technologies and Climate Policy

As we navigate the intersection of modern technologies and climate policy, it becomes evident that the challenges posed by climate change demand a strategic and integrated response. The synergy between technological advancements and policy imperatives are essential, emphasizing the critical need for their seamless integration.

Synergy Between Technology and Policy: The confluence of modern technologies and climate policy represents a strategic alliance poised to address the multifaceted challenges posed by climate change in agriculture (Aryal et al., 2020). This synergy is not merely a collaboration but a dynamic framework where advancements in technology offer the means to implement and enhance policy objectives. Gupta and Vegelin's (2016) call for integrated approaches resonates, emphasizing that technological innovations must align with policy goals to achieve effective climate change mitigation and adaptation in agriculture. The synergy becomes a strategic interplay, where the precision of technology finds purpose and direction within the broader context of climate policy objectives.

Importance of Integrating Technological Advancements with Policy Measures: The significance of seamlessly integrating technological advancements with policy measures cannot be overstated. While technology provides the tools for precision agriculture, climate monitoring, and adaptive strategies, effective policies serve as the guiding principles that harness the potential of these tools (Sarkar et al., 2025). Ritchie and Dowlatabadi's (2018) research underscores the importance of aligning technological advancements with policy frameworks for sustainable agricultural practices. Their work emphasizes that technological interventions, when synchronized with well-designed policies, can significantly enhance the resilience of agricultural systems. In essence, the integration ensures that technological innovations are strategically deployed to achieve overarching policy goals, fostering a harmonized and impactful response to climate challenges in agriculture (Meena et al., 2015).

Connecting Climate Change Impacts to Technological Solutions: The narrative of climate change impacts on agriculture serves as a backdrop for understanding the role of modern technologies. Unpredictable rainfall, heat stress, and increased disease severity, all consequences of climate change, necessitate advanced tools for monitoring and management (Balasundram et al., 2023). The examples of RS, ML,

IoT, and AI, discussed in detail in the previous section, become practical illustrations of how technology can be harnessed to address the specific challenges induced by climate change (Durán-Sandoval et al., 2023).

Setting the Stage for the Subsequent Exploration: With the understanding of the synergy established, the subsequent exploration promises an in-depth dive into the specific roles of modern technologies within the context of existing climate policies (Lipper & Zilberman, 2018). RS, ML, IoT, and AI are the protagonists in this narrative. Each technology will be scrutinized, not in isolation but within the intricate dance of existing climate policies. The aim is to dissect their applications, successes, and challenges, providing practical insights for policymakers, researchers, and practitioners (Calicioglu et al., 2019). This exploration goes beyond theoretical discourse; it delves into the practical intricacies of implementation (Kuhl., 2020). By examining the nuanced interactions of these technologies within the context of climate policies, we seek to unravel strategic pathways (Silici et al., 2021). The successes serve as guiding lights, while the challenges offer opportunities for refinement and innovation. This detailed exploration becomes a compass for those navigating the evolving landscape of sustainable agriculture in an era defined by climate uncertainty.

19.5 Policy Frameworks for Climate-Resilient Agriculture

Amidst the escalating challenges of climate change, the pivotal role of policy frameworks in steering agriculture towards resilience becomes unmistakable. This exploration unfolds with an overview of existing climate policies related to agriculture, delving into their gaps and areas for improvement, and culminating in the proposal of an integrated policy framework harmonizing with advancements in modern technologies.

A foundational understanding of the current landscape is crucial in shaping effective policy responses (Pahl-Wostl, 2007). Existing climate policies related to agriculture form a mosaic of strategies implemented at local, national, and international levels. Initiatives such as the Paris Agreement underscore the global commitment to building climate-resilient agricultural systems (Dröge, 2016). Nationally, countries have devised policies tailored to their specific climate challenges and agricultural landscapes. For instance, the Climate Smart Agriculture (CSA) initiative emphasizes the integration of climate considerations into farming practices, aiming for sustainable and adaptive agriculture (Ngoma, 2018).

While strides have been made in integrating climate considerations into agricultural policies, critical gaps and areas for improvement persist. One significant gap lies in the need for more granular and adaptive policies that account for regional variations in climate impacts. The disconnect between policy objectives and on-the-ground implementation poses another challenge. Additionally, policies often lack integration with cutting-edge technologies, limiting their effectiveness in the face of rapidly changing climate conditions (Lybbert & Sumner, 2010, 2012). These studies identify a gap in current policies related to adaptive measures in the face of shifting rainfall patterns. The study suggests the need for policies that actively incorporate data from advanced technologies for better adaptation.

In response to identified gaps, a proposal emerges for an integrated policy framework that harmonizes with technological advancements. This framework envisions policies that are adaptive, region-specific, and responsive to the dynamic nature of climate challenges (Chaterji et al., 2020). It recognizes the role of modern technologies as enablers for achieving policy objectives. The integration of RS in climate monitoring, ML in predictive modeling, IoT in real-time data gathering, and AI in adaptive management becomes a central theme (Shaikh et al., 2022). Such a framework not only addresses current gaps but also positions agriculture to thrive in the face of evolving climate uncertainties. The proposal for an integrated policy framework bridges the traditional gap between policy and technology (Araújo et al., 2023). It envisions policies that actively leverage the potential of modern technologies, transforming them from mere tools into integral components of climate-resilient agriculture. By integrating real-time data from advanced tools, policymakers gain insights into the effectiveness of implemented measures (Popescu et al., 2022). For instance, RS can provide feedback on the success of sustainable farming practices, and ML algorithms can inform adjustments to predictive models based on evolving climate patterns (Sun et al., 2019). The CSA initiative, with its emphasis on adaptive practices and sustainability, serves as a model for the integration of policy and technology to achieve climate-resilient agriculture (Ngoma, 2018; Kumar et al., 2022).

Therefore, the policy frameworks for climate-resilient agriculture are at a crossroads. The existing policies provide a foundation, but they require adaptation and augmentation to address current gaps and leverage the potential of modern technologies. The proposal for an integrated policy framework represents a forward-thinking approach—one that envisions a harmonious synergy between policy objectives and technological advancements. The aim is not only to build resilience in agriculture but to create a blueprint for sustainable and adaptive agricultural systems in a world shaped by climate uncertainty.

19.6 Practical Applications and Implementation

Navigating beyond theoretical realms, the need of the hour is tangible integration of advanced tools in agriculture. Offering practical insights, guidelines for seamless integration is required, underscoring the significance of scalability and adaptability across diverse agricultural settings, and tackle challenges with potential solutions.

As technology evolves, implementing advanced tools in agriculture requires a strategic approach. Guidelines for successful integration include establishing clear objectives, ensuring accessibility for end-users, and providing training programs. For instance, initiatives like the FAO's Digital Extension Guidelines offer a blueprint for incorporating technology into extension services, fostering knowledge transfer and capacity building (Salokhe et al., 2004). Scalability and adaptability are pivotal considerations in deploying advanced tools across different agricultural settings. A one-size-fits-all approach is impractical given the diversity of farming practices globally. The framework should be flexible enough to accommodate variations in climate, soil types, and socio-economic conditions. Successful case studies, such as the implementation of precision agriculture in diverse agroecosystems (Smith et al., 2021), underscore the importance of tailoring technological interventions to specific contexts.

The practical application of advanced tools in agriculture is not without challenges. Issues of data privacy, infrastructure limitations, and resistance to change can impede widespread adoption (Rathore et al., 2022). However, potential solutions exist. Collaborative efforts between governments, private sectors, and research institutions, as seen in the Agricultural Innovation Partnership (AIP), can address infrastructure gaps and foster a conducive environment for technology adoption (Jowi & Mbwette, 2017; Alfaro d'Alençon et al., 2023). Additionally, targeted awareness campaigns and capacity-building programs can mitigate resistance and facilitate smoother integration into existing agricultural practices. Therefore, the practical applications and implementation of advanced tools in agriculture necessitate a comprehensive and context-specific approach (De Olde et al., 2016). By adhering to guidelines, considering scalability and adaptability, and addressing challenges head-on with viable solutions, agriculture can harness the full potential of modern technologies. This serves as a guide for policymakers, researchers, and practitioners seeking to navigate the intricacies of incorporating advanced tools into real-world agricultural landscapes (Messerli et al., 2019).

19.7 Case Studies and Success Stories

The profound impact of modern technologies on agriculture is vividly illustrated through compelling case studies and success stories (Table 19.3). They serve as a tangible testament to the integration of technology for crop stress mitigation and its profound implications for both sustainable agriculture and climate policy.

19.7.1 Precision Agriculture in the Netherlands

In the Netherlands, precision agriculture represents a paradigm shift in farming practices. Through the strategic use of GPS-guided tractors and advanced sensors, farmers can precisely apply fertilizers and irrigation, optimizing resource utilization (Eastwood et al., 2017; Roep, 2019). This not only minimizes waste but also maximizes crop yield, contributing to the overall efficiency of agricultural operations. Precision agriculture aligns seamlessly with climate policy objectives by prioritizing resource efficiency. The reduction in fertilizer and water usage not only translates into economic benefits for farmers but also addresses environmental concerns. This resource optimization aligns with broader climate policy goals promoting responsible and sustainable agricultural practices.

Table 19.3 Details of cases studies where new technologies were used for sustainable agriculture and policymaking

Case studies	Tools used	Significance
Precision agriculture in the Netherlands	GPS-guided tractors, advanced sensors	Optimizing resource utilization, minimizing waste, maximizing crop yield, contributing to overall efficiency, aligning with climate policy objectives
ML for crop yield forecasting in the United States	Machine Learning, historical data, real-time climate conditions	Empowering farmers with insights for proactive decisions, strategic planning, resource allocation, contributing to climate policy goals related to food security and adaptive agriculture
Digital Extension Program in India	Mobile technology, real-time access to information, expert advice	Knowledge transfer, informed decision-making, promoting sustainable farming practices, aligning with climate policy goals for sustainable agriculture
Large-scale AI integration in Australia	AI algorithms, real-time data on plant physiology, climate conditions	Dynamic adjustment of farming strategies, enhancing agricultural resilience, aligning with climate policy goals for adaptive management in changing environmental conditions
Agroforestry Practices in Brazil	Integration of trees and shrubs into agricultural landscapes	Enhancing biodiversity, improving soil health, creating microclimates, mitigating climate impact on crop stress, aligning with climate policy goals for sustainable and adaptive agriculture
Drones for crop monitoring in Rwanda	Drones with high-resolution cameras and sensors	Revolutionizing crop monitoring, providing real-time data, identifying stress factors, promoting precision agriculture, reducing pesticide use, aligning with climate policy goals for sustainable practices
Smart irrigation systems in Israel	Sensors, weather data, algorithms	Optimizing irrigation schedules, conserving water resources, enhancing crop resilience, aligning with climate policy goals for water conservation and efficient resource use
Blockchain technology for supply chain transparency in Ghana	Blockchain technology in agricultural supply chains	Enhancing transparency, traceability, fair trade practices, reducing waste, promoting sustainable farming methods, aligning with climate policy goals for ethical and sustainable agriculture

19.7.2 ML for Crop Yield Forecasting in the United States

In the United States, a forward-thinking farming cooperative has embraced ML for accurate crop yield forecasting. By analyzing historical data and real-time climate conditions, farmers gain valuable insights into anticipated yields (Crane-Droesch, 2018; Shahhosseini et al., 2021). This empowers them to make proactive decisions, strategically planning planting schedules and resource allocation. The integration of ML in crop yield forecasting contributes significantly to climate policy objectives related to food security. Accurate forecasts enable better planning, reducing the risk of food shortages and aligning with climate policies emphasizing adaptive agriculture in the face of changing environmental conditions.

19.7.3 Digital Extension Program in India

In India, a digital extension program leveraging mobile technology has become a catalyst for knowledge transfer among farmers. Real-time access to agricultural information, expert advice, and market updates equips farmers with the knowledge needed to make informed decisions (Malik, 2023; Rao, 2007). This, in turn, promotes sustainable farming practices. The digital extension program aligns with climate policy objectives by fostering sustainable practices. Empowering farmers with knowledge enhances their adaptive capacity, enabling them to make informed decisions aligned with climate-resilient agriculture (Sarkar et al., 2014a, b). This grassroots approach contributes to the broader goals of climate policies related to sustainable farming.

19.7.4 Large-Scale AI Integration in Australia

Australia's agricultural initiative stands as a testament to the transformative power of AI in adaptive management. Real-time data on plant physiology, climate conditions, and historical performance are processed by AI algorithms to dynamically adjust farming strategies (Lockie et al., 2020; Taghikhahet al., 2022). This ensures that agricultural practices evolve in harmony with changing environmental dynamics. The integration of AI for adaptive management directly advances climate policy goals by enhancing agricultural resilience. Real-time adjustments based on climate data contribute to the adaptive strategies outlined in climate policies, ensuring that farming practices remain responsive to the challenges posed by climate change.

19.7.5 Agroforestry Practices in Brazil

In Brazil, agroforestry practices have emerged as a sustainable solution to mitigate the impact of climate change on crop stress. By integrating trees and shrubs into agricultural landscapes, farmers enhance biodiversity, improve soil health, and create microclimates that shield crops from extreme weather conditions (Levidow et al., 2021; Vosti et al., 1998). Agroforestry aligns with climate policy objectives by promoting biodiversity, sequestering carbon, and enhancing overall ecosystem resilience. The integration of trees into farming practices contributes to climate change mitigation efforts and supports the broader goals of sustainable and adaptive agriculture.

19.7.6 Drones for Crop Monitoring in Rwanda

In Rwanda, the utilization of drones for crop monitoring has revolutionized agricultural practices. Drones equipped with high-resolution cameras and sensors provide real-time data on crop health, identifying stress factors such as pests, diseases, and nutrient deficiencies (Sylvère & D'amour, 2020; Dinkar et al., 2024; Mujawamariya, 2021). This proactive monitoring enables farmers to implement targeted interventions. The use of drones for crop monitoring contributes to climate policy goals by promoting precision agriculture and reducing the need for indiscriminate pesticide use. This technology aids in sustainable farming practices, aligning with climate policies focused on minimizing environmental impact.

19.7.7 Smart Irrigation Systems in Israel

Israel's adoption of smart irrigation systems exemplifies a technological approach to address water scarcity and crop stress. These systems utilize sensors, weather data, and algorithms to optimize irrigation schedules, ensuring that crops receive the right amount of water (Gulati et al., 2021; Tal, 2019). This not only conserves water resources but also enhances crop resilience. Smart irrigation systems align with climate policy objectives related to water conservation and efficient resource use. By minimizing water wastage and maximizing irrigation efficiency, this technology contributes to climate-resilient agriculture and supports broader sustainability goals.

19.7.8 Blockchain Technology for Supply Chain Transparency in Ghana

In Ghana, the integration of blockchain technology in agricultural supply chains enhances transparency and traceability. Farmers and consumers can track the journey of agricultural products, ensuring fair trade practices, reducing waste, and promoting sustainable farming methods (Bai et al., 2022; Boakye et al., 2023; Musah et al., 2019). Blockchain technology contributes to climate policy goals by fostering transparency and accountability in the agricultural supply chain. This supports initiatives focused on sustainable and ethical farming practices, aligning with broader climate policy objectives.

In essence, these case studies underscore not only the technological advancements mitigating crop stress but also how these innovations align with and contribute to the broader objectives outlined in climate policies. From resource optimization to adaptive agriculture, these examples showcase the multifaceted contributions of modern technologies to building a resilient and sustainable agricultural future.

19.8 Building Agricultural Resilience

The integration of technologies and policy measures underscores the transformative potential to build resilience in agriculture against the challenges posed by climate change. These insights, drawn from diverse case studies and success stories, offer valuable recommendations for stakeholders, including policymakers, researchers, and practitioners.

Synthesis of Key Findings: The integration of modern technologies, such as precision agriculture, ML, drones, smart irrigation systems, and blockchain, with effective policy measures presents a dynamic framework for addressing crop stress and building resilience in agriculture. The case studies from the Netherlands, the United States, India, Australia, Brazil, Rwanda, Israel, and Ghana collectively highlight the versatility and effectiveness of technological interventions in mitigating the impact of climate change on agricultural systems.

19.8.1 Recommendations for Building Resilience

Holistic Technology Integration: Stakeholders are encouraged to adopt a holistic approach to technology integration, considering the diversity of agricultural landscapes (Sayer et al., 2013). Tailoring technological solutions to specific regional and crop-related challenges ensures effective resilience building.

Capacity Building: Policymakers should prioritize capacity-building initiatives to empower farmers with the necessary knowledge and skills to leverage advanced technologies (Loiruck, 2013). Training programs and digital literacy initiatives can enhance the adoption and impact of technological interventions.

Policy Framework Enhancement: Policymakers play a pivotal role in enhancing existing policy frameworks to accommodate and incentivize the adoption of sustainable agricultural technologies (North & Smallbone, 2006). Aligning policies with climate goals ensures a supportive environment for the implementation of adaptive and resilient farming practices.

International Collaboration: Given the global nature of climate change, international collaboration is paramount (Aldy et al., 2003). Policymakers and researchers should engage in knowledge exchange and collaborative efforts to share best practices, technologies, and policies that contribute to building agricultural resilience worldwide.

Research Investment: Continuous investment in research is essential to further refine and advance agricultural technologies (Sims et al., 2010). Researchers should focus on developing innovative solutions, addressing emerging challenges, and conducting longitudinal studies to assess the long-term impact of technological interventions on crop stress mitigation and resilience.

19.8.2 The Role of Stakeholders

Policymakers: Policymakers play a central role in shaping the regulatory environment. By crafting policies that support sustainable agricultural practices, incentivize technology adoption, and address the unique challenges faced by farmers, policymakers contribute significantly to building resilience in agriculture (Chatrchyan et al., 2017; Soriano et al., 2023). They need to actively engage with the scientific community, local governments, and international organizations to create comprehensive policies that foster innovation and address systemic challenges.

Researchers: Researchers are at the forefront of innovation, contributing through continuous exploration of cutting-edge technologies and their applications (Álvarez-Mingote et al., 2020; Soriano et al., 2023). Beyond developing solutions, researchers must communicate their findings effectively to policymakers and practitioners. Interdisciplinary collaboration is key, as researchers from diverse fields can bring complementary expertise to address complex agricultural challenges. Long-term studies and monitoring are essential to provide evidence-based insights that guide decision-making.

Practitioners: Farmers and agricultural practitioners are pivotal stakeholders in building resilience. Their intimate knowledge of local conditions and hands-on experience with technology implementation make them crucial contributors. Farmers should actively participate in training programs, share their experiences, and provide feedback to policymakers and researchers (Kangogo et al., 2020; Soriano et al., 2023). Creating platforms for knowledge exchange among practitioners can facilitate the adoption of best practices and enhance the collective resilience of the farming community.

Therefore, the success of building agricultural resilience lies in the collaborative efforts of all stakeholders. Policymakers, researchers, and practitioners must work together to create a harmonious ecosystem that fosters innovation, supports sustainable practices, and ensures the adaptability of agriculture to a changing climate.

19.9 Conclusion

In the intricate tapestry of agricultural evolution, the convergence of cutting-edge technologies with strategic climate policies emerges as a promising and transformative pathway to address the complex challenges thrust upon agriculture by a changing climate. This chapter orchestrates a seamless narrative that traverses the integration of RS, ML, IoT, and AI in agriculture, unraveling profound insights into their individual and collective roles. From the menacing specter of unpredictable rainfalls and intensifying droughts to the warming embrace of rising temperatures, these climate change-induced stressors have cast shadows over global food security and the sustainability of crop production. The challenges posed by climate change have been vividly illustrated through a lens focused on the tangible impacts on plant growth, yield, disease severity, and the overall resilience of agricultural ecosystems.

Our exploration has revealed a profound intersection of sustainable agriculture, climate policy, and modern technologies, offering a beacon of hope in the face of adversity. The challenges posed by climate change have been vividly illustrated through a lens focused on the tangible impacts on plant growth, yield, disease severity, and the overall resilience of agricultural ecosystems. With the expansive capabilities of RS, it plays a pivotal role as a data acquisition powerhouse. It acts as the vigilant eye from above, capturing the nuances of agricultural landscapes and climatic patterns, laying the foundation for informed decision-making in the face of climate uncertainties. In the context of ML, where algorithms are finely tuned and deeply analytical, ascend to predict crop yields with remarkable precision. By deciphering the intricate dance between historical climate data, soil quality, and crop performance, ML emerges as a potent tool empowering farmers with foresight. This predictive prowess becomes a cornerstone for resource optimization, enabling strategic decisions on planting, irrigation, and harvesting. The adaptability fostered by ML stands as a bulwark against the multifaceted challenges posed by climate stress. IoT behaves as an intricate web, where sensors and devices form a dynamic network across agricultural landscapes. This interconnected system, akin to a nervous system, continuously collects and disseminates real-time data on soil moisture levels, machinery performance, and climate conditions. The IoT emerges as an active and dynamic information hub, providing stakeholders with a comprehensive understanding of the current agricultural environment. Automated responses triggered by IoT in the face of environmental stress, such as initiating irrigation or activating shading mechanisms, epitomize proactive measures aligned with the overarching goals of climate-resilient agriculture. With the integration of AI into adaptive management within agriculture, real-time decision-making is possible. The prowess of AI lies not only in processing vast datasets encompassing plant physiological responses, climate data, and historical metrics but also in its context-aware decision-making capabilities. The algorithms navigate the experiences of different plant types under stress, ensuring that interventions, be they related to irrigation, fertilization, or pest control, dynamically evolve in real time with changing environmental conditions. The real-time evolution of management strategies becomes the hallmark of AI, resonating with the complexities of climate stress.

The symbiotic relationship between technological advancements and policy imperatives emerges as a recurring theme, extending beyond theoretical discourse. The call for integrated approaches becomes more than a theoretical proposition; it becomes the lynchpin of a dynamic framework where technology acts as the instrument for policy implementation and enhancement. The technological innovations, thus, become purposeful tools harmonizing with policy goals, facilitating effective climate change mitigation and adaptation in agriculture. The exploration extends to the global landscape of climate policies related to agriculture, revealing a mosaic of strategies at various levels. While progress is evident, critical gaps persist, calling for an integrated policy framework that adapts to the dynamic nature of climate challenges. The proposed framework envisions policies that actively leverage the potential of modern technologies, transforming them from tools into integral components of climate-resilient agriculture.

Transitioning to a practical perspective, the narrative unfolds practical guidelines for the integration of advanced tools in agriculture. The importance of clear objectives, accessibility, and training is underscored. Scalability and adaptability take center stage, recognizing the vast diversity inherent in global agriculture. Challenges, from data privacy concerns to resistance against technological change, are not merely acknowledged but met with collaborative solutions. This comprehensive approach sets the stage for the practical application of advanced tools in agriculture. The practical implications are evident from the compelling case studies and success stories from around the globe. Each case study serves as a tangible testament to the profound impact of modern technologies on agriculture. From precision agriculture in the Netherlands, ML for crop yield forecasting in the United States, to digital extension programs in India, the stories illustrate how these innovations align with and contribute to broader climate policy objectives. The multifaceted contributions, ranging from resource optimization and adaptive agriculture to sustainable farming practices, paint a vivid picture of the transformative power of technology in building a resilient and sustainable agricultural future.

Synthesizing these key findings, the integration of modern technologies with effective policy measures emerges as a dynamic framework for addressing crop stress and building resilience in agriculture. The recommendations span a spectrum of strategic considerations. Holistic technology integration, with due consideration for the diversity of agricultural landscapes, is advocated. Capacity building takes precedence as policymakers are urged to prioritize initiatives that empower farmers with the knowledge and skills to leverage advanced technologies. Policy framework enhancement becomes a focal point, emphasizing the need for policies that incentivize the adoption of sustainable agricultural technologies. Recognizing the global nature of climate change, international collaboration is deemed paramount. Policymakers and researchers are encouraged to engage in knowledge exchange and collaborative efforts to share best practices, technologies, and policies globally. Finally, continuous research investment is underscored as pivotal for refining and advancing agricultural technologies. In this collaborative landscape, stakeholders assume distinct roles in shaping the trajectory of agricultural resilience. Policymakers, as central architects, craft regulatory environments conducive to sustainable agricultural practices and technology adoption. Researchers, at the forefront of innovation, contribute through continuous exploration and effective communication of findings. Practitioners, the farmers and agricultural stakeholders, stand as crucial contributors, leveraging their intimate knowledge for effective technology implementation and providing valuable feedback to policymakers and researchers. The success of building agricultural resilience, as the narrative emphasizes, lies in the harmonious collaboration of all stakeholders.

In conclusion, the collaborative vision of climate-resilient agriculture becomes not just a theoretical aspiration but a tangible reality—a testament to human ingenuity and adaptability in the face of evolving environmental dynamics. The multifaceted exploration, from theoretical foundations to practical applications, serves as a comprehensive guide for policymakers, researchers, and practitioners navigating the intricate landscape of sustainable agriculture. It offers a blueprint for building a resilient, adaptive, and sustainable agricultural ecosystem, poised to thrive in an era defined by climate uncertainty.

Acknowledgments The authors would like to extend their heartfelt gratitude to their respective research laboratories for providing invaluable support and resources throughout the journey of working on this chapter. The experience gained in these labs has been instrumental in enhancing our understanding of the field and has contributed significantly to the quality of this work. We are also thankful to our colleagues and mentors for their guidance and insightful discussions, which have enriched our perspectives. This chapter would not have been possible without the collective efforts and support of all those who have contributed to our academic and professional growth.

References

Adhikari, S., Joshi, A., Chandra, A. K., Bharati, A., Sarkar, S., Dinkar, V., Kumar, A., & Singh, A. K. (2023). SMART plant breeding from pre-genomic to post-genomic era for developing climate-resilient cereals. In *Smart plant breeding for field crops in post-genomics era* (pp. 41–97). Springer.

Ahmed, R. A., Hemdan, E. E. D., El-Shafai, W., Ahmed, Z. A., El-Rabaie, E. S. M., & Abd El-Samie, F. E. (2022). Climate-smart agriculture using intelligent techniques, blockchain and Internet of Things: Concepts, challenges, and opportunities. *Transactions on Emerging Telecommunications Technologies, 33*(11), e4607.

Aldy, J. E., Ashton, J., Baron, R., Bodansky, D., Charnovitz, S., Diringer, E., Heller, T. C., Pershing, J., Shukla, P., & Tubiana, L. (2003). *Beyond Kyoto: Advancing the international effort against climate change*. Pew Center on Global Climate Change Arlington.

Alfaro d'Alençon, P., Boanada-Fuchs, A., Greene, M., Krishnamurthy, S., Michelutti, E., Moretto, L., & Smit, W. (2023). *Engaging urban research in policy making: Towards equitable economic growth. A partnership between Cities Alliance, AURI, REDEUS_LAC and N-AERUS*. Universitätsverlag der TU Berlin.

Álvarez-Mingote, C., Moore, A., & McNamara, P. (2020). Assessing the role of stakeholder platforms as drivers of resilient communities: The case of Malawi. *The Journal of Agricultural Education and Extension, 26*(1), 75–95.

Araújo, S. O., Peres, R. S., Ramalho, J. C., Lidon, F., & Barata, J. (2023). Machine learning applications in agriculture: Current trends, challenges, and future perspectives. *Agronomy, 13*(12), 2976.

Aryal, J. P., Sapkota, T. B., Rahut, D. B., & Jat, M. L. (2020). Agricultural sustainability under emerging climatic variability: The role of climate-smart agriculture and relevant policies in India. *International Journal of Innovation and Sustainable Development, 14*(2), 219–245.

Baath, G. S., Sarkar, S., Northup, B. K., Sapkota, B. R., Gowda, P. H., & Flynn, K. C. (2023). Summer pulses as sources of green manure and soil cover in the US Southern Great Plains. *Crop and Environment, 2*(2), 66–74.

Baath, G. S., Sarkar, S., Sapkota, B. R., Flynn, K. C., Northup, B. K., & Gowda, P. H. (2024). Forage yield and nutritive value of summer legumes as affected by row spacing and harvest timing. *Farming System, 2*(1), 100069.

Bai, C., Quayson, M., & Sarkis, J. (2022). Analysis of Blockchain's enablers for improving sustainable supply chain transparency in Africa cocoa industry. *Journal of Cleaner Production, 358*, 131896.

Balasundram, S. K., Shamshiri, R. R., Sridhara, S., & Rizan, N. (2023). The role of digital agriculture in mitigating climate change and ensuring food security: An overview. *Sustainability, 15*(6), 5325.

Balota, M., & Sarkar, S. (2020). *Transpiration of Peanut in the field under Rainfed production.* Paper presented at the American Peanut Research and Education Society Annual Meeting, Virtual, July 14–15, 2020.

Balota, M., Sarkar, S., Cazenave, A., Burow, M., Bennett, R., Chamberlin, K., Wang, N., White, M., Payton, P., & Mahan, J. (2021, July). *Vegetation Indices enable indirect phenotyping of Peanut physiologic and agronomic characteristics.* Paper presented at the American Peanut Research and Education Society Annual Meeting, Virtual.

Bello, O., Ganiyu, O., Wahab, M., Afolabi, M., Oluleye, F., Ig, S., Mahmud, J., Azeez, M., & Abdulmaliq, S. (2012). Evidence of climate change impacts on agriculture and food security in Nigeria. *International Journal of agriculture and Forestry, 2*(2), 49–55.

Bennett, R., Chamberlin, K., Morningweg, D., Wang, N., Sarkar, S., Balota, M., Burow, M., Chagoya, J., & Pham, H. (2021, July). *Response to drought stress in a subset of the US peanut mini-core evaluated in three states.* Paper Presented at the American Peanut Research and Education Society Annual Meeting, Virtual.

Bita, C. E., & Gerats, T. (2013). Plant tolerance to high temperature in a changing environment: scientific fundamentals and production of heat stress-tolerant crops. *Frontiers in Plant Science, 4*, 273.

Boakye, E. A., Zhao, H., & Ahia, B. N. K. (2023). Blockchain technology prospects in transforming Ghana's economy: A phenomenon-based approach. *Information Technology for Development, 29*(2–3), 348–377.

Calicioglu, O., Flammini, A., Bracco, S., Bellù, L., & Sims, R. (2019). The future challenges of food and agriculture: An integrated analysis of trends and solutions. *Sustainability, 11*(1), 222.

Calzadilla, A., Rehdanz, K., Betts, R., Falloon, P., Wiltshire, A., & Tol, R. S. (2013). Climate change impacts on global agriculture. *Climatic Change, 120*, 357–374.

Chandrashekharaiah, P., Kodgire, S., Sanyal, D., & Dasgupta, S. (2021). *Battling climate change: improving crop productivity and quality by increasing photosynthetic efficiency, deploying microbiome metagenomics, and effectively utilizing digital technology* (pp. 637–660). Climate Change and the Microbiome: Sustenance of the Ecosphere.

Chapu, I., Okello, D. K., Okello, R. C. O., Odong, T. L., Sarkar, S., & Balota, M. (2022). Exploration of alternative approaches to phenotyping of late leaf spot and groundnut rosette virus disease for groundnut breeding. *Frontiers in Plant Science, 13*, 912332.

Chaterji, S., DeLay, N., Evans, J., Mosier, N., Engel, B., Buckmaster, D., & Chandra, R. (2020). Artificial intelligence for digital agriculture at scale: Techniques, policies, and challenges. *arXiv preprint arXiv:2001.09786.*

Chatrchyan, A. M., Erlebacher, R. C., Chaopricha, N. T., Chan, J., Tobin, D., & Allred, S. B. (2017). United States agricultural stakeholder views and decisions on climate change. *Wiley Interdisciplinary Reviews: Climate Change, 8*(5), e469.

Crafts-Brandner, S. J., & Salvucci, M. E. (2002). Sensitivity of photosynthesis in a C4 plant, maize, to heat stress. *Plant Physiology, 129*(4), 1773–1780.

Crane-Droesch, A. (2018). Machine learning methods for crop yield prediction and climate change impact assessment in agriculture. *Environmental Research Letters, 13*(11), 114003.

Dadios, E., Vicerra, R., Lauguico, S., Bandala, A., Concepcion, R., & Sybingco, E. (2022). *Automation and control for adaptive management system of urban agriculture using computational intelligence* (p. 1). Automation and Control: Theories and Applications.

De Olde, E. M., Oudshoorn, F. W., Sørensen, C. A., Bokkers, E. A., & De Boer, I. J. (2016). Assessing sustainability at farm-level: Lessons learned from a comparison of tools in practice. *Ecological Indicators, 66*, 391–404.

Dedieu, J., Lessard-Fontaine, A., Ravazzani, G., Cremonese, E., Shalpykova, G., & Beniston, M. (2014). Shifting mountain snow patterns in a changing climate from remote sensing retrieval. *Science of the Total Environment, 493*, 1267–1279.

Dinkar, V., Sarkar, S., Pandey, S., & Antred, S. H. (2024). Plant stress phenotyping: Current status and future prospects. *Advances in Agronomy, 188*, 247.

Dröge, S. (2016). *The Paris Agreement 2015: Turning point for the international climate regime.* Retrieved from https://www.econstor.eu/bitstream/10419/253166/1/2016RP04.pdf

Durán-Sandoval, D., Uleri, F., Durán-Romero, G., & López, A. M. (2023). Food, climate change, and the challenge of innovation. *Encyclopedia, 3*(3), 839–852.

Eastwood, C., Klerkx, L., & Nettle, R. (2017). Dynamics and distribution of public and private research and extension roles for technological innovation and diffusion: Case studies of the implementation and adaptation of precision farming technologies. *Journal of Rural Studies, 49*, 1–12.

Farber, D. A. (2006). Basic compensation for victims of climate change. *University of Pennsylvaniya Law Review, 155*, 1605.

Garrett, K., Bebber, D., Etherton, B., Gold, K., Plex Sulá, A., & Selvaraj, M. G. (2022). Climate change effects on pathogen emergence: Artificial intelligence to translate big data for mitigation. *Annual Review of Phytopathology, 60*, 357–378.

Geli, H., Prihodko, L., Randall, J., Tran, S. C., Cao, H., Misra, S., Boucheron, L., Razzaghi, T., Sandoval, S., & Daniel, D. (2019). Climate adaptive smart systems for future agricultural and rangeland production. A White Paper on Artificial Intelligence Applications in Agriculture. Web access: https://aces-cesfas.nmsu.edu/documents/artificial-intelligence-in-agriculture%2D%2D-white-paper-with-summary%2D%2D-v6.pdf. Accessed 10 Jan 2020.

Gulati, A., Zhou, Y., Huang, J., Tal, A., Juneja, R., & Tal, A. (2021). *Israeli agriculture—Innovation and advancement* (pp. 299–358). Innovations in Indian, Chinese and Israeli Agriculture.

Gupta, J., & Vegelin, C. (2016). Sustainable development goals and inclusive development. *International Environmental Agreements: Politics, Law and Economics, 16*, 433–448.

Hamilton, E. W., Heckathorn, S. A., Joshi, P., Wang, D., & Barua, D. (2008). Interactive effects of elevated CO2 and growth temperature on the tolerance of photosynthesis to acute heat stress in C3 and C4 species. *Journal of Integrative Plant Biology, 50*(11), 1375–1387.

Hazmy, A. I., Hawbani, A., Wang, X., Al-Dubai, A., Ghannami, A., Yahya, A. A., Zhao, L., & Alsamhi, S. H. (2023). Potential of Satellite-Airborne Sensing Technologies for Agriculture 4.0 and Climate-Resilient: A Review. *IEEE Sensors Journal.*

Högy, P., Poll, C., Marhan, S., Kandeler, E., & Fangmeier, A. (2013). Impacts of temperature increase and change in precipitation pattern on crop yield and yield quality of barley. *Food Chemistry, 136*(3–4), 1470–1477.

Jabbari, A., Humayed, A., Reegu, F. A., Uddin, M., Gulzar, Y., & Majid, M. (2023). Smart farming revolution: Farmer's perception and adoption of smart iot technologies for crop health monitoring and yield prediction in jizan, Saudi Arabia. *Sustainability, 15*(19), 14541.

Jagadish, S. K., Pal, M., Sukumaran, S., Parani, M., & Siddique, K. H. (2020). Heat stress resilient crops for future hotter environments. *Plant Physiology Reports, 25*, 529–532.

Jha, U. C., Bohra, A., & Singh, N. P. (2014). Heat stress in crop plants: its nature, impacts and integrated breeding strategies to improve heat tolerance. *Plant Breeding, 133*(6), 679–701.

Jowi, J. O., & Mbwette, T. (2017). The role of research centres and networks. In *Regionalization of African higher education* (pp. 161–174). Brill.

Kangogo, D., Dentoni, D., & Bijman, J. (2020). Determinants of farm resilience to climate change: The role of farmer entrepreneurship and value chain collaborations. *Sustainability, 12*(3), 868.

Karthikeyan, L., Chawla, I., & Mishra, A. K. (2020). A review of remote sensing applications in agriculture for food security: Crop growth and yield, irrigation, and crop losses. *Journal of Hydrology, 586*, 124905.

Killi, D., Bussotti, F., Raschi, A., & Haworth, M. (2017). Adaptation to high temperature mitigates the impact of water deficit during combined heat and drought stress in C3 sunflower and C4 maize varieties with contrasting drought tolerance. *Physiologia Plantarum, 159*(2), 130–147.

Kuhl, L. (2020). Technology transfer and adoption for smallholder climate change adaptation: opportunities and challenges. *Climate and Development, 12*(4), 353–368.

Kumar, V., Sharma, A., Soni, J. K., & Pawar, N. (2017). Physiological response of C3, C4 and CAM plants in changeable climate. *The Pharma. Innovations, 6*(9, Part B), 70.

Kumar, P., Tokas, J., Kumar, N., Lal, M., & Singal, H. (2018). Climate change consequences and its impact on agriculture and food security. *International Journal of chemical studies, 6*(6), 124–133.

Kumar, N., Haak, D. C., Dunne, J. C., & Balota, M. (2022). Multilocation evaluation of Virginia and Runner-type peanut cultivars for yield and grade in Virginia–Carolina region. *Agronomy, 12*(12), 3206.

Kyei-Mensah, C., Kyerematen, R., & Adu-Acheampong, S. (2019). Impact of rainfall variability on crop production within the Worobong Ecological Area of Fanteakwa District, Ghana. *Advances in Agriculture, 2019*, 1–7.

Levidow, L., Sansolo, D., & Schiavinatto, M. (2021). Agroecological innovation constructing socionatural order for social transformation: two case studies in Brazil. Tapuya: Latin American Science, Technology and Society, 4(1), 1843318.

Levizzani, V., & Cattani, E. (2019). Satellite remote sensing of precipitation and the terrestrial water cycle in a changing climate. *Remote Sensing, 11*(19), 2301.

Lipper, L., & Zilberman, D. (2018). A short history of the evolution of the climate smart agriculture approach and its links to climate change and sustainable agriculture debates. In *Climate smart agriculture: building resilience to climate change* (pp. 13–30).

Liu, W., Sun, W., Huang, J., Wen, H., & Huang, R. (2021). Excessive rainfall is the key meteorological limiting factor for winter wheat yield in the middle and lower reaches of the Yangtze River. *Agronomy, 12*(1), 50.

Lockie, S., Fairley-Grenot, K., Ankeny, R., Botterill, L., Howlett, B., Mcbratney, A., Probyn, E., Sorrell, T., Sukkarieh, S., & Woodhead, I. (2020). *The future of agricultural technologies*: Australian Council of Learned Academies (ACOLA).

Loiruck, D. (2013). *Capacity building to farmers for maximizing agricultural output; a case study of Elakunoto Group at Mlangarini Ward in Arumeru District.* (Masters). The Open University of Tanzania, Retrieved from https://repository.out.ac.tz/1005/1/DISSERTATION_-_DANIEL_FINAL.pdf. Available from The Open University of Tanzania. https://repository.out.ac.tz/database. (1005).

Luck, J., Spackman, M., Freeman, A., & Tre˛ bicki, P., Griffiths, W., Finlay, K., & Chakraborty, S. (2011). Climate change and diseases of food crops. *Plant Pathology, 60*(1), 113–121.

Lüttger, A. B., & Feike, T. (2018). Development of heat and drought related extreme weather events and their effect on winter wheat yields in Germany. *Theoretical and Applied Climatology, 132*, 15–29.

Lybbert, T., & Sumner, D. (2010). Agricultural technologies for climate change mitigation and adaptation in developing countries: policy options for innovation and technology diffusion. *Retrieved from.* https://www.files.ethz.ch/isn/117246/agricultural-technologies-for-climate-change-mitigation-and-adaptation-in-developing-countries_web.pdf

Lybbert, T. J., & Sumner, D. A. (2012). Agricultural technologies for climate change in developing countries: Policy options for innovation and technology diffusion. *Food Policy, 37*(1), 114–123.

Malik, S. A. (2023). Linking climate-smart agriculture to farming as a service: mapping an emergent paradigm of datafied dispossession in India. In *Climate Change and Critical Agrarian Studies* (pp. 545–567). Routledge.

Meena, M., Singh, K., & Meena, H. (2015). Gendered Approach to Climate Resilient Agriculture: Technology & Policy-led Options. *Journal of AgriSearch, 2*(3), 206–211.

Mehmood, M. Z., Ahmed, M., Afzal, O., Aslam, M. A., Zoq-ul-Arfeen, R., Qadir, G., Komal, S., Shahid, M. A., Awan, A. A., & Awale, M. A. (2022). *Internet of Things (IoT) and sensors technologies in smart agriculture: applications, opportunities, and current trends* (pp. 339–364). Building Climate Resilience in Agriculture.

Messerli, P., Murniningtyas, E., Eloundou-Enyegue, P., Foli, E. G., Furman, E., Glassman, A., Hernández Licona, G., Kim, E. M., Lutz, W., & Moatti, J.-P. (2019). *Global sustainable development report 2019: the future is now–science for achieving sustainable development.* Retrieved from https://pure.iiasa.ac.at/id/eprint/16067/1/24797GSDR_report_2019.pdf

Mujawamariya, M. G. (2021). Crop conditions monitoring using IOT and transfer learning: Case study Maize in Rwanda. (Masters). University of Rwanda, . Retrieved from http://154.68.126.42/bitstream/handle/123456789/1634/MUJAWAMARIYA%20Marie%20Grace.pdf?sequence=1&isAllowed=y Available from University of Rwanda http://154.68.126.42/database.

Murray, V., & Ebi, K. L. (2012). IPCC special report on managing the risks of extreme events and disasters to advance climate change adaptation (SREX). *Journal of Epidemiology and Community Health, 66*(9), 759–760.

Musah, S., Medeni, T. D., & Soylu, D. (2019). *Assessment of role of innovative technology through Blockchain technology in Ghana's Cocoa beans food supply chains.* Paper presented at the 2019 3rd International Symposium on Multidisciplinary Studies and Innovative Technologies (ISMSIT) October 11–13, 2019.,

Newton, A. C., Johnson, S. N., & Gregory, P. J. (2011). Implications of climate change for diseases, crop yields and food security. *Euphytica, 179*, 3–18.

Ngoma, H., Angelsen, A., Carter, S., & Roman-Cuesta, R. M. (2018). Climate-smart agriculture. *Transforming REDD, 175.*

North, D., & Smallbone, D. (2006). Developing entrepreneurship and enterprise in Europe's peripheral rural areas: Some issues facing policy-makers. *European Planning Studies, 14*(1), 41–60.

Pahl-Wostl, C. (2007). Transitions towards adaptive management of water facing climate and global change. *Water Resources Management, 21*, 49–62.

Patil, S., Kumar, K., Jakhar, D. S., Rai, A., Borle, U., & Singh, P. (2016). Studies on variability, heritability, genetic advance and correlation in maize (Zea mays L.). *International Journal of Agriculture, Environment and Biotechnology, 9*(6), 1103–1108.

Popescu, T., & Saulescu, I. (2022). Harnessing the power of big data to drive evidence-based policy making in the public sector. *Journal of Big-Data Analytics and Cloud Computing, 7*(4), 45–57.

Rai, A., Sharma, V., & Heitholt, J. (2020). Dry bean [phaseolus vulgaris L.] growth and yield response to variable irrigation in the arid to semi-arid climate. *Sustainability, 12*(9), 3851.

Rai, A., Sarkar, S., & Jha, P. K. (2022). Deficit irrigation: An optimization strategy for a sustainable agriculture. In *Soil-water, agriculture, and climate change: exploring linkages* (pp. 163–181). Springer.

Rao, N. (2007). A framework for implementing information and communication technologies in agricultural development in India. *Technological Forecasting and Social Change, 74*(4), 491–518.

Rathore, B., Gupta, R., Biswas, B., Srivastava, A., & Gupta, S. (2022). Identification and analysis of adoption barriers of disruptive technologies in the logistics industry. *The International Journal of Logistics Management, 33*(5), 136–169.

Ritchie, J., & Dowlatabadi, H. (2018). Defining climate change scenario characteristics with a phase space of cumulative primary energy and carbon intensity. *Environmental Research Letters, 13*(2), 024012.

Roep, D. (2019). The recontextualization of precision agriculture to the Dutch crop farming sector. *Development.*

Ruane, A. C., & Rosenzweig, C. (2019). Climate change impacts on agriculture. In *Agriculture & food systems to 2050: Global trends, challenges and opportunities* (pp. 161–191). World Scientific.

Sage, R. F., & Kubien, D. S. (2007). The temperature response of C3 and C4 photosynthesis. *Plant, Cell & Environment, 30*(9), 1086–1106.

Salokhe, G., Pastore, A., Richards, B., Weatherley, S., Aubert, A., Keizer, J., Nadeau, A., Katz, S., Rudgard, S., & Mangstl, A. (2004). FAO's role in information management and dissemination-challenges, innovation, success, lessons learned. *IAALD Quarterly Bulletin, 49*(3–4), 73–83.

Sarkar, S. (2020). *Development of high-throughput phenotyping methods and evaluation of morphological and physiological characteristics of peanut in a sub-humid environment.* (Doctoral). Virginia Tech, Blacksburg, VA. Retrieved from https://vtechworks.lib.vt.edu/server/api/core/bitstreams/2070d20e-6d2b-4a0b-9cd2-d32f3fcb6d1d/content. Available from Virginia Tech Libraries VTechWorks database.

Sarkar, S., & Islam, M. (2016). Bird's-foot trefoil response to planting method and harvesting frequency. In *Wyoming Agricultural Experiment Station Field Days Bulletin 2016* (pp. 113–114). University of Wyoming.

Sarkar, S., & Islam, M. (2017). Effect of planting method, harvesting frequency, and cultivars on yield of bird's-foot trefoil. In *Wyoming agricultural experiment station field days bulletin 2017* (pp. 104–105). University of Wyoming.

Sarkar, S., & Jha, P. K. (2020). Is precision agriculture worth it? Yes, may be. *Journal of Biotechnology and Crop Science, 9*(14), 4–9.

Sarkar, S., Mishra, D., & Ghadei, K. (2014a). Collective marketing–A hope for the farmers. *Indian Journal of Crop Ecology, 2*(2).

Sarkar, S., Mishra, D., & Ghadei, K. (2014b). Collective marketing in Purvanchal Region of India–A case study of feasibility. *Indian Journal of Crop Ecology, 2*(2).

Sarkar, S., Cazenave, A. B., Oakes, J., McCall, D., Thomason, W., Abbot, L., & Balota, M. (2020). High-throughput measurement of peanut canopy height using digital surface models. *The Plant Phenome Journal, 3*(1), e20003.

Sarkar, S., Cazenave, A.-B., Oakes, J., McCall, D., Thomason, W., Abbott, L., & Balota, M. (2021a). Aerial high-throughput phenotyping of peanut leaf area index and lateral growth. *Scientific Reports, 11*(1), 21661.

Sarkar, S., Ramsey, A. F., Cazenave, A.-B., & Balota, M. (2021b). Peanut leaf wilting estimation from RGB color indices and logistic models. *Frontiers in Plant Science, 12*, 658621.

Sarkar, S., Oakes, J., Cazenave, A.-B., Burow, M. D., Bennett, R. S., Chamberlin, K. D., Wang, N., White, M., Payton, P., & Mahan, J. (2022a). Evaluation of the US peanut germplasm mini-core collection in the Virginia-Carolina region using traditional and new high-throughput methods. *Agronomy, 12*(8), 1945.

Sarkar, S., Rai, A., & Jha, P. K. (2022b). Remote sensing and high-throughput techniques to phenotype crops for drought tolerance. In *Soil-water, agriculture, and climate change: exploring linkages* (pp. 107–129). Springer.

Sarkar, S., Shekoofa, A., McClure, A., & Gillman, J. D. (2022c). Phenotyping and quantitative trait locus analysis for the limited transpiration trait in an upper-mid south soybean recombinant inbred line population ("Jackson"×"KS4895"): High throughput aquaporin inhibitor screening. *Frontiers in Plant Science, 12*, 779834.

Sarkar, S., Leyton, J. M. O., Noa-Yarasca, E., Adhikari, K., Hajda, C. B., & Smith, D. R. (2025). Integrating remote sensing and soil features for enhanced machine learning-based corn yield prediction in the Southern US. *Sensors (Basel, Switzerland), 25*(2), 543.

Sarkar, S., & Jha, P. K. (2025). Application of AI/ML in water resource management to resolve transboundary water conflict. In *Navigating the nexus: Hydrology, agriculture, pollution and climate change* (Vol. 1, pp. 431–455). Cham: Springer Nature Switzerland.

Sayer, J., Sunderland, T., Ghazoul, J., Pfund, J.-L., Sheil, D., Meijaard, E., Venter, M., Boedhihartono, A. K., Day, M., & Garcia, C. (2013). Ten principles for a landscape approach to reconciling agriculture, conservation, and other competing land uses. *Proceedings of the National Academy of Sciences, 110*(21), 8349–8356.

Shahhosseini, M., Hu, G., Huber, I., & Archontoulis, S. V. (2021). Coupling machine learning and crop modeling improves crop yield prediction in the US Corn Belt. *Scientific Reports, 11*(1), 1606.

Shaikh, T. A., Rasool, T., & Lone, F. R. (2022). Towards leveraging the role of machine learning and artificial intelligence in precision agriculture and smart farming. *Computers and Electronics in Agriculture, 198*, 107119.

Shukla, P. R., Skea, J., Calvo Buendia, E., Masson-Delmotte, V., Pörtner, H. O., Roberts, D., Zhai, P., Slade, R., Connors, S., & Van Diemen, R. (2019). *IPCC, 2019: Climate Change and Land: an IPCC special report on climate change, desertification, land degradation, sustainable land management, food security, and greenhouse gas fluxes in terrestrial ecosystems.* Retrieved from https://spiral.imperial.ac.uk/bitstream/10044/1/76618/2/SRCCL-Full-Report-Compiled-191128.pdf.

Shukla, R., Dubey, G., Malik, P., Sindhwani, N., Anand, R., Dahiya, A., & Yadav, V. (2021). Detecting crop health using machine learning techniques in smart agriculture system. *Journal of Scientific & Industrial Research, 80*(08), 699–706.

Silici, L., Rowe, A., Suppiramaniam, N., & Knox, J. W. (2021). Building adaptive capacity of smallholder agriculture to climate change: evidence synthesis on learning outcomes. *Environmental Research Communications, 3*(12), 122001.

Sims, R. E., Mabee, W., Saddler, J. N., & Taylor, M. (2010). An overview of second generation biofuel technologies. *Bioresource Technology, 101*(6), 1570–1580.

Smith, H. E., Sallu, S. M., Whitfield, S., Gaworek-Michalczenia, M. F., Recha, J. W., Sayula, G. J., & Mziray, S. (2021). Innovation systems and affordances in climate smart agriculture. *Journal of Rural Studies, 87*, 199–212.

Soriano, B., Garrido, A., Bertolozzi-Caredio, D., Accatino, F., Antonioli, F., Krupin, V., Meuwissen, M. P., Ollendorf, F., Rommel, J., & Spiegel, A. (2023). Actors and their roles for improving resilience of farming systems in Europe. *Journal of Rural Studies, 98*, 134–146.

Sun, A. Y., & Scanlon, B. R. (2019). How can Big Data and machine learning benefit environment and water management: a survey of methods, applications, and future directions. *Environmental Research Letters, 14*(7), 073001.

Sylvère, N., & D'amour, R. J. (2020). Updates on modern agricultural technologies adoption and its impacts on the improvement of agricultural activities in Rwanda: A review. *International Journal of Innovative Science and Research Technology, 5*(12), 222–229.

Symeonaki, E. G., Arvanitis, K. G., & Piromalis, D. D. (2019). *Cloud computing for IoT applications in climate-smart agriculture: A review on the trends and challenges toward sustainability*. Paper presented at the Innovative Approaches and Applications for Sustainable Rural Development: 8th International Conference, HAICTA 2017, Chania. https://doi.org/10.1007/978-3-030-02312-6_9

Taghikhah, F., Erfani, E., Bakhshayeshi, I., Tayari, S., Karatopouzis, A., & Hanna, B. (2022). Artificial intelligence and sustainability: solutions to social and environmental challenges. In *Artificial intelligence and data science in environmental sensing* (pp. 93–108). Elsevier.

Tal, A. (2019). *Israeli agricultural innovation: assessing the potential to assist smallholders*. Retrieved from https://www.syngentafoundation.org/sites/g/files/kgtney976/files/media/document/2019/11/15/israeli_agricultural_innovation_assessing_the_potential_to_assist_smallholders.pdf

Thornton, P. K., Ericksen, P. J., Herrero, M., & Challinor, A. J. (2014). Climate variability and vulnerability to climate change: A review. *Global Change Biology, 20*(11), 3313–3328.

Tobal, A., & Mokhtar, S. A. (2014). Weeds identification using evolutionary artificial intelligence algorithm. *Journal of Computer Science, 10*(8), 1355–1361.

Vosti, S., Witcover, J., Oliveira, S., & Faminow, M. (1998). Policy issues in agroforestry: technology adoption and regional integration in the western Brazilian Amazon. In P. K. R. Nair & C. R. Latt (Eds.), *Directions in tropical agroforestry research* (pp. 195–222). Springer.

Wang, D., Heckathorn, S. A., Barua, D., Joshi, P., Hamilton, E. W., & LaCroix, J. J. (2008). Effects of elevated CO2 on the tolerance of photosynthesis to acute heat stress in C3, C4, and CAM species. *American Journal of Botany, 95*(2), 165–176.

Yang, J., Gong, P., Fu, R., Zhang, M., Chen, J., Liang, S., Xu, B., Shi, J., & Dickinson, R. (2013). The role of satellite remote sensing in climate change studies. *Nature Climate Change, 3*(10), 875–883.

Zecca, F. (2019). The use of internet of things for the sustainability of the agricultural sector: The case of climate smart agriculture. *International Journal of Civil Engineering and Technology, 10*(3).

Socioeconomic Implications of Heat and Drought Stress on Indian Agriculture

20

Banda Sainath, Md Asadullah, Ujjwal Kumar, and Anirban Mukherjee

Abstract

The escalating threat of heat and drought stress, fueled by climate change, is wreaking havoc on India's agricultural sector, with far-reaching socioeconomic consequences. This chapter comprehensively examines the multifaceted impacts of these climatic stressors on crop yields, livestock productivity, food security, livelihoods, and rural communities. Evidence reveals alarming yield reductions across major crops: rice (10–40% due to heat, 20–50% due to drought), wheat (6–35% heat, 15–40% drought), maize (10–25% heat, 25–45% drought), and pulses (15–30% heat, 30–60% drought). Livestock, a crucial component of India's agricultural economy, also suffers, with milk yield losses of 10–30% in cattle and 8–25% in buffaloes. These losses translate into economic burdens exceeding billions of dollars annually. The socioeconomic ramifications extend beyond production losses. Increased food prices, malnutrition, farmer indebtedness, and forced migration are some of the devastating consequences. Heat and drought further exacerbate social vulnerabilities, especially for women, and intensify conflicts over dwindling resources. The health of rural populations is also at risk, with rising cases of heat-related illnesses and mental health issues. Mitigating these impacts necessitates a multipronged approach, encompassing drought-resistant crops, efficient irrigation, climate-smart agriculture, carbon farming, early warning systems, and targeted support for vulnerable communities. Addressing socioeconomic disparities and promoting gender equity is paramount to building long-term resilience in Indian agriculture.

B. Sainath (✉) · U. Kumar · A. Mukherjee
ICAR Research Complex for Eastern Region, Patna, Bihar, India
e-mail: sainat.banda@icar.gov.in

M. Asadullah
MBA (Sustainable Management), Nalanda University, Rajgir, Bihar, India

R. Dubey et al. (eds.), *Drought and Heat Stress in Agriculture*,
https://doi.org/10.1007/978-981-96-5735-3_20

Keywords

Heat and drought stress · Food security · Socioeconomic impact · Livestock · Yield loss

20.1 Introduction

Irregularities in climatic hazards, such as escalating floods, prolonged droughts, and erratic temperature fluctuations, have intensified in recent times, casting an ominous shadow over global agricultural systems. These amplified climatic events disrupt the delicate balance of agricultural production, hinder productivity, and trigger far-reaching socioeconomic consequences. These climatic events as a direct consequence of human-induced climate change, posing an existential threat to global food security and agricultural systems (IPCC, 2021).

Drought stands out as a particularly complex and devastating phenomenon (Hoque et al., 2021), inflicting significant damage on agricultural landscapes and economies worldwide. Its insidious nature, characterized by prolonged periods of water scarcity and soil moisture depletion, can decimate crop yields, erode livelihoods, and exacerbate food insecurity. Similarly, the escalating threat of heat stress, driven by rising temperatures and intensified solar radiation, poses a grave challenge to agricultural productivity (IPCC, 2021). Heat stress can impair plant physiological processes, disrupt reproductive development, and ultimately reduce crop yields, particularly for vulnerable staples like rice and wheat.

In regions heavily reliant on agriculture, such as South Asia, the confluence of these climatic hazards poses an existential threat to livelihoods, food security, and economic stability. The intricacies of these challenges necessitate a comprehensive understanding of their multifaceted impacts on agricultural systems, rural communities, and broader socioeconomic landscapes. This chapter endeavors to unravel the complexities of drought and heat stress, exploring their intricate interplay with socioeconomic factors and illuminating the pathways toward building resilience in the face of an increasingly volatile climate. In South Asia, a region characterized by its heavy reliance on agriculture for sustenance and economic stability, the escalating threat of heat and drought stress casts a long shadow. As climate change intensifies, these environmental stressors are no longer isolated events; they are a persistent and growing menace, threatening the delicate balance of agricultural productivity and the livelihoods of millions. The region's vulnerability is heightened by its dependence on rain-fed agriculture, a limited adaptive capacity among smallholder farmers, and underlying socioeconomic disparities (IPCC, 2021). The increasing frequency and intensity of heatwaves pose a significant risk to crop yields, particularly for staple crops like rice and wheat (Im et al., 2017). These heatwaves can disrupt critical growth stages, reduce pollination, and ultimately lead to substantial yield losses. This not only threatens food security in a region where a large portion of the population relies on agriculture for both sustenance and livelihoods but also exacerbates existing socioeconomic inequalities by disproportionately impacting marginalized communities who lack the resources and infrastructure to adapt to these climatic shocks (Thornton et al., 2014). India, with its diverse climate and extensive agricultural zones, is particularly vulnerable to the escalating impacts of climate change (Mall et al., 2006). The country's vulnerability is further compounded by its heavy reliance on rain-fed agriculture, limited adaptive capacity among smallholder farmers, and underlying socioeconomic disparities (IPCC, 2021). These factors make India particularly susceptible to the adverse effects of climatic hazards like droughts and heatwaves. Across India, erratic rainfall patterns and prolonged dry spells are on the rise, further exacerbating the vulnerability of rain-fed agricultural systems (Guhathakurta & Rajeevan, 2008).

The socioeconomic implications of these climatic challenges are immense, as nearly 58% of Indians rely on agriculture as their main source of income (World Bank, 2021), and the agricultural sector contributes significantly to the national GDP (GOI, 2024). India has witnessed a series of severe droughts in recent decades, including those in 2002, 2009, 2014, 2015, and 2016 (Mishra & Singh, 2010). These droughts have resulted in substantial economic losses, with the 2002 drought estimated to have cost the country $3 billion (World Bank, 2004) and the 2014–2015 drought exceeding $4.5 billion in damages (Dev & Sharma, 2016). The cascading effects of heat and drought stress extend far beyond crop production, permeating various facets of life in South Asia. Water scarcity, amplified by these climatic stressors, disrupts irrigation systems, limits livestock productivity, and threatens the availability of clean drinking water for both humans and animals. This can trigger a domino effect of adverse outcomes, including forced migration, conflicts over dwindling resources, and long-term socioeconomic instability (Lesk et al., 2016). Additionally, the health and well-being of rural populations are jeopardized by increased exposure to heat-related illnesses, malnutrition due to crop failures, and psychological distress resulting from the loss of livelihoods and displacement (Watts et al., 2018). The multifaceted nature of these challenges necessitates a comprehensive and multipronged approach to building resilience in the agricultural sector, particularly in vulnerable regions like India and South Asia. This includes investing in drought-resistant crop varieties, developing efficient irrigation systems, promoting sustainable agricultural practices, and empowering farmers with the knowledge and resources to adapt to a changing climate. Moreover, addressing the underlying socioeconomic disparities that exacerbate vulnerability to climatic shocks is crucial for ensuring equitable and sustainable development in the region. By understanding the complex interplay of climatic, environmental, and socioeconomic factors, we can pave the way for targeted interventions and adaptive strategies that safeguard the livelihoods of millions and ensure the long-term sustainability of agriculture in South Asia.

20.2 Impacts on Agricultural Productivity and Food Security in India

Figure 20.1 illustrates the intricate and interconnected nature of the socioeconomic impacts stemming from heat and drought stress, particularly in the context of agriculture-dependent regions like South Asia. The central node, “Impacts of Heat and Drought Stress,” is the driving force behind a web of consequences that ripple through various dimensions of vulnerability.

The diagram distinguishes between two types of impacts:

- *Unidirectional Impact (Dotted Lines):* This represents the direct, causal relationship between heat and drought stress and specific vulnerabilities. For example, the dotted line from the central node to “Food Insecurity” indicates that these climatic stressors directly reduce food production, leading to food shortages and price hikes. Similarly, the dotted lines to “Livelihood Insecurity” and “Economic Insecurity” highlight the direct impacts on farmers’ livelihoods and income due to crop failure, livestock losses, and increased production costs.
- *Bidirectional Relationship (Solid Lines):* The solid lines represent the complex interplay and interconnectedness between the different dimensions of vulnerability. For instance, the solid line between “Economic Insecurity” and “Social Vulnerability” signifies that economic hardship can exacerbate existing social inequalities and trigger conflicts over resources, while pre-existing social vulnerabilities can make communities more susceptible to the economic shocks of heat and drought. Similarly, the bidirectional relationship between “Food Insecurity” and “Livelihood Insecurity” indicates that food shortages can drive people to

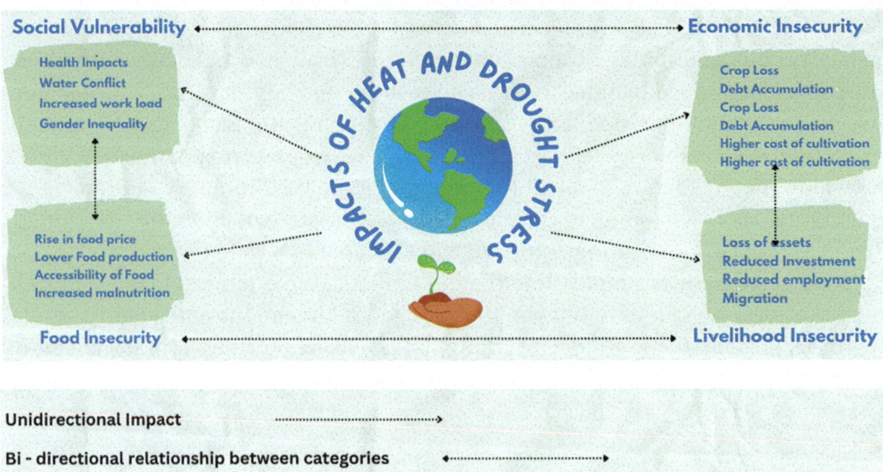

Fig. 20.1 Interconnected nature of the socioeconomic impacts of heat and drought stress. (Source: Author’s Compilation)

adopt risky livelihood strategies, further deepening their vulnerability, and vice versa.

20.3 Significance for the Indian Context

This diagram is particularly relevant for understanding the socioeconomic impacts of heat and drought stress in India, where a significant portion of the population relies on agriculture for their livelihoods (World Bank, 2021). The country's heavy dependence on rain-fed agriculture, coupled with limited adaptive capacity among smallholder farmers and underlying socioeconomic disparities, makes it highly vulnerable to these climatic stressors (IPCC, 2021). The diagram illustrates how these stressors not only threaten food security and livelihoods but also amplify social inequalities and trigger conflicts over scarce resources, as evidenced by the series of severe droughts that have plagued India in recent decades (Mishra & Singh, 2010).

Understanding these complex interconnections is crucial for developing effective adaptation and mitigation strategies. While immediate measures like drought-resistant crops and improved irrigation are essential, the diagram emphasizes the need for a holistic approach that addresses the underlying socioeconomic vulnerabilities and strengthens the resilience of communities in the face of an increasingly volatile climate.

20.3.1 Yield Losses

Heat and drought stress pose significant threats to India's agricultural productivity, resulting in substantial yield losses across various crops. Studies (Table 20.1) have shown that heat stress alone can cause yield reductions of 10–40% in rice (Peng et al., 2004), 6–35% in wheat (Challinor et al., 2014), 10–25% in maize (Lobell et al., 2011), and 15—30% in pulses like chickpea (Aggarwal et al., 2000). Drought stress, on the other hand, can lead to even more severe yield losses, with reductions

Table 20.1 Climate change-induced yield losses for major crops in India

Crop type	Crop/variety	Yield loss (%)	Source
Cereals	Rice	10–40	Peng et al. (2004)
	Wheat	6–35	Challinor et al. (2014)
	Maize	10–25	Lobell et al. (2011)
Pulses	Chickpea	15–30	Aggarwal et al. (2000)
	Pigeon pea	20–40	Rathore et al. (2015)
Oilseeds	Groundnut	25–50	Basu et al. (2016)
	Mustard	15–35	Kumar et al. (2010)
Vegetables	Tomato	30–50	Bhandari et al. (2021)
	Onion	20–40	Chattopadhyay et al. (2013)
Fruits	Mango	10–25	Kumar et al. (2019)
	Banana	15–30	Ravi et al. (2013)

of 20–50% in rice (Mall et al., 2006), 15–40% in wheat (Asseng et al., 2011), 25–45% in maize (Lobell et al., 2011), and 30–60% in pulses (Kumar et al., 2016). These figures highlight the significant impact of climate change-induced stressors on India's agricultural output and underscore the urgent need for adaptation and mitigation strategies.

Heat and drought stress pose significant threats to India's agricultural productivity, resulting in substantial yield losses across various crops. Studies have consistently shown that heat stress, characterized by elevated temperatures beyond the optimal range for plant growth, can trigger a cascade of detrimental physiological responses in crops, ultimately leading to reduced yields. For instance, research by Peng et al. (2004) indicates that heat stress alone can cause yield reductions of 10–40% in rice, a staple food crop for millions in India. The physiological disruptions caused by heat stress include impaired photosynthesis, increased respiration, and accelerated crop development, all of which contribute to lower grain yield and quality. Similarly, wheat, another crucial crop for the country's food security, faces yield losses of 6–35% due to heat stress (Challinor et al., 2014). The negative effects of heat stress on wheat include reduced grain filling, decreased grain size, and shortened growth duration, leading to significant reductions in overall productivity.

Maize and pulses, important components of India's agricultural landscape, are also not immune to the adverse effects of heat stress. Research by Lobell et al. (2011) suggests that maize yields can be reduced by 10–25% due to heat stress, primarily due to impaired pollination and accelerated senescence. Pulses, a major source of protein in the Indian diet, are particularly vulnerable to heat stress during reproductive stages, resulting in flower and pod abortion, reduced seed set, and yield losses of 15–30% (Aggarwal et al., 2000). These findings underscore the far-reaching consequences of heat stress on India's agricultural sector and the livelihoods of millions of farmers who depend on these crops for their sustenance and income.

20.3.2 Differential Impacts on Agricultural Regions and Farming Systems

The impacts of heat and drought stress are not uniform across India, varying significantly depending on agroecological zones and farming systems. Rain-fed regions, which constitute a substantial portion of India's agricultural landscape, are particularly vulnerable to drought due to their reliance on unpredictable rainfall patterns (Guhathakurta & Rajeevan, 2008). In these regions, prolonged dry spells and inadequate rainfall can severely hamper crop growth and lead to significant yield losses. On the other hand, regions with high temperatures, such as the northwestern plains, are more susceptible to heat stress, which can negatively impact crop physiology and reduce yields (Im et al., 2017). Irrigated areas, although relatively less vulnerable to drought, can still experience yield reductions during periods of extreme water scarcity or when groundwater levels decline due to overextraction (Shah, 2010).

20.4 Implications for Food Availability, Accessibility, and Affordability

The yield losses caused by heat and drought stress have profound implications for food security in India. Reduced agricultural production can lead to decreased food availability, which in turn can drive up food prices and make it difficult for vulnerable populations to access nutritious food (Dev & Sharma, 2016). This can exacerbate existing inequalities and disproportionately affect the poor and marginalized communities who are already struggling with food insecurity. Furthermore, the decline in agricultural productivity can also lead to job losses and economic hardship for farmers, particularly smallholders who lack the resources to cope with these climatic shocks (Thornton et al., 2014). This can create a vicious cycle of poverty and food insecurity, with long-term consequences for both rural and urban populations.

Heat and drought stress significantly exacerbate the vulnerability of farmers, particularly smallholder farmers and those in rain-fed regions, pushing them further into precarious socioeconomic conditions. Several key implications can be identified:

20.4.1 Economic Insecurity

- *Crop Losses:* Drought and heat-induced yield reductions directly translate to significant income losses for farmers, particularly in rain-fed regions heavily reliant on agriculture (Thornton et al., 2014). A global study published in *Nature Climate Change* by Asseng et al., (2015) found that a 1 °C increase in temperature during the growing season could lead to a 6% decline in global wheat yields, further emphasizing the economic vulnerability of farmers in the face of climate change. In India, where a significant portion of the population relies on agriculture for their livelihoods, crop losses due to these climatic stressors can be devastating. A study by the World Bank (2018) estimated that climate change could reduce India's agricultural GDP by 1.5% to 2.8% by 2050, underscoring the long-term economic consequences of these environmental stressors (Fig. 20.2).
- *Increased Production Costs:* Drought and heat stress not only reduce agricultural yields but also increase production costs, putting additional financial strain on farmers. Water scarcity necessitates increased investments in irrigation, often through the use of groundwater pumping, which can be expensive and energy-intensive (Shah, 2010). Farmers may also have to invest in drought-resistant crop varieties, which often come with a higher price tag than traditional varieties. Furthermore, the increased incidence of pests and diseases under heat and drought stress can necessitate additional expenditure on pesticides and other crop protection measures (Lobell et al., 2011). These increased production costs, coupled with reduced yields and lower market prices due to oversupply during times of drought, can significantly erode farmers' profit margins and exacerbate

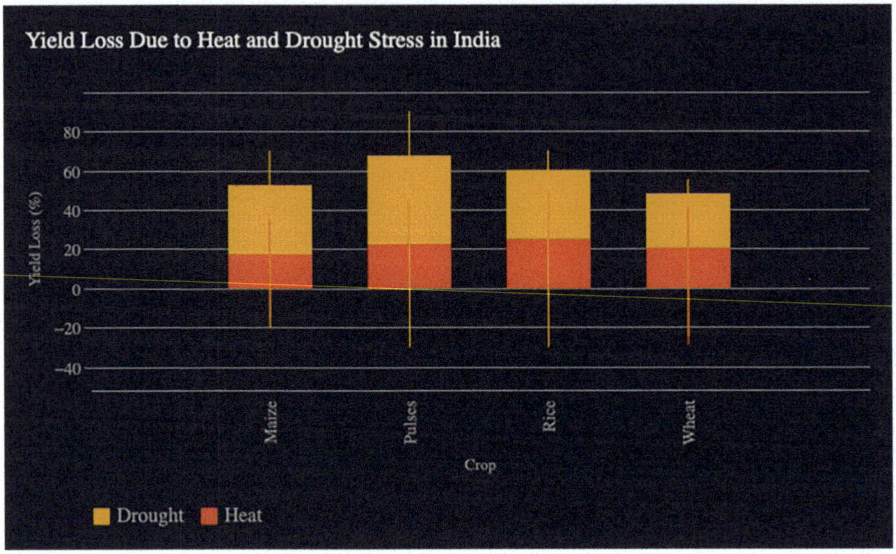

Fig. 20.2 Estimated yield loss (%) of major crops in India due to heat and drought stress

their economic vulnerability. Research published in *Agricultural Systems* by Aggarwal et al. (2000) indicates that the cost of adapting to climate change for Indian farmers could be substantial, further highlighting the economic burden faced by these vulnerable communities.

20.4.2 Livelihood Insecurity

Loss of Assets Drought and heat stress pose a grave threat to farmers' assets, particularly in regions heavily reliant on agriculture. Livestock, a critical source of income and sustenance for many farmers, are particularly vulnerable to these climatic stressors. Water scarcity and heat stress can lead to increased livestock mortality, reduced productivity (e.g., milk production), and lower market prices (Morton, 2007). This loss of livestock can have a devastating impact on farmers' livelihoods, as it not only deprives them of a valuable income source but also erodes their resilience and ability to recover from future shocks. In addition to livestock, drought and heat can also damage other productive assets such as farm machinery, irrigation infrastructure, and stored crops, further exacerbating farmers' economic vulnerability (IPCC, 2021). A study by Meze-Hausken (2004) found that drought-induced livestock losses can have long-lasting effects on household income and well-being, highlighting the importance of livestock in rural livelihoods and the vulnerability of this sector to climate change.

Reduced Employment Opportunities The decline in agricultural productivity due to heat and drought stress has a ripple effect on employment opportunities within the sector. As crop yields decrease and farm incomes dwindle, farmers may be forced to lay off agricultural laborers or reduce their working hours, leading to widespread unemployment and underemployment in rural areas (World Bank, 2021). Moreover, the decline in agricultural activities can also impact nonfarm rural employment, as demand for goods and services in rural areas decreases (Kumar et al., 2020). This can lead to a downward spiral of economic hardship, with limited alternative livelihood options available for those who lose their jobs in the agricultural sector (Morton, 2007). A study by Gray and Mueller (2012) on the impact of drought in Sub-Saharan Africa found that agricultural production shocks can lead to significant reductions in rural employment and income, highlighting the interconnectedness of livelihoods and agricultural productivity.

Forced Migration In extreme cases, when drought and heat stress decimate agricultural livelihoods and erode coping mechanisms, farmers may be forced to abandon their homes and migrate in search of alternative sources of income and sustenance (Lesk et al., 2016). This forced migration, often to urban areas or other rural regions, can disrupt social networks, fragment families, and lead to the loss of cultural identity (Hunter et al., 2015). Furthermore, migrants may face challenges in finding employment and accessing basic services in their new destinations, leading to further impoverishment and vulnerability (Chandra et al., 2018). The phenomenon of climate-induced migration is a growing concern in South Asia, with millions of people already displaced due to environmental factors (Gemenne, 2011). Addressing the root causes of this migration, such as enhancing agricultural resilience and providing livelihood diversification opportunities, is crucial for mitigating the long-term socioeconomic consequences of climate change in the region.

20.4.3 Food Insecurity

Reduced Food Production Heat and drought stress, through their detrimental impact on crop yields, directly threaten food availability at both household and community levels. This is particularly pronounced in regions heavily reliant on subsistence farming, where families depend on their own harvests for sustenance. When crops fail due to extreme weather events, these families face immediate food shortages and are forced to rely on expensive market purchases or food aid (Thornton et al., 2014). The Intergovernmental Panel on Climate Change (IPCC, 2021) has highlighted the increased risk of crop failures due to climate change, particularly in developing countries where adaptive capacity is limited. A study by the Food and Agriculture Organization (FAO, 2018) estimated that climate change could reduce global food production by up to 30% by 2050, further exacerbating food insecurity challenges. In India, this is evident in the declining yields of major crops like rice, wheat, and pulses due to heat and drought stress (Challinor et al., 2014; Lesk et al., 2016).

Increased Food Prices The reduced availability of food caused by heat and drought stress can lead to a surge in food prices, as basic economic principles of supply and demand come into play (Dev & Sharma, 2016). This price volatility disproportionately affects vulnerable populations, such as low-income households and those living in remote areas with limited access to markets. A study by the World Food Programme (WFP, 2018) found that climate-related disasters can trigger sharp increases in food prices, sometimes doubling or tripling the cost of staple foods. This can have devastating consequences for food security, as families are forced to make difficult choices between purchasing food and meeting other essential needs, such as healthcare and education. Additionally, research by Bellemare (2015) indicates that food price spikes can lead to social unrest and political instability, further highlighting the interconnectedness of food security, economic stability, and societal well-being.

Malnutrition The lack of access to adequate and diverse food, exacerbated by heat and drought-induced crop failures and price hikes, can lead to malnutrition, especially among vulnerable groups such as children and women (Watts et al., 2018). Malnutrition, in all its forms—including undernutrition, micronutrient deficiencies, and overweight/obesity—has devastating consequences for human health and development. In children, malnutrition can stunt growth, impair cognitive development, and increase vulnerability to infectious diseases (Black et al., 2013). Women, particularly pregnant and lactating women, are also at high risk of malnutrition due to their increased nutritional needs. A study by Grace et al. (2018) highlighted the disproportionate impact of climate change on women's health and nutrition, emphasizing the need for gender-sensitive approaches to address food insecurity in the context of climate change. Furthermore, research by Lloyd et al. (2011) suggests that climate change-induced food insecurity can have intergenerational consequences, as malnutrition in early life can lead to long-term health problems and reduced economic productivity in adulthood. Addressing malnutrition requires a multifaceted approach that encompasses not only improving food production and access but also addressing underlying social and economic inequalities that contribute to vulnerability.

20.4.4 Social Vulnerability

Social Vulnerability and Health Impacts of Heat and Drought Stress:

Increased Gender Inequality Women in agrarian communities, who often play a pivotal role in agricultural production and household management, are disproportionately burdened by the impacts of heat and drought stress. As water scarcity intensifies and crop yields decline, women face increased workloads in fetching water, tending to livestock, and managing household chores (Thornton et al., 2014).

This additional labor burden can have detrimental effects on women's health, nutrition, and overall well-being. Furthermore, gender disparities in access to resources and decision-making power can limit women's ability to adapt to climatic shocks, further exacerbating their vulnerability. A study by Alston and Whittenbury (2013) found that women in drought-prone regions of Africa experienced greater declines in income and food security compared to men, highlighting the gendered dimensions of climate change impacts on livelihoods.

Exacerbation of Social Conflicts Heat and drought stress can act as a catalyst for social conflict, particularly in regions with limited resources and weak governance structures. Competition for scarce resources, such as water and grazing land, can escalate into tensions and disputes between individuals, households, and communities (Lesk et al., 2016). A study by Hsiang et al. (2013) found a significant correlation between climatic events like drought and the onset of civil conflict in Sub-Saharan Africa. In India, where water scarcity is a perennial challenge, conflicts over water resources have become increasingly common, with farmers resorting to protests and legal battles to secure their access to irrigation (Shah, 2010). These conflicts not only disrupt social cohesion but also hinder collective efforts to adapt to and mitigate the impacts of climate change.

20.4.5 Health Impacts

- *Heat-Related Illnesses:* The escalating temperatures associated with climate change pose a direct threat to the health of farmers and agricultural workers who spend long hours working outdoors. Exposure to extreme heat can lead to a range of heat-related illnesses, including heatstroke, heat exhaustion, and dehydration (Watts et al., 2018). These conditions can be debilitating and even fatal, particularly for individuals with underlying health conditions or limited access to healthcare. A study by Kjellstrom et al. (2018) estimated that heat stress could reduce work capacity by up to 50% in some parts of South Asia, highlighting the significant health and economic impacts of rising temperatures.
- *Mental Health Issues:* The chronic stress and anxiety associated with crop losses, financial hardship, and displacement due to climate change can take a toll on the mental health of farmers and their families. Studies have shown a strong correlation between drought and an increased risk of depression, anxiety, and suicidal ideation among farmers (Berry et al., 2013). The loss of livelihoods, social disruption, and uncertainty about the future can contribute to psychological distress and exacerbate existing mental health vulnerabilities (Cianconi et al., 2020). Furthermore, the stigma associated with mental illness and limited access to mental health services in rural areas can further compound the challenges faced by farmers in coping with the psychological impacts of climate change.

20.4.6 Impacts on India's Livestock and Dairy Sectors

The detrimental effects of heat and drought stress are not confined to crop production alone; they also extend to India's vital livestock and dairy sectors. Elevated temperatures and water scarcity create a dual challenge for livestock, impacting their health, productivity, and overall well-being.

Heat stress directly affects livestock by disrupting their physiological processes, leading to reduced feed intake, impaired growth, and decreased reproductive efficiency (Sejian et al., 2010). High temperatures can also trigger heatstroke, a life-threatening condition that can lead to significant mortality in livestock populations. In dairy animals, heat stress is particularly detrimental, as it not only affects milk production but also impairs milk quality (Kadzere et al., 2002). Studies have shown that heat stress can reduce milk yield by 10–30%, with even greater losses observed in high-yielding breeds (Bohmanova et al., 2007).

Drought, on the other hand, exacerbates the challenges faced by livestock farmers by reducing the availability of fodder and water. This can lead to malnutrition, dehydration, and increased susceptibility to diseases in livestock (Morton, 2007). The economic impact of drought on livestock can be severe, as farmers face increased costs for feed and water, while also experiencing reduced income due to lower milk production and livestock mortality (Table 20.2).

20.4.7 Economic Impact of Heat and Drought on Livestock

The economic losses associated with heat and drought stress in the livestock sector are substantial. A study by the National Dairy Development Board (NDDB, 2012) estimated that heat stress alone costs the Indian dairy industry over ₹15,000 crores (approximately $2 billion USD) annually. These losses stem from reduced milk production, increased veterinary costs, and higher mortality rates. Drought further amplifies these economic impacts, as it limits fodder availability, increases feed prices, and necessitates additional expenditure on water procurement and transportation.

These cascading effects of heat and drought stress not only threaten the livelihoods of millions of livestock farmers but also jeopardize India's food security. Livestock plays a crucial role in providing milk, meat, and eggs, which are essential sources of protein and other nutrients for the population. The decline in livestock productivity due to climate change can lead to higher prices for animal products, making them less affordable for vulnerable populations.

Table 20.2 Estimated milk production losses in India due to heat stress

Livestock type	Milk loss (%)	Source
Cattle	10–30	Bohmanova et al. (2007)
Buffaloes	8–25	Das et al. (2016)

20.4.8 Fisheries and Aquaculture

The fishing and aquaculture sector, which provides employment and sustenance to millions in coastal and inland regions of India, is also adversely affected by climate change. Rising sea temperatures and ocean acidification disrupt marine ecosystems and fish populations (Brander, 2010). In inland regions, drought-induced water scarcity and increased evaporation can lead to the drying up of ponds and lakes, affecting fish production and livelihoods of those dependent on aquaculture (Kumar et al., 2019). The cascading impacts of heat and drought stress on multiple agricultural and allied sectors highlight the interconnectedness of these systems and the need for a comprehensive approach to building climate resilience across the entire food production landscape.

20.4.9 Mitigating Heat and Drought Stress: Adaptation and Mitigation Strategies in Indian Agriculture

The escalating impacts of heat and drought stress on Indian agriculture necessitate a multifaceted approach encompassing both short-term adaptation measures and long-term mitigation strategies. These interventions should aim to reduce vulnerability, enhance resilience, and promote sustainable agricultural practices.

1. *Crop Diversification and Drought-Resistant Varieties:* A shift towards crop diversification and the cultivation of drought-tolerant varieties can significantly enhance agricultural resilience. Traditional crop varieties, often adapted to local climatic conditions, can offer valuable traits for drought and heat tolerance. Additionally, advancements in agricultural biotechnology have led to the development of genetically modified crops with improved resilience to abiotic stresses. Research by Varshney et al. (2011) highlights the potential of genomics-assisted breeding to develop climate-resilient crop varieties for India.
2. *Efficient Water Management:* Water scarcity is a critical challenge exacerbated by heat and drought stress. Implementing efficient irrigation techniques, such as drip irrigation and sprinkler irrigation, can significantly reduce water wastage and improve water-use efficiency (Shah, 2010). Furthermore, promoting rainwater harvesting and groundwater recharge can help augment water resources and ensure their availability during dry spells. A study by Giordano et al. (2015) emphasizes the importance of integrated water management approaches that consider the interconnectedness of surface water, groundwater, and rainwater resources.
3. *Climate-Smart Agriculture (CSA):* CSA encompasses a range of practices that aim to sustainably increase agricultural productivity, enhance resilience to climate change, and reduce greenhouse gas emissions (FAO, 2013). This includes practices such as conservation agriculture, agroforestry, integrated pest management, and improved livestock management. These practices can enhance soil health, improve water-use efficiency, and reduce vulnerability to extreme weather events (Kumar et al., 2022).

4. *Carbon Farming and Carbon Credits:* Carbon farming, a practice that sequesters carbon in soil through sustainable agricultural practices, has emerged as a promising strategy for mitigating climate change and improving soil health (Lal, 2004). By adopting practices like cover cropping, reduced tillage, and organic amendments, farmers can enhance soil organic carbon, which not only improves soil fertility and water retention but also contributes to climate change mitigation by sequestering atmospheric carbon dioxide. Carbon credits, generated through these practices, can provide additional income streams for farmers, incentivizing the adoption of sustainable agricultural practices. Smallholder farmers can participate in the carbon credit programmes by forming groups or Farmer Producer Organisations (FPOs) (Sainath et al., 2025)
5. *Empowering Farmers with Information and Resources:* Providing farmers with timely and accurate information about weather forecasts, market prices, and climate-smart agricultural practices can empower them to make informed decisions and adapt to changing conditions (Thornton et al., 2014). This can be achieved through extension services, mobile phone-based information systems, and farmer training programs (Atheequlla et al., 2021). Additionally, providing access to credit and insurance can help farmers manage risks and invest in adaptation measures.
6. *Community Involvements:* Farmers Producers Organizations (FPOs) in India play a critical role in reducing the vulnerability of agriculture to heat and drought stress, addressing significant socioeconomic implications (Sainath & Mukherjee, 2024). These farmer-run institutions promote the adoption of climate-resilient practices such as drought-resistant crops, efficient water management, and soil conservation (Mukherjee et al., 2020). FPOs enhance access to financial resources through better credit terms and insurance schemes, allowing investments in essential infrastructure like irrigation systems. They also improve market access by aggregating produce, ensuring fair prices, and reducing exploitation by intermediaries (Anand et al., 2023). This collective approach increases farmers' incomes and economic stability. Additionally, FPOs foster a strong community support network, offering resources and mutual assistance during crises, and empowering farmers to advocate for policy changes (Mukherjee et al., 2018). Through these efforts, FPOs secure farmers' livelihoods, ensuring socioeconomic stability and sustainable agricultural development in India.
7. *Policy Interventions and Institutional Support:* Robust policy frameworks that incentivize sustainable agricultural practices, promote climate-smart technologies, and protect farmers from price volatility are crucial for building resilience in the agricultural sector (World Bank, 2018). Investments in research and development, extension services, and rural infrastructure are also essential for enhancing adaptive capacity and supporting farmers in their efforts to cope with the challenges of climate change.

By implementing a combination of these adaptation and mitigation strategies, India can build a more resilient agricultural sector that can withstand the impacts of heat and drought stress and ensure food security for its growing population.

20.5 Conclusion

The escalating threat of heat and drought stress in India, exacerbated by climate change, has unveiled a multifaceted crisis with far-reaching socioeconomic consequences. The evidence presented in this chapter paints a stark picture of the vulnerability of India's agricultural sector and the interconnected livelihoods it supports. Crop yield losses due to heat and drought stress are substantial and widespread, threatening food security for a nation with a rapidly growing population. These losses are not uniform, with rain-fed regions and marginalized communities disproportionately bearing the brunt of these climatic shocks.

The economic implications of heat and drought stress are profound. Reduced agricultural productivity translates to income losses for farmers, increased food prices for consumers, and a significant strain on the national economy (Dev & Sharma, 2016). The cyclical nature of debt accumulation further exacerbates economic insecurity, trapping farmers in a vicious cycle of poverty and vulnerability (Morton, 2007). Moreover, the increased production costs associated with coping with these stressors further erode farmers' profit margins and threaten their livelihoods.

The impacts of heat and drought extend beyond the economic realm, permeating the social fabric of rural communities. Livelihood insecurity manifests in the loss of assets, reduced employment opportunities, and forced migration (Lesk et al., 2016). The burden of these stressors falls disproportionately on women, who often bear the brunt of increased workloads and face additional challenges in accessing resources and decision-making power (Alston & Whittenbury, 2013). Furthermore, the competition for scarce resources like water and grazing land can escalate into social conflicts, further destabilizing communities already grappling with the impacts of climate change (Hsiang et al., 2013).

The health and well-being of rural populations are also jeopardized by the escalating frequency and intensity of heatwaves and droughts. Increased exposure to extreme heat leads to a rise in heat-related illnesses and mortality, particularly among agricultural workers who spend long hours outdoors (Watts et al., 2018). The psychological toll of crop failures, economic hardship, and displacement can also manifest in mental health issues like depression and anxiety (Berry et al., 2013; Cianconi et al., 2020).

In the face of this complex and interconnected crisis, the need for urgent and comprehensive action is clear. A multipronged approach is required to build resilience in the agricultural sector and mitigate the socioeconomic impacts of heat and drought stress. This includes investing in drought-resistant crop varieties, developing efficient irrigation systems, promoting sustainable agricultural practices, and empowering farmers with the knowledge and resources to adapt to a changing climate (Thornton et al., 2014; Varshney et al., 2011). Policy interventions that address underlying socioeconomic disparities, strengthen social safety nets, and promote climate-resilient livelihoods are also essential (World Bank, 2018).

This chapter has highlighted the critical need for a paradigm shift in India's agricultural sector towards a more sustainable, resilient, and equitable model. By adopting innovative technologies, embracing climate-smart practices, and addressing the underlying vulnerabilities of farmers and their communities, India can navigate the challenges of climate change and secure a sustainable future for its agricultural sector and the millions who depend on it. The path to resilience is paved with proactive and collaborative action, grounded in a deep understanding of the complex interplay between climate, environment, and society.

References

Aggarwal, P. K., Bandyopadhyay, S. K., Pathak, H., Kalra, N., Chander, S., & Kumar, S. (2000). Analysis of yield trends of the rice-wheat system in north-western India. *Outlook on Agriculture, 29*(4), 259–268.

Alston, M., & Whittenbury, K. (2013). Does climate change affect women and men differently? Evidence from rural Africa. In *Gender and climate change: An introduction* (pp. 83–106). Routledge.

Anand, S., Ghosh, S., & Mukherjee, A. (2023). Effectiveness of Farmer Producer Organizations (FPOs) at different growth stages in transitioning to secondary agriculture. *Indian Journal of Extension Education, 59*(3), 90–96.

Asseng, S., Ewert, F., Martre, P., Rötter, R. P., Lobell, D. B., Cammarano, D., et al. (2011). Rising temperatures reduce global wheat production. *Nature Climate Change, 1*(2), 143–147.

Asseng, S., Ewert, F., Martre, P., Rötter, R. P., Lobell, D. B., Cammarano, D., ... & Zhu, Y. (2015). Rising temperatures reduce global wheat production. *Nature Climate Change, 5*(2), 143–147.

Atheequlla, G. A., Mukherjee, A., Roy, M. L., & Chandra, N. (2021). Mobile based agro advisory service and farmer's willingness to pay: A case study in Bageshwar District of Uttarakhand. *Journal of Community Mobilization, 16*(3), 976–986.

Basu, B., Badole, S., & Gummagolmath, K. C. (2016). Drought in India: A study of its trends, spatiotemporal patterns, and impacts. In *Drought and its impact on agriculture in India* (pp. 21–54). Academic Press.

Bellemare, M. F. (2015). Rising food prices, food price volatility, and social unrest. *American Journal of Agricultural Economics, 97*(1), 1–21. https://doi.org/10.1093/ajae/aau038

Berry, H. L., Bowen, K., & Kjellstrom, T. (2013). Climate change and mental health: A causal pathways framework. *International Journal of Public Health, 58*(2), 123–132.

Bhandari, R., Neupane, N., & Adhikari, D. P. (2021). Climatic change and its impact on tomato (Lycopersicum esculentum L.) production in plain area of Nepal. *Environmental Challenges, 4*, 100129.

Black, R. E., Victora, C. G., Walker, S. P., Bhutta, Z. A., Christian, P., De Onis, M., ... & Uauy, R. (2013). Maternal and child undernutrition and overweight in low-income and middle-income countries. *The Lancet, 382*(9890), 427–451.

Bohmanova, J., Misztal, I., & Cole, J. B. (2007). Temperature-humidity indices describing livestock environment: A review. *Journal of Animal Science, 85*(13), 3574–3587.

Brander, K. M. (2010). Impacts of climate change on fisheries. *Journal of Marine Systems, 79*(3–4), 389–402.

Challinor, A. J., Watson, J., Lobell, D. B., Howden, S. M., Smith, D. R., & Chhetri, N. (2014). A meta-analysis of crop yield under climate change and adaptation. *Nature Climate Change, 4*(4), 287–291.

Chandra, N., Roy, M. L., Mukherjee, A., Jethi, R., & Joshi, K. (2018). A study of migration pattern in Kumaun Hills and associated socio-economic factors. *Journal of Community Mobilization., 13*(1), 107–112.

Chattopadhyay, S., Dutta, S., & Aggarwal, P. K. (2013). Impact of climate change on Indian agriculture: A dynamic simulation study. *Mitigation and Adaptation Strategies for Global Change, 18*(3), 321–342.

Cianconi, P., Betrò, S., & Janiri, L. (2020). The impact of climate change on mental health: A systematic descriptive review. *Frontiers in Psychiatry, 11*, 74.

Das, R., Upadhyay, R. C., & Ashutosh. (2016). Heat stress in buffaloes: A review. *Journal of Animal Physiology and Animal Nutrition, 100*(1), 1–13.

Dev, S. M., & Sharma, A. (2016). Drought and its impact on agriculture in India. In *Advances in agronomy* (Vol. 138, pp. 169–216). Academic Press.

FAO. (2013). *Climate-smart agriculture sourcebook*. Food and Agriculture Organization of the United Nations.

Food and Agriculture Organization of the United Nations (FAO). (2018). *The Future of Food and Agriculture – Alternative Pathways to 2050*. Rome: FAO. Retrieved from https://www.fao.org/3/i8429en/i8429en.pdf

Gemenne, F. (2011). Climate-induced population displacements in a 4°C+ world. *Philosophical Transactions of the Royal Society A: Mathematical, Physical and Engineering Sciences, 369*(1934), 182–204.

Giordano, M., et al. (2015). Integrated water resources management: A review of concepts, challenges, and opportunities. *Water Resources Management, 29*(11), 3817–3832.

Grace, K., Nagle, N. N., Burgert-Brucker, C. R., Rutzick, S., Van Riper, D. C., Dontamsetti, T., & Croft, T. (2018). Integrating environmental context into DHS analysis while protecting participant confidentiality: A new remote sensing method. *Population and Development Review, 45*(1), 197.

Government of India. (2024). Agriculture sector has registered an average annual growth rate of 4.18 Per cent over the last five years: economic survey. Press Information Bureau. https://pib.gov.in/PressReleasePage.aspx?PRID=2034943

Gray, C., & Mueller, V. (2012). Drought and population mobility in rural Ethiopia. *World Development, 40*(1), 134–145.

Guhathakurta, P., & Rajeevan, M. (2008). Trends in rainfall pattern over India. *International Journal of Climatology, 28*(11), 1453–1469.

Hoque, M. A., Rahaman, M. M., & Bhuiyan, M. A. R. (2021). Impact of drought on agriculture and food security: A review. *Journal of Agriculture and Food Research, 5*, 100168.

Hsiang, S. M., Burke, M., & Miguel, E. (2013). Quantifying the influence of climate on human conflict. *Science, 341*(6151), 1235367.

Hunter, L. M., Luna, J. K., & Norton, R. M. (2015). Environmental dimensions of migration. *Annual Review of Sociology, 41*, 377–397.

Im, E. S., Pal, J. S., & Eltahir, E. A. (2017). Deadly heat waves projected in the densely populated agricultural regions of South Asia. *Nature Climate Change, 7*(4), 364–370.

Intergovernmental Panel on Climate Change (IPCC). (2021). Climate Change 2021: The Physical Science Basis. In *Contribution of Working Group I to the Sixth Assessment Report of the Intergovernmental Panel on Climate Change*. Cambridge University Press.

Kadzere, C. T., Murphy, M. R., Silanikove, N., & Maltz, E. (2002). Heat stress in lactating dairy cows: A review. *Livestock Production Science, 77*(1), 59–91.

Kjellstrom, T., Freyberg, C., Lemke, B., Otto, M., & Briggs, D. (2018). Estimating the occupational health impact of climate change: A methods overview. *International Journal of Environmental Research and Public Health, 15*(6), 1124.

Kumar, A., & Kumar, P. (2010). Impact of elevated temperature on Brassica juncea productivity. *Journal of Agronomy and Crop Science, 196*(1), 70–76.

Kumar, N., & Kumar, S. (2019). Impact of climate change on Indian agriculture and allied sectors: A review. *Journal of Integrative Agriculture, 18*(8), 1671–1686.

Kumar, S., Sai Krishna, A., & Singh, R. (2016). Drought vulnerability assessment of chickpea growing environments of India. *Agricultural and Forest Meteorology, 226*, 180–193.

Kumar, U., Raman, R. K., Kumar, A., Singh, D. K., Mukherjee, A., Singh, J., & Bhatt, B. P. (2020). Return migration of labours in Bihar due to COVID-19: Status and strategies of deployment in agricultural sector. *Journal of Community Mobilization, 15*(1), 192–200.

Kumar, A., Mukherjee, A., & Choudhary, A. K. (2022). Climate Smart Village and its Role in Mitigating Climate Change and Improving the Farm Profitability. *Climate Smart Technologies for Improving Farm Productivity, 85*.

Lal, R. (2004). Soil carbon sequestration impacts on global climate change and food security. *Science, 304*(5677), 1623–1627.
Lesk, C., Rowhani, P., & Ramankutty, N. (2016). Influence of extreme weather disasters on global crop production. *Nature, 529*(7584), 84–87.
Lloyd, S. J., Kovats, R. S., & Chalabi, Z. (2011). Climate change, crop yields, and undernutrition: Development of a model to quantify the impact of climate scenarios on child undernutrition. *Environmental Health Perspectives, 119*(12), 1817–1823. https://doi.org/10.1289/ehp.1003311
Lobell, D. B., Schlenker, W., & Costa-Roberts, J. (2011). Climate trends and global crop production since 1980. *Science, 333*(6042), 616–620.
Mall, R. K., Singh, R., Gupta, A., Srinivasan, G., & Rathore, L. S. (2006). Impact of climate change on Indian agriculture: A review. *Climatic Change, 78*(2–4), 445–478.
Meze-Hausken, E. (2004). Contrasting climate variability and extremes between semi-arid regions in northern and southern Africa: environmental and societal impacts. *Climate Research, 27*(1), 33–53. Ministry of Agriculture & Farmers Welfare, Government of India. (2023). Annual Report 2022–2023.
Mishra, A. K., & Singh, V. P. (2010). A review of drought concepts. *Journal of Hydrology, 391*(1–2), 202–216.
Morton, J. F. (2007). The impact of climate change on smallholder and subsistence agriculture. *Proceedings of the National Academy of Sciences, 104*(50), 19680–19685.
Mukherjee, A., Singh, P., Ray, M., Satyapriya, S., & Burman, R. R. (2018). Enhancing farmers income through farmers' producers companies in India: Status and roadmap. *The Indian Journal of Agricultural Sciences, 88*(8), 1151–1161.
Mukherjee, A., Singh, P., Maity, A., Shubha, K., & Burman, R. R. (2020). Enhancing livelihood security of dairy farmers through farmers' producer company: a diagnostic study of Bundelkhand region. *Range Management and Agroforestry, 41*(1), 156–167.
National Dairy Development Board (NDDB). (2012). *Annual Report 2011–12*. Anand.
Peng, S., Huang, J., Sheehy, J. E., Laza, R. C., Visperas, R. M., Zhong, X., et al. (2004). Rice yields decline with higher night temperature from global warming. *Proceedings of the National Academy of Sciences, 101*(27), 9971–9975.
Rathore, L. S., Rao, V. U. M., & Singh, S. K. (2015). Impact of climate change on crop productivity in India. *Current Science, 109*(3), 535–545.
Ravi, I., Singh, A. K., & Sharma, S. B. (2013). Impact of climate change on banana productivity in India: A review. *Agricultural and Food Science, 22*(1), 103–118.
Sainath, B., & Mukherjee, A. (2024). Farmers producers organizations: Growing together for agricultural prosperity. *Agritech, 1*, 19–22.
Sainath, B., Kuymar, A.,Mukherjee, A., Dubey, R.,Kumar,U.,SIngh, D. K.,Raman, R.K.,Das, A. (2025) Carbon credits: opportunities, challenges and policy options. *ICAR RCER Technical Bulletin No. R-85/Patna-51* ICAR Research Complex for Eastern Region, Patna. pp. 56. https://www.researchgate.net/publication/389359258_Carbon_Credits_Opportunities_Challenges_and_Policy_Options
Sejian, V., et al. (2010). Effects of climate change on the productivity and reproduction of livestock in India. *Climate Change, 101*(1–2), 171–191.
Shah, T. (2010). *Taming the anarchy: Groundwater governance in South Asia*. Routledge.
Thornton, P. K., Ericksen, P. J., Herrero, M., & Challinor, A. J. (2014). Climate variability and vulnerability to climate change: A review. *Global Change Biology, 20*(11), 3313–3328.
Varshney, R. K., et al. (2011). Drought tolerance in chickpea (Cicer arietinum L.): status, genetics, and breeding. In *Chickpea breeding and management* (pp. 287–320). CABI.
Watts, N., Adger, W. N., Ayeb-Karlsson, S., Bai, Y., Byass, P., Campbell-Lendrum, D., et al. (2018). The Lancet Countdown on health and climate change: from 25 years of inaction to a global transformation for public health. *The Lancet, 391*(10120), 581–630.
World Bank. (2004). *India Country Overview*, 2004.
World Bank. (2018). *South Asia's hotspots: The impact of temperature and precipitation changes on living standards*. World Bank.
World Bank. (2021). World Development Indicators.

The manufacturer's authorised representative in the EU is Springer Nature Customer Service Centre GmbH, Europaplatz 3, 69115 Heidelberg, Germany. If you have any concerns regarding our products, please contact ProductSafety@springernature.com

Printed and bound by CPI Group (UK) Ltd, Croydon, CR0 4YY
30/09/2026
02225299-0010